Optimization: Insights and Applications

PRINCETON SERIES IN APPLIED MATHEMATICS

EDITORS

Ingrid Daubechies, *Princeton University*
Weinan E, *Princeton University*
Jan Karel Lenstra, *Eindhoven University*
Endre Süli, *University of Oxford*

TITLES IN THE SERIES

Chaotic Transitions in Deterministic and Stochastic Dynamical Systems: Applications of Melnikov Processes in Engineering, Physics, and Neuroscience by Emil Simiu

Selfsimilar Processes by Paul Embrechts and Makoto Maejima

Self-Regularity: A New Paradigm for Primal-Dual Interior-Point Algorithms by Jiming Peng, Cornelis Roos, and Tamás Terlaky

Analytic Theory of Global Bifurcation: An Introduction by Boris Buffoni and John Toland

Entropy by Andreas Greven, Gerhard Keller, and Gerald Warnecke

Auxiliary Signal Design for Failure Detection by Stephen L. Campbell and Ramine Nikoukhah

Thermodynamics: A Dynamical Systems Approach by Wassim M. Haddad, VijaySekhar Chellaboina, and Sergey G. Nesesov

Optimization: Insights and Applications by Jan Brinkhuis and Vladimir Tikhomirov

The Princeton Series in Applied Mathematics publishes high quality advanced texts and monographs in all areas of applied mathematics. Books include those of a theoretical and general nature as well as those dealing with the mathematics of specific applications areas and real-world situations.

Optimization: Insights and Applications

Jan Brinkhuis
Vladimir Tikhomirov

PRINCETON UNIVERSITY PRESS
PRINCETON AND OXFORD

Copyright © 2005 by Princeton University Press

Published by Princeton University Press,
41 William Street, Princeton, New Jersey 08540

In the United Kingdom: Princeton University Press,
3 Market Place, Woodstock, Oxfordshire OX20 1SY

All Rights Reserved

Library of Congress Cataloging-in-Publication Data

Brinkhuis, Jan.
  Optimization : insights and applications / Jan Brinkhuis, Vladimir Tikhomirov.
    p. cm. — (Princeton series in applied mathematics)
  Includes bibliographical references and index.
  ISBN-13 : 978-0-691-10287-0 (alk. paper)
  ISBN-10 : 0-691-10287-2 (alk. paper)
  1. Econometrics — Mathematical models. I. Tikhomirov, V. M. (Vladim Mikhailovich), 1934- II. Title. III. Series.

   HB141.B743 2005
   330`.01`5195—dc22   2005046598

British Library Cataloging-in-Publication Data is available

The publisher would like to acknowledge the authors of this volume for providing the camera-ready copy from which this book was printed.

Printed on acid-free paper.

pup.princeton.edu

Printed in the United States of America

10 9 8 7 6 5 4 3 2 1

Je veux parler du r ê v e. Du rêve, et des visions qu'il nous souffle - impalpables comme lui d'abord, et réticentes souvent à prendre forme. [...] Quand le travail est achevé, ou telle partie de travail, nous en présentons le résultat tangible sous la lumière la plus vive que nous pouvons trouver, nous nous en réjouissons, [...] Une vision, sans nom et sans contours d'abord, ténue comme un lambeau de brumes, a guidé notre main et nous a maintenus penchés sur l'ouvrage, sans sentir passer les heures.

I want to talk about the d r e a m. About the dream, and the visions that it whispers to us, intangible like itself at first, and often reluctant to take shape. [...] When the work is finished, or a certain part of the work, then we present the tangible result of it under the most lively light that we can find, we enjoy ourselves about it, [...] A vision, without name and without shape at first, tenuous as a shred of mist, has guided our hand and has kept us bent over the work, without feeling the hours.

A. Grothendieck

Contents

Preface		xi
0.1	Optimization: insights and applications	xiii
0.2	Lunch, dinner, and dessert	xiv
0.3	For whom is this book meant?	xvi
0.4	What is in this book?	xviii
0.5	Special features	xix

Necessary Conditions: What Is the Point? 1

Chapter 1. Fermat: One Variable without Constraints 3

1.0	Summary	3
1.1	Introduction	5
1.2	The derivative for one variable	6
1.3	Main result: Fermat theorem for one variable	14
1.4	Applications to concrete problems	30
1.5	Discussion and comments	43
1.6	Exercises	59

Chapter 2. Fermat: Two or More Variables without Constraints 85

2.0	Summary	85
2.1	Introduction	87
2.2	The derivative for two or more variables	87
2.3	Main result: Fermat theorem for two or more variables	96
2.4	Applications to concrete problems	101
2.5	Discussion and comments	127
2.6	Exercises	128

Chapter 3. Lagrange: Equality Constraints 135

3.0	Summary	135
3.1	Introduction	138
3.2	Main result: Lagrange multiplier rule	140
3.3	Applications to concrete problems	152
3.4	Proof of the Lagrange multiplier rule	167
3.5	Discussion and comments	181
3.6	Exercises	190

Chapter 4. Inequality Constraints and Convexity — 199

- 4.0 Summary — 199
- 4.1 Introduction — 202
- 4.2 Main result: Karush-Kuhn-Tucker theorem — 204
- 4.3 Applications to concrete problems — 217
- 4.4 Proof of the Karush-Kuhn-Tucker theorem — 229
- 4.5 Discussion and comments — 235
- 4.6 Exercises — 250

Chapter 5. Second Order Conditions — 261

- 5.0 Summary — 261
- 5.1 Introduction — 262
- 5.2 Main result: second order conditions — 262
- 5.3 Applications to concrete problems — 267
- 5.4 Discussion and comments — 271
- 5.5 Exercises — 272

Chapter 6. Basic Algorithms — 273

- 6.0 Summary — 273
- 6.1 Introduction — 275
- 6.2 Nonlinear optimization is difficult — 278
- 6.3 Main methods of linear optimization — 283
- 6.4 Line search — 286
- 6.5 Direction of descent — 299
- 6.6 Quality of approximation — 301
- 6.7 Center of gravity method — 304
- 6.8 Ellipsoid method — 307
- 6.9 Interior point methods — 316

Chapter 7. Advanced Algorithms — 325

- 7.1 Introduction — 325
- 7.2 Conjugate gradient method — 325
- 7.3 Self-concordant barrier methods — 335

Chapter 8. Economic Applications — 363

- 8.1 Why you should not sell your house to the highest bidder — 363
- 8.2 Optimal speed of ships and the cube law — 366
- 8.3 Optimal discounts on airline tickets with a Saturday stayover — 368
- 8.4 Prediction of flows of cargo — 370
- 8.5 Nash bargaining — 373
- 8.6 Arbitrage-free bounds for prices — 378
- 8.7 Fair price for options: formula of Black and Scholes — 380
- 8.8 Absence of arbitrage and existence of a martingale — 381
- 8.9 How to take a penalty kick, and the minimax theorem — 382
- 8.10 The best lunch and the second welfare theorem — 386

Chapter 9. Mathematical Applications — 391

- 9.1 Fun and the quest for the essence — 391

CONTENTS ix

9.2 Optimization approach to matrices ... 392
9.3 How to prove results on linear inequalities ... 395
9.4 The problem of Apollonius ... 397
9.5 Minimization of a quadratic function: Sylvester's criterion and Gram's formula ... 409
9.6 Polynomials of least deviation ... 411
9.7 Bernstein inequality ... 414

Chapter 10. Mixed Smooth-Convex Problems ... 417

10.1 Introduction ... 417
10.2 Constraints given by inclusion in a cone ... 419
10.3 Main result: necessary conditions for mixed smooth-convex problems ... 422
10.4 Proof of the necessary conditions ... 430
10.5 Discussion and comments ... 432

Chapter 11. Dynamic Programming in Discrete Time ... 441

11.0 Summary ... 441
11.1 Introduction ... 443
11.2 Main result: Hamilton-Jacobi-Bellman equation ... 444
11.3 Applications to concrete problems ... 446
11.4 Exercises ... 471

Chapter 12. Dynamic Optimization in Continuous Time ... 475

12.1 Introduction ... 475
12.2 Main results: necessary conditions of Euler, Lagrange, Pontryagin, and Bellman ... 478
12.3 Applications to concrete problems ... 492
12.4 Discussion and comments ... 498

Appendix A. On Linear Algebra: Vector and Matrix Calculus ... 503

A.1 Introduction ... 503
A.2 Zero-sweeping or Gaussian elimination, and a formula for the dimension of the solution set ... 503
A.3 Cramer's rule ... 507
A.4 Solution using the inverse matrix ... 508
A.5 Symmetric matrices ... 510
A.6 Matrices of maximal rank ... 512
A.7 Vector notation ... 512
A.8 Coordinate free approach to vectors and matrices ... 513

Appendix B. On Real Analysis ... 519

B.1 Completeness of the real numbers ... 519
B.2 Calculus of differentiation ... 523
B.3 Convexity ... 528
B.4 Differentiation and integration ... 535

Appendix C. The Weierstrass Theorem on Existence of Global Solutions ... 537

C.1	On the use of the Weierstrass theorem	537
C.2	Derivation of the Weierstrass theorem	544

Appendix D. Crash Course on Problem Solving — 547

D.1	One variable without constraints	547
D.2	Several variables without constraints	548
D.3	Several variables under equality constraints	549
D.4	Inequality constraints and convexity	550

Appendix E. Crash Course on Optimization Theory: Geometrical Style — 553

E.1	The main points	553
E.2	Unconstrained problems	554
E.3	Convex problems	554
E.4	Equality constraints	555
E.5	Inequality constraints	556
E.6	Transition to infinitely many variables	557

Appendix F. Crash Course on Optimization Theory: Analytical Style — 561

F.1	Problem types	561
F.2	Definitions of differentiability	563
F.3	Main theorems of differential and convex calculus	565
F.4	Conditions that are necessary and/or sufficient	567
F.5	Proofs	571

Appendix G. Conditions of Extremum from Fermat to Pontryagin — 583

G.1	Necessary first order conditions from Fermat to Pontryagin	583
G.2	Conditions of extremum of the second order	593

Appendix H. Solutions of Exercises of Chapters 1–4 — 601

Bibliography — 645

Index — 651

Preface

Our first aim has been to write an interesting book, [...] we can hardly have failed completely, the subject-matter being so attractive that only extravagant incompetence could make it dull. [...] it does not demand any great mathematical knowledge or technique.

G. H. Hardy and E. M. Wright

Das vorliegende Buch, aus Vorlesungen entstanden, die ich mehrfach in [...] gehalten habe, setzt sich zum Ziel, den Leser, ohne irgendwelche [...] Kenntnisse vorauszusetzen, in das Verständnis der Fragen einzuführen, welche gegenwärtig den Gipfel der Theorie [...] bilden. [...] Für den Kenner der Theorie werden immerhin vielleicht einige Einzelheiten von Interesse sein.

The present book, arisen from lectures, which I have given several times in [...], aims to introduce the reader, without assuming any [...] knowledge, into the understanding of the questions that form at present the summit of the theory. [...] For the experts of the theory, some details will perhaps be of interest.

E. Hecke

Wonder en is gheen wonder.

A miracle is not a miracle.

S. Stevin

Geometry draws the soul toward truth.

Plato

Solange ein Wissenszweig Ueberfluß an Problemen bietet, ist er lebenskräftig; [...]. Wie überhaupt jedes menschliche Unternehmen Ziele verfolgt, so braucht der mathematische Forscher Probleme. [...] Eine mathematische Theorie ist nicht eher als volkommen anzusehen, als du sie dem ersten Manne erklären könntest, den du auf der Straße triffst.

As long as a branch of science offers an abundance of problems, so long is it alive; [...] Just as every human undertaking pursues certain objectives, so also mathematical research requires its problems. [...] A mathematical theory is not to be considered complete until you can explain it to the first person whom you meet on the street.

D. Hilbert

Quand une situation, de la plus humble à la plus vaste, a été comprise dans les aspects essentiels, la démonstration de ce qui est compris (et du reste) tombe comme un fruit mûr à point. Alors que la démonstration arrachée comme un fruit encore vert à l'arbre de la connaissance laisse un arrière-goût d'insatisfaction, une frustration de notre soif, nullement apaisée.

When a situation, from the most humble to the most immense, has been understood in the essential aspects, the proof of what is understood (and of the remainder) falls like a fruit that is just ripe. Whereas the proof snatched like an unripe fruit from the tree of knowledge leaves an after-taste of dissatisfaction, a frustration of our thirst, not at all silenced.

A. Grothendieck

PREFACE

0.1 OPTIMIZATION: INSIGHTS AND APPLICATIONS

We begin by explaining the title.

- *Optimization.* It is our great wish to explain in one book all aspects of continuous optimization to a wide circle of readers. We think that there is some truth in the words: *"if you want to understand something, then you should try to understand everything."* We intend our self-contained text to be stimulating to the activity of solving optimization problems and the study of the ideas of the methods of solution. Therefore, we have made it as entertaining and informal as our wish to be clear and precise allowed. The emphasis on insights (by means of pictures) and applications (of a wide variety) makes this book suitable as a textbook (for various types of courses ranging from introductory to advanced). There are many books on this subject, but our book offers a novel and consistent approach. The contribution to the field lies in the collection of examples, but as well in the presentation of the theory.

  The focus is on solving optimization problems with a finite number of continuous variables analytically (that is, by a formula). Here our ambition is "to tell the whole story." The main message here is that the art of solving has almost become a craft, by virtue of a universal strategy to solve all optimization problems that can be solved at all.

  In addition, we give two brief introductions, to classical and modern numerical methods of optimization, and to dynamic optimization. Discrete and stochastic optimization fall outside the scope of our book.

- *Insights.* The overarching point of this book is that most problems —when viewed "correctly"—may be solved by the direct application of the theorems of Fermat, Lagrange, and Weierstrass. We have pursued an intensive quest to reach the essence of all theoretical methods. This has led to the surprising outcome that the proofs of all methods can be fully understood by means of simple geometric figures. Writing down the rigorous analytical proofs with the aid of these figures is a routine job. Advanced readers and experts might be interested in the simple alternative proofs of the Lagrange principle and of the Pontryagin maximum principle, given at the end of the book. They might also be surprised to see that all methods of optimization can be unified into one result, *the principle of Fermat-Lagrange.*

- *Applications.* We introduce the reader to mathematical optimization with continuous variables and apply the methods to a substantial and highly varied list of classical and practical problems. Many of these applications are really surprising, even for experts.

Concerning the painting on the front cover, its choice is inspired by the organization of our book, to be explained below.

0.2 LUNCH, DINNER, AND DESSERT

This book can be viewed as consisting of three parts: lunch, dinner, and dessert. We assume that you already have had some breakfast. We offer some snacks for those who did not have enough for breakfast.

Lunch. Lunch takes no effort to prepare, at least not in Moscow and Rotterdam, where we wrote this book. It is a light, simple and enjoyable meal. There is a rich choice of things to put on your sandwiches. You leave the table refreshed and often with a taste for more. This is all right, as it was not the purpose to silence all hunger and thirst.

Dinner. Dinner requires careful preparation, a visit to the shops, and some concentrated work in the kitchen. It is a substantial, refined, and tasty meal. Once at the table, you have not much choice, but have to "eten wat de pot schaft" ($\approx$ eat what is being served). You leave the table satisfied, that is, with the feeling that you have eaten enough. Although dinner and the accompanying wine are of excellent quality, you know that you have to learn to appreciate it. This and the idea that it will restore your energy will help you to finish your dinner.

Dessert. For dessert there is a wealth of things in the fridge and in the fruit basket to choose from. To eat dessert you need no ulterior motive. It is just fun, pure pleasure, and delicious. Of course you can combine as many of the choices as you want at home; we felt very much at home both in Moscow and in Rotterdam!

Food for thought in plain prose
- **Breakfast.** Course on vectors, matrices, differentiation and continuity (we assume that you have taken such a course).

- **Snacks.** Three short refreshment courses, appendices A (on vectors and matrices), B (on differentiation), C (on continuity), and the introductory chapter "Necessary Conditions: What Is the Point?"

- **Lunch.** Introductory course on optimization methods for those who are mainly interested in the applications, but moreover want to learn something about optimization: Chapters 1, 2, 3, 4, 6, 11. All proofs are optional, as well as chapter 5 and appendix D, "Crash Course on Problem Solving".

- **Dinner.** Advanced course on optimization for those who want full insights into all aspects of the subject. Chapters 5, 7, 10, 12, appendix G, "Conditions of Extremum from Fermat to Pontryagin", and all proofs in the "lunch sections" and in appendices E and F, two crash courses on optimization theory (one in geometrical style and one in analytical style).

- **Dessert.** Applications of optimization methods: Chapters 8, 9, all concrete problems and exercises throughout the book, and the use of software based on the numerical methods from Chapters 6, 7.

- **Appetizer.** Most chapters begin with a summary, motivating the contents of the chapter.

Royal road

The pharaoh of Egypt once asked the Greek geometer Euclid whether there was no easier way to learn geometry than by studying all the volumes of Euclid's textbook *"the Elements."* His legendary answer was: *"there is no royal road to geometry"*.

Fortunately, there exists a shortcut to the dessert. This route is indicated in each section under the heading "royal road."

Three sorts of desserts

Each concrete application of substance belongs to at least one of the following three classes.

1) **Pragmatic applications.** Pragmatic applications usually represent a trade off between two opposite effects. Moreover, there are countless examples of economic problems, for example, problems of minimizing cost, or maximizing profit or social welfare.

2) **Optimality of the world.** Most—or all—laws of physics can be viewed as though nature would be optimizing. For example, light behaves essentially as though it "chooses the fastest route." In economics there are similar examples. One of these concerns the basic

principle governing markets: the price for which supply equals demand is the price for which total social welfare is maximal.

3) **Mathematical applications.** The justification for mathematical applications is less straightforward. There is—and always has been—a tendency to reach the essence of things: scientific curiosity compels some people to go to the bottom of the matter. For example, if you have obtained an upper or lower bound that suffices for practical purposes, then nevertheless you can go on to search for the sharpest one.

0.3 FOR WHOM IS THIS BOOK MEANT?

- **Short answer.** A basic mathematical knowledge is sufficient to understand the topics covered in the book and to master the methods. This makes the book very useful for nonspecialists. On the other hand, more advanced readers and even experts might be surprised to see how all main results can be grounded on the so-called Fermat-Lagrange theorem. The book can be used for a wide range of courses on continuous optimization, from introductory to advanced, for any field for which optimization is relevant. The minimum goal would be to master the main tricks of finding analytical solutions of optimization problems. These are explained in a few pages in appendix D, "Crash Course on Problem Solving." The maximal goal would be to get to the bottom of the subject, to get a complete understanding of all insights and to study all applications.

- **Detailed answer.** This book is based on our research—resulting in many novel details—and on our teaching experience. Parts of this book have been tested in courses given at the departments of Mathematics and Economics of the Universities of Moscow and Rotterdam. The participants of these courses were students in Economics, Econometrics and Mathematics, Ph.D. students in Economics and Management Science, and Master's students in Maritime Economics. The aim is to introduce the reader, without requiring any previous knowledge of optimization, to the state of the art of solving concrete optimization problems. It is meant for beginners as well as for advanced readers and even for experts. The prerequisites are as follows:

Minimal. Interest to learn about optimization. A—vague—memory of the following concepts: linear equations, vectors, matrices, limits, differentiation, continuity, and partial derivatives. This memory can be refreshed quickly by means of the appendices.

Recommended. Interest in learning about optimization, more than just "the tricks." A good working knowledge of linear algebra and differential calculus.

Moreover, we think that experts will find something of interest in this book as well, if only in the collection of concrete problems, the simple proofs—here we recommend appendix G, "Conditions of Extremum from Fermat to Pontryagin"—and various finer points; moreover, there is the chapter about unification of all methods of optimization by the principle of Fermat-Lagrange (Chapter 10).

We have two types of reader in mind.

Beginners. Anyone with an interest in optimization who has ever taken a course on linear equations and differential calculus is invited to lunch. There we explain and apply all successful methods of optimization, using some simple pictures. We provide detailed proofs of all statements, using only elementary arguments. Readers who wish to understand all proofs and discover that they cannot follow some, because their grasp of matrices, limits, continuity, or differentiation is not sufficient, can refresh their memory by reading the appendices.

Having invested some time in familiarizing themselves with the methods of optimization, they can collect the fruits of this by amusing themselves with some of the applications to their favorite subject. For dessert, they will find the full analysis of some optimization problems from economics, mathematics, engineering, physics, and medicine. All are practical, beautiful, and/or challenging. Another interesting possibility is to try one's hand on some of the many exercises, ranging from routine drills to problems that are challenging but within reach of the tools provided at lunch.

Advanced readers. The other type of reader we hope to interest in our book has followed a mathematics course at the university level. They can come straight to dinner and make their choices of the desserts. Each item of the dessert is wonderful in its own way and needs no recommendation. We try to offer them a fuller and deeper insight into the state of the art of optimization. They might be surprised to see that the entire body of knowledge can be based on a small number of fundamental ideas, and how geometrically intuitive these ideas are.

The main dinner course is the principle of Fermat-Lagrange. This unifying principle is the basis of all methods to find the solution of an optimization problem. Finding the solution of an optimization problem might appear at first sight like finding a needle in a haystack. The principle of Fermat-Lagrange makes it possible to carry out this seemingly impossible task.

Beginners, advanced readers—and experts. Both types of reader can also profit from that part that is not specially written for them. On the one hand, we like to think that advanced readers—and even experts in optimization theory—will find something new in the lunch and in the short proofs of all necessary conditions of optimization theory—including the formidable Pontryagin maximum principle of optimal control—in appendix G, "Conditions of Extremum from Fermat to Pontryagin." It is worth pointing out as well that all the ideas of dinner are already contained in lunch, sometimes in embryonic form. On the other hand, readers of our lunch who have a taste for more are invited for dinner.

0.4 WHAT IS IN THIS BOOK?

Four-step method. Our starting point is a collection of concrete optimization problems. We stick to strict rules of admission to this collection. The problem should belong to at least one of three types: it should serve a *pragmatic aim*, represent a *law of optimality*, or be an *attempt to reach the essence* of some matter. We offer a universal four-step method that allows us to solve all problems that can be solved at all analytically, that is, by a formula.

Simple proofs based on pictures. We give insights into all the ingredients of this four-step method. The intuition for each of the theoretical results can be supported by simple geometric figures. After this it is a routine job to write down precise proofs in analytical style. We do more: we clarify all related matters, such as second order conditions, duality, the envelope theorem, sensitivity, shadow prices, and the unification of necessary conditions. These provide additional insights.

Simple structure theory. The structure itself of the "whole building" is also simple: everything follows from the *tangent space theorem* and the *supporting hyperplane theorem*. The first result allows us to profit from the *"smoothness,"* that is, the differentiability, of the given problem; the second one from the *"convexity."* In essence, the first (resp. second) result gives the existence of a linear approximation under assumption of smoothness (resp. convexity). Both results will be derived from the *Weierstrass theorem*, a geometrically intuitive existence result. Thus the state of the art of solving analytically concrete *finite-dimensional* or *static* optimization problems appears to be completely satisfactory: we do not know of any open questions.

Numerical methods. However, for many pragmatic optimization problems the best you can hope for is a *numerical* solution. Here the main issue is: what can you expect from algorithms and what not? We make clear why there will never be one algorithm that can solve all nonlinear optimization problems efficiently. The state of the art is that you should try to model your problem as a *convex* optimization problem; then you have a fighting chance to find a numerical solution of guaranteed quality, using *self-concordant barrier methods*—also called *interior point methods*—or using *cutting plane methods* such as the *ellipsoid method*.

Dynamic optimization. Finally, we will give a glimpse of *dynamic* optimization, that is, of the calculus of variations, optimal control, and dynamic programming. Here we enter the domain of *infinite-dimensional* optimization. We emphasize that we give in appendix G, "Conditions of Extremum from Fermat to Pontryagin," short novel proofs from an advanced point of view of all necessary conditions of optimization from the Fermat theorem to the Pontryagin maximum principle.

0.5 SPECIAL FEATURES

- **Four-step method.** We recommend writing all solutions of concrete optimization problems in the same brief and transparent way, in four steps:

    1. model the problem and establish existence of global solutions,

2. write down the equation(s) of the first-order necessary conditions,
3. investigate these equations,
4. write down the conclusion.

Thus, you do not have to be a Newton to solve optimization problems. The original solutions of optimization problems often represent brilliant achievements of some of the greatest scientists, like Newton. The four-step method turns their high art into a craft. The main aim of this book is to teach this craft, giving many examples.

- **Existence of solutions.** An essential moment in the four-step method is the verification of the existence of a global solution. This is always done in the same way: by using the Weierstrass theorem. Even when the most delicate assumption of this theorem—boundedness of the set of admissible points—does not hold, we can use this theorem to establish the required existence, thanks to the concept of "coercivity."

- **Applications.** An abundant and highly varied collection of completely solved optimization problems is offered. This corresponds to our view that the development of optimization methods arises from the challenges provided by concrete optimization problems. This is expressed by the words of Hilbert in one of the epigraphs of this preface.

  The methods of optimization are universal and can be applied to many different fields, as we illustrate by our choice of applications. In particular, we offer an abundance of economic applications, ranging from classical ones, such as Nash bargaining, to a model on time (in)consistency of Kydland and Prescott, the winners of the Nobel Prize in Economics in 2004. A reason for this emphasis is that one of the authors works at a School of Economics. We recommend that instructors who adopt this book for courses outside economics, econometrics, management science, and applied mathematics, add applications from their own fields to the the present collection. The full solutions of the exercises from the first four chapters are given at the end of the book.

- **User-friendliness.** Our main point is to make clear that solving optimization problems is becoming more and more a convenient *craft* for a wide circle of practitioners, whereas originally

PREFACE

it was an unattainable *art* of a small circle of experts. For some it will be a pleasant surprise that the role of all formal aspects, such as definitions, is very modest. For example, the definitions of the central concepts continuity and differentiability are never used in the actual problem solving. Instead, one uses for each of these a user-friendly calculus.

- **Finer points and tricks.** Many finer points and tricks, about problem solving as well as about the underlying methods, some of them novel, are offered throughout the book. For example, we clarify the source of the secret power of the Lagrange multiplier method: it is the idea of reversing the order of the two tasks, elimination and differentiation. This turns the hard task, elimination, from a nonlinear problem into a linear one. In particular, we emphasize that the power of this method does not come from the use of multipliers.

- **Historical perspective.** We emphasize background information and anecdotes, in order to make the book more readable and enjoyable for everyone. Thus one can read about how some of the methods of optimization were discovered. For this we have made use of the MacTutor History of Mathematics archive (www-history.mcs.st-andrews.ac.uk/history/index.html).

- **Comprehensiveness.** Simple proofs from scratch are given for all results that can be used in the analysis of finite-dimensional optimization problems. Thus anyone who might initially be surprised at the strength of these "miraculous" results will be led to agree with the words of Simon Stevin in one of the epigraphs of this preface: *Wonder en is gheen wonder.*

- **Pictures.** The idea of the proof of each result in this book has a simple geometric sense. That is, all ideas can be fully understood by means of simple pictures. This holds in particular for the proof of the central result, the Lagrange multiplier rule. We recall that this result is usually derived from the implicit function theorem, by means of a calculation, and that the proofs given in textbooks for this latter theorem are relatively technical and not very intuitive. The role of pictures in this book is in the spirit of the words of Plato chosen as one of the epigraphs of this preface: *Geometry draws the soul toward truth.*

- **Unification.** A new unification of all known necessary conditions that are used to solve concrete problems is given, to be called the *principle of Fermat-Lagrange*. This appears to mark, in our opinion, the natural limit of how far conditions of multiplier type—that can be used to solve concrete problems—can be pushed.

- **Structure of theory.** The structure itself of the theory underlying the four-step method for smooth-convex problems turns out to be simple as well. This theory is based in a straightforward way on three basic results, the Weierstrass theorem, the tangent space theorem, and the supporting hyperplane theorem.

  Concerning the proofs of these three results, the situation is as follows. The Weierstrass theorem is so plausible that it hardly needs a proof, unless one wants to delve into the foundations of the real number system. For the other two results there are "universal proofs," which extend to the infinite-dimensional case and therefore make it possible to extend the four-step method to the calculus of variations and to optimal control.

  However, the main topic of the present book is finite-dimensional optimization and in the finite-dimensional case each of these two results can be derived in a straightforward way from the Weierstrass theorem and the Fermat theorem, using the four-step method. We have chosen to present these last proofs in the main text.

  The sense of the three basic results is also clear. The Weierstrass theorem establishes the existence of global solutions of optimization problems; it is one of the many ways to express the "completeness" property of the real numbers. The tangent space theorem (resp. supporting hyperplane theorem) makes it possible to profit from the smoothness (resp. convexity) properties of optimization problems; it is a statement about the approximation of nonlinear smooth (resp. convex) "objects" by linear objects.

- **Role of second order conditions.** For each concrete problem, we begin by establishing the *existence* of a solution. This always can be done easily, and once this is done, it only remains to find the solution(s) using the first order conditions. This has the great advantage that there is no need to use the relatively heavy-handed technique of second order conditions to complete the analysis. The role of second order conditions is just that these give some *insight*, as will be shown.

- **Role of constraint qualifications.** For concrete problems with constraints, it is usually easy to show that λ_0, the Lagrange multiplier of the objective function, cannot be zero. This turns out to be more convenient than using a constraint qualification in order to prevent λ_0 from being zero. Again, the role of constraint qualifications is to give some insight.

- **Proofs for advanced readers.** We want to draw the attention of advanced readers and experts once again to appendix G, "Conditions of Extremum from Fermat to Pontryagin." In this appendix we offer novel proofs for all necessary first and second order conditions of finite- as well as infinite-dimensional optimization problems. These proofs are simpler than the proofs in the main text, but they are given from a more advanced point of view. For experts it might be interesting to compare these proofs to the proofs in the literature. For example, a short proof of the Lagrange multiplier rule is given, using Brouwer's fixed point theorem; this proof requires weaker smoothness assumptions on the problem than the usual one. Another example is a very short transparent proof of the Pontryagin maximum principle from optimal control, using standard results on ordinary differential equations.

- **Optimality of classical algorithms.** We show that the classical optimization algorithms are all optimal in some sense. That is, we show that these algorithms are essentially solutions of some optimization problem.

- **Simplified presentation of an advanced algorithm.** We simplify the presentation of the technical analysis of the celebrated self-concordant barrier method of Nesterov and Nemirovskii by means of the device of "restriction to a line." We offer as well a novel, coordinate-free presentation of the v-space approach to interior point methods for LP.

- **Bellman equation.** We tell a simple tale about a boat that represents the proof of the Bellman equation, the central result of continuous time dynamic programming.

- **Main text and extras.** The main text consists of the explanation of the methods, and their application to concrete problems. In addition there are examples, exercises, proofs, and texts giving insights and background information.

- **Easy access to the material.** We have tried to facilitate access to the material in various ways:
  1. The royal road.
  2. An introduction to necessary conditions.
  3. A summary at the beginning of most chapters.
  4. A "plan" at the beginning and a "conclusion" at the end of each section.
  5. A crash course on problem solving.
  6. Two crash courses on optimization theory, one in analytical style and one in geometrical style.
  7. Appendices on linear algebra, on real analysis, and on existence of solutions.
  8. A brief sketch of all aspects of—continuous variable—optimization methods, at the end of Chapter 1.
  9. Short proofs for advanced readers of necessary and sufficient first and second order conditions in appendix G, "Conditions of Extremum from Fermat to Pontryagin."
  10. Our websites contain material related to this book such as a list of any corrections that will be discovered and references to implementations of optimization algorithms.
  11. The index points to all statements of interest.
- **Uniform structure.** All basic chapters have a uniform structure: summary, introduction, main result, applications to concrete problems, proof main result, discussion and comments, exercises.

Acknowledgments. We would like to thank Vladimir Protasov for his contributions and for sharing his insights. We extend our thanks to Jan Boone, who provided a number of convincing economic applications. We are especially grateful to Jan van de Craats, who produced the almost hundred figures, which take a central position.

We were fortunate that friends, colleagues, and students were willing to read early drafts of our book and to comment upon it: Joaquim Gromicho, Bernd Heidergott, Wilfred Hulsbergen, Marieke de Koning, Charles van Marrewijk, Mariëlle Non, Ben Tims, Albert Veenstra, and Shuzhong Zhang.

We also thank the Rijksmuseum Amsterdam for permission to use the painting on the cover: *Stilleven met kaassoorten*, (Still life with cheeses) by the Dutch painter Floris van Dijck (1575-1651).

Necessary Conditions: What Is the Point?

The aim of this introduction to optimization is to give an informal first impression of *necessary conditions*, our main topic.

Six reasons to optimize

Exercises. Suppose you have just read for the first time about a new optimization method, say, the Lagrange multiplier method, or the method of putting the derivative equal to zero. At first sight, it might not make a great impression on you. So why not take an example: find positive numbers x and y with product 10 for which $3x + 4y$ is as small as possible. You try your luck with the multiplier rule, and after a minor struggle and maybe some unsuccessful attempts, you succeed in finding the solution. That is how the multiplier rule comes to life! You have conquered a new method.

Puzzles. After a few of these numerical examples, you start to lose interest in these exercises. You want more than applying the method to problems that come from nowhere and are leading nowhere. Then it is time for a puzzle. Here is one.

To save the world from destruction, agent 007 has to reach a skiff 50 meters off-shore from a point 100 meters farther along a straight beach and then disarm a timing device. The agent can run along the shore at 5 meters per second, swim at 2 meters per second, and disarm a timing device in 30 seconds. Can 007 save the world if the device is set to trigger destruction in 73 seconds?

The satisfaction of solving such a puzzle—by putting a derivative equal to zero in this case—is already much greater.

Test of strength. Now, someone comes with shocking news. The multiplier rule is worth nothing, according to him. For example, if, in the example above, you express y in terms of x by rewriting $xy = 10$ as $y = 10/x$, and substitute this into $3x + 4y$, then you get $3x + 40/x$. Putting the derivative equal to zero, you find the optimal x and so, using $y = 10/x$, the optimal y. You see that you do not need the multiplier method at all, and this solution is even simpler!? Does the multiplier rule have the right to exist?

Time for a test of strength: how can we put down four sticks to form a quadrangle of maximal area? The sticks may have different sizes. This is not just a run-of-the-mill problem. In ancient times, the Greek geometers knew this problem already. They tried very hard to solve it, but without success. With the multiplier rule, you can solve it without any difficulty. We do not know of any other way to solve it.

Insight. After a while you have played enough. You would like to do something useful with the optimization tools. Here is a suggestion. Some fruitful principles of economics supported by much empirical evidence tell us that certain things are automatically optimal. An example is the "invisible hand" of Adam Smith: individual self-interest can lead to common interest. Another example is the law of "comparative advantage," leading to the conclusion that free trade is always optimal, even for weak trading partners. In physics, all laws can be "understood" by optimization. The best-known example is that the law of Snellius on the refraction of light can be derived from the principle that light always chooses the path that takes the shortest time. Here we quote a centuries-old phrase of Euler: "Nothing happens in the world without a law of optimization having a hand in it."

Practical applications. Many things are not optimal automatically. Fortunately there are many experts (consultants, economic advisors, econometricians, engineers) who can give advice. Here are some possible questions. What is the optimal tax system? What is the best way to stimulate the creation of new jobs? What is the best way to stimulate research and development? How to achieve optimal stability of a bridge? How to achieve the highest possible return without unhealthy risks? What is the best way to organize an auction?

Turn art into craft. Many optimization problems of interest have been solved by some of the greatest minds, such as Newton. Their ingenious solutions can be admired, as beautiful *art* can be. The optimization methods that are now available allow you to learn the *craft* to solve all these problems—and more—by yourself. You don't have to be a Newton to do this! We hope that our book will help you to discover this, as well as the many attractions of optimization problems and their analysis.

Chapter One

Fermat: One Variable without Constraints

> When a quantity is the greatest or the smallest, at that moment its flow is neither forward nor backward.
>
> *I. Newton*

- How to find the maxima and minima of a function f of one variable x without constraints?

1.0 SUMMARY

> You can never be too rich or too thin.
>
> *W. Simpson, wife of Edward VIII*

One variable of optimization. The epigraph to this summary describes a view in upper-class circles in England at the beginning of the previous century. It is meant to surprise, going against the usual view that somewhere between too small and too large is the optimum, the "golden mean." Many pragmatic problems lead to the search for *the golden mean* (or *the optimal trade-off* or *the optimal compromise*). For example, suppose you want to play a computer game and your video card does not allow you to have optimal quality ("high resolution screen") as well as optimal performance ("flowing movements"); then you have to make an optimal compromise. This chapter considers many examples where this golden mean is sought. For example, we will be confronted with the problem that a certain type of vase with one long-stemmed rose in it is unstable if there is too little water in it, but as well if there is too much water in it. How much water will give optimal stability? Usually, the reason for such optimization problems is that a trade-off has to be made between two effects. For example, the height of houses in cities like New York or Hong Kong is determined as the result of the following trade-off. On the one hand, you need many people to share the high cost of the land on which the house is built. On the other hand, if you build the house very high, then the specialized costs are forbidding.

Derivative equal to zero. All searches for the "golden mean" can be modeled as problems of optimizing a function f of one variable x, minimization (maximization) if $f(x)$ represents some sort of cost (profit). The following method, due to Fermat, usually gives the correct answer: "put the derivative of f equal to zero," solve the equation, and – if the optimal x has to be an integer – round off to the nearest integer. This is well known from high school, but we try to take a fresh look at this method. For example, we raise the question why this method is so successful. The technical reason for this is of course the great strength of the available calculus for determining derivatives of given functions. We will see that in economic applications a conceptual reason for this success is the *equimarginal rule*. That is, rational decision makers take a marginal action only if the marginal benefit of the action exceeds the marginal cost; they will continue to take action till marginal benefit equals marginal cost.

Snellius's law. The most striking application of the method of Fermat is perhaps the derivation of the law of Snellius on the refraction of light on the boundary between two media—for example, water and air. This law was discovered empirically. The method of Fermat throws a striking light on this technical rule, showing that it is a consequence of the simple principle that light always takes the fastest path (at least for small distances).

1.1 INTRODUCTION

Optimization and the differential calculus. The first general method of solution of extremal problems is due to Pierre de Fermat (1608–1665). In 1638 he presented his idea in a letter to the prominent mathematicians Gilles Persone de Roberval (1602–1675) and Marin Mersenne (1588–1648). Scientific journals did not yet exist, and writing a letter to learned correspondents was a usual way to communicate a new discovery. Intuitively, the idea is that the tangent line at the highest or lowest point of a graph of a function is horizontal. Of course, this tangent line is only defined if the graph has no "kink" at this point.

The exact meaning became clear later when Isaac Newton (1642/43–1727) and Gottfried von Leibniz (1646–1716) invented the elements of classical analysis. One of the motivations for creating analysis was the desire of Newton and Leibniz to find general approaches to the solution of problems of maximum and minimum. This was reflected, in particular, in the title of the first published work devoted to the differential calculus (written by Leibniz, published in 1684). It begins with the words "Nova methodus pro maximis et minimis ...."

The Fermat theorem. In his letter to Roberval and Mersenne, Fermat had—from our modern point of view—the following proposition in mind, now called the (one-variable) Fermat theorem (but he could express his idea only for polynomials): if $\widehat{x}$ is a point of local minimum (or maximum) of f, then *the main linear part of the increment is equal to zero*. The following example illustrates how this idea works.

Example 1.1.1 *Verify the idea of Fermat for the function $f(x) = x^2$.*

Solution. To begin with, we note that the graph of f is a parabola that has its lowest point at $x = 0$.

Now let us see how the idea of Fermat leads to this point. Let $\widehat{x}$ be a point of local minimum of f. Let x be an arbitrary point close to $\widehat{x}$; write h for the increment of the argument, $x - \widehat{x}$, that is, $x = \widehat{x} + h$. The increment of the function,

$$f(x) - f(\widehat{x}) = (\widehat{x} + h)^2 - \widehat{x}^2 = 2\widehat{x}h + h^2,$$

is the sum of the *main linear part* $2\widehat{x}h$ and the *remainder* h^2.

This terminology is reasonable: the graph of the function $h \to 2\widehat{x}h$ is a straight line through the origin and the term h^2, which remains, is negligible—in comparison to h—for h small enough. That h^2 is negligible can be illustrated using decimal notation: if $h \approx 10^{-k}$

("k-th decimal behind point"), then $h^2 = 10^{-2k}$ ("$2k$-th decimal behind point"). For example, if $k = 2$, then $h = 1/100 = .01$, and then the remainder $h^2 = 1/10000 = .0001$ is negligible in comparison to h.

That the main linear part is equal to zero means that $2\widehat{x}h$ is zero for each h, and so $\widehat{x} = 0$.

Royal road. If you want a shortcut to the applications in this chapter, then you can read the statements of the Fermat theorem 1.4, the Weierstrass theorem 1.6, and its corollary 1.7, as well as the solutions of examples 1.3.4, 1.3.5, and 1.3.7 (solution 3); thus prepared, you are ready to enjoy as many of the applications in sections 1.4 and 1.6 as you like. After this, you can turn to the next chapter.

1.2 THE DERIVATIVE FOR ONE VARIABLE

1.2.1 What is the derivative?

To solve problems of interest, one has to combine the idea of Fermat with the differential calculus. We begin by recalling the basic notion of the differential calculus for functions of one variable. It is the notion of *derivative*. The following experiment illustrates the geometrical idea of the derivative.

"Zooming in" approach to the derivative. Choose a point on the graph of a function drawn on a computer screen. Zoom in a couple of times. Then the graph looks like a straight line. Its slope is the derivative. Note that a straight line through a given point is determined by its slope.

Analytical definition of the derivative. For the moment, we restrict our attention to the analytical definition, due to Auguste Cauchy (1789–1857). This is known from high school: it views the derivative $f'(\widehat{x})$ of a function f at a number $\widehat{x}$ as the limit of the quotient of the increments of the function, $f(\widehat{x} + h) - f(\widehat{x})$, and the argument, $h = (\widehat{x} + h) - \widehat{x}$, if h tends to zero, $h \to 0$ (Fig. 1.1).

Now we will give a more precise formulation of this analytical definition. To begin with, note that the derivative $f'(\widehat{x})$ depends only on the behavior of f at a neighborhood of $\widehat{x}$. This means that for a sufficiently small number $\varepsilon > 0$ it suffices to consider $f(x)$ only for x with $|x - \widehat{x}| < \varepsilon$.

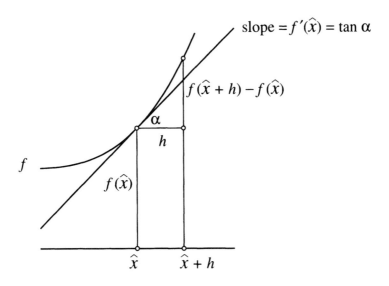

Figure 1.1 Illustrating the definition of differentiability.

We call the set of all numbers x for which

$$|x - \hat{x}| < \varepsilon$$

the ε-neighborhood of $\hat{x}$. This can also be defined by the inequalities

$$\hat{x} - \varepsilon < x < \hat{x} + \varepsilon,$$

or by the inclusion $x \in (\hat{x}-\varepsilon, \hat{x}+\varepsilon)$. The notation $f : (\hat{x}-\varepsilon, \hat{x}+\varepsilon) \to \mathbb{R}$ denotes that f is defined on the ε-neighborhood of $\hat{x}$ and takes its values in $\mathbb{R}$, the real numbers.

Definition 1.1 Analytical definition of the derivative. *Let $\hat{x}$ be a real number, $\varepsilon > 0$ a positive number, and $f : (\hat{x} - \varepsilon, \hat{x} + \varepsilon) \to \mathbb{R}$ a function of one variable x. The function f is called differentiable at $\hat{x}$ if the limit*

$$\lim_{h \to 0} \frac{f(\hat{x} + h) - f(\hat{x})}{h}$$

exists. Then this limit is called the (first) derivative of f at $\hat{x}$ and it is denoted by $f'(\hat{x})$.

For a linear function $f(x) = ax$ one has $f'(x) = a$ for all $x \in \mathbb{R}$, that is, for all numbers x of the real line $\mathbb{R}$. The following example is more interesting.

Example 1.2.1 *Compute the derivative of the quadratic function $f(x) = x^2$ at a given number $\hat{x}$.*

Solution. For each number h one has

$$f(\hat{x} + h) - f(\hat{x}) = (\hat{x} + h)^2 - \hat{x}^2 = 2\hat{x}h + h^2$$

and so

$$\frac{f(\hat{x} + h) - f(\hat{x})}{h} = 2\hat{x} + h.$$

Taking the limit $h \to 0$, we get $f'(\hat{x}) = 2\hat{x}$.

Analogously, one can show that for the function $f(x) = x^n$ the derivative is $f'(x) = nx^{n-1}$. It is time for a more challenging example.

Example 1.2.2 *Compute the derivative of the absolute value function $f(x) = |x|$.*

Solution. The answer is obvious if you look at the graph of f and observe that it has a kink at $x = 0$ (Fig. 1.2). To begin with, you see

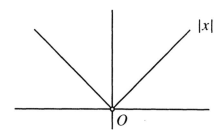

Figure 1.2 The absolute value function.

that f is not differentiable at $x = 0$. For $x > 0$ one has $f(x) = x$ and so $f'(x) = 1$, and for $x < 0$ one has $f(x) = -x$ and so $f'(x) = -1$. These results can be expressed by one formula,

$$|x|' = \frac{x}{|x|} = \operatorname{sgn}(x) \text{ for } x \neq 0.$$

FERMAT: ONE VARIABLE WITHOUT CONSTRAINTS

Without seeing the graph, we could also compute the derivative using the definition. For each nonzero number h one has

$$\frac{f(\widehat{x}+h) - f(\widehat{x})}{h} = \frac{|\widehat{x}+h| - |\widehat{x}|}{h}.$$

- For $\widehat{x} < 0$ this equals, provided $|h|$ is small enough,

$$\frac{(-\widehat{x}-h) - (-\widehat{x})}{h} = -1,$$

and this leads to $f'(\widehat{x}) = -1$.

- For $\widehat{x} > 0$ a similar calculation gives $f'(\widehat{x}) = 1$.

- For $\widehat{x} = 0$ this equals 1 for all $h > 0$ and -1 for all $h < 0$. This shows that $f'(0)$ does not exist.

The geometrical sense of the differentiability of f at $\widehat{x}$ is that the graph of f makes no "jump" at $\widehat{x}$ and has no "kink" at $\widehat{x}$. The following example illustrates this.

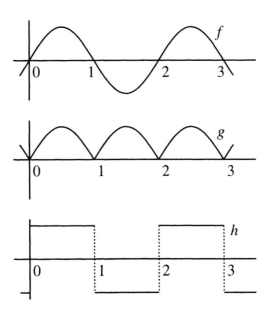

Figure 1.3 Illustrating differentiability.

Example 1.2.3 *In Figure 1.3, the graphs are drawn of three functions $f, g,$ and h. Determine for each of these functions the points of differentiability.*

Solution. The functions g and h are not differentiable at all integers, as the graph of g has "kinks" at these points and the graph of h makes "jumps" at these points. The functions are differentiable at all other points. Finally, f is differentiable at all points.

Sometimes other notation is used for the derivative, such as $\frac{d}{dx}f(\widehat{x})$ or $\frac{df}{dx}(\widehat{x})$. One writes $D^1(\widehat{x})$ for the set of functions of one variable x for which the first derivative at $\widehat{x}$ exists.

1.2.2 The differential calculus

Newton and Leibniz computed derivatives by using the definition of the derivative. This required great skill, and such computations were beyond the capabilities of their contemporaries. Fortunately, now everyone can learn how to calculate derivatives. *The differential calculus* makes it possible to determine derivatives of functions in a routine way. This consists of a list of basic examples such as

$$(x^r)' = rx^{r-1}, \ r \in \mathbb{R}; \ (a^x)' = a^x \ln a, \ a > 0;$$

$$(\sin x)' = \cos x; \ (\cos x)' = -\sin x; \ (\ln x)' = 1/x;$$

and rules such as the *product rule*

$$(f(x)g(x))' = f'(x)g(x) + f(x)g'(x)$$

and the *chain rule*

$$(f(g(x)))' = f'(g(x))g'(x).$$

In the $\frac{d}{dx}$ notation for the derivative this takes on a form that is easy to memorize,

$$\frac{dz}{dx} = \frac{dz}{dy}\frac{dy}{dx},$$

where z is a function of y and y a function of x.

Modest role of definitions. Thus the *definition* of the derivative plays a very modest role: to compute a derivative, you use a user-friendly *calculus*, but not the definition. Users might be pleasantly surprised to find that a similar observation can be made for most definitions.

The use of the differential calculus will be illustrated in the analysis of all concrete optimization problems.

1.2.3 Geologists and sin 1°

If you want an immediate impression of the power of the differential calculus, then you can read about the following holiday experience. This example will provide insight, but the details will not be used in the solution of concrete optimization problems.

Once a colleague of ours met in the mountains a group of geologists. They got excited when they heard he was a mathematician, and what was even stranger is the sort of questions they started asking him. They needed the first four or five decimals of ln 2 and of sin 1°. They even wanted him to explain how they could find these decimals by themselves. "Next time, we might need sin 2° or some cosine and then we won't have you."

Well, not wanting to be unfriendly, our friend agreed to demonstrate this to them on a laptop. However, the best computing power that could be found was a calculator, and this did not even have a sine or logarithm button. Excuse enough not to be able to help, but for some reason he took pity on them and explained how to get these decimals even if you have no computing power. The method is illustrated below for sin 1°.

The geologists explained why they needed these decimals. They possessed some logarithmic lists about certain materials, but these were logarithms to base 2 and the geologists needed to convert these into natural logarithms; therefore, they had to multiply by the constant ln 2. This explains ln 2. They also told a long, enthusiastic story about sin 1°. It involved their wish to make an accurate map of the site of their research, and for this they had to measure angles and do all sorts of calculations.

Example 1.2.4 *How to compute the first five decimals of* sin 1° *?*

Solution. Often the following observation can be used to compare two functions f and g on an interval $[a, b]$ (cf. exercise 1.6.38).

If $f(a) = g(a)$ and $f'(x) \leq g'(x)$ for all x, then $f(x) \leq g(x)$ for all x.

Now we turn to $\sin 1° = \sin \frac{\pi}{180}$, going over to radians.
- **First step.** To begin with, we establish that

$$0 \leq \sin x \leq x$$

for all $x \in [0, \frac{1}{2}\pi]$; that is, we "squash" $\sin x$ in between 0 and x, for all x of interest to the geologists. To this end, we apply the observation above. The three functions 0, $\sin x$, and x have the same value at $x = 0$, and the inequality between their derivatives,

$$0 \leq \cos x \leq 1,$$

holds clearly for all $x \in [0, \frac{1}{2}\pi]$.

- **Second step.** In the same way one can establish that

$$1 - \frac{1}{2}x^2 \leq \cos x \leq 1$$

for all $x \in [0, \frac{1}{2}\pi]$, using the inequality that we have just established. Indeed, the three functions $1 - \frac{1}{2}x^2$, $\cos x$, and 1 have the same value at $x = 0$, and the inequality between their derivatives,

$$-x \leq -\sin x \leq 0,$$

is essentially the inequality that we have just established. This is no coincidence: the expression $1 - \frac{1}{2}x^2$ is chosen in such a way that its derivative is $-x$ and that its value at $x = 0$ equals $\cos 0 = 1$.

- **Third step.** The next step gives

$$x - \frac{1}{6}x^3 \leq \sin x \leq x,$$

again for all $x \in [0, \frac{1}{2}\pi]$. We do not display the verification, but we note that we have chosen the expression $x - \frac{1}{6}x^3$ in such a way that

$$\frac{d}{dx}\left(x - \frac{1}{6}x^3\right) = 1 - \frac{1}{2}x^2$$

and $(x - \frac{1}{6}x^3)_{x=0} = \sin 0$.

- **Approximation is adequate.** Now let us pause to see whether we already have a sufficient grip on the sine to help the geologists. We have

$$\frac{\pi}{180} - \frac{1}{6}\left(\frac{\pi}{180}\right)^3 \leq \sin 1° \leq \frac{\pi}{180}.$$

Is this good enough? A rough calculation on the back of an envelope, using $\pi \approx 3$, gives

$$\frac{1}{6}\left(\frac{\pi}{180}\right)^3 \approx \frac{1}{6}\left(\frac{1}{60}\right)^3 = \frac{1}{6}\frac{1}{216000} \approx 10^{-6}.$$

That is, we have squashed our constant $\sin 1°$ in between two constants that we can compute and that differ by about 10^{-6}. This suffices; we have more than enough precision to get the required five decimals,

$$\sin 1° \approx \frac{\pi}{180} \approx 0.01745,$$

provided we know the first few decimals of $\pi \approx 3.1416$.

The calculation of $\ln 2$ can be done in essentially the same way, but requires an additional idea; we postpone this calculation to exercise 1.6.37.

Historical comments. What we have just done amounts to rediscovering the *Taylor polynomials* or *Taylor approximations* and the *Taylor series*, the method of their computation, and their use—for the functions sine and cosine at $x = 0$. Taylor series were discovered in 1712 by Brook Taylor (1685–1731), inspired by a conversation in a coffeehouse. However, Taylor's formula was already known to Newton before 1712, and it gave him great pleasure that he could use it to calculate everything. In 1676 Newton wrote to Oldenburg about his feelings regarding his calculations:

"it is my shame to confess with what accuracy I calculated the elementary functions $\sin, \cos, \log$ *etc."*

In some of the exercises you will be invited to a further exploration of this method of approximating functions (exercises 1.6.35–1.6.39).

Behind the sine button. You can check the answer of the example above by pushing the sine button of your calculator. There is no need to know how the calculator computes the sine of angles like $1°$. However, for the sake of curiosity, what is behind this button? Maybe tables of values of the sine are stored in the memory? It could also be that there is some device inside that can construct rectangular triangles with given angles and measure the sides? In reality, it is neither. What is behind the sine button (and some of the other buttons) is the method of the Taylor polynomials presented above.

Conclusion of section 1.2. The derivative is the basic notion of differential calculus. It can defined as the limit of the quotient of increments. However, to *calculate* it one should use lists of basic examples, and rules such as the product rule and the chain rule.

1.3 MAIN RESULT: FERMAT THEOREM FOR ONE VARIABLE

Let $\widehat{x}$ be a number and f a function of one variable, defined on an open interval containing $\widehat{x}$, say, the interval of all x with $|x-\widehat{x}| < \varepsilon$ for some positive number $\varepsilon > 0$. The function $f : (\widehat{x}-\varepsilon, \widehat{x}+\varepsilon) \to \mathbb{R}$ might also be defined outside this interval, but this is not relevant for the present purpose of formulating the Fermat theorem. It is sometimes convenient to consider minima and maxima simultaneously. Then we write extr (*extremum*). This sense of the word "extremum" did not exist until the nineteenth century, when it was invented by Paul David Gustav Du Bois-Reymond (1831–1889).

Definition 1.2 *Main problem: unconstrained one-variable optimization.* *The problem*

$$f(x) \to \text{extr} \qquad (P_{1.1})$$

is called a one-variable optimization problem without constraints.

The main task of problem $(P_{1.1})$ is to find the points of global extremum, but we will also find the points of local extremum, as a byproduct of the method. We will now be more precise about this. The concept of local extremum is useful, as the available method for finding the extrema makes use of differentiation and this is a *local* operation, using only the behavior near the solution. A minimum (resp. maximum) is from now on often called a *global* minimum (resp. maximum) to emphasize the contrast with a local minimum (resp. maximum). A point is called a point of local minimum (resp. maximum) if it is is a point of global minimum (resp. maximum) on a sufficiently small neighborhood. Now we display the formal definition of local extremum.

Definition 1.3 *Local minimum and maximum.* *Let $f : A \to \mathbb{R}$ be a function on some set $A \subseteq \mathbb{R}$. Then $\widehat{x} \in A$ is called a point of local minimum (resp. maximum) if there is a number $\varepsilon > 0$ such that $\widehat{x}$ is* a global minimum (resp. maximum) of the restriction of $f : A \to \mathbb{R}$ to the set of $x \in A$ that lie in the ε-neighborhood of $\widehat{x}$, that is, for which $|x - \widehat{x}| < \varepsilon$ (Fig. 1.4).

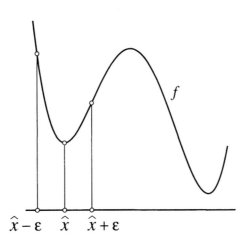

Figure 1.4 Illustrating the definition of local minimum.

An alternative term for global (local) minimum/maximum is *absolute* (*relative*) minimum/maximum.

Now we are ready to formulate the Fermat theorem.

Theorem 1.4 *Fermat theorem—necessary condition for one-variable problems without constraints.* *Consider a problem of type* $(P_{1.1})$. *Assume that the function* $f : (\widehat{x} - \varepsilon, \widehat{x} + \varepsilon) \to \mathbb{R}$ *is differentiable at* $\widehat{x}$. *If* $\widehat{x}$ *is a local extremum of* $(P_{1.1})$, *then*

$$f'(\widehat{x}) = 0. \tag{1.1}$$

The Fermat theorem follows readily from the definitions. We display the verification for the case of a minimum; then the case of a maximum will follow on replacing f by $-f$.

Proof. We have $f(\widehat{x} + h) - f(\widehat{x}) \geq 0$ for all numbers h with $|h|$ sufficiently small, by the local minimality of $\widehat{x}$.

- If $h > 0$, division by h gives $(f(\widehat{x} + h) - f(\widehat{x}))/h \geq 0$; letting h tend to zero gives $f'(\widehat{x}) \geq 0$.

- If $h < 0$, division by h gives $(f(\widehat{x}+h) - f(\widehat{x}))/h \leq 0$ ("dividing a nonnegative number by a negative number gives a nonpositive number"); letting h tend to zero gives $f'(\widehat{x}) \leq 0$.

Therefore, $f'(\widehat{x}) = 0$. □

Remark. This proof can be reformulated as follows. If the derivative of f at a point $\bar{x}$ is positive (resp. negative), then f is increasing (resp. decreasing) at $\bar{x}$; therefore, it cannot have a local extremum at $\bar{x}$. This formulation of the proof shows its algorithmic meaning. For example, if we want to minimize f, and have found $\bar{x}$ with $f'(\bar{x}) > 0$, then there exists $\bar{\bar{x}}$ slightly to the left of $\bar{x}$ that is "better than $\bar{x}$" in the sense that $f(\bar{\bar{x}}) < f(\bar{x})$.

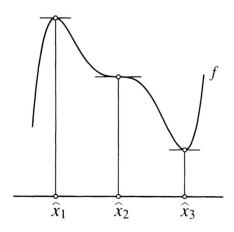

Figure 1.5 Illustrating the definition of stationary point.

Stationary points. Solutions of the equation $f'(x) = 0$ are called *stationary points* of the problem $(P_{1.1})$ (Fig. 1.5).

Trade-off. Pragmatic optimization problems always represent a trade-off between two opposite effects. Usually such a problem can be modeled as a problem of optimizing a function of one variable. We give a numerical example of such a trade-off, where one has to minimize the sum ("representing total cost") of two functions of a positive variable x ("representing the two effects"); one of the terms is an increasing and the other one a decreasing function of x. We will give an application of this type of problem to skyscrapers (exercise 1.6.14).

FERMAT: ONE VARIABLE WITHOUT CONSTRAINTS

Example 1.3.1 *Find the solution $\hat{x}$ of the following problem:*

$$f(x) = x + 5x^{-1} \to \min, \ x > 0.$$

Solution. By the Fermat theorem, $\hat{x}$ is a solution of the equation $f'(x) = 0$. Computation of the derivative of f gives $f'(x) = 1 - 5x^{-2}$. The resulting equation $1 - 5x^{-2} = 0$ has a unique positive solution $x = \sqrt{5}$. Therefore, the choice $\hat{x} = \sqrt{5}$ gives the optimal trade-off.

Remark. Unconstrained one-variable problems can be defined in the following more general and more elegant setting as follows. We need the concept of an open set.

A set V in $\mathbb{R}$, that is, a set V of real numbers, is called *open* if there exists for each number $v \in V$ a sufficiently small number $\varepsilon > 0$ such that the open interval $(v - \varepsilon, v + \varepsilon)$ is contained in V. For example, each open interval (a, b) is an open set. $V \in \mathcal{O}(\mathbb{R})$ means that V is an open set in $\mathbb{R}$; $V \in \mathcal{O}(\hat{x}, \mathbb{R})$ means that V is an open set in $\mathbb{R}$ containing $\hat{x}$, and such a set is called a *neighborhood* of $\hat{x}$ in $\mathbb{R}$.

Let $V \in \mathcal{O}(\mathbb{R})$ and $f : V \to \mathbb{R}$. Then the problem

$$f(x) \to \min \qquad (P_{1.1})$$

is called a *problem without constraints*. A point $\hat{x} \in V$ is called a point of *local minimum* if there exists $U \subset V$, $U \in \mathcal{O}(\hat{x}, \mathbb{R})$ such that $\hat{x}$ is a global minimum of the restriction of f to U.

1.3.1 Existence: the Weierstrass theorem

The Fermat theorem does not give a *criterion* for optimality. The following example illustrates this.

Example 1.3.2 *The function $f(x) = x^3$ has a stationary point. This is not a point of local extremum. Show this.*

Solution. Calculate the derivative $f'(x) = 3x^2$ and put it equal to zero, $3x^2 = 0$. This stationarity equation has a unique solution, $x = 0$. However, this is not a point of local extremum, as

$$f(x) = x^3 > 0 = f(0) \quad \text{for all } x > 0$$

and

$$f(x) = x^3 < 0 = f(0) \quad \text{for all } x < 0.$$

Therefore, we need a theorem to complement the Fermat theorem. Its formulation requires the concept of *continuity* of a function.

Definition 1.5 *Continuous function. A function $f : [a,b] \to \mathbb{R}$ on a closed interval $[a,b]$ is called continuous if*

$$\lim_{x \to \bar{x}} f(x) = f(\bar{x})$$

for all $\bar{x} \in [a,b]$ (that is, for $a \leq \bar{x} \leq b$).

The geometric sense of the concept of continuity is that the graph of the function $f : [a,b] \to \mathbb{R}$ makes no "jumps." That is, "it can be drawn without lifting pencil from paper."

How to verify continuity. To verify that a function is continuous on a closed interval $[a,b]$, one does not use the definition of continuity. Instead, this is done by using the fact that the basic functions such as

$$x^r, \ c^x, \ \sin x, \ \cos x, \ \ln x$$

are continuous if they are defined on $[a,b]$, and by using certain rules, for example, $f+g$, $f-g$ and fg are continuous on $[a,b]$ if f and g are continuous on $[a,b]$. In appendix C two additional rules are given.

We formulate the promised theorem for the case of minimization (of course it holds for maximization as well).

Theorem 1.6 *Weierstrass theorem. If a function $f : [a,b] \to \mathbb{R}$ on a closed interval is continuous, then the problem*

$$f(x) \to \min, \ a \leq x \leq b,$$

has a point of global minimum.

This fact is intuitively clear, but if you want to prove it, you are forced to delve into the foundations of the real numbers. For this we refer to appendix C.

In many applications, the variable does not run over a closed interval. Usually you can reduce the problem to the theorem above. We give an example of such a reduction. A function $f : \mathbb{R} \to \mathbb{R}$ on $\mathbb{R}$ is called *coercive (for minimization)* if

$$\lim_{|x| \to +\infty} f(x) = +\infty.$$

Corollary 1.7 *If a function $f : \mathbb{R} \to \mathbb{R}$ is continuous and coercive for minimization, then the problem $f(x) \to \min$, $x \in \mathbb{R}$, has a point of global minimum.*

Proof. We choose a number $M > 0$ large enough that $f(x) > f(0)$ for all x with $|x| > M$, as we may, by the coercivity of f. Then we can add to the given problem the constraints

$$-M \leq x \leq M$$

without changing the—possibly empty—solution set. In particular, this does not change the solvability status. It remains to note that the resulting problem,

$$f(x) \to \min, \ -M \leq x \leq M,$$

has a point of global minimum, by the Weierstrass theorem. □

Often you need variants of this result, as the following completion of the analysis of example 1.3.1 illustrates.

Example 1.3.3 *Show that the problem*

$$f(x) = x + 5x^{-1} \to \min, \ x > 0,$$

has a solution.

Solution. Observe that

$$\lim_{x \to +\infty} f(x) = +\infty \quad \text{and} \quad \lim_{x \downarrow 0} f(x) = +\infty.$$

Therefore, we can choose $M > 0$ large enough that

$$f(x) > f(1) \quad \text{if } x < 1/M \text{ or } x > M.$$

Then we can add to the given problem the constraints

$$1/M \leq x \leq M$$

without changing the—possibly empty—solution set. In particular, this does not change the solvability status. It remains to note that the resulting problem,

$$f(x) \to \min, \ 1/M \leq x \leq M,$$

has a point of global minimum, by the Weierstrass theorem.

Learning by doing. Such variants will be used without comment in the solution of concrete problems. Here, and in similar situations, we have kept the following words of Gauss in mind: Sed haec aliaque artificia practica, quae ex usu multo facilius quam ex praeceptis ediscuntur, hic tradere non est instituti nostri. (*But it is not our intention to treat of these details or of other practical artifices that can be learned more easily by usage than by precept.*)

Conclusion. Using only the Fermat theorem does not lead to certainty that an optimum has been found. The recommended method of completing the analysis of a problem is by using the Weierstrass theorem. This allows us to establish the existence of solutions of optimization problems.

1.3.2 Four-step method

In order to be brief and transparent, we will use a four-step method to write down the solutions of all problems in this book:

1. Model the problem and establish existence of global solutions.

2. Write down the equation(s) of the first order necessary conditions.

3. Investigate these equations.

4. Write down the conclusion.

What (not) to write down. We will use shorthand notation, without giving formal definitions for the notation, if the meaning is clear from the context. Moreover, we will always be very selective in the choice of what to display. For example, to establish existence of global solutions, one needs to verify the continuity of the objective function (cf. theorem 1.6). Usually, a quick visual inspection suffices to make sure that a given function is continuous. How to do this is explained above, and more fully in appendix C. We recommend, *not to write down* the verification. In particular, in solving each optimization problem in this book, we have checked the continuity of the objective function, but in the text there is no trace of this verification.

Now we illustrate the four-step method for example 1.3.3 (=example 1.3.1).

Example 1.3.4 *Solve the problem*

$$f(x) = x + 5x^{-1} \to \min, \ x > 0$$

by the four-step method.

FERMAT: ONE VARIABLE WITHOUT CONSTRAINTS

Solution

1. $f(0+) = +\infty$ and $f(+\infty) = +\infty$, and so existence of a global solution $\widehat{x}$ follows, using the Weierstrass theorem.

2. Fermat: $f'(x) = 0 \Rightarrow 1 - 5x^{-2} = 0$.

3. $x = \sqrt{5}$.

4. $\widehat{x} = \sqrt{5}$.

Logic of the four-step method. The Fermat theorem (that is, steps 2 and 3 of the four-step method) leads to the following conclusion: *if* the problem has a solution, then it must be $x = \sqrt{5}$. However, it is not possible to conclude from this implication alone that $x = \sqrt{5}$ is the solution of the problem. Indeed, consider again example 1.3.2: the Fermat method leads to the conclusion that if the problem $f(x) = x^3 \to$ min has a solution, then it must be $x = 0$. However, it is clear that this problem has no solution. This is the reason that the four-step method includes establishing the existence of a global solution (in step 1). By combining the two statements "there exists a solution" and "if a solution exists, then it must be $x = \sqrt{5}$," one gets the required statement: the unique solution is $x = \sqrt{5}$.

Why the four-step method? This problem could have been solved in one of the following two alternative ways. They are more popular, and at first sight they might seem simpler.

1. If we use the simple and convenient technique of the sign of the derivative, known from high school, then the Weierstrass theorem is not needed.

2. Another easy possibility is to use the second order test.

The great thing about the four-step method is that it is *universal*. It works for all optimization problems that can be solved analytically.

1. The sign of derivative test breaks down as soon as we consider optimization problems with two or more variables, as we will from chapter two onward.

2. The second order test can be extended to tests for other types of problems than unconstrained one-variable problems, but these tests are rather complicated. Moreover, these tests allow us to check *local* optimality only, whereas the real aim is to check

global optimality. The true significance of the second order tests is that they give a valuable insight, as we will make clear in section 1.5.1, and more fully in chapter five. We recommend not using them in problem solving.

The need for comparing candidate solutions. If step 3 of the four-step method leads to more than one candidate solution, then the solution(s) can be found by a comparison of values. The following example illustrates this.

Example 1.3.5   *Solve the problem*
$$f(x) = x/(x^2+1) \to \max$$
by the four-step method.

Solution

1. $\lim_{|x| \to +\infty} f(x) = 0$ and there exists a number $\bar{x}$ with $f(\bar{x}) > 0$, in fact $f(x) > 0$ for all $x > 0$, and so existence of a global solution $\hat{x}$ follows, using the Weierstrass theorem (after restriction to the closed interval $[-N, N]$ for a sufficiently large N).

2. Fermat: $f'(x) = 0 \Rightarrow ((x^2+1) - x(2x))/(x^2+1)^2 = 0$.

3. $x^2 + 1 - 2x^2 = 0 \Rightarrow x^2 = 1 \Rightarrow x = \pm 1$. Compare values of f at the candidate solutions 1 and -1:
$$f(-1) = -\frac{1}{2} \text{ and } f(1) = \frac{1}{2}.$$

4. $\hat{x} = 1$.

Variant of the coercivity result. In this example we see an illustration of one of the announced variants of corollary 1.7. It can be derived from the Weierstrass theorem in a similar way as corollary 1.7 has been derived from it.

Logic of the four-step method. The Fermat theorem (that is, steps 2 and 3 of the four-step method) leads to the following conclusion: the set of solutions of the problem is included in the set $\{1, -1\}$, to be called the set of candidate solutions. However, this statement does not exclude the possibility that the problem has no solution. By comparing the f-values of the candidates, one is led to the following more precise statement: if there exists a solution, then it must be $x = 1$. By combining this with the statement that a solution exists, which is established in step 1, one gets the required statement: there is a unique solution $x = 1$.

Four-step method is universal. We will see that the four-step method is the universal method for solving analytically all concrete optimization problems of all types. A more basic view on this method is sketched in the remark following the proof of the fundamental theorem of algebra (theorem 2.8).

1.3.3 Uniqueness; value of a problem

The following observation is useful in concrete applications. It can be derived from the Weierstrass theorem.

Corollary 1.8 *Uniqueness. Let f be a function of one variable, for which the second derivative $f'' = (f')'$ is defined on an interval $[a, b]$. If*

$$f'(a) < 0, \; f'(b) > 0, \; \text{and} \; f''(x) > 0, \; a < x < b,$$

then the problem

$$f(x) \to \min, \; a < x < b,$$

has a unique point of global minimum.

The following example illustrates how this observation can be used.

Example 1.3.6 Solve the problem $f(x) = x^2 - 2\sin x \to \min$.

Solution

1. The existence and uniqueness of a global solution $\hat{x}$ follow from the remark above. Indeed,

   $$f''(x) = 2(1 + \sin x) > 0$$

   for all x except at points of the form $\frac{3}{2}\pi + 2k\pi, \; k \in \mathbb{Z}$, where it is 0. Moreover,

   $$f'(x) = 2x - 2\cos x,$$

   and so $f'(a) < 0$ and $f'(b) > 0$ for a sufficiently small and b sufficiently large.

2. Fermat: $f'(x) = 0 \Rightarrow 2(x - \cos x) = 0$.

3. The stationarity equation cannot be solved analytically.

4. There is a unique point of global minimum. It can be characterized as the unique solution of the equation $x = \cos x$.

Push the cosine button. Here is a method to find the solution of this problem numerically: if you repeatedly push on the cosine button of a calculator (having switched from degrees to radians to begin with), then the numbers on the display stabilize at 0.73908. This is the unique solution of the equation $x = \cos x$ up to five decimals.

Value problem. Finally, we mention briefly the concept of *value* of an optimization problem, which is useful in the analysis of some concrete problems. For a solvable optimization problem it can be defined as the value of the objective function at a point of global solution. The concept of value can be defined, more generally, for all optimization problems. The precise definition and the properties of the value of a problem are given in appendix B. We will write $S_{\min}$ (resp. $S_{\max}$) for the value of a minimization (resp. maximization) problem.

1.3.4 Illustrations of the Fermat theorem

What explains the success of the Fermat theorem? It is the combination with the power of the differential calculus, which makes it easy to calculate derivatives of functions. Thus, unconstrained optimization problems are reduced to the easier task of solving equations. We give two illustrations.

Example 1.3.7 *Consider the quadratic problem*

$$f(x) = ax^2 + 2bx + c \to \min$$

with $a \neq 0$. If $a > 0$ it has a unique solution $\widehat{x} = -b/a$ and minimum value

$$(ac - b^2)/a.$$

If $a < 0$ it has no solution. Verify these statements.

For those who know matrices and determinants: one can view the minimal value as the quotient of determinants of two symmetric matrices of size 2×2 and 1×1:

$$\begin{pmatrix} a & b \\ b & c \end{pmatrix} \text{ and } (a).$$

The resulting formula can be generalized to quadratic polynomials $f(x_1, \ldots, x_n)$ of n variables, giving the quotient of the determinants of two symmetric matrices of size $(n+1) \times (n+1)$ and $n \times n$, as we will see in theorem 9.4

Solution 1. One does not need the Fermat theorem to solve this problem: one can use one's knowledge of parabolas. It is well known that for $a > 0$ the graph of f is a *valley parabola*, which has a unique point of minimum

$$\widehat{x} = -b/a.$$

Therefore, its minimal value is

$$f(\widehat{x}) = (ac - b^2)/a.$$

For $a < 0$, the graph of f is a *mountain parabola*, and so there is no point of minimum.

Solution 2. The solution of the problem follows also immediately from the formula that you get for $a \neq 0$ by "completing the square,"

$$ax^2 + 2bx + c = a(x + b/a)^2 + (ac - b^2)/a.$$

Solution 3. However, it is our aim to illustrate the Fermat theorem.

1. To begin with,

$$f(x) = x^2(a + 2b/x + c/x^2);$$

the first factor of the right-hand side tends to $+\infty$ and the second one to a for $|x| \to +\infty$. If $a < 0$, it follows that the problem has value $-\infty$ and so it has no point of global minimum. If $a > 0$, it follows that f is coercive and so a solution $\widehat{x}$ exists.

2. Fermat: $f'(x) = 0 \Rightarrow 2(ax + b) = 0$.

3. $x = -b/a$.

4. If $a > 0$, then there is a unique minimum, $\widehat{x} = -b/a$; if $a < 0$, then $S_{\min} = -\infty$.

1.3.5 Reflection

The real fun. Four types of things have come up in our story: concrete problems worked out by some calculations, theorems, definitions, and proofs. Each of these has its own flavor and plays its own role. To be honest, the real fun is working out a concrete problem. However, we use a tool to solve the problem: the Fermat theorem. In the formulation of this result, the concepts "derivative" and "local extremum" occur.

Need for a formal approach. Therefore, we feel obliged to give precise definitions of these concepts and a precise proof of this result. Reading and digesting the proof might be a challenge (although it must be said that the proof of the Fermat theorem does not require any ideas; it is essentially a mechanical verification using the definitions). It might also require some effort to get used to the definitions.

Modest role of the formal approach. You might of course choose not to look very seriously at this definition and this proof. We have already emphasized the modest role of the *definition* of the derivative for practical purposes. In the same spirit, reading the proof gives you insight into the Fermat theorem, but this insight is not required for the solution of concrete problems.

These remarks apply to all worked-out problems, theorems, definitions, and proofs.

1.3.6 Comments: physical, geometrical, analytical, and approximation approach to the derivative and the Fermat theorem

Initially, there were three—equivalent—approaches to the notion of derivative and the Fermat theorem: *physical, geometrical*, and *analytical*. To these a fourth one was added later: *by means of approximation*. The analytical approach, due to Cauchy, is the one we have used to define the derivative.

Physical approach. The *physical* approach is due to Newton. Here is how he formulated the two basic problems that led to the differential and integral calculus.

" *To clarify the art of analysis, one has to identify some types of problems. [...]*

1. Let the covered distance be given; it is demanded to find the velocity of movement at a given moment.

2. Let the velocity of movement be given; it is demanded to find the length of the covered distance."

That is, for Newton the derivative is a *velocity*: the first problem asks to find the derivative of a given function. To be more precise, let a function $s(t)$ represent the position of an object that moves along a straight line and let the variable t represent time. Then the derivative $s'(t)$ is the *velocity* $v(t)$ at time t. We give a classical illustration.

Example 1.3.8 *Let us relate two facts that are well known from high school, but are often learned separately. Galileo Galilei (1564–1642), the astronomer who defended the heliocentric theory, is also known for his experiments with stones, which he dropped from the tower of Pisa. His experiments led to two formulas,*

$$s(t) = \frac{1}{2}gt^2$$

for the distance covered in the vertical direction after t seconds, and

$$v(t) = gt$$

for the velocity after t seconds. Here g is a constant ($g \approx 10$). Show that these two formulas are related.

Solution. This was one of the examples of Newton for illustrating his insight that velocity is always the derivative of the covered distance. This insight threw a new light on the formulas of Galileo:

the derivative of the expression $\frac{1}{2}gt^2$ is precisely gt.

Thus the formulas of Galileo are seen to be related.

Newton formulated the Fermat theorem in terms of velocity by the following phrase, which we used as the epigraph to this chapter:

"when a quantity is the greatest or the smallest, at that moment its flow is neither forward nor backward."

Geometrical approach. The *geometrical* approach is due to Leibniz and well known from high school. He interpreted the derivative as the *slope of the tangent line* (Fig. 1.1). This is $\tan \alpha$, the *tangent* of the angle α between the tangent line and the horizontal axis. Then the Fermat theorem means that the tangent line at a point of local extremum is horizontal, as we have already mentioned.

Weierstrass. The *approximation* approach was given much later, by Karl Weierstrass (1815–1897). He was one of the leaders in rigor in analysis and is known as the "father of modern analysis." In addition he is considered to be one of the greatest mathematics teachers of all time. He began his career as a teacher of mathematics at a secondary school. This involved many obligations: he was also required to teach physics, botany, geography, history, German, calligraphy, and even gymnastics. During a part of his life he led a triple life, teaching during the day, socializing at the local beer hall during the evening, and doing mathematical research at night. One of the examples of his rigorous approach was the discovery of a function that, although continuous, has no derivative at any point. This counter-intuitive function caused dismay among analysts, who depended heavily on their intuition for their discoveries.

The reputation of his university teaching meant that his classes grew to 250 pupils from all around the world. His favorite student was Sofia Kovalevskaya (1850–1891), whom he taught privately since women were not allowed admission to the university. He was a confirmed bachelor and kept his thoughts and feelings always to himself. To this rule, Sofia was the only exception. Here is a quotation from a letter to her from their lifelong correspondence:

"dreamed and been enraptured of so many riddles that remain for us to solve, on finite and infinite spaces, on the stability of the world system, and on all the other major problems of the mathematics and the physics of the future [...] you have been close [...] throughout my life [...] and never have I found anyone who could bring me such understanding of the highest aims of science and such joyful accord with my intentions and basic principles as you."

Approximation approach. Another fruit of the work of Weierstrass on the foundations of analysis is his *approximation* approach to the derivative. One can split the increment $f(\widehat{x}+h) - f(\widehat{x})$ in a unique way as a sum of a term that depends on h in a linear way, called the *main linear part*, and a term that is negligible in comparison to h if h tends to zero, called the *remainder*. Then the main linear part equals the product of a constant and h. This constant is defined to be the derivative $f'(\widehat{x})$.

This is *the universal approach*, as we will see later (definition 2.2 and section 12.4.1), whenever we have to extend the notion of derivative. Moreover, it is the most fruitful approach to the derivative for economic applications. We will give the formal definition of the approximation definition of the derivative. If r is a function of one variable, then we write $r(h) = o(h)$ ("small Landau o") if

$$\lim_{h \to 0} \frac{|r(h)|}{|h|} = 0.$$

For example, $r(h) = h^2$ is $o(h)$, and we write $h^2 = o(h)$. The sense of this concept is that $|r(h)|$ is negligible in comparison to $|h|$, for $|h|$

sufficiently small—to be more precise, $|r(h)|/|h|$ is arbitrary small for $|h|$ small enough.

Definition 1.9 *Approximation definition of the derivative. Let $\widehat{x}$ be a real number and f a function of one variable x. The function f is called differentiable at $\widehat{x}$ if it is defined, for some $\varepsilon > 0$, on the ε-neighborhood of $\widehat{x}$, that is,*

$$f : (\widehat{x} - \varepsilon, \widehat{x} + \varepsilon) \to \mathbb{R},$$

and if, moreover, there exist a number $a \in \mathbb{R}$ and a function $r : (-\varepsilon, +\varepsilon) \to \mathbb{R}$, for which $r(h) = o(h)$, such that

$$f(\widehat{x} + h) - f(\widehat{x}) = ah + r(h)$$

for all $h \in (-\varepsilon, +\varepsilon)$. Then a is called the (first) derivative of f at $\widehat{x}$ and it is denoted by $f'(\widehat{x})$.

We illustrate this definition for example 1.1.1.

Example 1.3.9 *Compute the derivative of the quadratic function $f(x) = x^2$ at a given number $\widehat{x}$, using the approximation definition.*

Solution. For each number h, one has

$$f(\widehat{x} + h) - f(\widehat{x}) = (\widehat{x} + h)^2 - \widehat{x}^2 = 2\widehat{x}h + h^2,$$

so this can be split up as the sum of a linear term $2\widehat{x}h$ and a small Landau o term h^2. The coefficient of $2\widehat{x}h$ is $2\widehat{x}$; therefore,

$$f'(\widehat{x}) = 2\widehat{x}.$$

As a further illustration of the approximation definition, we use it to write out the proof of the Fermat theorem 1.4 (again for the case of minimization):

Proof. $0 \leq f(\widehat{x} + h) - f(\widehat{x}) = f'(\widehat{x})h + o(h)$ and so, dividing by h and letting h tend to zero from the right (resp. left) gives $0 \leq f'(\widehat{x})$ (resp. $0 \geq f'(\widehat{x})$). Therefore, $f'(\widehat{x}) = 0$. □

Conclusion. There are four—equivalent—approaches to the notion of derivative and the Fermat theorem, each shedding additional light on these central issues.

1.3.7 Economic approach to the Fermat theorem: equimarginal rule

Let us formulate the economic approach to the Fermat theorem. A rational decision maker takes a marginal action only if the marginal benefit of the action exceeds the marginal cost. Therefore,

in the optimal situation the marginal benefit and the marginal cost are equal.

This is called the *equimarginal rule*. To see the connection with the Fermat theorem, note that if we apply the Fermat theorem to the maximization of profit, that is, of the difference between benefit and cost, then we get the *equimarginal* rule ("marginal" is another word for derivative).

The tradition in economics of carrying out a *marginal analysis* was initiated by Alfred Marshall (1842–1924). He was the leading figure in British economics (itself dominant in world economics) from 1890 until his death in 1924. His mastery of mathematical methods never made him lose sight that the name of the game in economics is to understand economic phenomena, whereas in mathematics the depth of the result is the crucial point.

Conclusion section. All optimization problems that can be solved analytically can be solved by one and the same four-step method. For problems of one variable without constraints, this method is based on a combination of the Fermat theorem with the Weierstrass theorem.

1.4 APPLICATIONS TO CONCRETE PROBLEMS

1.4.1 Original illustration of Fermat

Fermat illustrated his method by solving the geometrical optimization problem on the largest area of a right triangle with given sum a of the two sides that make a right angle.

Problem 1.4.1 *Solve the problem of Fermat if the given sum a is 10.*

Solution. The problem can be modeled as follows

$$f(x) = \frac{1}{2}x(10-x) = -\frac{1}{2}x^2 + 5x \to \max, \ 0 < x < 10.$$

This is a special case of example 1.3.7, if we rewrite our maximization problem as a minimization problem by replacing $f(x)$ by $-f(x)$. It follows that it has a unique solution, $\hat{x} = 5$. That is, the right triangle for which the sides that make a right angle both have length 5 is the unique solution of this problem.

Remark. The solution of the problem of Fermat in its general form, that is, for $a > 0$,

$$\frac{1}{2}x_1 x_2 \to \max, \quad x_1 + x_2 = a, \quad x_1, x_2 > 0,$$

is $\widehat{x}_1 = \widehat{x}_2 = a/2$.

Corollary 1.10 *Inequality of the geometric-arithmetic means. The following inequality holds true for all positive numbers x_1, x_2:*

$$\sqrt{x_1 x_2} \leq (x_1 + x_2)/2.$$

Proof. Choose arbitrary positive numbers x_1, x_2. Write $a = x_1 + x_2$; then by the solution of the Fermat problem

$$\sqrt{x_1 x_2} \leq \sqrt{(a/2)^2} = a/2$$

and, by $a/2 = (x_1 + x_2)/2$, we have established the required inequality

$$\sqrt{x_1 x_2} \leq (x_1 + x_2)/2.$$

□

1.4.2 A rose in a lemonade glass

The next problem illustrates how the search for the golden mean between too little and too much leads to an optimization problem.

Problem 1.4.2 *A girl gets a beautiful long-stemmed rose from her friend. She does not have a suitable vase, but then she thinks of one of her lemonade glasses. This glass has the form of a cylinder and is very narrow—the diameter is about 4 cm—and it is quite long for a glass, almost 20 cm. She puts the rose in the glass, but the rose makes the glass tumble and she catches the glass just in time.*

Then she puts some water in the glass and tries again. This is definitely better, but still not good enough. She does not give up and adds more water. Now all is well; the glass with water and rose are stable. However, to be on the safe side, she adds even more water. To her surprise, this makes things worse: the glass again becomes unstable and almost tumbles.

Can you calculate how much water gives optimal stability?

Modeling a problem always requires some technical knowledge about the subject you are dealing with. For example, in this problem we are confronted with the stability of an object. Here the height of the *center of gravity* of the object plays a role.

Technical background on centers of gravity

- **Center of gravity.** Now we provide the required technical background for the next problem. The center of gravity of an object is a point in space. Intuitively, it lies somewhere in the middle of the mass of the object. The precise definition is not needed here.

- **A formula.** Often, it is of interest to know the height of the center of gravity of an object that is made up of two objects, say, of weights m_1 and m_2 g (=gram) and with centers of gravity at heights h_1 and h_2 cm. This height is the *weighted average* of h_1 and h_2, where h_i has weight m_i, $i = 1, 2$. This is defined by the following formula:

$$(m_1/(m_1 + m_2))h_1 + (m_2/(m_1 + m_2))h_2.$$

- **Qualitative insight into the formula.** You can get a first insight into this formula by means of "exaggeration": if m_1 is much larger than m_2, then the height of the center of gravity of the total object is almost equal to that of the first object. More generally, the larger m_1 compared to m_2, the more the center of gravity is pulled toward the center of gravity of the first object.

- **Stability and center of gravity.** The role of the height of the center of gravity for the stability of an object is that the higher (lower) this height is, the less (more) stable the object is.

- **Application of the formula to stability.** For example, suppose that you are climbing a stepladder. Then there is the problem that you feel less and less stable, especially if you are "heavy." Let us consider this problem from the point of view of the concept of center of gravity. The formula above shows that the height of the center of gravity of you and the stepladder, taken together, goes up, when you climb. Moreover, the heavier you are, the faster this center goes up, and so the higher it is when you are on top of the stepladder. Thus, the formula above throws light on this decreasing stability effect, and this insight is readily seen to lead to practical solutions to the problem.

Solution
A simplified model. We model the glass with the rose by a cylinder of height 20 cm, of weight 100 g, and with a bottom area of 10 cm². That is, we omit the rose and ignore the heavy bottom of the glass. This simple model suits our purpose. Indeed, such a cylinder is not very stable, just like the glass with rose. Our model without the rose will give us some insight into what happens to the center of gravity if we add water.

The optimization problem in words. We want to find out how much water we should add if we want the center of gravity to be as low as possible—for optimal stability. A more refined model should include the heavy bottom and the rose: the rose is tilting over and this affects the stability as well.

The model in formulas. In the initial situation, the center of gravity is rather high, at a height half of that of the glass, 10 cm.

If we add water to a height of h cm, then the center of gravity of the water is at a height of $\frac{1}{2}h$ cm, the volume of the water is $10h$ cm³, and so its weight is $10h$ g (we recall that 1 cm³ of water weighs 1 g).

Now, we compute the height $g(h)$ of the center of gravity of cylinder and water, taken together, using the formula above, with

$$m_1 = 100,\ h_1 = 10\ (\text{``glass''}) \text{ and } m_2 = 10h,\ h_2 = h/2\ (\text{``water''}).$$

This gives that the height of the center of gravity of cylinder and water equals the weighted average:

$$g(h) = \frac{100}{100 + 10h} 10 + \frac{10h}{100 + 10h}\left(\frac{1}{2}h\right) = \frac{1000 + 5h^2}{100 + 10h}.$$

If we plot the graph of the function g, we see that it starts in $h = 0$ at level 10 ("empty glass"), then with increasing h it first goes down, and then goes up to its final level 10 at $h = 20$ ("glass filled with water"). To be precise, we can proceed as follows.

Solving the optimization problem. The problem is modeled as follows:

1. $g(h) = (1000 + 5h^2)/(100 + 10h) \to \min,\ 0 \leq h \leq 20$.

Existence of a global minimum $\widehat{h}$ follows from Weierstrass.

2. Fermat : $g'(h) = 0 \Rightarrow$
$$(10h(100 + 10h) - (1000 + 5h^2)10)/(100 + 10h)^2 = 0.$$

3. $50(h^2 + 20h - 200) = 0$. The only positive root of this equation is

$$h = 10\sqrt{3} - 10 \approx 7.3.$$

Thus the list of candidate minima consists of the stationary point $10\sqrt{3} - 10$ and the two boundary points 0 and 20.

Compare:

$$g(0) = 10, \ g(20) = 10, \ g(10\sqrt{3} - 10) = 10\sqrt{3} - 10 \approx 7.3.$$

4. $\widehat{h} = 10\sqrt{3} - 10$. That is, the conclusion of this numerical example is that the cylinder should be filled slightly more than a third with water. This lowers the center of gravity from 10 cm to about 7.3 cm.

Explanation of a "coincidence"

The "coincidence." In this example we have found that in the optimal situation the level of the water $\widehat{h}$ equals $g(\widehat{h})$, the height of the center of gravity of glass and water: both are equal to $10\sqrt{3} - 10 \approx 7.3$. This might be a surprising outcome at first sight.

How does center of gravity change? However, it is no coincidence: it is always true, for an intuitive physical reason. Indeed, let us analyze the effect of adding a bit of water to a glass that is already partially filled with water. The key observation is that if you add to a first object a second object, then the effect on the height of the center of gravity depends on where you add it: it is lowered if you add it below the original center of gravity and raised if you add it above it.

[**How does average mark change?** We note in passing that this effect is known to all high school students who are keeping track of their average grade: this average goes up (down) precisely if the latest grade is higher (lower) than the last average.]

"Coincidence" explained. Applying this to our lemonade glass, we see that in the beginning we add water below the center of gravity of glass and water, and therefore the center of gravity of glass and water goes down, starting from level 10, while the level of the water rises, starting from level zero. There will come a moment that these two levels are equal. If we were to add water after that moment, this water would end up above the center of gravity of glass and water, and therefore, the center of gravity of glass and water would go up again.

FERMAT: ONE VARIABLE WITHOUT CONSTRAINTS

1.4.3 Saving the World

Problem 1.4.3 *To save the world from destruction, agent 007 must reach a skiff 50 meters off shore from a point 100 meters away along a straight beach, and then disarm a timing device. The agent can run along the shore on the shifting sand at 5 meters per second, swim at 2 meters per second, and disarm a timing device in 30 seconds. Assuming that the device is set to trigger destruction in 74 seconds, is it possible for 007 to succeed?*

Solution. Anyone but 007 would run 100 meters along the beach. This would take 20 seconds. After this, he would throw himself into the water and swim the 50 meters, which would take 25 seconds. This leaves 74-20-25=29 seconds to disarm the bomb. Too bad for the world; this is not enough. Agent 007 does not run the full 100 meters, but x meters less; then he jumps into the water and swims in a straight line to the skiff (Fig. 1.6).

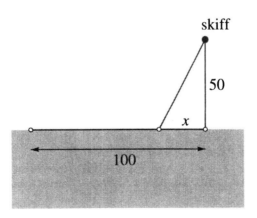

Figure 1.6 Saving the world.

1. Then he has

$$f(x) = 74 - \frac{100-x}{5} - \frac{\sqrt{x^2 + 50^2}}{2}$$

seconds to disarm the bomb. Is this enough? To begin with, 007 realizes that it is not necessary for his purpose to check that a global solution $\widehat{x}$ exists (however, we note in passing that f is coercive for maximization, using that $\sqrt{x^2 + 50^2} \approx |x|$ for sufficiently large $|x|$).

2. Suppose that he calculates in a flash the derivative of f in order to choose x:

$$f'(x) = \frac{1}{5} - \frac{1}{4}\frac{2x}{\sqrt{x^2 + 50^2}}.$$

Putting this derivative equal to zero, he is led to the equation

$$10x = 4\sqrt{x^2 + 50^2}.$$

3. Squaring gives $100x^2 = 16(x^2 + 50^2)$; this equation has one positive solution only, $\widehat{x} \approx 21.82$. This is the choice of 007. A calculation gives $f(\widehat{x}) \approx 31.09$, so more than 30.

4. Just enough to save the world.

1.4.4 Optimal price for a car

We will solve the following economic problem, with a marginal analysis. Suppose you want to sell one object, for example, a car or a house, to a potential buyer. What is the best price to ask? If you ask a high price he might not buy it; if you ask a low price, you will have only a small profit from the transaction.

You can find the price that gives the optimal trade-off between these two effects, if you know the *distribution* of the price the buyer is prepared to pay. For example, the buyer's price might lie between two known bounds and you might consider its distribution to be *uniform*; that is, all prices between these bounds have the same chance of being the buyer's price, as far as you know.

Problem 1.4.4 *You want to sell someone your car, which is worth 9,000 euros to you. You don't know how much he is prepared to pay for it. You assume that it is at least 10,000 euros and not more than 15,000 euros; apart from this you consider all possibilities as equally likely. To be more precise, you assume a uniform distribution for the buyer's price. What price should you ask if you want the highest expected profit?*

FERMAT: ONE VARIABLE WITHOUT CONSTRAINTS

Solution. Suppose you plan to ask a price x (in units of 1,000 euros). You may assume that $10 \leq x \leq 15$, as other prices are certainly not optimal. The probability that this price is accepted is

$$(15 - x)/(15 - 10) = (15 - x)/5,$$

and your profit will be $x - 9$. Therefore, your expected profit is the product

$$f(x) = (x - 9)(15 - x)/5.$$

- Before you carry out your plan, you do a marginal analysis: if you made an incremental change $\triangle x$ in the price, this would give the following incremental change in the expected profit:

$$f(x+\triangle x) - f(x) = (x+\triangle x-9)(15-(x+\triangle x))/5 - (x-9)(15-x)/5.$$

Expanding the brackets and neglecting $(\triangle x)^2$ gives

$$((24 - 2x)/5)\triangle x.$$

This shows that if the expression $(24 - 2x)/5$ is positive (respectively negative), then you can increase your expected profit by asking a slightly higher (resp. lower) price: by making a small incremental positive (resp. negative) change $\triangle x$.

You can continue to make incremental changes till you come to a price x for which $(24 - 2x)/5$ is zero. Then, no incremental change leads to an improvement; that is, you have arrived at a maximum (strictly speaking a *local* maximum). The condition that we have obtained, $(24 - 2x)/5 = 0$, gives $x = 12$. The conclusion is that you should ask 12,000 euros for your car.

- Thus we have given an example of a marginal analysis. If our aim had been to give a quick solution of the problem, we could have argued as follows. The graph of f is a mountain parabola that intersects the horizontal axis in two points, $x = 9$ and $x = 15$. Therefore, its top lies at $x = 12$, precisely in between.

- Alternatively, we could have used that the present problem is a special case of example 1.3.7.

- Finally, we display the solution with the four-step method.

  1. Consider the problem

  $$f(x) = (x - 9)(15 - x)/5 = (-x^2 + 24x - 135)/5 \to \max,$$

$$10 \leq x \leq 15.$$

Existence of a global solution $\widehat{x}$ follows from Weierstrass.

2. Fermat: $f'(x) = 0 \Rightarrow (-2x+24)/5 = 0$.
3. $x = 12$.
4. $\widehat{x} = 12$.

1.4.5 How high does an arrow go?

To solve the following problem, it is helpful to use the physical interpretation of the Fermat theorem, as suggested by Newton's phrase. We will solve this problem, using the law of conservation of energy.

Problem 1.4.5 *Someone shoots an arrow straight up into the air. How high will it go if he manages to give it an initial velocity of v_0 meters per second?*

Solution. Let the height of the arrow after t seconds be $s(t)$. Then the derivative $s'(t)$ is the velocity at time t. We neglect the length of the person, that is, $s(0) = 0$, and we neglect the resistance of the air. What will we see when the arrow shoots up?

It will lose speed as it goes up. At its highest point it will seem to come to a standstill for a fraction of a second. In the words of Newton: "when a quantity is the greatest or the smallest, at that moment its flow is neither forward nor backward." This observation is precisely the Fermat theorem:

at the point $\widehat{t}$ of maximum of $s(t)$, the velocity $s'(\widehat{t})$ is zero.

Now we use the law of conservation of energy and the observation that only two types of energy play a role here, kinetic energy ("energy of movement") $\frac{1}{2}ms'(t)^2$ and potential energy ("energy of position") $mgs(t)$; here m is the mass of the arrow and g is the constant of gravity, which is approximately 10. The law of conservation of energy gives that the sum

$$\frac{1}{2}ms'(t)^2 + mgs(t)$$

has the same value at all moments t. Note that at $t = 0$, there is only kinetic energy $\frac{1}{2}mv_0^2$—as $s(0) = 0$—and at $t = \widehat{t}$, there is only potential energy $mgs(\widehat{t})$—as $s'(\widehat{t}) = 0$. Therefore, the following equality holds true

$$\frac{1}{2}mv_0^2 = mgs(\hat{t}).$$

It follows that the maximal height reached by the arrow is

$$v_0^2/2g.$$

In particular, this height does not depend on the mass m of the arrow.

We will not complete the analysis of this problem, as this not of interest for our purpose: it is obvious from observation that the answer we have obtained is the global maximum.

1.4.6 Path of light and optimization

Light chooses the quickest path. The *principle of least time*, the first *variational principle*, is also due to Fermat (1662). According to this principle, light "chooses" the path (from point to point) that minimizes the time it takes to travel. To be more precise, this *principle* states that the path taken is only *stationary* and not necessarily a global or even a local minimum. Therefore, *the principle of stationary time* might have been a better terminology.

Application 1: Snellius's law. This principle can, for example, be applied to the problem of the refraction of a ray of light at the boundary between two media where light has different velocities, v_1 and v_2. This leads to the formula

$$\frac{\sin \alpha}{\sin \beta} = \frac{v_1}{v_2}$$

(Fig. 1.7), which is now known as the law of Snellius.

In 1621 Willebrord van Roijen Snell (1580–1626) had discovered empirically that the quotient $\sin \alpha / \sin \beta$, called the *refraction index*, is the same for each ray. Fermat's idea to introduce the velocities of light and to formulate his variational principle gave an explanation of this empirical discovery.

However, at least as important to convince people of the value of this principle is that it predicts new things.

Application 2: relation between refraction indices. For example, suppose we do experiments to measure for each pair (i,j) chosen from three media, say glass (1), water (2), and air (3), the

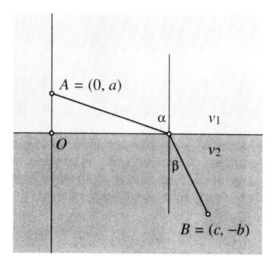

Figure 1.7 Refraction of a ray of light.

refraction index n_{ij}. Then the Fermat principle predicts the following relation between the results of these measurements:

$$n_{23} = \frac{n_{13}}{n_{12}}.$$

Indeed,

$$n_{23} = \frac{v_2}{v_3} = \frac{v_1/v_3}{v_1/v_2} = \frac{n_{13}}{n_{12}}.$$

Using only the discovery of Snell, such a prediction could not be made.

Application 3: angles of refraction and speeds of light. Another argument for this principle is that after measuring *angles* of refraction of light between the media water and air, we can predict the outcome of another experiment: a measurement of the *speed* of light in water and air will show that the first is less than the second. That these angles predict something about these speeds is a spectacular and convincing success of the variational principle of Fermat. It was only much later that velocities of light in different media could be measured accurately enough to verify these predictions. Exercises 1.6.7–1.6.13 are related to the principle of least time; in some of these we will give additional consequences of this principle.

Problem 1.4.6 *The law of Snellius. Derive the law of Snellius for the refraction of light at the boundary between two media from the*

FERMAT: ONE VARIABLE WITHOUT CONSTRAINTS

variational principle of Fermat that light "always takes the quickest path."

Solution. We begin by modeling the problem. Let the two media be separated by the horizontal coordinate axis, and let us assume that the velocity of light in the upper medium is v_1 and in the lower v_2. A ray of light traveling from the point $A = (0, a)$ to the point $B = (c, -b)$ (Fig. 1.7) breaks on the boundary of these media. The problem is to find the break-point $\hat{x}$.

1. This point is, according to the principle of Fermat, a stationary point of the problem,

$$f(x) = \frac{\sqrt{a^2 + x^2}}{v_1} + \frac{\sqrt{b^2 + (c-x)^2}}{v_2} \to \min.$$

Existence and uniqueness of a global solution $\hat{x}$ follow from corollary 1.8, as f is coercive (note that $f(x) \approx |x|/v_1 + |x|/v_2$ for sufficiently large $|x|$) and as the following inequality holds true:

$$f''(x) = \frac{a^2}{v_1(a^2+x^2)^{1\frac{1}{2}}} + \frac{b^2}{v_2(b^2+(c-x)^2)^{1\frac{1}{2}}} > 0$$

for all x.

2. Fermat: $f'(x) = 0 \Rightarrow$

$$\frac{x}{v_1\sqrt{a^2+x^2}} - \frac{c-x}{v_2\sqrt{b^2+(c-x)^2}} = 0.$$

3. It would be hard to solve the stationarity equation. However, what can be done—and what is more interesting anyway—is to rewrite it as a more meaningful equation. Indeed, the stationarity equation is precisely the law of Snellius, as

$$\sin\alpha = \frac{x}{\sqrt{a^2+x^2}} \text{ and } \sin\beta = \frac{c-x}{\sqrt{b^2+(c-x)^2}},$$

by the high school definition of the sine in terms of a right triangle. Therefore, we get $\sin\alpha/v_1 = \sin\beta/v_2$, and this can be rewritten as

$$\sin\alpha / \sin\beta = v_1/v_2.$$

We discard the case that $a = b = 0$: then the ray of light follows the boundary.

4. The problem has a unique stationary point; it can be characterized by the law of Snellius.

Historical comment. This quick derivation of the law of Snellius from Fermat's variational principle makes use of the Fermat theorem. You might think that this derivation must be due to Fermat himself. However, it was given for the first time by Leibniz, one of the inventors of the differential calculus. Fermat gave another, much more complicated derivation, which did not use the Fermat theorem. The reason for this is that Fermat could only apply his theorem to polynomial functions, for which he possessed the equivalent of differentiation. This illustrates once more the point that

the Fermat theorem needs the differential calculus in order to lead to a powerful method.

A remark is in order here. We did not *solve* the stationarity equation in the last problem, as this is not so simple. The best that can be done here is to rewrite it as a "meaningful" equation, the law of Snellius.

Conclusion to section 1.4. We have seen that it is possible to apply the method of Fermat to problems of

- geometry (problem 1.4.1),
- algebra (corollary 1.10),
- mechanics (problem 1.4.2),
- security (problem 1.4.3),
- economics (problem 1.4.4),
- natural science (problems 1.4.5 and 1.4.6).

Invitation to chapters eight and nine. At this point you could continue with some problems in chapters eight and nine.

- The problem "Why not to sell your house to the highest bidder" in section 8.1 gives one of the surprises of *auction theory*; moreover, it makes use of the envelope theorem.
- Problem 8.2.1 "Optimal speed of ships and the cube law," is an application to maritime economics, based on the cube rule, which is used by practitioners.

- Problem 9.1.1, "Which number is bigger, e^π or π^e?" illustrates the distinction between the desire "to reach the essence of things" and the wish to crack a puzzle just for the fun of it.

1.5 DISCUSSION AND COMMENTS

1.5.1 All methods of continuous optimization in a nutshell

It is possible to sketch all questions and methods of optimization that we want to present, by discussing these for one-variable problems. We will omit precise assumptions and will make use of some notions that will be defined later in a precise way, such as that of a *convex function*, meaning that each segment with endpoints on the graph of the function lies nowhere below this graph. If, moreover, no points of these segments lie on the graph except the endpoints, then the function is called *strictly convex*.

Base. *What makes it all work?* It is, in the first place, the method of linear approximation: this consists of the *differential calculus* and a powerful result, the *tangent space theorem*. The differential calculus concerns the concept derivative and the rules to calculate derivatives. The tangent space theorem allows us to calculate the linear approximation to the solution set of a system of nonlinear equations at one of the solutions. You might know this result already in a different but equivalent form: as *the implicit function theorem*. The tangent space theorem is more convenient for the purposes of optimization.

In the second place, convex functions play some role here as well. These functions may be nondifferentiable at some points, but even at these points it is possible to linearize, in a different—parallel— sense. Here the method of linear approximation consists of the *convex calculus* and a powerful result, *the supporting hyperplane theorem*.

The convex calculus consists of the definition of the subdifferential and the rules to calculate subdifferentials. To begin with, convex functions f have at all points $\widehat{x}$ subgradients, that is, numbers α for which

$$f(x) - f(\widehat{x}) \geq \alpha(x - \widehat{x})$$

for all x. The subdifferential $\partial f(\widehat{x})$ is defined to be the closed interval consisting of all these numbers α. The subdifferential plays the same role for convex functions as the derivative for smooth functions. The supporting hyperplane theorem allows us to define the concept linear approximation to nonlinear convex sets at one of the boundary points of the set. Here we have emphasized the parallel between the smooth and the convex theory. However, in the present book, the use we make of the convex theory is restricted to the supporting hyperplane theorem. This causes no inconvenience, as the other results of the convex theory are easy corollaries of this theorem.

In the third place, we have existence results; the main one will be given below.

Necessary conditions. *What are the conditions that a point of local minimum has to satisfy?* The main condition is the equation $f'(\hat{x}) = 0$ at a point of differentiability and the inclusion $0 \in \partial f(\hat{x})$ if f is convex. This latter condition is also sufficient for global minimality, and then this point of minimum is unique if f is strictly convex. Moreover, if f is twice differentiable at $\hat{x}$, then there is in addition a second order condition, $f''(\hat{x}) \geq 0$, for local minimality of f at $\hat{x}$.

Sufficient conditions. *Which conditions are sufficient for local minimality?* These are the first order condition, $f'(\hat{x}) = 0$, together with the second order condition, $f''(\hat{x}) > 0$.

Thus we see that the difference between the second order necessary and sufficient conditions is only a "tiny line."

Existence. *Does there exist a global solution?* Without doing any calculation, one can be sure that a continuous function f on a closed interval $[a, b]$ attains its global minimum. A variant of this result is that a coercive, continuous function on the whole real line has a global minimum. Continuity means that the graph of f can be drawn without lifting pencil from paper. Coercivity (for minimization) means that $\lim_{|x| \to +\infty} f(x) = +\infty$.

Uniqueness. *Is the global solution unique?* A strictly convex function can have at most one point of global minimum. A function f for which the second derivative f'' exists and only takes positive values is strictly convex.

Sensitivity. *How sensitive is the optimal value for small changes in the data of the problem?* To compute this sensitivity we may assume that the solution itself does not change (*the envelope theorem*). To express this result analytically, consider a *perturbation*, that is, a family of problems of one variable x,

$$f(x, a) \to \min, \qquad (P_a)$$

depending on the parameter a. Let $x(a)$ be the solution of (P_a) for each a, which we assume to exist and to be unique. Let $S(a)$ be the value of (P_a), that is,

$$S(a) = f(x(a), a).$$

Let $\bar{a}$ be a specific number and write $\bar{x} = x(\bar{a})$. Then the derivative of the function

$$a \to S(a) = f(x(a), a)$$

at $a = \bar{a}$ is equal to the partial derivative of f with respect to a at the point $(x, a) = (\bar{x}, \bar{a})$, that is,

$$S'(\bar{a}) = \frac{\partial f}{\partial a}(\bar{x}, \bar{a}).$$

This is the promised result:

to compute $S'(\bar{a})$, that is, the sensitivity of the optimal value for small changes in a, we may assume that the solution $\bar{x}$ itself does not change.

This result allows us to calculate the number $S'(\bar{a})$, which we want to get hold of, in a simple way: *it does not require knowledge of the function S, which is often hard to get hold of.*

A precise version of this result will be given in section 1.11. The number $S'(\bar{a})$ represents, in embryonic form, the central concept of a "Lagrange multiplier." The envelope theorem is a useful tool in economics, for example in mechanism design (cf. [53],[51]). We have selected one application: the envelope theorem will be used in the solution of the auction problem "Why you should not sell your house to the highest bidder" in section 8.1.

Fighting chance. How nice should a problem be in order to give us a "fighting chance" to solve it? If we want to solve the problem analytically—"by a closed formula"—then the problem has to be smooth, convex, or mixed smooth-convex to begin with. The problems with an *analytical* solution form a small minority, but an interesting one.

The class of problems that can be solved *numerically*—"by an efficient algorithm that gives a guarantee of quality for the resulting approximate solution"—is much larger; it corresponds, roughly speaking, to the class of convex problems. In fact, convex problems can "usually" be solved efficiently by *self-concordant barrier algorithms*—see chapter seven—and/or by the *ellipsoid method*—see chapter six.

Algorithms. How to find a minimum numerically? We will discuss three fundamental algorithms to find a point of minimum $\hat{x}$ of a function f on an interval $[a, b]$.

Bisection method. Assume that f is *unimodular*, that is, that it is decreasing (resp. increasing) to the left (resp. right) of $\hat{x}$. The bisection method starts with computing the derivative of f at the midpoint of the interval $[a,b]$ (Fig. 1.8). If this derivative is positive (resp. negative), then $\hat{x}$ lies in the left (resp. right) half of the interval. Continuing in this way, you halve ("bisect") the length of the suspected interval each time, narrowing down the position of $\hat{x}$, the solution of the problem, more and more, until you have reached a degree of accuracy that you are satisfied with. If you are so lucky as to come across a point where the derivative is zero, then this point must be $\hat{x}$, and so you can stop.

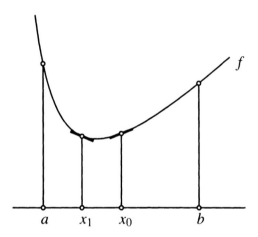

Figure 1.8 Illustrating the bisection method.

At first sight, it is not obvious that the bisection method leads to an approximation of guaranteed quality of $f(\hat{x})$, the *value* of the problem, as well. However, we will see that such a guarantee can be given if f is convex (theorem 6.3).

Newton method. The idea of the Newton method is to *linearize*. Assume that we can make a reasonable guess x_0 at the solution $\hat{x}$ of the equation $f'(x) = 0$. Then we can replace the left-hand side of this equation by

$$f'(x_0) + f''(x_0)(x - x_0),$$

the *linear approximation* at the current approximation x_0, and hope that the solution x_1 of the resulting linear equation is a better guess. A geometric illustration of a few steps of the algorithm show that this hope is not unreasonable (Fig. 1.9). In fact, the improvement turns out to be spectacular, *provided x_0 is sufficiently close to $\hat{x}$*: after each step the error is reduced to the square of the previous error, roughly speaking. This means that the number of correct decimals doubles with each step! This is called *quadratic convergence*. The formula

$$x_1 = x_0 - f''(x_0)^{-1} f'(x_0)$$

describes the recursion in an analytical way. Note that the Newton

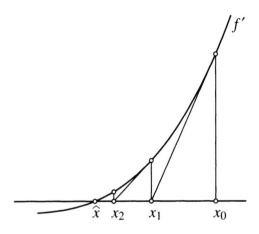

Figure 1.9 Illustrating the Newton method.

method is essentially a method to solve any equation $f(x) = 0$ (replace f' by f in Fig. 1.9), and not just a method to solve a stationary equation $f'(x) = 0$.

Then the recursion formula is

$$x_1 = x_0 - f'(x_0)^{-1} f(x_0).$$

The first time Newton mentioned this method was in the letter to Oldenburg where he also wrote the words about his enthusiasm for calculating many decimals of the elementary functions, which we quoted at the beginning of this section. In this letter he illustrated the method for some polynomials.

Modified Newton method. The modified Newton method is similar to the Newton method, only now each time you take a linear approximation with the same slope, so that you get parallel lines (Fig. 1.10).

This is in contrast to the Newton method, where the slope is updated each time. If this slope is approximately equal to $f''(\widehat{x})$, then the algorithm converges with satisfactory speed: after each step the number of correct decimals grows at least by a fixed number. This is called *linear convergence*. The recursion formula is

$$x_1 = x_0 - a^{-1} f'(x_0),$$

where a is the chosen slope.

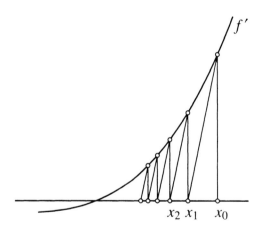

Figure 1.10 Modified Newton algorithm for solving $f'(x) = 0$.

Modified Newton method as a powerful key. The modified Newton method is more than just an algorithm. It provides a universal technique to prove all the deep results of optimization theory, from the Lagrange multiplier method (chapter three) to the Pontryagin maximum principle (chapter twelve). The geometric meaning of this technique can be demonstrated for the inverse function theorem for the one-variable case. This theorem states that a continuously differentiable function $y = f(x)$, for which $f(0) = 0$ and $f'(0) \neq 0$, has a continuously differentiable inverse function in a neighborhood of $x = 0$. That is, roughly speaking, for each $y \approx 0$ there is a unique $x \approx 0$ for which $f(x) = y$. We write $g(y)$ for this unique x, and then g is the inverse function f^{-1} of f. This inverse theorem turns out to be the heart of the matter: once you have it, all deep results of optimization theory follow readily.

Of course, the one-variable case is an immediate consequence of the easy fact that a continuous function assumes for each two values all intermediate values ("intermediate value theorem"). However, this proof does not extend to more than one variable.

Now we give the idea of the proof. For each $y \approx 0$ we apply the modified Newton method to the equation $f(x) = y$, that is, to the equation $f(x) - y = 0$, starting at the point $x = 0$, and choosing as constant slope the number $f'(0)$. This is a reasonable choice: one has $f'(0) \approx f'(x)$ if $x \approx 0$, by continuity of f'.

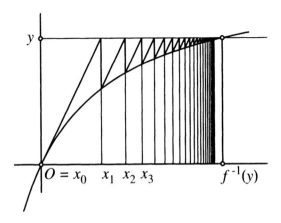

Figure 1.11 Modified Newton algorithm for solving $f(x) = y$.

In Figure 1.11 it is seen that this leads to a sequence that converges to the unique solution of $f(x) = y$, as required. Having seen this, you can try to turn this inspiring picture into a rigorous analytical proof. This turns out to be essentially a routine exercise. This is a spectacular illustration of the words of Plato from one of the epigraphs of the preface: *"geometry draws the soul toward truth."*

Newton and modified Newton compared. It is a natural idea to use the Newton method itself to prove the inverse function theorem (Fig. 1.12). However, this does not work. Figure 1.13 gives a first impression of the reason for this failure. Another thing that can be said here is that there is a trade-off between speed and reliability: the modified Newton method has "traded" speed for reliability.

Other keys

1. **Weierstrass and Fermat.** Another proof of the inverse function theorem is based on Weierstrass and Fermat, the main tools

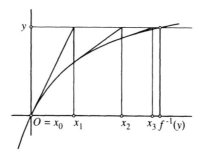

Figure 1.12 The Newton algorithm for solving $f(x) = y$.

used throughout this book. This proof is not universal; it only holds in finite dimensions. This is the technique of proof that we have chosen in the main text, as it makes use of the two tools that we use all the time, and also as we think it gives a better insight. To be specific, we use this technique to prove the Lagrange multiplier rule. This technique is not usually chosen in textbooks; see, however, [70]. The idea of the proof of the inverse function theorem using this technique is as follows. For each $y \approx 0$ we minimize $h(x) = |f(x) - y|$ for $x \approx 0$, say, $-\varepsilon \leq x \leq \varepsilon$ for a small (but not too small) $\varepsilon > 0$. Then the following statements hold true:

- This minimization problem has a global solution $\hat{x}$ by Weierstrass. Moreover, one can verify that it is unique.
- The boundary points are not points of global minimum. Indeed, it turns out that $h(\varepsilon) > f(0)$ and $h(-\varepsilon) > f(0)$.
- The points of differentiability of h are not points of global minimum. Indeed, for such points Fermat would be applicable, and this is seen to lead to a stationarity equation that has no solution.

In all, it follows that the problem has a unique global solution $\hat{x}$ and that h is not differentiable at this point. The only possibility that h is not differentiable at $\hat{x}$ is that $|f(\hat{x}) - y| = 0$, as the absolute value function $z \mapsto |z|$ is only not differentiable at $z = 0$, where it has a "kink." Thus we get $f(\hat{x}) = y$, as required.

2. **Brouwer fixed point principle.** The Brouwer fixed point principle states that a continuous mapping of a closed n-dimen-

FERMAT: ONE VARIABLE WITHOUT CONSTRAINTS

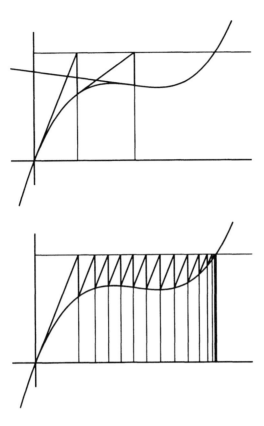

Figure 1.13 The Newton algorithm fails to converge (top figure), but the modified algorithm (bottom figure) still converges.

sional ball to itself leaves at least one point fixed (cf. [52]). This result holds in fact for any bounded, closed, convex set and not only for balls. It can be used to prove a strong version of the inverse function theorem. In appendix G, "Conditions of Extremum from Fermat to Pontryagin," it is shown how the methods of optimization can be based on the Brouwer fixed point principle. An advantage of this advanced approach is that it leads to stronger results.

3. **Contractions.** Often, the proof of the inverse function theorem is based on *the contraction principle*. This can be seen to be a reformulation of the proof using the modified Newton method. We now explain the idea of the contraction principle. If you take a map of Rotterdam and throw it on the ground somewhere in

Rotterdam, then there is exactly one point on the map that lies exactly on top of the point of Rotterdam that it represents. This even holds if the map is folded. This principle is illustrated in Figure 1.14 for a living room and a floor plan of this living room that is lying on the floor of the same room. The idea of the proof is as follows. Take any point x_1 on the floor. Find the point on the map that represents x_1 and take the point on the floor that lies directly under it. Call this point x_2. Then repeat this step, starting now from x_2. This leads to a point x_3 on the floor. Continuing this, we are led to an infinite sequence $x_1, x_2, x_3, \ldots$. This sequence is seen to converge to the desired unique point on the floor that lies directly under the point on the plan that it represents. Again, it is a routine exercise to turn this sketch into a rigorous analytical proof.

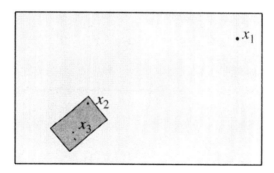

Figure 1.14 The plan of the living room (scale 1 : 5) lying on the floor of the same room. The sequence $x_1, x_2, x_3, \ldots$ converges to the point on the plan that lies exactly on top of the point of the floor that it represents.

How to find extrema on boundary. What to do, if the point of extremum is a boundary point and so the derivative need not be equal to zero? The idea of the following algorithm is that of a *barrier*: this allows us to reduce the task of finding a boundary point of extremum to the task of finding successively a series of interior points of extremum of certain auxiliary optimization problems.

Barrier methods (also called interior point methods): impressionistic view. This barrier method will be illustrated below for the following type of problem

$$f(x) \to \min, \ a \leq x \leq b.$$

This is no more than an illustration. For these problems the situation is so simple that one does not need barriers to solve them: there are only two boundary points and so one can inspect them "one by one." What will you see if you apply the barrier method to such a simple situation? The sequence of successive approximations will in the beginning "wander aimlessly" in the interval (a, b), that is, in the *interior* of the admissible set, but after a while it will "make up its mind" and tend to one of the endpoints: to a or to b. As soon as you see to which endpoint it tends, you can stop the algorithm. This endpoint is the solution of the given optimization problem.

For problems of finding the extrema of functions of two or more variables, the boundary can be very complicated, and then the barrier method is much more efficient than searching for a point of extremum among a huge collection of boundary points.

Barrier methods (also called interior point methods): the idea. Assume that $f : [a, b] \to \mathbb{R}$ is a differentiable strictly convex function and let $\widehat{x}$ be the unique solution of the problem (P). Then we define for each number $0 \leq \mu \leq 1$ the unconstrained problem

$$\mu f(x) - (1 - \mu)(\ln(x - a) + \ln(b - x)) \to \min, \ a < x < b, \quad (P_\mu)$$

and let $x(\mu)$ be the unique solution of the problem. If we let μ run from 0 to 1, then $x(\mu)$ runs from $x(0) = \frac{1}{2}(a + b)$ to $x(1) = \widehat{x}$. Now we try to follow $x(\mu)$ from its known start to its unknown finish. How to do this?

Path following using Newton. Suppose we have computed a point $x(\mu_1)$. Then we choose a number μ_2 which is bigger than μ_1 and compute $x(\mu_2)$ with the Newton method for the problem (P_{μ_2}) with starting point $x(\mu_1)$. The crucial problem is *how to choose* μ_2. There is a trade-off for each step: a *small* increase of the parameter μ gives a reliable—but small—improvement, as the starting point for the Newton method is close to its aim $x(\mu_2)$; a *large* increase gives a great— but unreliable—improvement, as we aim for a point on the curved line which is close to the finish $\widehat{x}$. It turns out that a satisfactory trade-off between reliability and speed is possible. This leads to *linear* convergence of the algorithm; that is, each step we gain a fixed number of digits of the decimal expansion of $\widehat{x}$.

Conclusion. All topics of the present book have been illustrated for one variable optimization problem.

1.5.2 Envelope theorem

The *envelope theorem* gives information on how sensitive the value of a problem is to a small perturbation. It is a basic and useful tool, for example, in economics. The intuitive version given in section 1.5.1 is adequate for informal purposes. For a full insight, we will formulate—and prove—a precise version as well. We will make use of some notions and results that are well known, but that will, strictly speaking, only be given in later chapters. In the next chapter we will give a simpler formulation of the envelope theorem, by making use of vector notation.

Setup. Consider the following setup. Let two open intervals U, V, a function f of two variables $x \in U$ and $a \in V$, and two numbers $\bar{x} \in U$, $\bar{a} \in V$ be given. We consider the family of problems

$$f(x, a) \to \min, \ x \in U, \qquad (P_a)$$

where the parameter a runs over V. We view this family as a *perturbation* of the problem $(P_{\bar{a}})$. That is, we want to consider the sensitivity of the value of the problem $(P_{\bar{a}})$ to small changes of the parameter a.

Let $S(a)$ denote the *value* of the problem (P_a), for all $a \in V$. The function S is called the *value function* of the perturbation

$$\{(P_a) : a \in V\}.$$

In terms of S, our question is: how does S change near $a = \bar{a}$? If S is differentiable, this means that we are interested in the derivative $S'(\bar{a})$.

Definitions. We need to make some definitions.

- A function g of one variable x is *continuously differentiable* at a point b if there exists $\varepsilon > 0$ such that the derivative $g'(x)$ exists for all x with $|x - b| < \varepsilon$ and if, moreover, g' is continuous at the point b. Then one writes $g \in C^1(b)$.

- A function g of one variable x is *twice continuously differentiable* at a point b if there exists $\varepsilon > 0$ such that its first and second derivatives, $g'(x)$ and $g''(x)$, exist for all x with $|x - b| < \varepsilon$ and if, moreover, $g''(x)$ is continuous at $x = b$. Then one writes $g \in C^2(b)$.

- A function g of *two* variables x_1, x_2 is *twice continuously differentiable* at $x_1 = b_1$, $x_2 = b_2$ if all first and second order partial derivatives, that is,

$$g_{x_1}(x_1, x_2), \ g_{x_2}(x_1, x_2), \ g_{x_1 x_1}(x_1, x_2),$$

$$g_{x_2x_2}(x_1,x_2),\ g_{x_1x_2}(x_1,x_2),\ g_{x_2x_1}(x_1,x_2)$$

exist, for all x_1, x_2 with $|x_i - b_i| < \varepsilon$, $i = 1, 2$, and are continuous at $x_1 = b_1$, $x_2 = b_2$. Then one writes $g \in C^2(b_1, b_2)$. This implies that

$$g_{x_1x_2}(b_1,b_2) = g_{x_2x_1}(b_1,b_2).$$

Sufficient conditions. Note that the first and second order sufficient conditions for the problem $(P_{\bar{a}})$ and the point $\bar{x}$ have the following form: $f_x(\bar{x}, \bar{a}) = 0$ and $f_{xx}(\bar{x}, \bar{a}) > 0$.

We need the concept of a *neighborhood* in order to formulate the next result. The main example of a neighborhood of a point p in $\mathbb{R}^2$ is an open disk with center p, that is,

$$\{x \in \mathbb{R}^2 : |x - p| < \varepsilon\}.$$

However, the class of open disks is not flexible enough for many purposes. A neighborhood of a point $p \in \mathbb{R}^2$ is a set W in $\mathbb{R}^2$ that contains w and for which there exists, for all $w \in W$, a sufficiently small number $\varepsilon > 0$ such that the open disk with center w and radius ε is contained in W. In other words, W contains p and is a union of open disks.

Theorem 1.11 Envelope theorem. *Consider the setup given above. Assume that $f \in C^2(\bar{x}, \bar{a})$ and, moreover, that the first and second order sufficient conditions hold for the problem $(P_{\bar{a}})$ and the point $\bar{x}$.*

Then we can achieve, by reducing the neighborhoods U of $\bar{x}$ and V of $\bar{a}$, that problem (P_a) is strictly convex and has a unique point of global minimum $x(a)$, for all $a \in V$. Then $x(\cdot) \in C^1(\bar{a})$, $x(\bar{a}) = \bar{x}$, and $S(\cdot) \in C^2(\bar{a})$. Moreover,

$$S(\bar{a}) = f(\bar{x}, \bar{a}),$$

$$S'(\bar{a}) = f_a(\bar{x}, \bar{a}),$$

and

$$S''(\bar{a}) = f_{xx}(\bar{x}, \bar{a})^{-1}(f_{xx}(\bar{x}, \bar{a})f_{aa}(\bar{x}, \bar{a}) - f_{xa}(\bar{x}, \bar{a})^2).$$

We will give a sketch of the proof.

Proof.

Simplification of set-up. We assume that $\bar{x} = 0$, $\bar{a} = 0$, and $f(0,0) = 0$, as we may without restricting the generality of the argument. We choose open intervals U, V in $\mathbb{R}$ containing 0 such that, for all $a \in V$, the function $x \to f(x,a)$ is strictly convex on U, using that $f_{xx}(0,0) > 0$ and f_{xx} is continuous, and so $f_{xx}(x,a) > 0$ for all (x,a) close to zero.

Application of the inverse function theorem. Reducing U and V if necessary, we may assume that the C^1-mapping

$$(x,a) \to (f_x(x,a), a),$$

defined for $x \in U$ and $a \in V$, has an inverse mapping Ψ, which is C^1. This follows from the inverse function theorem B.3, which is given in appendix B. (We note in passing that this is the point where we need the greater flexibility of the class of neighborhoods, compared to the class of open disks with given center.) The assumption of the inverse mapping theorem can be seen to hold, using $f_{xx}(0,0) > 0$.

The solution $x(a)$ of (P_a). Let $x(a) \in U$ be the first coordinate of $\Psi(0, a)$ for all $a \in V$. Then

$$f_x(x(a), a) = 0,$$

that is, $x(a)$ is a stationary point of the function $x \to f(x,a)$. Therefore, $x(a)$ is the unique point of minimum of the function $x \to f(x,a)$ on U, using the strict convexity of this function.

Formula for $S(\bar{a})$. Therefore, $S(a) = f(x(a), a)$, by definition of the value function S; substitution of $a = \bar{a}$ gives

$$S(\bar{a}) = f(\bar{x}, \bar{a}).$$

Formula for $S'(\bar{a})$. Moreover, by the chain rule,

$$S'(a) = \frac{d}{da} f(x(a), a) = f_x(x(a), a) x'(a) + f_a(x(a), a),$$

which is $f_a(x(a), a)$, as $f_x(x(a), a) = 0$. That is, $S'(a) = f_a(x(a), a)$. Substitution of $a = \bar{a}$ gives

$$S'(\bar{a}) = f_a(\bar{x}, \bar{a}).$$

Formula for $S''(\bar{a})$. Continuing in this way, we get, by differentiating the following relations at $\bar{a}$,

$$S'(a) = f_a(x(a), a), \ f_x(x(a), a) = 0;$$

then, eliminating $x'(a)$, and substituting $a = \bar{a}$, we get the required expression for $S''(\bar{a})$ as well:

$$S''(\bar{a}) = f_{xx}(\bar{x},\bar{a})^{-1}(f_{xx}(\bar{x},\bar{a})f_{aa}(\bar{x},\bar{a}) - f_{xa}(\bar{x},\bar{a})^2).$$

□

Conclusion. To calculate the sensitivity of the value of a problem to small perturbations, we may assume that the solution does not change.

1.5.3 Duality theory

Duality theory as supplement to the envelope theorem. The envelope theorem has a supplement, called *duality theory*, if the function f is not only smooth but also satisfies an additional condition (essentially, strict convexity near $(\bar{x}, \bar{a})$). Duality theory is of great interest, and here we have the opportunity to observe it in embryonic form, where a picture can illuminate it. We will give an intuitive description of duality theory—say, for minimization problems—in the present situation. We note that it is not standard practice to give the idea of duality theory as a local result in the smooth case. Usually, it is only given as a global result in the convex case.

Setup. We associate to the perturbation $(P_a)_{a \in V}$ of the given minimization problem $(P_{\bar{a}})$ a *maximization* problem with a different variable, but with the same "local value" as $(P_{\bar{a}})$. Moreover, $S'(\bar{a})$ is a strict local solution of the associated problem. The associated problem is called the *dual* problem and, in this context the given problem is called the *primal* problem. The essence of duality theory can be given by means of a simple figure (Fig. 1.15 with $\bar{a} = 0$), as we will see.

Assumptions. It remains to define the dual problem. We make an additional assumption, for convenience of exposition,

$$f_{xx}(\bar{x},\bar{a})f_{aa}(\bar{x},\bar{a}) - f_{xa}^2(\bar{x},\bar{a}) > 0$$

(the sense of this condition is that it implies, together with $f_{aa} > 0$, that f is strictly convex near $(\bar{x}, \bar{a})$, as is shown in appendix B). The function S is strictly convex near $a = \bar{a}$, by the expression for $S''(\bar{a})$ in the envelope theorem and the continuity of S'' at $\bar{a}$.

Definition of dual problem. The following facts are geometrically obvious (Fig. 1.15) and can be verified analytically as well. For each number z close to $S'(\bar{a})$, there is a unique number $a(z)$ near $\bar{a}$

such that
$$S'(a(z)) = z.$$
Take the tangent line to the graph of S at $a = a(z)$ and let $\varphi(z)$ denote the second coordinate of the point of intersection of this tangent line with the vertical line $a = \bar{a}$ in the x, a-plane. The *dual problem* is defined to be the problem to maximize the function φ.

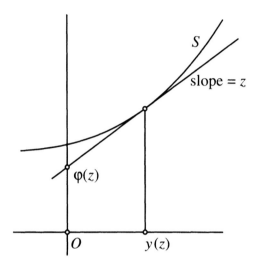

Figure 1.15 Illustrating the idea of duality theory.

Duality theory. Duality theory consists of the following observations on Figure 1.15 (with $\bar{a} = 0$):
The values of the primal and dual problem are equal. The derivative of the value function S at $\bar{a}$ is the solution of the dual problem ("shadow price interpretation" of the solution of the dual problem).

Primal problems can be maximization problems. In a similar way one can define the dual of a perturbation of a maximum problem: this will be a minimum problem, and then the primal problem is a maximization problem.

Conclusion to section 1.5. We give a *local* realization of one of the central ideas of optimization theory, duality theory (the usual realization, to be given in chapter four is *global*). This realization makes again use of the concept perturbation of an optimization problem. We show that local duality theory is a supplement of the envelope theorem.

1.6 EXERCISES

The first three exercises illustrate certain specific phenomena. The first one illustrates that a point of minimum or maximum of f need not be a solution of $f'(x) = 0$ if f is considered on a closed interval $[a,b]$. This is due to the possibility of a boundary solution. The method of Fermat is flexible enough to deal with such problems.

Exercise 1.6.1 *Solve the problem*

$$f(x) = x(x+6) \to \min, \ 0 \le x \le 1.$$

Solution

1. Existence of a global solution $\widehat{x}$ follows from Weierstrass.

2. Fermat: $f'(x) = 0 \Rightarrow 2x + 6 = 0$.

3. $x = -3 \notin [0,1]$. Therefore, the candidate minima are 0 and 1, the boundary points.

   Compare values: $f(0) = 0$ and $f(1) = 7$.

4. $\widehat{x} = 0$.

In the next two exercises we illustrate that the method of Fermat can also handle "kinks." A more challenging problem involving "kinks" is given in exercise 1.6.33.

Exercise 1.6.2 *Solve the problem* $f(x) = \sqrt{|x^2 - 1|} - x \to$ extr.

Solution

1. $f(-\infty) = +\infty$ and $f(+\infty) = 0-$ ($0-$ denotes convergence to 0 from the negative side), so existence of a global minimum $\widehat{x}_{\min}$ follows from Weierstrass, and it follows that a global maximum does not exist.

2. Fermat: $f'(x) = 0 \Rightarrow$

$$\frac{1}{2}|x^2 - 1|^{-\frac{1}{2}}|x^2 - 1|^{-1}(x^2 - 1)2x - 1 = 0,$$

using the chain rule twice and using the following calculus rules: $|x|' = |x|^{-1}x$ and $(x^r)' = rx^{r-1}$ for $r = \frac{1}{2}$ and $r = 2$.

3. Squaring leads to $x^2 = |x^2 - 1|$. This equation has two roots, $\frac{1}{2}\sqrt{2}$ and $-\frac{1}{2}\sqrt{2}$, but only the second one satisfies the original equation. Therefore, the candidate extrema are the stationary point $-\frac{1}{2}\sqrt{2}$ and the two points of nondifferentiability ± 1.

   Compare values: $f(-\frac{1}{2}\sqrt{2}) = \sqrt{2}$, $f(-1) = 1$ and $f(1) = -1$.

4. $\widehat{x}_{\min} = 1$ and $S_{\max} = \infty$.

Exercise 1.6.3 *Solve the problem*

$$f(x) = \sqrt{x^2 - x^4} \to \text{extr}, \quad -1 \leq x \leq 1.$$

Solution

1. Existence of a global maximum $\widehat{x}_{\max}$ and minimum $\widehat{x}_{\min}$ follows from Weierstrass.

2. Fermat: $f'(x) = 0 \Rightarrow \frac{1}{2}(x^2 - x^4)^{-\frac{1}{2}}(2x - 4x^3) = 0$.

3. $x = \pm\frac{1}{2}\sqrt{2}$. Therefore, the candidate extrema are the two stationary points $\pm\frac{1}{2}\sqrt{2}$, the two boundary points ± 1, and the point of nondifferentiability 0.

   Compare values: $f(\pm 1) = f(0) = 0$ and $f(\pm\frac{1}{2}\sqrt{2}) = \frac{1}{2}$.

4. $\text{argmin} = \{0, \pm 1\}$ and $\text{argmax} = \{\pm\frac{1}{2}\sqrt{2}\}$.

We give two numerical run-of-the-mill problems.

Exercise 1.6.4 $f(x) = \sqrt{x^2 + 1} - \frac{1}{2}x \to \min$.

Exercise 1.6.5 $f(x) = e^{x_1 x_2} \to \text{extr}, \quad x_1 + x_2 = 1$.

Market clearing prices are socially optimal

- **Price is market clearing price.** One of the basic laws of economics is that prices are determined by the clearing of markets, that is, by equality of supply and demand. For example, if a hairdresser charges his customers a very high price, he might not attract enough customers. If he asks a low price, he will have more clients than he can serve. These two forces will determine his price.

FERMAT: ONE VARIABLE WITHOUT CONSTRAINTS

- **Market clearing price is optimal.** The optimization problem in the next exercise leads to the valuable—and cheerful—conclusion that the clearing price is in a sense the best one for everybody:

it maximizes the social welfare ($\approx$ total pleasure) of producers and consumers.

To be more precise, we assume that there is a market where producers and consumers of a certain good come together. Each producer has a minimal price at which he is prepared to sell; each consumer has a maximal price at which he is prepared to buy.

- **Social welfare.** Now suppose we could choose a price and impose it on the market. Then each producer who is tempted to sell will calculate his *surplus*: that is, the difference between what he gets for his products and what he would have got if he had sold at his bottom price. Each consumer who is tempted to buy will also calculate his *surplus*: the difference between what he would have paid if he had bought at his ceiling price and what he has actually paid. The sum of all producer and consumer surpluses is called the total *social welfare*; it can be viewed as the total pleasure of producers and consumers of their transactions.

- **Modeling.** We model the situation as follows. Let a market for a certain good be described by a supply function $Q = S(P)$ and a demand function $Q = D(P)$. That is, if the price is P then the quantity of goods, supplied by all producers for which P is not lower than their bottom price, equals $S(P)$; the definition of the demand function is similar. For simplicity we assume that the graphs of the supply and demand functions are straight lines.

Exercise 1.6.6 *Show that the social welfare is maximal precisely if the price is the market clearing price.*

Hint. Show that the producer and consumer surpluses can be interpreted as areas of regions in the plane, as indicated in Figure 1.16.

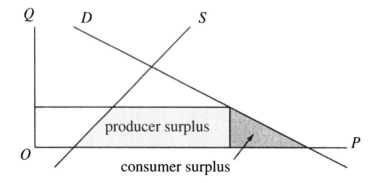

Figure 1.16 Optimality market clearing prices.

The following seven problems are related to the principle of least time.

Exercise 1.6.7 *The caring neighbor.* *A man lives 50 meters from the river and wants to bring an old lady who lives 100 meters from the river, but 300 meters further down, a bucket of water from the river (Fig. 1.17). He wants to walk to some point of the river, put water in his bucket, and walk to the house of the old lady.*

1. *For which point is the length of his walk minimized?*

2. *For which point is the total time of his walk minimized if he walks only half as fast with a full bucket as with an empty one?*

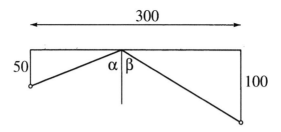

Figure 1.17 The caring neighbor.

Hint: try to relate for the length-optimal (time-optimal) point the angles α and β from Figure 1.17.

This reduces the first question to the problem of finding the shortest path between two given points: the answer is the straight line connecting the points.

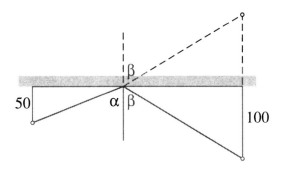

Figure 1.18 Old lady reflected.

Remark. The first question of this exercise has the following physical interpretation as well: the question how light is reflected by a mirror reduces, by the principle of least time, to the same type of optimization problem. Both questions can be reduced to a situation where the law of Snellius can be applied, by means of a reflection. This is illustrated in Figure 1.18.

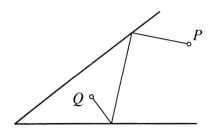

Figure 1.19 Path connecting P and Q touching two lines.

Exercise 1.6.8 *Repeated reflection.* *Two points lie between two lines. What is the shortest path between these points that touches each of the lines (Fig. 1.19)?*

Solution. A geometric solution of this problem is suggested by Figure 1.20.

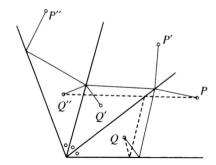

Figure 1.20 Geometric solution by two reflections.

By applying two successive reflections to the angular region between the lines, the problem is reduced to the problem of finding the shortest path between two given points: the answer is the straight line connecting the points. This leads to the solution of the original problem, by means of two reflections.

Exercise 1.6.9 *Catching a fish. If you try to catch a fish in a pond with your hands, then you have the handicap that it is not at the position where you see it (Fig. 1.21). Explain this using the law of*

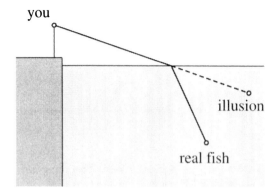

Figure 1.21 Catching a fish.

Snellius. How can you take the law of Snellius into account when you try to catch a fish with your hands?

Exercise 1.6.10 *It might seem hard to believe that 007 could carry out the calculations of the Saving the World problem fast enough. The explanation is that 007 knows the law of Snellius and has used this law instead of the Fermat theorem in order to save time. Check that the solution of the Saving the World problem follows this law.*

In the next three exercises we give consequences of Fermat's principle of least time that are more advanced.

Exercise 1.6.11 * *Setting sun. When we see the sun setting, it is already below the horizon. Explain this using the principle of Fermat and the facts that the earth's atmosphere is thin at the top and dense at the bottom, and that light travels slower in air than in vacuum.*

Exercise 1.6.12 * *Mirages on hot roads. If you drive on a hot road, you often see a mirage. You see "water" on the road, but when you get there, it is as dry as the desert. Explain this using the principle of Fermat and the fact that the air is very hot just above a hot road and therefore, it is expanded and so it is less dense.*

Legend. Fata Morgana, also known as Morgan le Fay, was a fairy enchantress, skilled in the art of changing shape. She was King Arthur's sister and learned many of her skills from Merlin the Magician. She lived in a marvelous castle under the sea. Sometimes the enchantress made this castle appear reflected up in the air, causing seamen who mistook it for a safe harbor to be lured to their deaths.

Exercise 1.6.13 * *Fata morgana. What is the explanation of the following phenomenon, called fata morgana? Traveling in a desert you see an oasis in the distance, but when you come closer you discover that it was an illusion.*

Trade-off: too expensive or too high. In some places in the world, like Manhattan and Hong Kong, the land is so expensive that only by building skyscrapers can you keep the price of an apartment affordable. However, you should not build too high, as this would lead to very high specialized costs. In the next exercise we look for an optimal trade-off. It requires some care, as the solution has to be an integer.

Exercise 1.6.14 *Skyscrapers. The cost of building an apartment block, x floors high, is made up of three components:*

1. ten million dollars for the land,

2. a quarter of a million dollars per floor,

3. specialized costs of $10,000x$ dollars per floor.

How many floors should the block contain, if the average cost per floor is to be minimized?

Here are three more problems where the solutions are required to be integers.

Exercise 1.6.15 *Solve the problem $\sqrt[n]{n} \to$ extr, where n runs over the natural numbers.*

How to round off. The following "method" to deal with problems where the solution has to be an integer leads often—but not always— to the optimal solution: model the problem, ignore that the variable x has to be an integer, then find the global minimum using the Fermat theorem, and finally round off to the nearest integer. The reason that it often works is that graphs tend to be more or less symmetric near local extrema. The next problem considers the special case of cubic polynomials.

Exercise 1.6.16 *Rounding off*

1. *Consider the problem*

$$f(x) = 50x^3 - 35x^2 - 12x + 4 \to \min,$$

where x runs over the nonnegative integers. Show that the above "method" involves rounding off from 0.6 to 1 and leads to a point of value $7 = f(1)$, whereas the optimal value of this problem is much lower, $4 = f(0)$.

2. **For general cubic polynomials, the situation is as follows. Let*

$$f(x) = x^3 + ax^2 + bx + c,$$

let $\widehat{x}$ be a local minimum of f lying between the consecutive integers n and $n+1$, and assume that there is no other stationary point between n and $n+1$.

Verify that "rounding off gives the right answer" if $\widehat{x} < n + \frac{1}{2}$ and also if $\widehat{x} > n + \frac{2}{3}$. That is,

$$f(n) < f(n+1) \text{ if } \widehat{x} < n + \tfrac{1}{2}, \text{ and } f(n+1) < f(n) \text{ if } x > n + \tfrac{2}{3}.$$

FERMAT: ONE VARIABLE WITHOUT CONSTRAINTS

Exercise 1.6.17 *Sum with maximal product*

1. Write 100 as the sum of a collection of positive real numbers such that their product is maximal.

2. The same problem as in 1, but now the numbers are required to be positive integers.

The following exercise gives another illustration—in addition to problem 1.4.4 ("optimal price for a car")—of how a marginal analysis can be used in economics.

Exercise 1.6.18 Standby ticket. *Flying a 200-seat plane across the ocean costs an airline 100,000 dollars. Therefore, the average cost per passenger is 500 dollars. Suppose the airplane is about to take off with ten empty seats. A last-minute passenger is waiting at the gate, willing to pay 300 dollars for a seat. A trainee argues that this makes no sense as this is less than 500 dollars. However, in reality it turns out that the airline is prepared to sell.*

Explain this by a marginal analysis, comparing marginal benefit and cost. Take into account your estimate of the additional cost of taking this passenger. What is the optimization problem here?

Here is an example in the same spirit. This one illustrates the meaning of the equimarginal rule ($\approx$ the Fermat theorem) without any high school algebra.

Exercise 1.6.19 *Given below are three columns of, respectively, total output, total revenue, and total cost for a firm.*

1	15	7
2	29	14
3	41	22
4	51	31
5	60	42
6	66	55

How much output, if any, should this firm produce? At first glance, the more the better. At second glance we realize that the cost should not be left out of consideration. What is the optimization problem here?

Trade-off monopolist. A monopolist can choose what price to charge for his product. He faces the following trade-off. If he asks a high price, not many customers will buy his product; if he wants to sell his product to many customers, he has to ask a low price and will make a small profit only at each transaction. Part 1. of the following example illustrates this.

How to play the game. Now consider another problem as well: you want a share of the profits of the monopolist—for example, because you are the inventor of the product. Then you should take the profit-maximizing behavior of the monopolist into account. You are, as it were, playing a game against the monopolist: both you and the monopolist are optimizing, each one having your own objective function. This is illustrated by part 2. of the following example.

Exercise 1.6.20 *Optimal royalties. An artist has recorded a dvd. His record company can produce the dvd with no fixed cost and a variable cost of 5 dollars per dvd. The company's marketing department determines that the demand for the dvd is given by the two columns below, the first giving the price in dollars and the second the number of dvds.*

P	D
24	10,000
22	20,000
20	30,000
18	40,000
16	50,000
14	60,000

Here is a clear trade-off between high sales and a large profit per dvd sold.

1. **Trade-off monopolist.** Which one of the six dvd prices will maximize profit? What will be the profit?

2. ***Game: optimal royalties.** How many royalties—the amount of money he will get for each dvd sold—would you, as his agent, advise the artist to demand from his record company?

3. ***Game: selling the royalties.** Is it a good idea to sell the royalty rights to the company for a fixed amount?

4. ***Continuous prices.** Would it make any difference in part 2, if we allowed all prices, not just the six in the first column?

You can view this problem as a game between two optimizing agents, the record company and the artist.

Hint for the agent. Of course, you will begin by asking the artist what he really wants. If the objective is to maximize, for whatever reason, the number of dvds that will be sold, then the artist should not ask any royalties. This is because any fee will increase the cost per dvd for the record company, this profit-maximizing company will be led to set a higher dvd price, and this will lead to a lower demand. We assume that the artist answers your question by saying that he is mainly interested in one thing: money.

Exercise 1.6.21 Labor or leisure. *Each day an individual must decide how to allocate her 24 hours between labor and leisure. She can choose to supply L hours of labor in order to earn money to buy consumer goods C. The remaining hours, Z, constitute her leisure time. Hence $0 \leq L \leq 24$ and $Z = 24 - L$. The individual takes as given the wage rate w (dollars per hour) and the price of consumer goods, so that $C = wL$. The pleasure she has— from consumption and leisure time—is modeled by the utility function*

$$u(C, Z) = \frac{1}{3} \ln C + \frac{2}{3} \ln Z$$

(if you want to avoid logarithms, you can instead take the utility function $u(C, Z) = C^{\frac{1}{3}} Z^{\frac{2}{3}}$).

She chooses L so as to maximize the value of her utility function. How many hours a day will she work and how does this depend on the wage rate w?

If you try to guess, without calculation, the effect of the wage rate, then you could make a case that a higher wage rate will be an incentive to work longer. However, you can also make a plausible case that the effect is opposite.

Does a medicine help? Here is a medical application. How can you test whether a medicine really helps? This is done as follows. We give a simplified description of the method. You take a large group of patients and give some the medicine and some a placebo. Then you check—for each patient—whether his condition improves or not. If the condition of each patient who got the medicine improves and all the others don't, then you have strong evidence that the medicine helps.

However, tests do not always lead to such a clear-cut outcome. Sometimes there is a positive probability p that the condition of the patient also improves without any medicines. Moreover, usually it is not a certainty that a patient improves after having received the medicine. We write rp for the probability that the condition improves after the medicine. Then the factor r gives the effect of the medicine: if $r > 1$, then the medicine is of help.

This makes clear that it is of the greatest interest to know or at least to estimate r. The following exercise illustrates the *maximum likelihood method* for estimating r, based on the outcomes of the test above. The following exercise gives a justification for the following widely used estimate:

$$r = \left(\frac{n_1}{n}\right) \bigg/ \left(\frac{m_1}{m}\right).$$

The justification will be that this estimate is, in a sense, the best one.

Exercise 1.6.22 *Testing medicines.* A group of m (resp. n) patients are given a placebo (resp. the medicine). In the group of patients who got a placebo (resp. the medicine), m_1 (resp. n_1) show an improvement of their condition, but the remaining m_2 (resp. n_2) do not. For which choices of p and r are the probabilities of these outcomes of the test maximal ("maximum likelihood")? The resulting r is the promised maximum likelihood estimate.

Remark. Usually one is not interested in p; it is called a *nuisance parameter*.

If you are moving, then you are sometimes confronted by problems of getting large pieces of furniture around corners. In the next exercise

FERMAT: ONE VARIABLE WITHOUT CONSTRAINTS

we will look at a simplified version of this problem, where the freedom to maneuver is restricted to a two-dimensional horizontal plane.

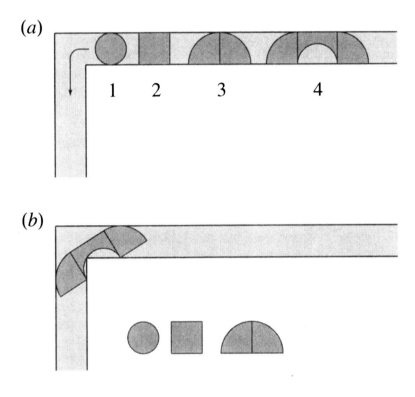

Figure 1.22 Shapes that go around the corner.

Exercise 1.6.23 *Around the corner.* What is the object in the plane, to be called a "piece of furniture," of maximal area, that you can get around the corner of the corridor (Fig. 1.22(a))?

First attempt. Your first guess might be a round table of diameter the width of the corridor (shape 1). Its area is $\frac{\pi}{4} \approx .78$ if we choose the width of the corridor as the unit of length.

Second attempt. Now suppose you are allowed to ask someone in the other corridor to help you by pulling. Then you can do better. A square table with sides equal to 1 (shape 2) has area 1.

Third attempt. To improve upon this candidate requires some inspiration. A table in the form of a half-disk with radius equal to the width of the corridor (shape 3) has area $\frac{\pi}{2} \approx 1.57$. This is a very good attempt. It might be a surprise that it is not the best one.

Best attempt known. You can do more than 40% better. A sofa of shape 4 can be moved around the corner as well, for each choice of the size of the radius of the inner circle, from 0 up to 1. Figure 1.22(b) can help to show this: you have to use the fact that from the position of each point on a half circle one sees the endpoints of the half circle under a right angle.

Final touch. Now one question remains: for which choice of radius of the inner half-circle of the sofa do we get a sofa of maximal area? Determine this optimal radius and the resulting area, and compare it with the runner-up, the half-circle table.

The secret of the cubic equations. Nicolo Tartaglia (1506–1557) lived in a time where mathematical contests were popular. At one contest his task was to solve 30 cubic equations

$$x^3 + px + q = 0$$

for different values of p and q. Shortly before the deadline, he discovered that his opponent was initiated into the secret of the solution of cubic equations. Through a tremendous effort, Tartaglia managed, on his own, to find the general method eight days before the deadline. He never published his method, but in a number of his works he announced various applications, one of these to a maximum problem.

Exercise 1.6.24 *Problem of Tartaglia.* Divide the number 8 into two parts such that the product of these two parts, multiplied with their difference, is maximal.

We extend this exercise.

Exercise 1.6.25 *Generalized problem of Tartaglia.* Find the local minima and maxima of a polynomial function of the third degree,

$$f(x) = ax^3 + bx^2 + cx + d,$$

with $a \neq 0$.

This is the only polynomial problem that can be solved in general; all textbooks and puzzle books are filled with special cases of this problem. In problem 8.2.1 we will give an application to maritime economics, and it can be used to solve several of the exercises of this section.

Euclid and the pharaoh. The most prominent mathematician of antiquity is Euclid of Alexandria (325BC–265BC). He is best known for the *Elements*, the first scientific monograph and textbook in human history. This textbook was used till the beginning of the twentieth century, and about 1,700 editions are known to exist. Once, the pharaoh Ptolemy Soter asked Euclid whether geometry could not be learned in an easier way than by reading the *Elements*, to which he replied that *there is no royal road to geometry.*

In the *Elements* there is one maximum problem.

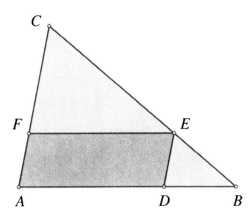

Figure 1.23 The problem of Euclid.

Exercise 1.6.26 **Problem of Euclid. In a given triangle ABC, inscribe a parallelogram ADEF (with $D \in AB$, $E \in BC$, $F \in AC$) of maximal area (Fig. 1.23).*

This problem of Euclid can be seen as the starting point of some interesting general developments, which will be discussed in problem 2.4.1, and in sections 7.2 and 9.5.

The geometrical problems in the next two exercises play a prominent role in the construction of algorithms—the *center of gravity method* and the *ellipsoid method*—as we will see in chapter six. The *center of gravity* of a triangle in the two-dimensional plane can be defined as

$$\left(\frac{1}{3}(a_1 + b_1 + c_1), \frac{1}{3}(a_2 + b_2 + c_2)\right)$$

if the vertices are (a_1, a_2), (b_1, b_2), and (c_1, c_2). It is a popular mistake to think that each line through the center of gravity divides the triangle into two parts of equal area. The following exercise shows that the two parts have, in general, different areas, although the difference is never very large.

Exercise 1.6.27 *Each of the two parts into which a given triangle is divided by a line through the center of gravity of the triangle has area between $\frac{4}{9} \approx 0.4444$ and $\frac{5}{9} \approx 0.5556$ times the area of the given triangle. These constants cannot be improved. Verify these statements.*

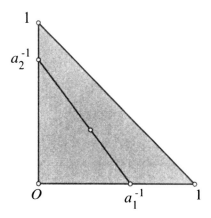

Figure 1.24 Center of gravity method.

Hint. In this exercise, we may assume without restriction of the generality of the argument that the triangle is the one with vertices $(0,0)$, $(1,0)$, and $(0,1)$ and that the line through the center of gravity intersects the two sides that make a right angle. This can be achieved by a linear change of coordinates. Then the center of gravity of the triangle is $(\frac{1}{3}, \frac{1}{3})$, and the line through the center of gravity has an equation of the form

$$a_1 x_1 + a_2 x_2 = 1 \quad \text{with} \quad 1 \leq a_1, a_2 \leq 2 \quad \text{and} \quad a_1 + a_2 = 3.$$

These simplifications are illustrated in Figure 1.24.

Ellipse. We define a *solid ellipse* in the two-dimensional plane to be the image of the closed unit disk under a linear transformation. That is, it is the set of all points

$$(r + ax + by, s + cx + dy),$$

where (x, y) runs over all solutions of the inequality $x^2 + y^2 \leq 1$ (the "unit disk") and a, b, c, d, r, s are given numbers satisfying the condition $ad - bc \neq 0$ (to exclude "degenerate" ellipses). The image of the center of the unit circle under the linear transformation, that is, the point (r, s), is called the *center* of the ellipse. For example, the set of solutions of the inequality

$$\frac{x^2}{a^2} + \frac{y^2}{d^2} \leq 1$$

is a solid ellipse with center $(0, 0)$. A *half-ellipse* of a solid ellipse is defined to be one of the two parts into which the solid ellipse is divided by a line through its center.

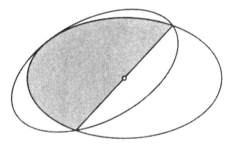

Figure 1.25 Ellipsoid method.

This constant cannot be improved. Verify these statements.

Exercise 1.6.28 *Each half-ellipse of a given solid ellipse can be contained in a solid ellipse with area not more than $\frac{4}{9}\sqrt{3} \approx 0.7698$ times the area of the given solid ellipse (Fig. 1.25).*

The statement in this exercise is the heart of the matter of the ellipsoid method (section 6.10). In the next chapter you will be invited to extend the results of the last two exercises to dimension n (exercises 2.6.9 and 2.6.10).

Kepler and the wine barrels. In retrospect, one already can see a trace of the Fermat theorem in the book *Nova stereometria doliorum vinariorum* ("A New Stereometry of Wine Barrels") by Johann Kepler (1571–1630), the well-known discoverer of the laws governing the elliptic orbits of the planets.

At the beginning of this book he describes an event in his life that occurred in the fall of 1613:

"In December of last year [...] I brought home a new wife at a time when Austria, having brought in a bumper crop of noble grapes, distributed its riches. [...] The shore in Linz was heaped with wine barrels that sold at a reasonable price. [...] That is why a number of barrels were brought to my house and placed in a row, and four days later the salesman came and measured all the tubs, without distinction, without paying attention to the shape, without any thought or computation. [...] I was astonished."

Why does one measurement suffice? Kepler thought it strange that by means of a single measurement (Fig. 1.26), one could deter-

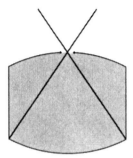

Figure 1.26 Measurement of the volume of a barrel by the salesman.

mine the volumes of barrels of different shape. He goes on: "As a bridegroom, I thought it proper to take up a new subject of mathematical studies and to investigate the geometric laws of a measurement so useful in housekeeping, and to clarify its basis if such exists."

Optimal shape of barrels. His analysis leads to the conclusion that barrels are made in such a shape that they have a certain maximality property; therefore small deviations in this shape have hardly any influence on the volume. To put it in modern terms, the derivative in an optimum is equal to zero. These are his own words:

Birth of the Fermat theorem. *"From this it is clear that, when making a barrel, Austrian barrel makers, as though guided by common and geometric sense, take [...] and [the barrel] will thus have maximal capacity even if one deviated somewhat from the exact rules during the making of the barrel, because figures close to the optimum change their capacity very little. [...] This is so because near a maximum the changes on both sides are in the beginning only imperceptible."*

This is historically the first description of the main idea of the Fermat theorem. Kepler considered a number of optimization problems inspired by this event, and solved them, but not by using the insight contained in these words. One of these problems is given in the following exercise.

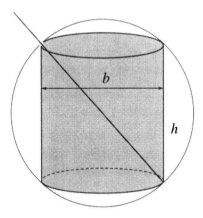

Figure 1.27 A wine barrel problem of Kepler.

Exercise 1.6.29 * *Problem of Kepler.*
What is the optimal ratio of the base diameter b to the height h of a cylinder that is inscribed in a given ball (Fig. 1.27), if we want the volume of the cylinder to be as large as possible? The sense of this exercise in terms of wine barrels is that we now suppose that the hole in the wine barrel ($\approx$ cylinder) is at a different position (cf. Figs. 1.26 and 1.27).

Greatest achievement of Archimedes

1. **Modern point of view.** Archimedes of Syracuse (287 BC–212 BC) is one of the greatest scientists of all time. Now his most famous theorem is Archimedes' principle, contained in his work *On Floating Bodies*.

2. **Contemporary point of view.** However, in his time he had gained a great reputation because of the machines he had invented. They were used in a very effective way as engines of war in the defense of Syracuse. This was attacked and besieged by the Romans under the command of Marcellus in 212 BC. As the contemporary Roman historian Plutarch writes, Archimedes

had been persuaded by his friend King Hiero II of Syracuse to build such machines:

"These machines [Archimedes] had designed and contrived, not as matters of any importance, but as mere amusements in geometry; in compliance with King Hiero's desire and request, some little time before, that he should reduce to practice some part of his admirable speculation in science, and by accommodating the theoretic truth to sensation and ordinary use, bring it more within the appreciation of the people in general."

Plutarch describes the effect of Archimedes' engines of war during the siege of Syracuse of 212 BC:

"when Archimedes began to ply his engines, he at once shot against the land forces all sorts of missile weapons, and immense masses of stone, which came down with incredible noise and violence; against which no man could stand; for they knocked down those upon whom they fell in heaps, breaking all their ranks and files."

"or else the ships, drawn by engines within, and whirled about, were dashed against steep rocks that stood jutting out under the walls, with great destruction of the soldiers that were aboard them. A ship was frequently lifted up to a great height in the air (a dreadful thing to behold) and was rolled to and fro, and kept swinging, until the mariners were all thrown out, when at length it was dashed against the rocks, or let fall.".

3. **Archimedes' own point of view.** Yet Archimedes, although he achieved fame by his mechanical inventions, considered these of little value compared to his other achievements. He considered that his most significant accomplishments were those concerning a cylinder circumscribing a sphere. On his request a representation of this was inscribed on his tomb, as is confirmed by Cicero, who was in Sicily in 75 BC. Cicero writes how he searched for Archimedes' tomb:

"and found it enclosed all around and covered with brambles and thickets; for I remembered certain doggerel lines inscribed, as I had heard, upon his tomb, which stated that a sphere along with a cylinder had been put on top of his grave. Accordingly, after taking a good look all around [...], I noticed a small column arising a little above the bushes, on which there was a figure of a sphere and a cylinder. [...] Slaves were sent in with sickles

[...] and when a passage to the place was opened we approached the pedestal in front of us; the epigram was traceable with about half of the lines legible, as the latter portion was worn away."

All descriptions of Archimedes are written in superlatives; let us again quote Plutarch:
"Archimedes possessed so high a spirit, so profound a soul, and such treasures of scientific knowledge,"

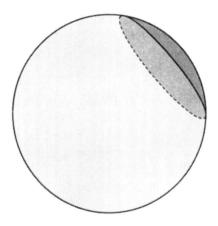

Figure 1.28 Problem of Archimedes.

A ruse of Archimedes. It is refreshing to realize, reading the preface of his work *On Spirals*, that Archimedes was no legend but a human being. He was in the habit of sending his friends in Alexandria statements of his latest theorems, but without giving proofs. Then sometimes other people claimed these results as their own. In this preface Archimedes writes that the next time he sent theorems, he included two that were false:
"so that those who claim to discover everything, but produce no proofs of the same, may be confuted as having pretended to discover the impossible."

In his work *On the Sphere and Cylinder*, Archimedes poses and solves the following maximum problem.

Exercise 1.6.30 *Problem of Archimedes. Find the maximum volume that spherical segments (one of the two parts of a ball that* is separated by a plane) with a given area of the curved part of its surface (its "lateral" surface) can have (Fig. 1.28). Here the ball is not fixed: its radius is allowed to vary. You may want to use the following formulas for the volume V of a spherical segment of height h of a sphere with radius R, and for the area A of the lateral surface, which were discovered by Archimedes:

$$V = \pi h^2(R - h/3), \quad A = 2\pi Rh.$$

One of the authors mentioned once in the first lecture of a course on optimization that he did not know whether this result extends to dimension n. The next lecture one of the students, Vladeen Timorin, handed him a manuscript giving the analysis of the problem for dimension n. If you want a challenge, you can try to model this extension to dimension n and to solve it.

Exercise 1.6.31 *Problem of Guillaume François Antoine Marquis de L'Hospital (1661–1704).* *Pose and solve the following problem: among all truncated cones inscribed in a given ball, find the cone for which the area of the curved part of its surface (its "lateral" surface) is maximal (Fig. 1.29).*

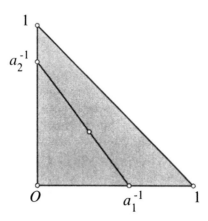

Figure 1.29 Problem of de l'Hospital.

Hint: the lateral surface of the truncated cone equals $\frac{1}{2}lp$, where l is the distance from the top to a point of the circle at the basis of the cone, $p = 2\pi r$ is the perimeter of the circle, and r is the radius of the circle. This fact follows from the well known formula for the area of a triangle, $\frac{1}{2} \times$ base $\times$ height, if we approximate the circle by polygons and so approximate the area of the lateral surface by a sum of areas of triangles.

The following extension is recommended as an exercise in modelling.

Exercise 1.6.32 * Generalized problem of de L'Hospital.
Pose and solve the problem of de L'Hospital in n-dimensional space.

We give another illustration that the Fermat theorem can handle kinks. It shows, moreover, that the concept *median* from statistics is in some sense optimal,

Exercise 1.6.33 *

1. Find the number(s) for which the average distance to the numbers $0, 1, 3,$ and 10 is minimal.

2. Show that a median of a finite sequence of numbers has among all real numbers minimal average distance to the numbers of this sequence.

Barrier methods: the simplest illustration. The following exercise illustrates a basic idea of the celebrated barrier algorithm for finding boundary solutions of optimization problems. A sketch of this algorithm was given at the end of section 1.5.1; it will be considered in more detail in section 7.3 (and, without mentioning barriers, in section 6.9). The interest of using barrier methods is in solving problems of n variables with inequality constraints.

The simplest such problem is

$$x \to \min, \ x \geq 0.$$

For this simple problem one does not need the barrier method: it is immediately clear what its solution is. However, this problem affords the opportunity to get a first impression of a basic idea of the barrier methods. The Fermat theorem is not valid at the point of minimum, and the basic effect of the barrier method is a reduction to a situation where the Fermat theorem can be used. Well, if we add to the objective function the barrier $-\ln x$, multiplied with "weight" $c > 0$, then we get a new optimization problem,

$$x - c \ln x \to \min, \ x > 0, \qquad (P_c)$$

for which the Fermat theorem is valid at its point of minimum $\widehat{x}_c$. If we let the weight c tend to 0, then the effect of the barrier term becomes less and less noticeable. This makes it plausible that $\widehat{x}_c$ tends to $\widehat{x}$ if we let the weight c tend to 0. The purpose of the next exercise is to verify that this is indeed the case. Thus a basic idea of the barrier method is then illustrated: the solution $\widehat{x}$ of the given problem is the limit for $c \downarrow 0$ of the solution $\widehat{x}_c$ of an auxiliary problem (P_c), and the solution $\widehat{x}_c$ of (P_c) does satisfy the Fermat theorem. To put it simply, barrier methods are methods to solve (P) numerically in the following way: one tries to compute—with the use of the Fermat theorem—the solution of (P_c) for smaller and smaller $c > 0$ until a satisfactory approximation of the solution of (P) is found.

Exercise 1.6.34 *Calculate the solution $\widehat{x}_c$ (resp. $\widehat{x}$) of the problem*

$$x - c \ln x \to \min, \ x > 0$$

(resp. $x \to \min, \ x \geq 0$) and verify that $\widehat{x}_c$ tends to $\widehat{x}$ if the weight c tends to 0.

Taylor polynomials. In example 1.2.4 on the geologists and $\sin 1°$, we did a few steps of a procedure that is known as the approximation of $\sin x$ by Taylor polynomials. In the following exercises, this procedure is explored further.

Exercise 1.6.35 *Continue the approximation procedure in order to get the next approximation of $\sin x$. How many decimals of $\sin 1°$ would this give?*

In the following exercise we show how to compute the value of e^x for some x as accurately as we want.

Exercise 1.6.36 **Use the idea of the approximation method to compute the first five decimals of $e = e^1$ and $\sqrt{e} = e^{\frac{1}{2}}$. You may start with the crude approximation $1 \leq e \leq 3$.*

The next exercise shows how to deal with an obstacle that arises from the second request of the geologists: to calculate the first five decimals of $\ln 2$.

Exercise 1.6.37 **One could try to use the idea of the approximation method to compute the first five decimals of $\ln(1+x)$ at $x = 1$. Which practical problem are you confronted with? Use the identity $\ln \frac{1}{2} = -\ln 2$ to solve this practical problem.*

The next exercise gives a physical interpretation of the basic fact from which we have derived in section 1.2.3 the method of approximating functions by Taylor polynomials.

Exercise 1.6.38 *Consider the following observation. If two persons start a race along a running track and one of them never runs slower than the other one, then that person will never be behind. Use the velocity interpretation of the derivative to translate this observation into a statement about two functions and their derivatives. Show that this statement holds true.*

Finally we consider the approximation of functions by Taylor polynomials in the general case. If a function on an interval I is n times continuously differentiable—that is, the n-th order derivative exists and is continuous on I—then we write $f \in C^n(I)$.

Exercise 1.6.39 *Taylor polynomials and remainder term. Let f be a function in $C^n([0, s], \mathbb{R})$ for which $C \leq f^{(n)}(x) \leq D$ for $0 \leq x \leq s$. Define the Taylor polynomial of f at $x = 0$ of degree n to be*

$$f(0) + f'(0)x + \frac{1}{2!}f''(0)x^2 + \cdots + \frac{1}{n!}f^{(n)}(0)x^n$$

and define the corresponding remainder term to be the difference

$$R(x) = f(x) - (f(0) + f'(0)x + \frac{1}{2!}f''(0)x^2 + \cdots + \frac{1}{n!}f^{(n)}(0)x^n).$$

Show that

$$\frac{C}{(n+1)!}x^{n+1} \leq R(x) \leq \frac{D}{(n+1)!}x^{n+1}$$

for $0 \leq x \leq s$.

The following exercise illustrates the power of the Weierstrass theorem. This theorem implies a result that allows to determine the course of the sign of functions. This result is tacitly used in the high school method to find the extrema of a function f: this method involves the determination of the course of the sign of f'.

Exercise 1.6.40 *Intermediate value theorem. If f is a continuous function on a closed interval $[a, b]$ for which the product $f(a)f(b)$ is negative, then it has a zero between a and b. Derive this result, called the intermediate value theorem, by analyzing the problem of finding the extrema of a primitive g of the given function f.*

However, this implication is like shooting with a cannon at a mosquito, as it is easier to prove the intermediate value theorem directly.

Chapter Two

Fermat: Two or More Variables without Constraints

Near a maximum the changes [···] are in the beginning only imperceptible.

J. Kepler

- How do we find the extrema of a function of two or more variables without constraints, that is, how can we solve a problem $f(x) \to$ extr, where f is a function of $x_1, \ldots, x_n$?

2.0 SUMMARY

Two or more variables of optimization. Sometimes an optimal choice is described by a choice of two of more numbers. This allows many interesting economic models. For example, we will consider the problem of choosing the best location from which to serve three clients. We consider as well the optimal location for two competing ice-cream sellers on a beach. This is a simple problem, but it leads to a fruitful insight: "the invisible hand" (=hand of the market), one of the central ideas in economics, does not always work. Other models give insights into taxes: some taxes can lead to a net loss for society and some taxes can be used to give an incentive to limit pollution. Another model, for Coca Cola and Pepsi Cola, gives some insight into the beneficial effects of competition.

Least squares approximations. A basic econometric problem is to determine relationships between variables. The simplest example is the problem of finding the linear relation between two variables that "best" approximates given data (x_i, y_i), $1 \leq i \leq k$. Then the choice of a linear relation $y = ax + b$ between two variables x and y amounts to the choice of two numbers, a and b. In the same spirit, we can look at the well-known complication that a system of linear equations can be contradictory, and so have no solutions. This need not be the last word on such a system: one can ask for the "best approximate solution." This turns out to be a fruitful and useful concept in other situations as well, and this has led to an entire subject, *approximation theory*.

Derivative equal to the zero vector. Vector concepts allow to extend the method of Fermat to all unconstrained problems of two or more variables: then the variable of optimization is modeled as a vector and the derivative is a vector and not a number. As a further example of the extension, consider the concept of an ε-neighborhood $(\hat{x} - \varepsilon, \hat{x} + \varepsilon)$ of a number $\hat{x}$ for a small number $\varepsilon > 0$, which plays a role in the definition of "point of local extremum" of a function of one variable. The extension to n variables is an open ball in n-dimensional space with center $\hat{x}$ and radius a small number $\varepsilon > 0$. The power of the method comes again from the differential calculus, in combination with the technique of partial derivatives.

The method of choice to complete the analysis—which aims to "find the *global* minima and maxima"—is to use a result that at first sight might look to be only of theoretical interest. This is the Weierstrass theorem, which establishes the existence of solutions, however, without giving information on their location. Other, less convenient methods to complete the analysis are "second order conditions" (see chapter five) and "direct verification of optimality." We note that the high school method of the sign of the derivative does not extend beyond the one-variable case.

Fermat-Torricelli point. One of the most remarkable applications is to the problem of the best location to serve three clients. This two-variable problem was originally posed as a geometrical problem by Fermat to Torricelli. This problem has a many-sided sense. You can see it not only as an economic problem, an *optimal location problem*, but also as a convincing test case for the strength of the n-dimensional Fermat method, and moreover, in several ways as a physical problem (soap films, mechanics). The n-dimensional Fermat theorem will be seen to solve this problem in one straightforward blow, but without this tool it would require a geometrical ruse to solve it.

2.1 INTRODUCTION

The theorems of Fermat and Weierstrass remain true for functions of two or more variables, by virtue of vector concepts and notation. For example, the essence of the method of Fermat will remain the same: in a local extremum the derivative is zero; however, *the derivative of a function of two or more variables is not a number, but a vector*. The power of vector concepts and notation is that they bring structure and organization to what otherwise would be messy formulas. This allows us to view a collection of objects—numbers, formulas or equations—conceptually as one single object—a vector, a vector function, or a vector equation.

Royal road. If you want a shortcut to the applications, then you can read formula (1) in section 2.2.4 for the derivative in terms of partial derivatives, the statements of the Fermat theorem 2.5, the Weierstrass theorem 2.6 and corollary 2.7, as well as the solutions of examples 2.3.1 and 2.3.2; thus prepared, you are ready for the applications in sections 2.4 and 2.6. After this, you can turn to the next chapter.

2.2 THE DERIVATIVE FOR TWO OR MORE VARIABLES

We are going to invest some effort in a solid foundation for the analysis of unconstrained problems with two or more variables, also called the n-dimensional case.

2.2.1 The length of a vector

For the definition of the derivative of a function of n variables we need the concept "length of a vector."

Dot product. We will use $\mathbb{R}^n$ as the notation for the space of all n-dimensional column vectors and $(\mathbb{R}^n)^T$ or $(\mathbb{R}^n)'$ for the space of all n-dimensional row vectors. Moreover, for each row $a \in (\mathbb{R}^n)^T$ and each column $x \in \mathbb{R}^n$ we will denote their matrix product

$$\sum_{i=1}^n a_i x_i = a_1 x_1 + \cdots + a_n x_n$$

by $a \cdot x$, the *dot product*.

Transpose. Often, we will write a column vector $x \in \mathbb{R}^n$ as a row vector in $(\mathbb{R}^n)^T$, for typographical reasons. The resulting row vector is called the *transpose* of x and it is denoted by x^T. Similarly, sometimes we will write a row vector $y \in (\mathbb{R}^n)^T$ as a column vector, call it the transpose of y, and denote it by y^T. More generally, for an $m \times n$ matrix A, its transpose $B = A^T$ is defined to be the $n \times m$ matrix obtained by writing the rows of A as columns, or equivalently, the columns of A as rows, that is, $b_{ji} = a_{ij}$ for $1 \leq i \leq m$, $1 \leq j \leq n$.

Length. The number

$$|x| = (x^T \cdot x)^{1/2} = \left(\sum_{i=1}^{n} x_i^2\right)^{1/2}$$

is called the *length* or the *Euclidean norm* or the *modulus* of the vector $x \in \mathbb{R}^n$. We note that

$$|x|^2 = \sum_{i=1}^{n} x_i^2,$$

that $|\alpha x| = |\alpha||x|$, and that

$$|x + y|^2 = |x|^2 + 2x^T \cdot y + |y|^2$$

for all $\alpha \in \mathbb{R}$ and $x, y \in \mathbb{R}^n$. The vector in the space of n-dimensional column vectors $\mathbb{R}^n$ that has all coordinates equal to zero is called the *origin* or the *zero vector* in $\mathbb{R}^n$ and is denoted by 0_n. Other vectors in $\mathbb{R}^n$ are called *nonzero vectors*. Note that the length of a nonzero vector is a positive number.

Theorem of Pythagoras in n-dimensional space. Two vectors $x, y \in \mathbb{R}^n$ with $x^T \cdot y = 0$ are called *orthogonal*. As a consequence of the definitions, we get the n-dimensional theorem of Pythagoras:

$$|x + y|^2 = |x|^2 + |y|^2$$

if x and y are orthogonal vectors in $\mathbb{R}^n$. For $n = 2$ ("the plane") and $n = 3$ ("three-dimensional space") the Euclidean norm is just the ordinary length of the vector x, as can be verified using the theorem of Pythagoras; for $n = 1$ ("the real line $\mathbb{R}$") we get $|x| = (x^2)^{\frac{1}{2}}$, which is equal to the absolute value of the number x, so our notation for the Euclidean norm does not clash with the usual notation for the absolute value. Moreover, for $n = 2$ and $n = 3$, "orthogonal" agrees with the usual meaning that the angle between the vectors is $\frac{1}{2}\pi$ ("a right angle").

2.2.2 Standard vector inequalities

For later use, we give two useful properties of the length.

Theorem 2.1 *The following inequalities hold true:*

1. **The Cauchy-Bunyakovskii-Schwarz inequality**

$$|x^T \cdot y| \leq |x||y|$$

for all $x, y \in \mathbb{R}^n$ *—in fact, the inequality holds strictly unless one of the two vectors x and y is a scalar multiple of the other.*

2. **The triangle inequality**

$$|x + y| \leq |x| + |y|$$

for all $x, y \in \mathbb{R}^n$.

Proof.

1. We assume that x is not the zero vector, as the case $x = 0_n$ is trivial. The problem

$$f(\alpha) = |\alpha x + y|^2 \to \min$$

has a solution, and its value is

$$\frac{|x|^2|y|^2 - (x^T \cdot y)^2}{|x|^2},$$

using

$$|\alpha x + y|^2 = (\alpha x + y)^T \cdot (\alpha x + y) = \alpha^2 |x|^2 + 2\alpha x^T \cdot y + |y|^2$$

and example 1.3.7. On the other hand, this value is nonnegative, as $f(\alpha)$ is defined as a square. Thus the inequality is proved. To prove the supplementary statement, we may assume $x \neq 0_n$ (otherwise the statement is obvious), and then we have to prove that the value of the problem $f(\alpha) \to \min$ is positive, if y is not a scalar multiple of x. This follows as $f(\alpha) = |\alpha x + y|^2$ does not take the value zero if y is not a scalar multiple of x.

2. One gets $|x+y|^2 \leq (|x|+|y|)^2$ on expanding, using $|v|^2 = v^T \cdot v$ (three times) and the Cauchy-Bunyakovskii-Schwarz inequality.

$\square$

Historical remark. The inequality

$$\left|\sum_{k=1}^n a_k b_k\right| \leq \left(\sum_{k=1}^n a_k^2\right)^{\frac{1}{2}} \left(\sum_{k=1}^n b_k^2\right)^{\frac{1}{2}}$$

is due to Cauchy. Viktor Yakovlevich Bunyakovskii (1804–1889) generalized it in 1859 to integrals

$$\left| \int a(t)b(t)dt \right| \leq \left(\int a^2(t)dt \right)^{\frac{1}{2}} \left(\int b^2(t)dt \right)^{\frac{1}{2}}.$$

Hermann Amandus Schwarz (1843–1921) generalized this to complex valued functions. This inequality and its generalizations are sometimes called the inequality of Cauchy or of Cauchy-Schwarz or of Bunyakovskii. This is one of the many examples where terminology is not always universal and does not always attribute results to their original discoverers.

2.2.3 Definition of the derivative of a function of n variables

Observe that the analytical, geometrical, and physical definitions of the derivative of a function of one variable cannot be extended literally to n variables. For example, the analytical definition cannot be generalized literally, as it is not possible to make sense of division by a *vector* $h \in \mathbb{R}^n$. We will extend the *approximation* definition 1.9 of the derivative. For this we have to make four changes.

1. We have to replace the modulus (or absolute value) of the number h by the modulus of the vector $h \in \mathbb{R}^n$—in order to extend the notions "neighborhood" and "small Landau o."

2. We have to replace the product $f'(\widehat{x})h$ of the numbers $f'(\widehat{x})$ and h by the dot product (that is, the matrix product) $f'(\widehat{x}) \cdot h$ of the row vector $f'(\widehat{x}) \in (\mathbb{R}^n)'$ and the column vector $h \in \mathbb{R}^n$. Note here that the simplest nontrivial function on $\mathbb{R}^n$ is a *linear function*, that is, a function of the type

$$h \to a \cdot h = a_1 h_1 + \cdots + a_n h_n,$$

where a is a nonzero row vector from $(\mathbb{R}^n)^T$. Now we can explain why it is convenient to have two notations for row vectors. If we view an n-dimensional row vector a as a linear function of n variables, then we write $a \in (\mathbb{R}^n)'$; if we view it just as a row vector, then we write $a \in (\mathbb{R}^n)^T$.

3. **Open balls.** We have to replace the open interval $(\widehat{x} - \varepsilon, \widehat{x} + \varepsilon)$ by the *open ball* $U_n(\widehat{x}, \varepsilon)$ in $\mathbb{R}^n$ with center $\widehat{x}$ and radius ε. This open ball is defined to be the set consisting of all $x \in \mathbb{R}^n$ for which $|x - \widehat{x}| < \varepsilon$. The terminology "open ball" is suggested

by the usual terminology for the case $n = 3$. For $n = 2$ the set $U_n(\widehat{x}, \varepsilon)$ is an *open disk* and for $n = 1$ it is the *open interval* $(\widehat{x} - \varepsilon, \widehat{x} + \varepsilon)$. The purpose of open balls with center $\widehat{x}$ is to serve as neighborhoods of the point $\widehat{x}$.

Now we recall again the words of Kepler from the epigraph to this chapter:

"Near a maximum the changes ... are in the beginning only imperceptible."

We have to give a precise definition of the concept that "changes are in the beginning only imperceptible."

4. **Small Landau** *o*. We have to extend the small Landau o notation to functions $r(h)$ of a variable vector $h \in \mathbb{R}^n$:

$$r(h) = o(h) \text{ if } \lim_{|h| \to 0} |h|^{-1} |r(h)| = 0.$$

The meaning of this property is that the length of the vector $r(h)$ is negligible ("imperceptible") compared to the length of the vector h, if h is sufficiently close to 0_n.

We are now ready to give the promised definition of the derivative. It extends the approximation definition of the derivative from one to n variables.

Definition 2.2 *Definition of the derivative.* Let $f : U_n(\widehat{x}, \varepsilon) \to \mathbb{R}$ *be a function of n variables, defined on an open ball with center $\widehat{x}$. Then $f : U_n(\widehat{x}, \varepsilon) \to \mathbb{R}$ is called differentiable at $\widehat{x}$ if there exist a row vector $f'(\widehat{x})$ in $(\mathbb{R}^n)'$, and a function $r : U_n(0_n, \varepsilon) \to \mathbb{R}$ for which $r(h) = o(h)$, such that*

$$f(\widehat{x} + h) - f(\widehat{x}) = f'(\widehat{x}) \cdot h + r(h)$$

for all $h \in U_n(0_n, \varepsilon)$.

Note that differentiability at $\widehat{x}$ implies continuity at $\widehat{x}$. Details on the concept continuity of functions of n variables are given in appendix C. The row vector $f'(\widehat{x})$ is called the *(first) derivative* of f at $\widehat{x}$; the column vector $f'(\widehat{x})^T$ is called the *gradient* of f at $\widehat{x}$. Its direction is the one of *steepest ascent* of f at $\widehat{x}$, as we will see (cf. exercise 2.6.15). One writes $D^1(\widehat{x})$ for the set of functions that are defined on some open ball with center $\widehat{x}$ and for which the first derivative at $\widehat{x}$ exists.

Definition in words. We recall the idea of the definition. The increment $f(\widehat{x} + h) - f(\widehat{x})$ is split up as the sum of two terms: *the main linear part* $f'(\widehat{x}) \cdot h$ and the *remainder term* $r(h)$, defined by

$$r(h) = (f(\widehat{x} + h) - f(\widehat{x})) - f'(\widehat{x}) \cdot h$$

(Fig. 2.1). The core of the matter is that this splitting up is char-

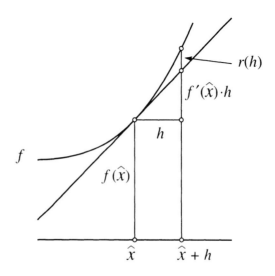

Figure 2.1 Illustrating the definition of differentiability.

acterized by the following properties of these terms: the dependence on h of the main linear part is *linear* and the dependence on h of the remainder term is $o(h)$, that is, it is negligible compared to the modulus of h, if h is sufficiently close to 0_n.

Uniqueness of the derivative. To be precise, we have to check that if $f : U_n(\widehat{x}, \varepsilon) \to \mathbb{R}$ is differentiable at $\widehat{x}$, then $f'(\widehat{x})$ is uniquely determined by definition 2.2. Well, if $a_1, a_2 \in (\mathbb{R}^n)^T$ with

$$f(\widehat{x} + h) = f(\widehat{x}) + a_i \cdot h + o(h), \ i = 1, 2,$$

then subtraction gives

$$0 = (a_1 - a_2) \cdot h + o(h),$$

and so $a_1 = a_2$, as required.

For a linear function $f(x) = a \cdot x$, one has $f'(x) = a$ for all $x \in \mathbb{R}^n$. Let us consider a quadratic example.

Example 2.2.1 *Show that the derivative of $f(x) = x_1^2 + x_2^2$ at $\widehat{x} = (\widehat{x}_1, \widehat{x}_2)^T$ is $2\widehat{x}^T$.*

Solution. Expanding the expression for the remainder term $r(h) = f(\widehat{x} + h) - f(\widehat{x}) - 2\widehat{x}^T \cdot h$ gives $r(h) = h_1^2 + h_2^2 = |h|^2$, and this is $o(h)$.

Remark. *Quadratic functions and matrix notation.* Let us consider, more generally, the class of quadratic functions of n variables $x_1, \ldots, x_n$. A function $f : \mathbb{R}^n \to \mathbb{R}$ is called *quadratic* if it is a polynomial function of $x_1, \ldots, x_n$ of degree at most 2. That is,

$$f(x) = \sum_{i,j=1}^{n} a_{ij} x_i x_j + 2 \sum_{k=1}^{n} b_k x_k + c.$$

The factor 2 is written as it leads to simpler formulas later on. Such a function can be written, using vector and matrix notation, as

$$f(x) = x^T A x + 2b \cdot x + c,$$

with $A = (a_{ij})_{ij}$ an $n \times n$ matrix, $b = (b_k)_k \in (\mathbb{R}^n)^T$, and $c \in \mathbb{R}$. An $n \times n$ matrix B is called *symmetric* if $b_{ij} = b_{ji}$ for all $i, j \in \{1, \ldots, n\}$, that is, if the matrix B is equal to its transpose B^T. We may assume without restriction that the matrix A is symmetric: indeed,

$$x^T A x = \frac{1}{2} x^T (A + A^T) x,$$

and the matrix $A + A^T$ is symmetric. In terms of coordinates, for each pair (i, j) the coefficient a_{ij} (resp. a_{ji}) of the product $x_i x_j$ (resp. $x_j x_i$) is replaced by the average of these two coefficients ("equal sharing").

Now we are ready to calculate the derivative of a quadratic function of n variables. This is useful for solving the many practical optimization problems for which the objective function is quadratic.

Problem 2.2.1 *Derivative of a quadratic function.* *Consider a quadratic function on $\mathbb{R}^n$, say,*

$$f(x) = x^T A x + 2b \cdot x + c,$$

with A a symmetric $n \times n$ matrix, $b \in (\mathbb{R}^n)^T$, and $c \in \mathbb{R}$. Then the derivative of $f : \mathbb{R}^n \to \mathbb{R}$ at a given point $\widehat{x}$ is

$$f'(\widehat{x}) = 2(\widehat{x}^T A + b).$$

Derive this formula.

Solution. For each vector $h \in \mathbb{R}^n$ we have $h^T A \widehat{x} = \widehat{x}^T A h$, by the definition of matrix multiplication, and by $A^T = A$. Therefore

$$\begin{aligned}&f(\widehat{x}+h) - f(\widehat{x}) \\ &= (\widehat{x}^T + h^T)A(\widehat{x}+h) + 2b \cdot (\widehat{x}+h) + c - \widehat{x}^T A \widehat{x} - 2b \cdot \widehat{x} - c \\ &= 2(\widehat{x}^T A + b) \cdot h + h^T A h.\end{aligned}$$

As $|h^T A h| \leq |h||Ah|$ by the Cauchy-Bunyakovskii-Schwarz inequality, it follows that $h^T A h = o(h)$. Therefore, the function $f : \mathbb{R}^n \to \mathbb{R}$ is differentiable at the point $\widehat{x}$, and the derivative $f'(\widehat{x})$ of the function $f : \mathbb{R}^n \to \mathbb{R}$ at the point $\widehat{x}$ is the row vector $2(\widehat{x}^T A + b)$.

Historical remark. After the invention by Newton and Leibniz of the derivative for functions of one variable, it took no less than two centuries before it was generalized to two or more variables: definition 2.2 is due to Weierstrass. Later it was rediscovered in a more general context by Maurice Fréchet (1878–1973).

Conclusion. The definition of the derivative of a function of n variables can be given in exactly the same way as the derivative of a function of one variable, provided we use the approximation definition and make use of vectors.

2.2.4 The differential calculus

Plan. We show how the calculation of the derivative of an n-variable function can be reduced to that of derivatives of one-variable functions, by means of *partial derivatives.*

It is recommended to calculate the derivatives of functions by means of the differential calculus, using *partial derivatives.* One defines $\frac{\partial f}{\partial x_i}(\widehat{x})$, the i-th partial derivative of a function $f : U_n(\widehat{x}, \varepsilon) \to \mathbb{R}$ at $\widehat{x}$, where $i \in \{1, \ldots, n\}$, to be the derivative of the function

$$g(x_i) = f(\widehat{x}_1, \ldots, \widehat{x}_{i-1}, x_i, \widehat{x}_{i+1}, \ldots, \widehat{x}_n)$$

at $x_i = \widehat{x}_i$, if this exists. That is, all coordinates except the i-th one are put equal to the coordinates of $\widehat{x}$, and then one takes from the

FERMAT: TWO OR MORE VARIABLES WITHOUT CONSTRAINTS

resulting function of one variable x_i the derivative at $x_i = \hat{x}_i$. Details on partial derivatives are given in appendix B. The main fact is that a function $f : U_n(\hat{x}, \varepsilon) \to \mathbb{R}$ is differentiable at $\hat{x}$ if all partial derivatives exist on $U_n(\hat{x}, \varepsilon)$ and are continuous at $\hat{x}$. This useful fact leads to a *reduction of the calculation of derivatives to the one-variable case* by virtue of the following expression of the derivative in terms of partial derivatives

$$f'(\hat{x}) = \left(\frac{\partial f(\hat{x})}{\partial x_1}, \ldots, \frac{\partial f(\hat{x})}{\partial x_n} \right). \tag{1}$$

This reduction is illustrated in the following calculation of the derivative of the modulus function.

Example 2.2.2 *Calculate the derivative of the modulus function*

$$f(x) = |x| = \sqrt{x_1^2 + \cdots + x_n^2}$$

at a nonzero point by means of partial derivatives. Show that

$$|x|' = |x|^{-1} x^T.$$

This extends the formula for the derivative of the absolute value function, which was derived in example 1.2.2 in two different ways: by means of Figure 2.1 and by means of the definition of the derivative. It could also have been derived by means of the differential calculus (chain rule and rule $(x^r)' = rx^{r-1}$) in the following way:

$$|x|' = ((x^2)^{\frac{1}{2}})' = \frac{1}{2}(x^2)^{-\frac{1}{2}} 2x = |x|^{-1} x.$$

It is this third possibility which is the most convenient one for the present purpose, the derivation of the extension of this formula to the n-dimensional case.

Solution. Calculation of the partial derivatives of f gives

$$\frac{\partial f}{\partial x_i}(x) = x_i / \sqrt{x_1^2 + \cdots + x_n^2}$$

for all $i \in \{1, \ldots, n\}$. These functions are seen to be continuous everywhere except at the origin 0_n, using the fact that the denominator is nonzero for $x \neq 0_n$. It follows that f is differentiable everywhere except at the origin, with

$$f'(x) = (x_1 / \sqrt{x_1^2 + \cdots + x_n^2}, \ldots, x_n / \sqrt{x_1^2 + \cdots + x_n^2}) = |x|^{-1} x^T.$$

As a motivation for the convenience of the differential calculus you could try to calculate the derivative of the modulus function directly from the definition of the derivative.

Warning. It is possible that all partial derivatives

$$\frac{\partial f(\widehat{x})}{\partial x_k}, \ 1 \leq k \leq n,$$

exist, but that $f : U_n(\widehat{x}, \varepsilon) \to \mathbb{R}$ is not differentiable at $\widehat{x}$, as the following example shows.

Example 2.2.3 *Investigate the differentiability properties at the origin of the function f of two variables defined by*

$$f(x) = x_1 x_2/(x_1^2 + x_2^2) \text{ if } x \neq 0_2 \text{ and } f(0_2) = 0.$$

Solution. The function f is not continuous at 0_2, and so certainly not differentiable. However, the partial derivative $\frac{\partial f}{\partial x_1}(0_2)$ equals by definition the derivative of the function g of one variable given by $g(x_1) = f(x_1, 0)$. One has $g(x_1) = 0$, both for $x_1 \neq 0$ and for $x_1 = 0$. Therefore, $\frac{\partial f}{\partial x_1}(0_2) = 0$. In the same way, $\frac{\partial f}{\partial x_2}(0_2) = 0$.

Conclusion. The approximation definition of the derivative of a function of one variable can be extended to the case of two or more variables (provided you take the right point of view). The resulting derivative is a vector, not a number. The *calculation* of this derivative proceeds by calculating the partial derivatives.

2.3 MAIN RESULT: FERMAT THEOREM FOR TWO OR MORE VARIABLES

Let $\widehat{x}$ be a point of $\mathbb{R}^n$ and $f : U_n(\widehat{x}, \varepsilon) \to \mathbb{R}$ a function of n variables, defined on an open ball with center $\widehat{x}$ and radius $\bar{\varepsilon} > 0$. The function $f : U(\widehat{x}, \bar{\varepsilon}) \to \mathbb{R}$ might also be defined outside this open ball, but this is not relevant for the present purpose.

Definition 2.3 Main problem: unconstrained n-variable optimization. *The problem*

$$f(x) \to \text{extr}, \qquad (P_{2.1})$$

is called an unconstrained n-variable or n-dimensional problem.

Definition 2.4 Local extrema. *One says that $\widehat{x}$ is a (point of) local minimum (maximum) of a function $f : U_n(\widehat{x}, \bar{\varepsilon}) \to \mathbb{R}$ if there exists a sufficiently small positive number $\varepsilon < \bar{\varepsilon}$ such that $\widehat{x}$ is a point of global minimum (maximum) of $f : U_n(\widehat{x}, \varepsilon) \to \mathbb{R}$.*

Now we give the extension of Fermat's method to the case of n variables. Usually, this result is called the *Fermat theorem* as well.

Theorem 2.5 *The Fermat theorem—necessary condition for unconstrained n-variable problems.* Consider a problem of type $(P_{2.1})$. Assume that $f : U_n(\widehat{x}, \bar{\varepsilon}) \to \mathbb{R}$ is differentiable at $\widehat{x}$. If $\widehat{x}$ is a point of local extremum of $(P_{2.1})$, then

$$f'(\widehat{x}) = 0_n^T. \tag{2.1}$$

Stationary points. The point $\widehat{x}$ is called *a point of stationarity* of the problem $(P_{2.1})$ (and also of the function $f : U_n(\widehat{x}, \bar{\varepsilon}) \to \mathbb{R}$) if it is a point of differentiability of f for which the derivative $f'(\widehat{x})$ is equal to the n-dimensional zero row vector 0_n^T.

Thus this result states that all points of differentiability that are local extrema are stationary points. Once again we quote the words of Kepler, which we have chosen as the epigraph of this chapter:

"*near a maximum the changes [...] are in the beginning only imperceptible.*"

These words capture the essence of this result.

The proof of the Fermat theorem for n variables is essentially the same as in the case of one variable. Again it suffices to give the proof in the case of minimization.

Proof. On the one hand, the increment $f(\widehat{x} + h) - f(\widehat{x})$ is, for sufficiently small $|h|$, nonnegative, by the local minimality of $\widehat{x}$. On the other hand, it is equal to the sum of the main linear part $f'(\widehat{x})$ and the remainder term, $r(h)$, with $r(h) = o(h)$, by the definition of the derivative. This gives

$$f'(\widehat{x}) \cdot h + r(h) \geq 0.$$

Replacing h by th with h arbitrary but fixed and $t > 0$ variable, we get

$$tf'(\widehat{x}) \cdot h + r(th) \geq 0.$$

Dividing by t and letting t tend to zero gives $f'(\widehat{x}) \cdot h \geq 0$. As h is an *arbitrary* vector in $\mathbb{R}^n$, it follows that $f'(\widehat{x}) = 0_n^T$. To see this, write $a = f'(\widehat{x})$ and note that $a \cdot e_i = a_i$, $1 \leq i \leq n$, where e_i is the vector in $\mathbb{R}^n$ with 1 on the i-th place and zero on all other places. $\square$

This is really a "one-line proof":

$$0 \leq f(\widehat{x} + h) - f(\widehat{x}) = f'(\widehat{x}) \cdot h + o(h).$$

That is, the sum of a linear function and a small Landau o function of the variable vector h is a nonnegative-valued function. This is only possible if the linear function is the zero function, and therefore, $f'(\widehat{x}) = 0_n^T$.

Crash course theory. This is a good moment to look up in appendices E and F, the two crash courses on optimization theory, the parts concerning this chapter.

Proof for advanced readers. Advanced readers and experts might prefer the presentation of the proof given in appendix E, "Conditions of Extremum from Fermat to Pontryagin."

Algorithmic character proof: reasonable descent. Exercise 2.6.14 emphasizes the algorithmic spirit of the proof. In fact the proofs of all first order necessary conditions for optimization problems have an algorithmic character: the proof shows that if these conditions do not hold at some point, then a *reasonable descent* can be given, that is, a nearby admissible point with a lower value of the objective function can be constructed (cf. exercises 3.6.17 and 4.6.18).

Compact sets. To extend the Weierstrass theorem to the case of n variables, we have to replace the "closed interval $[a, b]$" by a *nonempty compact* set in $\mathbb{R}^n$, that is by a *nonempty, closed,* and *bounded* set in $\mathbb{R}^n$. We will only give the formulation of the result here. If you are not familiar with the concepts "continuous function of n variables" and "closed and bounded set in $\mathbb{R}^n$," then you can consult appendix C, the Weierstrass theorem on existence of global solutions.

Theorem 2.6 *Weierstrass theorem for n variables. A continuous function on a nonempty compact set in $\mathbb{R}^n$ attains its global maxima and minima.*

Often we will use the following consequence. A continuous function $f : \mathbb{R}^n \to \mathbb{R}$ on the whole $\mathbb{R}^n$ is called *coercive* (for minimization) if

$$\lim_{|x| \to +\infty} f(x) = +\infty.$$

Corollary 2.7 *A continuous function $f : \mathbb{R}^n \to \mathbb{R}$ that is coercive for minimization has a global minimum.*

One has a similar result for maximization.

FERMAT: TWO OR MORE VARIABLES WITHOUT CONSTRAINTS

Example 2.3.1 *Solve the problem*
$$f(x) = |x|^2 = x_1^2 + x_2^2 \to \min, \ x \in \mathbb{R}^2.$$

Solution

1. Existence of a global solution $\widehat{x}$ follows from the coercivity of f.

2. Fermat: $f'(x) = 0_2^T \Rightarrow (2x_1, 2x_2) = (0, 0)$ by means of partial derivatives.

3. $x = 0_2$.

4. $\widehat{x} = 0_2$.

Example 2.3.2 *Solve the problem of finding the point in the plane $\mathbb{R}^2$ for which the sum of the squares of the distances to three given points in the plane, $a, b,$ and c, is minimal.*

Solution

1. This problem can be modeled as follows:
$$f(x) = |x - a|^2 + |x - b|^2 + |x - c|^2 \to \min, \ x \in \mathbb{R}^2.$$

   Existence of a global solution $\widehat{x}$ follows from the coercivity of f (note that $f(x) \approx 3|x|^2$ for sufficiently large $|x|$).

2. Fermat: $f'(x) = 0_n^T \Rightarrow 6x - 2a - 2b - 2c = 0_n$,

   using
   $$(|x - a|^2)' = (x^T \cdot x - 2a^T \cdot x + a^T \cdot a)' = 2(x - a)^T,$$
   applying the formula of problem 2.1.

3. $x = \frac{1}{3}(a + b + c)$.

4. $\widehat{x} = \frac{1}{3}(a + b + c)$, the "center of gravity" of the triangle with vertices $a, b,$ and c.

We often need variants of corollary 2.7. The idea of all these variants is that you show that the objective function f of the problem has a nonempty and bounded level set $\{x : f(x) \leq C\}$. We illustrate this in the following example (we mention in passing that it belongs to the problem type that will be considered in chapter four).

Example 2.3.3 *A consumer wants to buy a consumption bundle, consisting of n types of products that can be bought in any integral and nonintegral quantity (like sugar and rice). He wants to buy the cheapest bundle that gives him a certain amount of pleasure. Model this problem and show that the resulting optimization problem has a solution.*

Solution. The consumption bundle can be modeled by a nonnegative vector in n-space. The pleasure can be modeled by a continuous utility function $u : \mathbb{R}_+^n \to \mathbb{R}$, where $u(x)$ denotes the pleasure of the consumer of buying bundle x. Let $\bar{u}$ be the required minimum level of utility. We make the assumption that there exists a consumption bundle $\tilde{x}$ that gives the required pleasure, that is, $u(\tilde{x}) \geq \bar{u}$. Let the positive n-dimensional row vector $p \in (\mathbb{R}^n)^T$ be the price vector. This leads to the problem

$$f(x) = p \cdot x \to \min, \ x \in \mathbb{R}^n, \ x \geq 0_n, \ u(x) \geq \bar{u}.$$

The Weierstrass theorem cannot be applied immediately, as there is no reason why the admissible set should be bounded. However, note that the—possibly empty—solution set of the problem does not change if we add the constraint $p \cdot x \leq p \cdot \tilde{x}$. Indeed, the optimal bundle cannot be more expensive than the bundle $\tilde{x}$. By adding this constraint, we get a problem to which Weierstrass is applicable:

- the admissible set is nonempty as it contains $\tilde{x}$,

- the admissible set is closed as the functions $x \to p \cdot x$ and $x \to u(x)$ are continuous,

- the admissible set is bounded, as the inequalities $x \geq 0_n$ and $p \cdot x \leq p \cdot \tilde{x}$ imply $0 \leq p_i x_i \leq p \cdot \tilde{x}$, and so $0 \leq x_i \leq p_i^{-1} p \cdot \tilde{x}$, for $1 \leq i \leq n$.

It follows that a solution of the problem exists.

Conclusion to section 2.3. Unconstrained optimization problems of n variables can be solved in exactly the same way as those of one variable: the one-dimensional Fermat and Weierstrass theorem can be extended easily, using vectors. To extend the Weierstrass theorem, one has to replace "closed intervals $[a, b]$ in $\mathbb{R}$" by "nonempty compact (=closed and bounded) sets in $\mathbb{R}^n$."

2.4 APPLICATIONS TO CONCRETE PROBLEMS

2.4.1 Minimization of a quadratic function

Many concrete problems are a special case of the following problem type, which is the generalization of the problem of Euclid—exercise 1.6.26—to n variables (see examples 2.3.1 and 2.3.2): minimization of a quadratic function. Now we will solve this problem analytically under a suitable assumption, which implies that the problem has a unique solution. To begin with, we recall that a quadratic function can be written in matrix notation and that this involves a symmetric matrix (cf. the remark preceding problem 2.2.1). A symmetric $n \times n$ matrix A is called *positive definite* if $x^T A x > 0$ for all nonzero $x \in \mathbb{R}^n$. This property implies that A is invertible. We mention in passing that it is not hard to verify whether a given symmetric matrix is positive definite (cf. section 5.2.2).

Problem 2.4.1 *Let a quadratic function $f(x)$ of n variables be given. Write the function f in matrix notation, $f(x) = x^T A x + 2b \cdot x + c$, where A is a symmetric $n \times n$ matrix, $b \in (\mathbb{R}^n)^T$, and $c \in \mathbb{R}$. Consider the problem*

$$f(x) = x^T A x + 2b \cdot x + c \to \min.$$

Assume that A is positive definite. Show that the given problem has a unique global solution, $\hat{x} = -A^{-1}b^T$, and that the value of this problem equals

$$c - bA^{-1}b^T.$$

Solution

1. Existence of a global solution $\hat{x}$ follows from the coercivity of f. The details of the verification of the coercivity of f are as follows. Consider, to begin with, the auxiliary problem

   $$x^T A x \to \min, \ |x| = 1.$$

   The Weierstrass theorem can be applied to it as the unit sphere in $\mathbb{R}^n$ is nonempty and compact, that is, closed and bounded. Let us check this. The unit sphere is nonempty as it contains the point $(1, 0, \ldots, 0)^T \in \mathbb{R}^n$. It is closed as it is the solution set of the equation $|x| - 1 = 0$, and $|x| - 1$ is a continuous function. It is bounded as the equation $|x| = 1$ can be written as $x_1^2 + \cdots + x_n^2 = 1$, and this implies $-1 \leq x_i \leq 1$ for all i.

As a result, the auxiliary problem has a global solution. Therefore, its minimal value γ is positive, using the fact that A is positive definite. It follows that

$$x^T A x = |x|^2 (|x|^{-1} x)^T A (|x|^{-1} x) \geq \gamma |x|^2,$$

for all nonzero vectors $x \in \mathbb{R}^n$.

Now we are ready to verify the coercivity of f:

$$|f(x)| \geq |x^T A x| - 2|b.x| - |c| \geq \gamma |x|^2 - 2|b||x| - |c| \to +\infty$$

for $|x| \to +\infty$, as $\gamma > 0$.

2. Fermat: $f'(x) = 0_n^T \Rightarrow 2x^T A + 2b = 0_n^T$.

3. $x = -A^{-1} b^T$.

4. $\widehat{x} = -A^{-1} b^T$.

Finally, the minimal value of the problem is found to be $c - bA^{-1}b^T$ by substitution of the solution $\widehat{x}$ into the formula for f.

The discussion of this problem will be continued in sections 7.2 and 9.5

2.4.2 Selling ice cream on the beach

Let us consider an economic application of the Fermat theorem. Sometimes, the unregulated self-interest of individuals leads to the common good of the whole society. Then it is as though an invisible hand—"of the market"—is guiding these individuals. This "invisible hand" is one of the most fruitful insights of Adam Smith (1723–1790), the founding father of economics. The invisible hand does not always work, as the following problem shows. Sometimes government regulations are required to achieve a socially desirable objective.

Moreover, the effect that two political parties both tend to move toward the middle between left and right in their struggle for votes will be seen in action in the first half of this problem.

Problem 2.4.2 *Should the town council of a seaside resort tell the sellers of ice cream on its beach where to stand?*

Solution

Sellers choose location. Imagine you are on the beach and want to buy an ice cream. The sand is hot, so you don't want to walk too far, and therefore you head for the nearest ice cream seller. Now put yourself in the position of an ice cream seller. The tourists are evenly spread ("uniformly distributed") over the beach. You will choose a position on the beach where you are the closest seller for as many potential clients as possible. Keeping this in mind, what will happen on a 200-meter beach with two ice cream sellers who are free to choose a position for their businesses?

You will see that they will end up next to each other at the middle of the beach. Until this happens, the seller who is farther away from the middle will not be happy with his position, as he can improve the number of his potential clients by moving closer to the middle. This analysis can be written down more formally using the Fermat theorem, of course. At the final position, the ice cream sellers have an equal share of the market. Unfortunately, the average walking distance for the customers is rather long, 50 m.

Prescribe the location of the sellers. Now let us organize things differently. The ice cream sellers are told where to sell by someone who has the interest of the tourists in mind, say a representative of the council. She aims to keep the average walking distance to an ice cream minimal.

Modeling of the problem. Let us model this problem of the optimal locations for the sellers. Choose 200 m as a unit of length, the length of the beach, and let x, y, and z be the lengths of the three parts into which the beach is divided by the positions of the two sellers, so $z = 1 - x - y$ (Fig. 2.2).

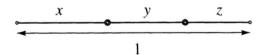

Figure 2.2 Two ice cream sellers on the beach.

Formula for the average distance. Now we calculate the average distance $f(x, y)$ from a tourist to the nearest ice cream seller, assuming that the tourists are distributed uniformly over the beach. This gives the following formula:

$$f(x,y) = x(x/2) + (y/2)(y/4) + (y/2)(y/4) + (1-x-y)((1-x-y)/2).$$

To see this, let us consider, for example, the first term of this expression, $x(x/2)$, the product of the factors x and $x/2$. It represents the contribution to the average made by the tourists on the part of the beach of length x to the left side of the first ice cream seller. Indeed, the fraction of the tourists on this part of the beach is x as they are uniformly distributed; this explains the factor x. The average walking distance to the first ice cream seller is $x/2$ for these tourists; this explains the factor $x/2$.

The other three terms can be explained in the same way. For example, the second term is the contribution to the average of the tourists who are on the beach between the two ice cream sellers and closer to the left one than to the right one. The expression for the average distance $f(x,y)$ can be simplified to

$$f(x,y) = x^2/2 + y^2/4 + (1-x-y)^2/2.$$

Solution of the optimization problem. It follows that the problem to choose the locations for the sellers can be modeled as follows:

1. $f(x,y) = x^2/2 + y^2/4 + (1-x-y)^2/2 \to \min$. The function f is quadratic with quadratic part

   $$x^2/2 + y^2/4 + (-x-y)^2/2,$$

   which is positive for all nonzero vectors (x,y). Therefore, we can apply the result of problem 2.4.1.

   It follows that there is a unique global minimum $(\hat{x}, \hat{y})$.

2. Fermat: $f'(x,y) = 0_2^T \Rightarrow$

   $$\frac{\partial f}{\partial x}(x,y) = x + (1-x-y)(-1) = 2x + y - 1 = 0,$$

   $$\frac{\partial f}{\partial y}(x,y) = y/2 + (1-x-y)(-1) = x + 3y/2 - 1 = 0.$$

3. The first equation gives $y = 1 - 2x$. Substitution in the second equation gives

   $$x + 3(1-2x)/2 - 1 = -2x + \frac{1}{2} = 0.$$

   Therefore, $x = \frac{1}{4}$ and $y = \frac{1}{2}$.

4. $(\widehat{x}, \widehat{y}) = (\frac{1}{4}, \frac{1}{2})$. That is, it is optimal from the point of view of the tourists to have one ice cream seller at a quarter along the beach and the other one at three quarters. Then the average walking distance is $f(\frac{1}{4}, \frac{1}{2}) = \frac{1}{8}$, that is, 25 m. In other words the rules of the council make it on average twice as easy to get ice cream. Moreover, the sellers are not worse off; they have again an equal share of the market.

This example makes clear that the "invisible hand" does not work in all situations. Then there is reason to worry.

Variants. If you want to interpret the poetry of this model in a literal sense, then you can embellish it by taking into account that the ice cream sellers might have different flavors, prices, and service, which might tempt some customers to walk past one seller to buy an ice cream from a competitor. Moreover, some people might not buy an ice cream if they have to walk too far to get it. For some of the resulting models you will find that the council does not have to interfere.

2.4.3 Should firms always maximize profits?

Profit maximization always best? In economics, the standard assumption is that firms maximize profits. One argument to defend this is the following. Suppose that firms do not maximize profits, then they would be driven from the market by firms that do. This argument we consider in more detail.

Competition: profit versus revenue maximization. Consider two firms that produce a homogeneous good in a market with linear demand of the form

$$p = 1 - q_1 - q_2,$$

where p is the price and q_i is the output level of firm i ($i = 1, 2$). Each firm produces with constant marginal costs equal to c. Firm 1 maximizes profits; that is, it chooses its output level q_1 to solve

$$\max_{q_1} (1 - q_1 - q_2 - c) q_1.$$

Firm 2 chooses its output level q_2 to maximize profits, but the managers at this firm also give some weight α to total revenue,

$$\max_{q_2} (1 - q_1 - q_2 - c) q_2 + \alpha (1 - q_1 - q_2) q_2.$$

That is, these managers give some weight to the empire they are building. At parties they can boast about the number of employees their company has. An interpretation of this objective function is that these managers maximize revenue under the constraint that profits do not fall below some appropriately chosen level π_2.

Problem 2.4.3 *Stakeholder vs. shareholder in the duopoly context. Which of these two companies makes the highest profits for $\alpha > 0$, the profit maximizing firm or the firm that maximizes revenue under a profit constraint?*

Solution. To answer this question we analyze the *Nash equilibrium* outcome, where each firm chooses its output level, taking the opponent's output level as given. The first order conditions (here these conditions are seen to be necessary and sufficient, using our knowledge of parabolas) for the firms can then be written as

$$1 - 2q_1 - q_2 - c = 0, \qquad (2.1)$$
$$1 - q_1 - 2q_2 - c + \alpha - \alpha q_1 - 2\alpha q_2 = 0. \qquad (2.2)$$

These can be solved for output and price levels as follows, where $\varphi = (1+\alpha)^{-1}(1-c+\alpha)$:

$$q_1 = \frac{2}{3}(1-c) - \frac{1}{3}\varphi, \qquad (2.3)$$

$$q_2 = \frac{2}{3}\varphi - \frac{1}{3}(1-c), \qquad (2.4)$$

$$p = 1 - \frac{1}{3}(1-c) - \frac{1}{3}\varphi. \qquad (2.5)$$

As a consequence, profits can be written as

$$\pi_1 = \left(\frac{2}{3}(1+c) - \frac{1}{3}\varphi\right)\left(\frac{2}{3}(1-c) - \frac{1}{3}\varphi\right), \qquad (2.6)$$

$$\pi_2 = \left(\frac{2}{3}(1+c) - \frac{1}{3}\varphi\right)\left(\frac{2}{3}\varphi - \frac{1}{3}(1-c)\right). \qquad (2.7)$$

Since φ is increasing in α, it is routine to verify that the profits of firm 1 are decreasing in α while the profits of firm 2 are increasing in α as long as $\varphi < 2$ and $\varphi < 5/4 + 3c/4$.

As a consequence, we see that the firm that does not maximize profits, in fact makes in equilibrium higher profits than the firm that does maximize profits! What is the intuition for this result?

2.4.4 Problem of Fermat and Evangelista Torricelli (1608–1647)

Problem 2.4.4 *Find the point in the plane for which the sum of the distances to three given points in the plane is minimal.*

This geometrical problem is illustrated in Figure 2.3.

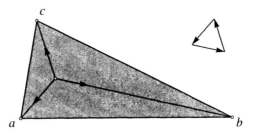

Figure 2.3 Problem of Fermat-Torricelli and geometric sense stationarity.

This problem has a many-sided sense.

1. **Geometrical problem.** It was posed as a geometrical problem by Fermat to Torricelli, who solved it. The latter was a student of Galilei, best known for Torricelli's vacuum, which he invented.

2. **Location problem.** This problem allows the following economic interpretation as well. What is the optimal location to serve three equally important clients, if you want to minimize the average distance to the clients?

3. **Soap films.** In addition it can be interpreted in terms of soap films. When is a configuration of three soap films that come together in a straight line stable? Here, nature behaves as though it were minimizing the total area of the soap films (this has minimal surface tension).

4. **Three holes in a table.** An even clearer physical interpretation is the following one. Drill three holes in a table, take three equal weights, attach them to three strings, pull the strings up through the holes in the table, and tie the three loose ends in a knot above the table. Then let the system go and see what happens. By a law of physics, the position of the knot will keep changing place as long as this leads to a lowering of the center of gravity of the system. At which position will the knot come to a rest?

Solution

1. This problem can be modeled as follows:
$$f(x) = |x - a| + |x - b| + |x - c| \to \min, \ x \in \mathbb{R}^2.$$
Here $a, b, c \in \mathbb{R}^2$ are given. Existence of a global solution $\widehat{x}$ follows from the coercivity of f (note that $f(x) \approx 3|x|$ for sufficiently large $|x|$).

2. Fermat: $f'(x) = 0_2^T \Rightarrow$
$$f'(x)^T = \frac{x-a}{|x-a|} + \frac{x-b}{|x-b|} + \frac{x-c}{|x-c|} = 0_2,$$
using example 2.2.2.

3. This equation allows a geometrical interpretation. We have three vectors in the plane of equal length 1 with sum the zero vector. Therefore, these vectors form a triangle with equal sides, and so with equal angles, when they are placed head-to-tail (Fig. 2.3). This shows that the angle between each two of these three vectors is equal.

Therefore the candidate minima are the stationary points described above (if these exist) and the points of nondifferentiability of f: a, b, and c. It is the unique solution of the problem.

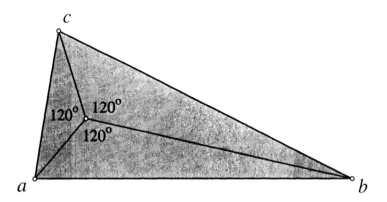

Figure 2.4 The Fermat-Torricelli point.

4. A more precise analysis would reveal that if one of the angles of the triangle with vertices a, b, and c is $\geq \frac{2}{3}\pi$, then the optimal

point is the vertex of this angle. Otherwise there is a unique point $\widehat{x}$ inside the triangle with vertices a, b, and c for which the angles $\angle a\widehat{x}b$, $\angle b\widehat{x}c$, and $\angle c\widehat{x}a$ are equal to $\frac{2}{3}\pi$, that is, $120°$. This point is called the *Fermat* or *Torricelli point* (Fig. 2.4).

2.4.5 Least squares line

Least squares is "the best method." A principal concern of economic and managerial theory and econometrics is the determination of relationships among variables. One method of establishing the coefficients in these relationships is the method of least squares. We will not display the formulas of this method; our main aim is to make clear that this method is in a sense the *best one*, and that it can be obtained by solving an optimization problem. The scope of the least squares method is very broad, but here we only will consider a numerical example of the most basic least squares problem. In section 2.4.9 we will consider some nonlinear problems (with linear coefficients) as well. Moreover, many nonlinear least squares problems can be viewed from an abstract point of view as a special case of the problem of finding the orthogonal projection of a vector in $\mathbb{R}^n$ onto a linear subspace of $\mathbb{R}^n$ spanned by a given finite collection of vectors in $\mathbb{R}^n$. This problem is considered in section 2.4.8.

Line through a cloud of points. The basic problem is to determine the numbers a and b for which the straight line $y = ax+b$ in some sense "best approximates" the relationship between two variables x and y based on the available data. That is, the problem is to draw through a given "cloud" of points in the plane the line that approximates the cloud in the best way. The most popular sense to interpret "best" is to take for each point in the cloud the distance to the line in the vertical direction, square it, take the sum of all these squares, and demand that this measure of the quality of the approximation be as small as possible. The resulting line is called the *least squares line*. The problem of finding the least squares line is illustrated in the following exercise.

Problem 2.4.5 *Let a sample set of data be available, say, the following five pairs* (x_i, y_i), $1 \leq i \leq 5$: *(1,2), (2,5), (4,7), (6,10), (8,14). Determine a and b such that*

$$f(a,b) = \sum_{i=1}^{5}(y_i - ax_i - b)^2,$$

the sum of the 5 squares of the differences between actual value y_i and predicted value $ax_i + b$, is minimal.

Solution

1. We consider the problem

   $$f(a,b) = |a(1,2,4,6,8) + b(1,1,1,1,1) - (2,5,7,10,14)|^2 \to \min.$$

   Existence of a global minimum $(\widehat{a}, \widehat{b})$ follows from the coercivity of f (note that

   $$f(a,b) \geq (|a(1,2,4,6,8) + b(1,1,1,1,1)| - |(2,5,7,10,14)|)^2,$$

   by the triangle inequality).

2. Fermat: $f'(a,b) = (0,0) \Rightarrow 242a + 42b - 424 = 0$ and $42a + 10b - 76 = 0$.

3. $a = \frac{131}{82} \approx 1.597$ and $b = \frac{73}{82} \approx 0.890$.

4. The line

   $$y = \frac{131}{82}x + \frac{73}{82} \approx 1.597x + 0.890$$

   is the unique least squares line for the given sample set of data.

Variants. This basic problem has a number of interesting variants and generalizations.

For example, "solve a system of linear equations that has no solutions," or, to be more precise, minimize the function $f(x) = |Ax - b|$ on $\mathbb{R}^n$ for a given $m \times n$ matrix A and a given vector $b \in \mathbb{R}^m$ (exercise 2.6.8). This problem contains problem 2.4.5 as a special case, with $m = 5$ and $n = 2$.

Another example is to find the numbers $a, b,$ and c for which the parabola $y = ax^2 + bx + c$ is the best approximation of a certain relationship between two variables x and y based on the data available. Here "best" is again meant in the least squares sense.

These variants can be handled in a similar way. We will continue this topic in sections 2.4.8 and 2.4.9.

Theorem 2.8 Fundamental theorem of algebra. *Each polynomial*

$$p(x) = a_0 + a_k x^k + \cdots + a_n x^n$$

in one complex variable x—with $n \geq 1$ and a_k and a_n not equal to zero—has at least one root.

Proof.

1. We consider the problem

$$f(x) = |p(x)| \to \min, \ x \in \mathbb{C}.$$

   This has a solution $\widehat{x}$, as $x \to |p(x)|$ is coercive:

$$|p(x)| = |x|^n |a_n + a_{n-1}\frac{1}{x} + \cdots + a_0 \frac{1}{x^n}|;$$

   the first factor tends to $+\infty$ and the second one to $|a_n|$ for $|x| \to +\infty$. We assume that $\widehat{x} = 0$, as we may without restricting the generality of the argument (of course, this changes the coefficients $a_0, \ldots, a_n$ and the number k, but we retain the notation). In order to prove the theorem, we have to show that $a_0 = 0$.

2. Let u be a solution of $a_0 + a_k x^k = 0$. Define $x_\alpha = \alpha u$ for $\alpha \geq 0$ and $g(\alpha) = f(x_\alpha)$. Then

$$g(\alpha) \geq g(0) \Rightarrow |a_0 + (\alpha u)^k + o(\alpha^k)| \geq |a_0|.$$

3. $|(1 - \alpha^k)a_0 + o(\alpha^k)| \geq |a_0| \Rightarrow a_0 = 0.$

4. The theorem is proved.

□

Remark. We have applied the four-step method in its most basic form here: in the second step we construct a suitable "variation" x_α, $\alpha \geq 0$, of the point $\widehat{x}$, and apply the local minimality property of $\widehat{x}$ to this variation. Usually we apply a user friendly version: we use instead a necessary condition in this step. This is essentially the same method, as necessary conditions are proved by means of variations. This basic form of the four-step method is developed in more detail in [13].

Historical comments. One of the first attempts to state the fundamental theorem of algebra was made by René Descartes (1596–1650) in 1637: *"You must know that every equation may have as many roots as its dimension. However, it sometimes happens that some of these roots are false or less than nothing."* The first proof was given by Carl Friedrich Gauss (1777–1855) in 1794. The idea of the proof presented above was suggested in the works of Jean Le Rond d'Alembert (1717–1783) in 1746–1748.

2.4.6 Basic property of symmetric matrices and Weierstrass

We give another illustration of the power of the Weierstrass theorem, by using it to establish the basic property of symmetric matrices. This is one of the most widely used results from matrix theory. Note that it is not obvious at first sight that this property has anything to do with optimization. However, in many textbooks on optimization this is derived using the Weierstrass theorem, the powerful Lagrange multiplier rule 3.3, and the technique of complete induction. In particular, the proof involves a sequence of optimization problems. We proceed in a different way; we propose and analyze *one* optimization problem, using only the Weierstrass theorem. This immediately gives the required property. We could not find this simple proof explicitly in the literature. In courses on linear algebra, this property is often not proved; at best, it is derived from the fundamental theorem of algebra.

An $n \times n$ matrix P is called *orthogonal* if $PP^T = I_n$ or, equivalently, if the columns of P are orthogonal to each other and have length one. The product of two orthogonal $n \times n$ matrices is again orthogonal, clearly.

Theorem 2.9 *Orthogonal diagonalization of symmetric matrices.* *For each symmetric $n \times n$ matrix A there exists an orthogonal $n \times n$ matrix P such that $P^T A P$ is a diagonal matrix.*

Proof. **A less ambitious task.** Let A be a symmetric $n \times n$ matrix that is not a diagonal matrix, with $n > 2$ (the case $n = 1$ is trivial and the case $n = 2$ can be settled by explicit formulas). We want to turn it into a diagonal matrix by transition to a matrix of the form $P^T A P$, with P an orthogonal matrix. To begin with, let us be less ambitious and try to make it only "a bit more like a diagonal matrix." To be more precise, we will measure how far a symmetric matrix B is removed from being diagonal by the sum $g(B)$ of the squares of the off-diagonal elements of B. Then our aim is, to begin with, to find an orthogonal matrix P such that

$$g(P^T A P) < g(A).$$

How to make a move toward a diagonal matrix. We choose $1 \leq i < j \leq n$, such that $a_{ij} \neq 0$, as we may, as A is not diagonal. For each $n \times n$ matrix M, we denote by M_{ij} the 2×2 submatrix of M that one gets by selecting the i-th and the j-th row and column of M. Now we consider the 2×2 matrix A_{ij}, and we choose an orthogonal 2×2 matrix Q for which $Q^T A_{ij} Q$ is diagonal, using the case $n = 2$ of the theorem. Then we let P denote the $n \times n$ matrix for which $P_{ij} = Q$, $p_{kk} = 1 \ \forall k \neq i, j$, and all other entries are zero; note that P is an orthogonal matrix. Then

$$g(P^T A P) = g(A) - 2a_{ij}^2 < g(A),$$

as one readily checks.

Conclusion of the proof. Now let us be more ambitious and ask for an orthogonal matrix P for which $P^T A P$ is as little removed from a diagonal matrix as possible. Such a matrix exists: the problem

$$f(P) = g(P^T A P) \to \min, \ P \text{ orthogonal}$$

has a solution $\widehat{P}$ by Weierstrass (note that the set of orthogonal matrices is a closed, bounded, and nonempty set in the space of all symmetric matrices). The resulting matrix $\tilde{A} = \widehat{P}^T A \widehat{P}$ is a diagonal matrix. To prove this, let us argue by contradiction. If it were not diagonal, then there would be an orthogonal matrix $\tilde{P}$ for which $g(\tilde{P}^T \tilde{A} \tilde{P}) < g(\tilde{A})$, as we have seen above. This would give $f(\tilde{P}\widehat{P}) < f(\widehat{P})$, contradicting the minimality of $\widehat{P}$. □

The idea of the proof above is algorithmic in spirit. In fact, a very efficient class of numerical methods for orthogonally diagonalizing a symmetric matrix is based on this idea: the class of Jacobi methods (cf. [27]).

Conclusion of section 2.4.6. We have illustrated the Weierstrass theorem by deriving the central property of symmetric matrices from it. In the same way, we will later illustrate two other optimization methods, the Lagrange multiplier rule and the second order conditions, by deriving this property using one of these two other methods (cf. sections 3.3.7 and 5.3.4).

2.4.7 Matrix norms and their main property

A basic and useful concept in the theory of linear equations is "the norm of a matrix." In the following problem we establish this concept and its main property, with the use of the Weierstrass theorem.

Problem 2.4.6 *Let A be an $m \times n$ matrix. Consider the problem*

$$f(x) = \frac{|Ax|}{|x|} \to \max, \ x \in \mathbb{R}^n, \ x \neq 0.$$

(i) Show that this problem has a global solution x_A.

Define the norm of A to be $\|A\| = f(x_A)$. Show that the following property holds true:

(ii) $\|A\| > 0$ if A is not the $m \times n$ zero matrix.

For example, the norm of a diagonal $n \times n$ matrix with diagonal elements $d_1, \ldots, d_n$ is the maximum of the absolute values $|d_1|, \ldots, |d_n|$.

We will continue the investigation of problem 2.7 in the case of symmetric matrices A in problems 3.7 and 5.3.

Solution. (i) We begin by showing the existence of a global solution. We have $f(\mu x) = f(x)$ for all $\mu \in \mathbb{R}$, $x \in \mathbb{R}^n$, clearly. In particular,

$$f(x) = f(|x|^{-1}x)$$

for all nonzero $x \in \mathbb{R}^n$. This shows that we can add the constraint $|x| = 1$ to the problem, without changing the existence question. But then the Weierstrass theorem can be applied, as the unit sphere in $\mathbb{R}^n$ is nonempty and compact; we have verified this in the solution of problem 2.2. As a result, the given problem has a global solution.

(ii) Let A be a nonzero $m \times n$ matrix. We pick a nonzero entry of A, say a_{ij}, and take $x = e_j$, the vector in $\mathbb{R}^n$ with j-th coordinate equal to 1 and all others equal to 0. This gives

$$f(e_j) = \frac{|Ae_j|}{|e_j|} = \left(\sum_{k=1}^m a_{kj}^2\right)^{\frac{1}{2}} \geq |a_{ij}| > 0,$$

and so $\|A\| > 0$ as $\|A\| = f(x_A) \geq f(e_j)$ by the global maximality property of x_A.

We will use the following consequences.

Corollary 2.10   1. Let $v_1, \ldots, v_m$ be linearly independent vectors in $\mathbb{R}^n$. Then there is a constant $C > 0$ for which

$$|y_1 v_1 + \cdots + y_m v_m| \geq C|y|,$$

for all $y = (y_1, \ldots, y_m)^T \in \mathbb{R}^m$.

2. Let A be an $m \times n$ matrix of rank m. Then there is a constant $C > 0$ for which

$$|A^T y| \geq C|y|$$

for all $y \in \mathbb{R}^m$.

Proof.

1. One can argue as above, applying the Weierstrass theorem to the problem

$$f(y) = |y_1 v_1 + \cdots + y_n v_m| \to \min, \ |y| = 1.$$

2. Using the result above, one gets, with A^T the $n \times m$ matrix with columns $v_1, \ldots, v_m$ and with $y = (y_1, \ldots, y_m)^T$, the following upper bound for $|y| = |(AA^T)^{-1}(AA^T)y|$,

$$|y| \leq \|(AA^T)^{-1}\| |A(A^T y)| \leq \|(AA^T)^{-1}\| \|A\| |A^T y|$$

Therefore the choice of constant

$$C = \|(AA^T)^{-1}\|^{-1} \|A\|^{-1}$$

satisfies the requirement. Here we use the fact from matrix theory that the $m \times m$ matrix AA^T is invertible as the columns of A^T are linearly independent.

□

In fact, the two statements of this corollary are essentially equivalent: this can be seen by letting A^T be the $n \times m$ matrix with columns $v_1, \ldots, v_m$.

Other useful properties of the matrix norm include

- $\|A + B\| \leq \|A\| + \|B\|$,
- $\|\alpha A\| = |\alpha| \|A\|$.

Both follow from the corresponding properties for vectors.

2.4.8 Orthogonal projection

In this section, we illustrate by two examples a nonstandard way of proving basic results of matrix theory (more examples will be given in sections 9.2 and 9.5). The traditional way to prove results on matrices is by means of the Gaussian elimination algorithm. In view of this, you might find it refreshing to see an alternative way: by optimization methods. Note that most results of matrix theory seem to have nothing to do with optimization.

Problem 2.4.7 *Existence and uniqueness of the orthogonal projection. Let a vector $v \in \mathbb{R}^m$ and m linear independent vectors $v_1, \ldots, v_m$ in $\mathbb{R}^n$ be given. Then there exists a unique vector ξ that is a linear combination of $v_1, \ldots, v_m$, and for which, moreover, the line through v and ξ is orthogonal to all vectors $v_1, \ldots, v_m$ (and so to all linear combinations of $v_1, \ldots, v_m$. Show this.*

This vector ξ is called *the orthogonal projection* of v on the space of all linear combinations of $v_1, \ldots, v_m$. Note that this problem contains least squares problems as special cases, such as the problem to find the least squares quadratic function $y = ax^2 + bx + c$ for a given collection of points (x_i, y_i), $1 \leq i \leq N$, in the plane.

Solution. Geometric intuition suggests that the linear combination of $v_1, \ldots, v_m$ that is closest to v is the desired vector ξ. Therefore, we consider the problem.

1. $f(y) = \left| v - \sum_{i=1}^{m} y_i v_i \right|^2 \to \min,\ y \in \mathbb{R}^m$.

   Existence of a global minimum $\widehat{y}$ follows from coercivity. (Choose $C > 0$ such that $|\sum_{i=1}^{m} y_i v_i| \geq C|y|$ for all $y \in \mathbb{R}^m$. Then

   $$\left| v - \sum_{i=1}^{m} y_i v_i \right| \geq \left| \sum_{i=1}^{m} y_i v_i \right| - |v| \geq C|y| - |v|,$$

   and this gives the coercivity of f.)

2. Fermat: $f'(y) = 0_m^T \Rightarrow$

   $$\frac{\partial f}{\partial y_i} = 2(v - \sum_{i=1}^{m} y_i v_i)^T \cdot v_i = 0,\ 1 \leq i \leq m.$$

3. The stationarity equations can be interpreted geometrically:

the line through v and $\sum_{i=1}^{m} y_i v_i$ is orthogonal to the vectors v_i, $1 \leq i \leq m$, and so to the space of all linear combinations of these vectors.

By 1, we know that there exists at least one stationary point. Now we verify that this point is unique. If $\bar{y}$ and $\bar{\bar{y}}$ are stationary points, then $\bar{y} - \bar{\bar{y}}$ is orthogonal to the space of all linear combinations of these vectors. As $\bar{y} - \bar{\bar{y}}$ is contained in this space, we get that this vector is orthogonal to itself. Therefore, it is the zero vector, that is, $\bar{y} = \bar{\bar{y}}$.

4. There exists a unique $\hat{y} \in \mathbb{R}^m$ for which the line through v and $\sum_{i=1}^{m} \hat{y}_i v_i$ is orthogonal to all vectors $v_1, \ldots, v_m$.

Corollary 2.11 *If all solutions of a system of linear equations in n variables satisfy another linear equation in these n variables, then this equation is a linear combination of the equations of the system.*

Proof. Let $a_i \in (\mathbb{R}^n)^T$, $0 \leq i \leq m$, and β_i, $0 \leq i \leq m$, be given and assume that all solutions $x \in \mathbb{R}^n$ of the linear system

$$a_i \cdot x + \beta_i = 0, \ 1 \leq i \leq m,$$

satisfy the linear equation $a_0 \cdot x + \beta_0 = 0$. Let $(\bar{a}, \bar{\beta}) \in (\mathbb{R}^{n+1})^T$ with $\bar{a} \in (\mathbb{R}^n)^T$ and $\bar{\beta} \in \mathbb{R}$, be the orthogonal projection of $(a_0, \beta_0) \in (\mathbb{R}^{n+1})^T$ on the linear span of (a_i, β_i), $1 \leq i \leq m$. To prove the corollary, it suffices to show that $(\bar{a}, \bar{\beta}) = 0_{n+1}^T$. Note that all solutions of the linear system

$$a_i \cdot x + \beta_i = 0, \ 1 \leq i \leq m,$$

satisfy $\bar{a} \cdot x + \bar{\beta} = 0$, by the assumption together with the fact that (a_0, β_0) and $(\bar{a}, \bar{\beta})$ differ by a linear combination of the vectors $(a_i, \beta_i) \in (\mathbb{R}^n)^T$, $0 \leq i \leq m$.

By the definition of orthogonal projection,

$$a_i \cdot (\bar{a})^T + \beta_i \bar{\beta} = 0, \ 1 \leq i \leq m.$$

That is, $x = \bar{\beta}^{-1}(\bar{a})^T$ is a solution of the linear system

$$a_i \cdot x + \beta_i = 0, \ 1 \leq i \leq m.$$

This implies that $x = \bar{\beta}^{-1}(\bar{a})^T$ is a solution of the linear equation $\bar{a} \cdot x + \bar{\beta} = 0$. Substitution and multiplication with the scalar $\bar{\beta}$ give

$$\bar{a} \cdot (\bar{a})^T + \bar{\beta}^2 = 0,$$

that is, $|\bar{a}|^2 + \bar{\beta}^2 = 0$. This shows that $(\bar{a}, \bar{\beta}) = 0_{n+1}^T$, as required. □

2.4.9 Polynomials of least deviation

Approximation theory. We cannot resist the temptation to give a glimpse of *approximation theory*, a fruitful subject, which considers the problem of approximating a complicated object by a simpler object. The basic example is the least squares approximation considered above, where a complicated set of points in the plane is approximated by a set of points that lie on a line. We will give the solutions of some problems of approximation that are more advanced than the least squares approximation, and explain their usefulness. The analysis is in each case a true gem.

Abstract point of view. All examples of approximation problems can be viewed in the following abstract way. Let X be a vector space in which a suitable concept of length (also called *norm*) $\|x\|_X$ is defined for all $x \in X$. Find a point in a given set S that lies closest to a given point p,

$$f(\xi) = \|p - \xi\|_X \to \min, \ \xi \in S.$$

L_q-**norms.** Now we consider the following class of approximation problems. For each q with $1 \leq q \leq +\infty$ we define a norm $\|\cdot\|_{L_q(I)}$, the L_q-*norm*, on polynomials of one variable as follows, where $I = [-1, 1]$:

$$\|y(\cdot)\|_{L_q(I)} = \left(\int_I |y(t)|^q dt \right)^{1/q} \ \text{for } 1 \leq q < +\infty,$$

and

$$\|y(\cdot)\|_{L_\infty(I)} = \max_{t \in I} |y(t)|.$$

We write $x_k(t) = t^k$, with k a nonnegative integer. Let $\mathcal{P}_k$ be the set of all polynomials of degree at most k in the variable t. Note that such polynomials have $k + 1$ coefficients, and so $\mathcal{P}_k$ can be identified with the vector space $\mathbb{R}^{k+1}$.

Polynomials of least deviation. Consider the problem

$$f_n(\xi(\cdot)) = \|x_n(\cdot) - \xi(\cdot)\|_{L_q(I)} \to \min, \ \xi(\cdot) \in \mathcal{P}_{n-1}. \quad (P_{nq})$$

This problem has a unique solution $\widehat{\xi}(\cdot)$, and the polynomial $T_{nq}(t) = t^n - \widehat{\xi}(t)$ is called *the polynomial of degree n with leading coefficient 1 of least deviation in $L_q(I)$*.

The cases $q = 2, \infty, 1$. The cases of most interest turn out to be $q = 2, \infty$, and 1. Fortunately, for these cases the explicit expressions of polynomials of least deviation are known. One can write $\cos ns$ as a polynomial expression in $\cos s$, using repeatedly the well known formulas

$$\cos(a+b) = \cos a \cos b - \sin a \sin b$$

and

$$\sin(a+b) = \sin a \cos b + \cos a \sin b.$$

For example, $\cos 2s = 2\cos^2 s - 1$ and $\cos 3s = 4\cos^3 s - 3\cos s$. The resulting polynomials can be defined alternatively in the following slick way: as $\cos n \arccos s$. For example, writing $s = \arccos t$, and so $t = \cos s$, the two expressions above give

$$\cos 2 \arccos s = 2s^2 - 1$$

and

$$\cos 3 \arccos s = 4s^3 - 3s.$$

One writes $n! = 1 \cdot 2 \cdots n$; this is called n *factorial*. Finally, we recall that $\frac{d^n}{(dt)^n} x(t)$ denotes the n-th derivative of $x(t)$ and that $\dot{x}(t)$ is an alternative notation for the first derivative $\frac{d}{dt} x(t)$.

Theorem 2.12 *Polynomials of least deviation.*

1. $T_{n2}(t) = \frac{n!}{(2n)!} \frac{d^n}{(dt)^n} (t^2 - 1)^n$ *(Rodrigues, 1816).*

2. $T_{n\infty}(t) = 2^{-(n-1)} \cos n \arccos t$ *(Chebyshev, 1854).*

3. $T_{n1}(t) = \dot{T}_{n+1\infty}(t)/(n+1)$ *(Korkin and Zolotarev, 1886).*

Here, we only will prove the formula of Rodrigues; the other two formulas will be proved in chapter nine (see the proofs of theorems 9.3 and 9.4). In exercises 2.11 and 2.12 you are invited to compute some polynomials $T_{n2}(t)$ without using the formula of Rodrigues.

Orthogonal polynomials. The polynomials $\{T_{n2}(t)\}_{n \geq 0}$ have the following property: they are *orthogonal* to each other, in the sense that

$$\langle T_{m2}, T_{n2} \rangle_{L_2(I)} = 0$$

if $m \neq n$. Here the inner product $\langle, \rangle_{L_2(I)}$ is defined by

$$\langle f, g \rangle_{L_2(I)} = \int_I f(t) g(t) dt.$$

Historical remark. The first system of orthogonal polynomials $\{P_n(t)\}_{n\geq 0}$, the so-called Legendre polynomials, was constructed by Legendre (1787). These polynomials are equal up to a scalar multiple to the polynomials considered by Rodrigues, but they are normalized by the equality $P_n(1) = 1$. Here is the expression for the Legendre polynomials:

$$P_n(t) = \frac{1}{2^n n!} \frac{d^n}{dt^n}(t^2 - 1)^n.$$

Now we are ready to prove the formula of Rodrigues.

Proof.

1. We consider the problem without constraints,

$$f(x) = \int_{-1}^{1} \left(t^n - \sum_{k=1}^{n} x_k t^{k-1} \right)^2 dt \to \min.$$

Existence of a global solution follows from the coercivity of f.

2. Fermat: $f'(x) = 0_n^T \Rightarrow$

$$\int_{-1}^{1} (t^n - \sum_{k=1}^{n} x_k t^{k-1}) t^{k-1} dt = 0, \ 1 \leq k \leq n. \quad (i)$$

This allows the following interpretation: the polynomial

$$t^n - \sum_{k=1}^{n} x_k t^{k-1}$$

is orthogonal—with respect to the inner product $\langle \, , \, \rangle_{L_2(I)}$—to the power functions t^{k-1}, $1 \leq k \leq n$, and so to all polynomials in t of degree at most $n-1$.

3. Observe that the polynomial

$$\frac{n!}{(2n)!} \frac{d^n}{(dt)^n}(t^2 - 1)^n$$

has degree n, that the coefficient of t^n is 1, and that it is orthogonal to all polynomials of degree at most $n-1$ (using partial integration).

4. The Rodrigues formula is verified.

Motivation for polynomials of least deviation. What is the point of these problems and their solutions? They turn out to be more than mere curiosities: they provide an answer to a fundamental problem of approximating complicated functions in an accurate and reliable way by simpler functions. Let us consider the contrast with the "usual" way of approximating functions. The simplest example of this is the approximation of a function $x(t)$ of one variable t that is differentiable at zero, by

$$x(0) + \dot{x}(0)t,$$

the linear approximation of $x(\cdot)$ at zero. This approximation is defined in such a way that it is an accurate approximation for t sufficiently close to zero. This can be generalized to approximation by Taylor polynomials, as we have seen in section 1.2.3 and exercise 1.37.

However, here we are interested in something else. What if we want to *approximate on a whole interval?*

Approximations on an interval. If one wants to approximate a function $x(\cdot)$ by a function of the form $t \to a + bt$ *on a whole interval*, say, $[-1, 1]$, then it is better to take the function for which

$$\max_{t \in [-1,1]} |x(t) - (a + bt)|$$

is as small as possible.

It is too much to ask, that a simple function $t \to a + bt$ will give an accurate approximation on a whole interval.

Therefore, to get better approximations, one could approximate instead with functions $a+bt+ct^2$, or more generally, $a_0+a_1t+\cdots+a_kt^k$.

Problem of large coefficients. A practical problem arises. The graphs of the functions $t^2, t^4, \ldots$ are rather similar, as are the graphs of the functions $t, t^3, t^5, \ldots$; because of this, the absolute values $|a_i|$ of the coefficients a_i of the approximations are often very large compared to the values taken by the given function $x(\cdot)$, as it turns out. This is not very convenient from a numerical point of view. Moreover, if an approximating polynomial is not accurate enough and you replace it by the approximating polynomial of one degree higher, you have to start all over again: all coefficients change completely.

Solution of the large coefficient problem. Therefore, one usually writes these approximations in the following alternative way:

$$b_0 + b_1p_1(t) + \cdots + b_kp_k(t),$$

where $p_r(t)$ is the polynomial of degree r with leading coefficient 1 for which

$$\max_{t \in [-1,1]} |p_r(t)|$$

is as small as possible. These are precisely the polynomials $T_{r\infty}(t)$. Then the numbers $|b_0|, |b_1|, \ldots, |b_k|$ are usually not very large.

Additional advantage. There is another advantage of expressing the approximating polynomial as a linear combination of the polynomials 1, $p_1(t)$, $p_2(t)$, ..., $p_k(t)$: if you want to go over from degree k to degree $k+1$, then the coefficients of 1, $p_1(t)$, ..., $p_k(t)$ do not change; therefore it suffices to compute one number, the coefficient of $p_{k+1}(t)$.

Source of the large coefficient problem. We will sketch the reason for this. The phenomenon of large coefficients can be illustrated by the following example, where we consider points in $\mathbb{R}^2$ instead of polynomials. The two vectors $(1,1)$ and $(.999, 1.001)$ are rather similar; if we write, for example, the vector $(1,-1)$ as

$$a(1,1) + b(.999, 1.001),$$

then we find $a = 1000$ and $b = -1000$. This suggests that it is better to replace the polynomials 1, t, t^2, ..., t^k by $k+1$ polynomials with mutually very different graphs.

Gram-Schmidt procedure. How to find such polynomials with mutually different graphs? This can be done in general by the following algorithm, called the Gram-Schmidt procedure. This procedure can be applied to vectors in $\mathbb{R}^n$ and a choice of norm on $\mathbb{R}^n$, and also to polynomial functions on an interval I of degree at most $r = n-1$ and a choice of norm $\|\cdot\|_{L_q(I)}$. Replace linearly independent vectors v_1, ..., v_n by vectors w_1, ..., w_n, where w_l is defined for each $l \in \{1, \ldots, n\}$ to be the smallest vector u for which $v_l - u$ can be expressed as a linear combination of v_1, ..., v_{l-1}. The resulting vectors w_1, ..., w_n are as dissimilar to each other as possible in the following sense: if you add to one of these vectors, say w_l, a linear combination of the previous ones, w_1, ..., w_{l-1}, then the norm increases; if $q = 2$ this means that the vectors are orthogonal.

If the Gram-Schmidt procedure is applied to the polynomials

$$1, \ t, \ t^2, \ \ldots, \ t^k,$$

and the choice of norm $\|\cdot\|_{L_q(I)}$, then the result is the polynomials $T_{iq}(t)$, $0 \leq i \leq k$. For the special case $q = \infty$, this concludes the motivation for the polynomials $T_{n\infty}(t)$. For the other polynomials $T_{nq}(t)$, the motivation is similar.

The Gram-Schmidt procedure carries the names of Jorgen Pederson Gram (1850–1916) and Erhard Schmidt (1876–1959), but it was first presented by Pierre-Simon Laplace (1749–1827).

Historical comments

Benjamin Olinde Rodrigues (1794–1851) has been a somewhat obscure figure in the world of mathematics. He showed great intuition in studying important problems, but he was remarkably poorly appreciated by his contemporaries. This extends to his work as a social reformer, where many of his views have much later become the accepted ones.

Dreams of a better world. The Rodrigues formula, given above, is contained in his thesis. After this, Rodrigues became wealthy as a banker, supporting the development of the French railway system. He was involved with Claude Henri Comte de Saint-Simon, a rich man who had dreams of a better world, improved by the scientific and social reform of mankind. Saint-Simon gathered some famous scientists around him in support of his beliefs, including Lagrange, but he also brought on board bankers.

Sentimental ideas. Rodrigues became one of the leaders of this movement. His views included that all races had *"equal aptitude for civilization in suitable circumstances"* and also that women had equal aptitude and that *"women will one day achieve equality without any restriction."* These views were much criticized even in his own progressive circle. It was argued that Rodrigues was being sentimental and that science proved that he was wrong.

Banker against interest. Although most of his writings were on politics and social reform, he also wrote pamphlets on banking and occasional papers on mathematics. Saint-Simon was apparently very persuasive, as under his influence the banker Rodrigues argued for lending without interest.

Other discoveries. He made at least two other interesting discoveries in mathematics, his formula for the composition of rotations and a "formula" for the number of permutations of n objects with a given number of inversions. All his discoveries were forgotten, and rediscovered much later by others. In the case of his fine work on permutations, Rodrigues's contribution was only brought to light in 1970.

Adrien-Marie Legendre (1752–1833) attracted attention for the first time by winning the 1782 prize on projectiles offered by the Berlin Academy. The actual task was the following one:

Determine the curve described by cannonballs and bombs, taking into consideration the resistance of the air; give rules for obtaining the ranges corresponding to different initial velocities and to different angles of projection.

Comets and least squares. Then he went on to publish papers on celestial mechanics such as "Recherches sur la figure des planètes," which contains the Legendre polynomials.

Later, he published a book on determining the orbits of comets. His method involved three observations taken at equal intervals, and he assumed that the comet followed a parabolic path, so that he ended up with more equations than there were unknowns. In an appendix of this book the first description is given of the celebrated least squares method of fitting a curve to available data.

Other work. Legendre also published papers and books on other subjects such as geometry, and in particular he was one of the pioneers of number theory.

In 1791 he became a member of the committee to standardize weights and measures. This committee worked on the introduction of the metric system and it measured the earth. After this he became one of the leaders of a project to produce logarithmic and trigonometric tables. Between 70 and 80 assistants worked for 10 years to complete this task.

Pafnuty Lvovich Chebyshev (1821–1894) was once called a splendid Russian mathematician, to which he objected that he was a worldwide mathematician. Indeed, almost every summer he traveled in Western Europe, where he had contact with many leading mathematicians and where he received many honors. For example, he was the first Russian since Peter the Great to be elected to the Paris Academy.

Wonderful lecturer. Many famous Russian mathematicians were taught and influenced by him, and the later rise of Russian mathematics is in no small measure due to him. The following quote from his student Dmitry Grave shows that he brought some fresh air from the world into his classroom:

"Chebyshev was a wonderful lecturer. His courses were very short. As soon as the bell sounded, he immediately dropped the chalk, and,

limping, left the auditorium. On the other hand he was always punctual and not late for classes. Particularly interesting were his digressions when he told us about what he had spoken outside the country. [...] Then the whole auditorium strained not to miss a word."

Theory and practice. Chebyshev had as a guiding principle for his work that *"the closer mutual approximation of the points of view of theory and practice brings the most beneficial results. [...] science finds a reliable guide in practice."* For example, his work on polynomials of least deviation has its origin in the steam engine.

Efficient steam engine. Here is how this came about. From his early childhood, Chebyshev loved to play with mechanisms. In later life he devised mechanisms for a wheelchair, a rowboat, and a mechanical calculator. In the summer of 1852, he traveled through Europe and, among other things made many tours of inspection of various types of machinery—windmills, water turbines, railways, iron works, and steam engines. One of the major challenges of steam engine designers was how to turn a linear motion into a circular one. The circular motion was what is wanted to run factory machinery. The linear motion was what is provided by the up and down movements of the piston rod. James Watt had found an approximate solution. The fact that this was only an approximation led to a certain amount of leakage and frictional wear.

Steam engine and the birth of approximation theory. Chebyshev got an idea to improve the quality of the approximation and so make the steam engine more efficient. He worked out his idea in a paper, which is now considered a mathematical classic: it laid the foundations for the new and fruitful subject of best approximation of functions by means of polynomials. He was the first to see for these polynomials the possibility of a general theory and its applications. However, the idea was never applied to steam engines: his suggested improvement was only theoretical. If carried out, it would have involved more moving parts with their own problems of inefficiency.

Other work. Chebyshev worked successfully on various subjects, including probability and number theory. For example, he proved one of the first results in what has since become a vast subject called analytic number theory. This result came close to the definitive one that if n is a number, the number of prime numbers less than n is approximately equal to $n/\ln n$. Moreover, he proved the Bertrand conjecture: there is always at least one prime number between n and $2n$ for $n > 3$.

Great love. As to Chebyshev's personal life, he lived alone in a large house with ten rooms, he was rich, spending little on everyday comforts, but he had one great love, buying property.

Conclusion to section 2.4. We have seen that it is possible to apply the method of Fermat and Weierstrass to problems of

- matrix theory and linear algebra (problems 2.4.1, 2.4.6 and 2.4.7 and theorem 2.9,),

- economics (problems 2.4.2 and 2.4.3),

- geometry, mechanics, and location theory (problem 2.4.4),

- least squares approximation (problem 2.4.5),

- algebra (theorem 2.8)

- approximation theory (theorem 2.12).

Invitation to chapters eight and nine and to the proof of the tangent space theorem. At this point, you could continue with some problem in chapters eight and nine and with the proof of the tangent space theorem in chapter three.

Problem 8.3.1 "Why do there exist discounts on airline tickets with a Saturday stayover" illustrates the concept price discrimination.

In chapter nine we illustrate that some results from vector and matrix theory, which seem to have nothing to do with optimization, can be proved using optimization methods. As an example, we show in problem 9.2.2 that a positive definite matrix is invertible. Other examples are the *Sylvester criterion* for positive definiteness, in terms of main minors (theorem 9.3), and the explicit formula for the distance between a point and the linear subspace spanned by a given system of vectors (theorem 9.4).

Perhaps the most striking applications of the Fermat theorem are the derivations of the explicit formulas for *the polynomials of least deviation* for the cases $q = \infty$ (theorem 9.5) and $q = 1$ (theorem 9.6).

Finally, in section 3.4.3, *the tangent space theorem* and then the Lagrange multiplier rule is proved using the Fermat theorem. We recall that the tangent space theorem is one of the two building blocks of the entire optimization theory, and in fact a central result of mathematical analysis.

2.5 DISCUSSION AND COMMENTS

The "one-dimensional" envelope theorem 1.11 can be extended from functions $f(x, a)$ of two variables to functions $F(x_1, \ldots, x_n, y_1, \ldots, y_m)$ of two vector variables $(x_1, \ldots, x_n)$ and $(y_1, \ldots, y_m)$, and the proof is a straightforward generalization of the one-dimensional case. We display only the part of the statement that is used most often.

Setup. Consider the following setup. Let two open sets $U \subseteq \mathbb{R}^n$, $V \subseteq \mathbb{R}^m$, a function $F : U \times V \to \mathbb{R}$ of two vector variables $x \in U$ and $y \in V$, and two points $\bar{x} \in U$, $\bar{y} \in V$ be given. We consider the family of problems

$$f(x,y) \to \min, \ x \in U, \quad (P_y)$$

where the parameter y runs over V. We view this family as a *perturbation* of the problem $(P_{\bar{y}})$. That is, we want to consider the sensitivity of the value of the problem $(P_{\bar{y}})$ to small changes of the parameter y.

Sufficient conditions. The first and second order sufficient conditions for the problem $(P_{\bar{y}})$ and the point $\bar{x}$ have the following form: $F_x(\bar{x}, \bar{y}) = 0_n^T$ and the $n \times n$ matrix $F_{xx}(\bar{x}, \bar{y})$ is positive definite.

The definitions of "open neighborhoods" and "C^2-functions" can be given in the same way as in theorem 1.11.

Theorem 2.13 Envelope theorem. *Let points $\bar{x} \in \mathbb{R}^n$, $\bar{y} \in \mathbb{R}^m$, positive numbers $\bar{\varepsilon}$, $\bar{\delta}$, and a function $F : U_n(\bar{x}, \bar{\varepsilon}) \times U_m(\bar{\delta}) \to \mathbb{R}$ be given.*

Assume that $F \in C^2(\bar{x}, \bar{y})$ and, moreover, that the first and second order sufficient conditions hold at the point $\bar{x}$ for the problem

$$F(x, \bar{y}) \to \min, x \in U_n(\bar{x}, \varepsilon).$$

There exist positive numbers $\varepsilon < \bar{\varepsilon}$ and $\delta < \bar{\delta}$ such that, for all $y \in U_n(\bar{y}, \delta)$, the problem

$$F(x, y) \to \min, \ x \in U_n(\bar{x}, \varepsilon)$$

is strictly convex and so has a unique point of minimum $x(y)$.

Then the function $x(\cdot) : U_n(\bar{y}, \delta) \to \mathbb{R}$ is continuously differentiable and $x(\bar{y}) = \bar{x}$. Moreover,

$$\frac{d}{dy} F(x(y), y) = F_y(x(y), y)$$

for all $y \in U_n(\bar{y}, \delta)$.

Conclusion to section 2.5. The envelope theorem for one-variable problems without constraints can be extended in a routine way to the case of two or more variables by virtue of vector notation.

2.6 EXERCISES

We begin with three numerical examples. A feature of interest of the third example is that there are many stationary points, one of which is a local but not a global maximum. We will reconsider this example in chapter five after we have given a systematic method to detect local extrema (see exercise 5.1).

Exercise 2.6.1 $x_1 x_2 + 50 x_1^{-1} + 20 x_2^{-1} \to \min,\ x_1, x_2 > 0$.

Exercise 2.6.2 $x_1^2 + x_2^2 + x_3^2 - x_1 x_2 + x_1 - 2x_3 \to \min$.

Exercise 2.6.3 $2x_1^4 + x_2^4 - x_1^2 - 2x_2^2 \to$ extr. *Find the stationary point that is a local maximum.*

Price discrimination. The idea of *price discrimination* is so attractive that it is hard for producers to resist its temptation. Producers would like to charge each individual customer the maximum price the customer is prepared to pay for the product. Then different customers would pay different prices for exactly the same product. Such open price discrimination, even if it could be realized, is forbidden by law.

However, various ways have been invented to achieve some sort of price discrimination that does not break any law, but on the contrary is beneficial to society. A simple and widespread example is given in the next exercise. This exercise and its successor require relatively extensive calculations.

Exercise 2.6.4 **Telecom market and price discrimination*

In the telecom market, retail customers are charged different prices from wholesale customers for the same service. Let P_1 and Q_1 denote the price and demand for retail customers and let the demand equation be

$$P_1 + Q_1 = 500.$$

Let P_2 and Q_2 denote price and demand for wholesale customers and let the demand equation be

$$2P_2 + 3Q_2 = 720.$$

The total cost is

$$TC = 50000 + 20Q,$$

where $Q = Q_1 + Q_2$.

(i) Determine the firm's pricing policy that maximizes profit under price discrimination, and calculate the maximum profit.

(ii) Suppose a new law makes it illegal to charge different prices in different markets. Compare profits before and after the new law. How much profit is lost?

(iii) Suppose instead that a tax of t per unit is imposed on the market for retail customers. Show that this has the following consequences. It has no influence on sales for wholesale customers. The amount sold for retail customers is lowered and the price in this market—so including the tax—goes up. The tax revenue for the state is less than the loss in profits, so the tax causes a net loss, called "deadweight loss."

The next exercise gives some insight into the beneficial effects of competition.

Exercise 2.6.5 * **Monopoly and duopoly.** Coca Cola and Pepsi Cola each determine the price for their own cola, and they each produce as much as is demanded. The demands are given by

$$q_c = (29 - 5p_c + 4p_p)_+$$

and

$$q_p = (16 + 4p_c - 6p_p)_+.$$

Here we write x_+ to be x if $x \geq 0$ and zero otherwise. Coca Cola has total costs $5 + q_c$ and Pepsi Cola $3 + 2q_p$.

1. **Monopoly.** Coca Cola and Pepsi Cola have decided to cooperate and to maximize their total profits $\pi_c + \pi_p$, thus creating a cola monopoly. Find the prices, production levels, and profits for each firm.

2. **Duopoly.** Suppose that Coca Cola breaks the cooperation and chooses its price to maximize its profit, taking the current price of Pepsi Cola as given. Show that the effect of this on the demand for Pepsi is rather drastic. Pepsi answers by maximizing its profits with the new price of Coca Cola as given, and so on. Show how the prices, production levels, and profits for Coca Cola and Pepsi Cola change over time. What happens in the long run?

3. **Beneficial effect of competition.** Compare the monopoly and duopoly solutions; that is, what is the effect of price competition on the production levels and prices?

Equilibrium. One point of this exercise is to illustrate how in the long run an *equilibrium* is reached. We also came across this concept in the analysis of problem 2.4.3.

Cournot. The use of the concept equilibrium in economics goes back to Antoine Augustin Cournot (1801–1877).

He studied mathematics and his doctoral thesis impressed Poisson, on whose recommendation Cournot became professor of mathematics at Grenoble. Later Cournot wrote of Poisson's opinion of his first papers: *"Poisson discovered in them a philosophical depth, and, I must honestly say, he was not altogether wrong. Furthermore, from them he predicted that I would go far in the field of pure mathematical speculation but (as I have always thought and have never hesitated to say) in this he was wrong."*

Cournot became rector of the University of Grenoble, and later inspector general of public education. He published *Recherches sur les principes mathématiques de la théorie des richesses*, which contains pioneering work on mathematical economics, in particular, supply and demand functions, and conditions for equilibria with monopoly and duopoly.

His definition of a market still is the basis for the one still in use:

"Economists understand by the term Market, not any particular market place in which things are bought and sold, but the whole of any region in which buyers and sellers are in such free intercourse with one another that the prices of the same goods tend to equality easily and quickly."

Nash. The concept equilibrium was fully developed much later by John Forbes Nash (born 1928). In 1948, while studying for his doctorate in mathematics at Princeton, he wrote a paper that was 45 years later to win him a Nobel prize for economics. During this period Nash established the mathematical principles of game theory. The Nash equilibrium is perhaps the most important concept of game theory. Its applications in economics are numerous: in analyzing problems as diverse as candidates' election strategies, the causes of war, agenda manipulation in legislatures, or the actions of interest groups, the central issue is always the search for the Nash equilibria.

Our next example is another one where the "invisible hand" does not lead to an optimal outcome, just as in the "ice-cream sellers on the beach" problem. The activities of a firm might have negative consequences for others, for example, pollution. One can give the firm an incentive to take this into account by means of a tax. This turns the costs to others ("externalities") into internal costs. This is illustrated by the following exercise.

Exercise 2.6.6 Externalities and taxes. *A soot-spewing factory that produces steel windows is next to a laundry. We will assume that the factory faces a prevailing market price $P = 40$. Its cost function is $C = X^2$, where X is window output.*

The laundry produces clean wash, which it hangs out to dry. The soot from the window factory smudges the wash, so that the laundry has to clean it again. This increases the laundry's costs. In fact, the cost function of the laundry is $C = Y^2 + 0.05X$, where Y is pounds of laundry washed. The demand curve faced by the laundry is perfectly horizontal at a price of 10 euro per pound.

(i) What outputs would maximize the sum of the profits of these two firms?
(ii) Will those outputs be set by a competitive market?
(iii) What per unit tax would we need to set on window production to obtain the outputs that maximize the profits of the two firms?

We have considered the problem of finding the optimal location to serve three equally important clients, if you want to minimize the average distance to the clients (problem 2.4.4). The following exercise considers the optimal location problem in the case that the clients are *not* equally important.

Exercise 2.6.7 *Minimize the weighted average distance of a point in the plane to three given points, with given weights.*

The extension of the physical interpretation in terms of a table with three holes is that you hang different weights on the strings.

The next example is one of the extensions of the problem of finding the least squares line through a cloud of points in the plane. It produces some sort of approximate solution for a system of linear equations even if this system has no solution. We recall that in the same spirit the least squares line shows how to draw a straight line that runs approximately through a given collection of points, if the points of this collection are not lying on a straight line.

Exercise 2.6.8 Least squares solution of a system of linear equations.
Find the least squares solution of a system of linear equations that has no solutions.

Let us be more precise. Let A be an $m \times n$ matrix of rank m and let b be an m-dimensional column vector. Call a vector $\bar{x}$ for which the residue $\bar{r} = A\bar{x} - b$ of the system $Ax = b$ has minimal modulus, a least squares solution of $Ax = b$.

Show that there is a unique least squares solution and that it can be given by the formula

$$\bar{x} = (A^T A)^{-1} A^T b.$$

The following two exercises are extensions to dimension n of exercises 1.6.27 and 1.6.28 on the center of gravity method and the ellipsoid method.

Exercise 2.6.9 * *Formulate and prove the n-dimensional version of exercise 1.6.27. Which surprising property do you observe if you let n tend to infinity?*

In the same spirit we reconsider exercise 1.28.

Exercise 2.6.10 * *Formulate and prove the n-dimensional version of exercise 1.6.28.*

In the next two exercises you are invited to derive the Rodrigues formula for the polynomials T_{n2} (given in theorem 2.12) in the cases $n = 1$ and $n = 2$.

Exercise 2.6.11 *Minimize $\int_{-1}^{1} (t^2 - x_2 t - x_1)^2 dt$.*

Exercise 2.6.12 *Minimize $\int_{-1}^{1} (t^3 - x_3 t^2 - x_2 t - x_1)^2 dt$.*

Global property of the collection of extrema. A differentiable function of one variable that is coercive for minimization has one more local minimum than it has local maxima, clearly. What is remarkable about this obvious observation is that it is a global fact: it is a property of the collection of all local minima and maxima. What is not so obvious is that this fact can be extended to functions of more than one variable. In the following exercise we consider the case of functions of two variables.

Exercise 2.6.13 *Hilly landscapes can be very complicated, but they all obey the following iron rule:*

the number of mountain passes is one less than the sum of the number of mountain tops and the number of mountain valleys.

Give a geometrical proof of this fact.

This result is the top of a "mathematical iceberg," called *Morse theory* [52]. We make the following assumptions:

1. the landscape is the graph of a differentiable function f of two variables,

2. f is coercive for maximization,

3. f has only finitely many stationary points $\widehat{x}$,

4. for each stationary point $\widehat{x}$ of f, the 2×2 matrix $f''(\widehat{x})$ is invertible.

A stationary point $\widehat{x}$ of f corresponds to a top (resp. valley, resp. pass) of the landscape if $f''(\widehat{x})$ is positive definite (resp. negative definite, resp. indefinite).

Solution. According to legend, there was once one more continent, Atlantis. However, long ago, it sank to the bottom of the ocean. If we could have witnessed this tragedy we would have understood why the theorem above is true.

Indeed, we can imagine that the graph of the function in the theorem represents the shape of Atlantis. Let the sea level be at level 0. We assume that initially all mountaintops, valleys, and mountain passes are above sea level.

Now pay attention to what happens if a stationary point goes under water.

- If it is a mountain top, then we lose an island.

- If it is a valley, we gain a lake.

- If it is a mountain pass, either we gain an island, as one island splits up into two islands, or we lose a lake, as two lakes become one lake.

These three observations are the core of the proof of the theorem.

It follows that the following relation between the number of tops, valleys, lakes, mountain passes and islands holds true at all moments τ:

$$t(\tau) + v(\tau) + l(\tau) = m(\tau) + i(\tau).$$

Indeed, this formula holds certainly as soon as Atlantis has disappeared completely. Therefore, the formula holds at all times: whenever one of the numbers on one side of the equality changes, then one of the numbers on the other side changes in the same way, as the three observations above make clear. In particular, it holds at the initial moment: then the number of islands was one, Atlantis, and the number of lakes was zero. This gives the conclusion of the theorem.

The last exercises are theoretical. To begin with, it is of interest to present the proof of the Fermat theorem in a more "algorithmic" spirit: as a result on "reasonable descent."

That is, we show how to find, close to a nonstationary point, a point at which the value is lower.

Exercise 2.6.14 *Let $\widehat{x} \in \mathbb{R}^n$ and $f \in D^1(\widehat{x})$. If $\widehat{x}$ is not a stationary point point of f, then we have*

$$f(\widehat{x} + \alpha \bar{x}) = f(\widehat{x}) - \alpha + o(\alpha),$$

where

$$\bar{x} = -|f'(\widehat{x})|^{-2} f'(\widehat{x})^T,$$

and so $x_\alpha = \widehat{x} + \alpha \bar{x}$ with $\alpha \geq 0$ is a "reasonable descent":

$$f(x_\alpha) < f(\widehat{x}) \quad \text{for} \quad \alpha > 0 \quad \text{sufficiently small}.$$

Therefore, $\widehat{x}$ is not a point of local minimum of f. Verify these statements.

Exercise 2.6.15 *Steepest ascent interpretation gradient. Let $\widehat{x} \in \mathbb{R}^n$ and $f \in D^1(\widehat{x})$ and assume that $\widehat{x}$ is not a stationary point of f. We consider the problem in what direction f makes the steepest ascent, starting from $\widehat{x}$, that is,*

$$g(h) = \lim_{t \downarrow 0}(f(\widehat{x} + th) - f(\widehat{x}))/t \to \max, \ h \in \mathbb{R}^n, \ |h| = 1.$$

Show that the "normalized gradient" $|f'(\widehat{x})|^{-1} f'(\widehat{x})^T$ is the unique solution of this problem.

Chapter Three

Lagrange: Equality Constraints

> Nous ne faisons ici qu'indiquer ces procédés dont il sera facile de faire l'application; mais on peut les réduire à ce principe général: Lorsqu'une fonction de plusieurs variables doit être un *maximum* ou *minimum*, et qu'il y a entre ces variables une ou plusieurs équations, il suffira d'ajouter à la fonction proposée les fonctions qui doivent être nulles, multipliées chacune par une quantité indéterminée, et de chercher ensuite le *maximum* ou *minimum* comme si les variables étaient indépendantes; les équations que l'on trouvera, combinées avec les équations données, serviront à déterminer toutes les inconnues.

> Here we only sketch these procedures and it will be easy to apply them, but one can reduce them to this general principle: If a function of several variables should be a *maximum* or *minimum* and there are between these variables one or several equations, then it will suffice to add to the proposed function the functions that should be zero, each multiplied by an undetermined quantity, and then to look for the *maximum* or the *minimum* as if the variables were independent; the equations that one will find, combined with the given equations, will serve to determine all the unknowns.
>
> *J.-L. Lagrange*

- How does one find the extrema of an n-dimensional problem with equality constraints; that is, how can one solve a problem

$$f_0(x) \to \text{extr}, \quad f_i(x) = 0, \quad 1 \leq i \leq m,$$

where f_i, $0 \leq i \leq m$, are functions of $x_1, \ldots, x_n$?

3.0 SUMMARY

Constraints. It is natural to want the best. However, usually there are constraints that make this impossible. Then you want to do the best you can, given the constraints. For example, a boy ("consumer") might not have enough money to buy unlimited ice cream and, at the same time, to play as many video games as he wants. A producer might have limited technological possibilities to make her products. It can also happen that you are a manager and have two objectives, you want your firm to make as much profit as possible, but at the same time you want a high total revenue because your bonus depends on it. A more sophisticated question is, how to determine the benefits of trade, if producers are at the same time consumers.

Inequalities. Many problems with constraints arise from the wish to establish an inequality. An example that illustrates both the need for inequalities and the relation with optimization arises from least squares approximation. Here the sum of the squares of the deviations $r_i = y_i - (ax_i + b)$ is taken as a measure of the quality of the choice of line $y = ax + b$. This is done for convenience, although the sum of the absolute values of these deviations is the most natural measure. The justification is given by inequalities between the two measures, which show them to be always comparable in size. The way to establish these inequalities is by keeping one of the measures fixed and finding the extremal values of the other measure under this inequality.

Perfect pendulum. A class of intriguing applications concern the problem of finding the distance from a point to a curved line. These problems were studied systematically in the seventeenth century. The most prominent example is the case when the curved line is a *cycloid*; this played a central role in the construction of the *perfect pendulum* by Huygens. However, such problems were already considered at the dawn of mathematics by Apollonius, for example, for ellipses; as a consequence Apollonius stumbled upon the mysterious *astroid*, a curve that governs the solution of the problem.

Multiplier method. To solve problems with constraints, an excellent method is available, *Lagrange's method of elimination of constraints*. This relates the variables of optimization at the point of optimum, using auxiliary numbers called *multipliers*. The multipliers are not only of technical help in finding the solution of a problem. They have a conceptual meaning as well: they measure the sensitivity of the optimal value of the problem for small changes in the constraints. In fact, for problems of cost minimization, the multipliers have an illuminating economic interpretation, as a sort of prices, called *shadow prices*. A special feature of our treatment is an accessible and intuitive proof of the validity of this method, which uses the intuitive theorem of Weierstrass. The proof is usually considered to be relatively complicated. To dispel any possible doubts whether the method of Lagrange is essential or whether the same results could be obtained by ordinary elimination followed by the Fermat theorem, we offer the following test case.

Four sticks. Make from four sticks a quadrangle with maximal area. It is easy to solve this problem using the multiplier method, but the authors know of no other method and even no geometrical ruse to crack this hard nut, which was known to the ancient Greek geometers, but which defied their ingenuity.

3.1 INTRODUCTION

The multiplier rule in the words of Lagrange. Often the freedom to make an optimal choice is restricted by *equality constraints*. The first general method of investigation of such problems was formulated by Joseph-Louis Lagrange (1736–1813). In 1797 ("Prairial an V", that is, in the ninth month (10 May–18 June) of the fifth year ("from 1792")) he wrote in his book *Théorie des fonctions analytiques* the words that we have chosen as the epigraph to this chapter.

The powerful method expressed by these words is the central result of this section. The idea behind it is so strong that it extends far beyond the problem type considered in this chapter, problems with equality constraints. Many successful necessary conditions have been discovered since Lagrange, for a myriad of totally different types of optimization problems. All of these are variants of Lagrange's method. Sometimes this is not obvious at all, and it was only realized in retrospect, as in the case of the Pontryagin maximum principle. This epoch-making result was discovered in the 1950s; we will discuss it briefly in chapter twelve.

We note that the method of Lagrange might look counterintuitive at first sight: instead of eliminating some of the variables, it brings new variables into the game!

Royal road. If you check that the statement of theorem 3.3 – with $\lambda_0 = 1$ – corresponds to Lagrange's words, then you are ready to read examples 3.2.3, 3.2.4, and 3.2.8, and the applications in sections 3.3 and 3.6.

Intuitive description of Lagrange's method: forcing constraints by a "carrot and stick" policy

The following intuitive economic description of this rule gives some insight, but it is not used for the solution of concrete problems. Someone wants to choose n quantities $x_1, \ldots, x_n$ in order to minimize a function f of n variables $x_1, \ldots, x_n$; you can think of f as representing some sort of cost—in the case of maximization you could think of profit. However, the choice is not free; certain equality constraints have to be satisfied:

$$f_i(x) = 0, \ 1 \leq i \leq m.$$

Maybe some "(benevolent) dictator" has ordered this. He can make sure that his orders are carried out without the use of force. He can use a "carrot and stick" method instead of force, letting *prices* do the work. To this end, for each of the m constraints $f_i(x) = 0$, $1 \leq i \leq m$, he chooses a suitable number λ_i, to be called the *shadow price* or Lagrange multiplier of the constraint. He imposes a fine

$$|\lambda_i f_i(x)| = \lambda_i f_i(x) \text{ if } \lambda_i f_i(x) > 0$$

and gives a reward

$$|\lambda_i f_i(x)| = -\lambda_i f_i(x) \text{ if } \lambda_i f_i(x) < 0$$

("carrot and stick policy"). This leads to the following total costs:

$$f_0(x) + \lambda_1 f_1(x) + \cdots + \lambda_m f_m(x).$$

If one minimizes these total costs, *without taking any constraints into account*, then the minimum is attained at the solution of the original constrained problem, provided the right choice of fines/rewards λ_i, $1 \leq i \leq m$, has been made.

Conclusion. Lagrange's method gives a possibility of "decentralization." Indeed, just by setting the right prices, the benevolent dictator can ensure that an optimizing decision-maker will observe, from his own free will, all the equality constraints the dictator wants to be observed. The dictator does not have to enforce the constraints.

In section 3.5.3, a more detailed description will be given of the economic interpretation of the Lagrange method.

The bad case. A warning is in place. Although the method as formulated by Lagrange has been successful in each "serious" application, it is possible to construct artificial examples where it fails. There is a simple way to "repair" this: one has to multiply the objective function as well by a Lagrange multiplier λ_0, and allow it to be zero ("the bad case"). In each "serious" application, it turns out to be possible to show—by a simple ad hoc argument—that λ_0 cannot be zero. This is more convenient than the following way, which enjoys some popularity: one can work with a "constraint qualification," a condition that implies that the literal version of the method is correct.

On the proof. Lagrange did not prove his method. The proof required the theory of implicit functions, which was developed a hundred years after the discovery of the multiplier rule.

Conclusion to section 3.1. The quotation from Lagrange that has been chosen as epigraph of this chapter describes one of the most powerful ideas of optimization theory. It can be interpreted for a cost minimization problem with equality constraints as a "carrot and stick" policy to force constraints, without the use of force. The idea is to let *prices* do the work.

3.2 MAIN RESULT: LAGRANGE MULTIPLIER RULE

3.2.1 Precise formulation of the multiplier rule

Our goal is to express the principle of Lagrange in a precise way. Let $\widehat{x}$ be a point of $\mathbb{R}^n$, $\varepsilon > 0$, and let $f_i : U_n(\widehat{x}, \varepsilon) \to \mathbb{R}$, $0 \leq i \leq m$, be functions of n variables, defined on an open ball with center $\widehat{x}$, such that $f_i(\widehat{x}) = 0$, $1 \leq i \leq m$. These functions might be defined outside the ball as well, but this is not relevant for the present purpose.

Definition 3.1 *Main problem: equality constrained optimization.* The problem

$$f_0(x) \to \text{extr}, \quad f_i(x) = 0, \quad 1 \leq i \leq m, \qquad (P_{3.1})$$

is called a finite-dimensional problem with equality constraints.

The following two examples illustrate this type of problem, and the possibility of solving such problems without Lagrange's method.

Example 3.2.1 *Solve, for a given constant $a > 0$, the problem*

$$f_0(x) = x_1 x_2 \to \max, \quad f_1(x) = x_1^2 + x_2^2 - a^2 = 0, \quad x_i > 0, \; i = 1, 2.$$

Solution.

- Below, we will solve this problem with the method of the Lagrange multiplier rule.

- Here we note that the problem has a trivial geometrical solution (Fig. 3.1), using that the area of the shaded triangle equals

$$\frac{1}{2} ah = \frac{1}{2} x_1 x_2,$$

  which reduces the problem to that of maximizing the height h.

- Alternatively, we can easily reduce to an unconstrained problem

$$x_1 \sqrt{a^2 - x_1^2} \to \max, \; 0 < x_1 < a,$$

  and solve this using the Fermat theorem.

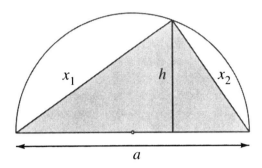

Figure 3.1 Geometrical solution.

- Another possibility is the ad hoc chain of inequalities,

$$x_1 x_2 \leq x_1 x_2 + \frac{1}{2}(x_1 - x_2)^2 = \frac{1}{2}(x_1^2 + x_2^2) = \frac{1}{2}a^2,$$

which holds for each admissible point, and which holds exactly if $x_1 = x_2$.

We restrict ourselves to writing down the answer: $\widehat{x}_1 = \widehat{x}_2 = \frac{a}{\sqrt{2}}$.

Example 3.2.2 Solve, for a given constant $a > 0$, the problem

$$f_0(x) = x_1 x_2 \to \max, \quad f_1(x) = x_1^6 + x_2^6 - a^6 = 0, \quad x_i > 0, \quad i = 1, 2.$$

Solution. This problem can be reduced in the same way to an unconstrained problem,

$$x_1 \sqrt[6]{a^6 - x_1^6} \to \max, \quad 0 < x_1 < a,$$

and this problem can be solved using the Fermat theorem, but if you do the calculations, you will see that these require some algebraic manipulations that are not so pleasant.

Admissible point. A point $x \in \mathbb{R}^n$ such that

$$f_i(x) = 0, \quad 1 \leq i \leq m,$$

is called *admissible* for problem $(P_{3.1})$.

Local minimum and maximum. The vector $\widehat{x}$ is called *a point of local minimum of the problem* $(P_{3.1})$ if it is a point of global minimum for the problem that you get by adding to the problem the constraint that x is contained in a sufficiently small open ball with center $\widehat{x}$. The definition of a *point of local maximum* is analogous.

We have to make stronger smoothness assumptions on the functions f_i, $1 \leq i \leq m$, that define the equality constraints, than just differentiability. This stronger concept, continuous differentiability at a point, is a very convenient one for three reasons.

1. It is not too strong: in—almost—all applications this property turns out to hold at all points of differentiability.

2. It is strong enough to exclude certain "abnormal" behavior that cannot be excluded if you only have differentiability. For example, the solution set of an equation $f(x_1, x_2) = 0$ near a point at which f is only differentiable, might not "look a curved line."

3. It is usually easy to check whether it holds for a given function: you have to check that all partial derivatives are continuous. In contrast, it is not so easy to check whether a function is differentiable and to compute the derivative, if the function is not continuously differentiable.

Definition 3.2 *Continuous differentiability at a point. A function $f : U_n(\widehat{x}, \varepsilon) \to \mathbb{R}$ of n variables, defined on an open ball with center $\widehat{x}$, is called continuously differentiable at the point $\widehat{x}$ if all partial derivatives*

$$\frac{\partial f}{\partial x_i}, \ 1 \leq i \leq n,$$

exist on a sufficiently small open ball with center $\widehat{x}$ and are continuous at $\widehat{x}$. Then we write $f \in C^1(\widehat{x})$.

The function

$$\mathcal{L}(x, \lambda) = \sum_{i=0}^{m} \lambda_i f_i(x)$$

with $\lambda = (\lambda_0, \ldots, \lambda_m)$ is called *the Lagrange function of* $(P_{3.1})$, and the variables λ_i, $0 \leq i \leq m$, are called *the Lagrange multipliers of* $(P_{3.1})$. We will call the row vector $\lambda \in (\mathbb{R}^{m+1})'$ *a selection of Lagrange multipliers*. We write $\mathcal{L}_x(x, \lambda)$ for the row vector of partial derivatives with respect to the x-variables,

$$\left(\frac{\partial \mathcal{L}}{\partial x_1}(x, \lambda), \ldots, \frac{\partial \mathcal{L}}{\partial x_n}(x, \lambda)\right).$$

Theorem 3.3 *The Lagrange multiplier rule—necessary condition for problems with equality constraints.* Consider a problem of type $(P_{3.1})$. Assume that this problem is smooth at $\widehat{x}$ in the following sense. The function $f_0 : U_n(\widehat{x}, \varepsilon) \to \mathbb{R}$ is differentiable at $\widehat{x}$ and the functions $f_i : U_n(\widehat{x}, \varepsilon) \to \mathbb{R}$, $1 \leq i \leq m$, are continuously differentiable at $\widehat{x}$.

If $\widehat{x}$ is a point of local extremum of the problem $(P_{3.1})$, then it is a stationary point of the Lagrange function of the problem for a suitable nonzero selection of Lagrange multipliers $\lambda \in (\mathbb{R}^{m+1})'$, that is,

$$\mathcal{L}_x(\widehat{x}, \lambda) = 0_n^T \Leftrightarrow \sum_{i=0}^m \lambda_i f_i'(\widehat{x}) = 0_n^T \Leftrightarrow \sum_{i=0}^m \lambda_i \frac{\partial f_i}{\partial x_j}(\widehat{x}) = 0, \quad 1 \leq j \leq n.$$
(3.1)

The conclusion of this rule can be formulated as follows:

the vectors $f_i'(\widehat{x})$, $0 \leq i \leq m$, are linearly dependent.

Stationary points. Points $\widehat{x}$ for which equation (3.1), *the Lagrange equation*, holds for some nonzero selection of Lagrange multipliers λ are called *stationary points* of the problem $(P_{3.1})$.

In all problems of interest one can take $\lambda_0 = 1$. Then we see that this theorem corresponds precisely to the words of Lagrange from the epigraph and that it gives $m + n$ equations (including the equality constraints) in as many unknowns.

To facilitate the memorizing of this system, you can write it as follows:

$$\frac{\partial \mathcal{L}}{\partial x_j} = 0, \ 1 \leq j \leq n, \ \frac{\partial \mathcal{L}}{\partial \lambda_i} = 0, \ 1 \leq i \leq m,$$

where the multiplier λ_0 is set equal to 1; that is, all partial derivatives of the function $\mathcal{L}(x, 1, \bar{\lambda})$ of $(x, \bar{\lambda})$ are set equal to zero.

The secret of the power of Lagrange's elimination method. Suppose we want to solve a problem of type $(P_{3.1})$. The Fermat method requires us to eliminate first and differentiate afterward. The elimination is the hard task: often it is even impossible, and when it is possible it often leads to complicated nonlinear expressions, which makes the other task—differentiation—more complicated. The clever idea behind the Lagrange method is to *reverse the order*: to differentiate first and eliminate afterward. This helps: it leads to an elimination problem for a system of *linear* equations, and, moreover, to a differentiation step that is as simple as possible. Thus the secret of the power of the Lagrange method is the reduction of the main task—the elimination—from a nonlinear problem to a linear one. There is an additional idea of the Lagrange method: the introduction of multipliers. This makes the method more convenient, but it is not essential.

Illustrations of these insights into Lagrange's elimination method will be given in sections 3.5.1 and 3.5.2.

Conclusion. The multiplier rule states that stationarity of the Lagrange function is a necessary condition for local optimality. The secret of its success lies in the idea of reducing the main task, a *nonlinear* elimination problem, to an easier task, a *linear* elimination problem.

3.2.2 First illustrations of the Lagrange multiplier rule

No need for sufficient conditions. We stress once again that the four-step method leads to a complete solution of problems with equality constraints, although the equations of the Lagrange method are only necessary conditions, and not sufficient conditions. The reason for this is that we establish each time the existence of global solutions. The logic behind this is exactly the same as for problems without constraints; this logic is explained in section 1.3.2.

We illustrate this rule with the problem from example 3.2.1.

Example 3.2.3 *Solve, for a given constant $a > 0$, the problem*

$$f_0(x) = x_1 x_2 \to \max, \ f_1(x) = x_1^2 + x_2^2 - a^2 = 0, \ x_i > 0, \ i = 1, 2.$$

Solution
1. Existence of a global solution $\hat{x}$ follows from the Weierstrass theorem. (We allow x_1 and x_2 to be zero, which makes no difference here: the quarter-circle $x_1^2 + x_2^2 = a^2$, $x_1, x_2 \geq 0$ is closed, bounded, and nonempty, and the function $x_1 x_2$ is continuous.)

2. Lagrange function

$$\mathcal{L}(x, \lambda) = \lambda_0 x_1 x_2 + \lambda_1 (x_1^2 + x_2^2 - a^2).$$

Lagrange: $\mathcal{L}_x(x,\lambda) = 0_2^T \Rightarrow$

$$\frac{\partial \mathcal{L}}{\partial x_1} = \lambda_0 x_2 + 2\lambda_1 x_1 = 0 \quad \text{and} \quad \frac{\partial \mathcal{L}}{\partial x_2} = \lambda_0 x_1 + 2\lambda_1 x_2 = 0.$$

We put $\lambda_0 = 1$, as we may ($\lambda_0 = 0$ would imply $\lambda_1 \neq 0$, and then the Lagrange equations would give $x_1 = x_2 = 0$, contradicting the equality constraint).

3. Elimination of λ_1 gives $x_1^2 = x_2^2$. Use of the admissibility conditions gives $x_1 = x_2 = \frac{1}{2}a\sqrt{2}$.

4. $\hat{x} = (\frac{1}{2}a\sqrt{2}, \frac{1}{2}a\sqrt{2})^T$.

As a second illustration, we consider the slightly more complicated problem from example 3.2.2. Here the Lagrange multiplier already gives a much simpler solution than the Fermat theorem.

Example 3.2.4 *Solve, for a given constant $a > 0$, the problem*

$$f_0(x) = x_1 x_2 \to \max, \quad f_1(x) = x_1^6 + x_2^6 - a^6 = 0, \quad x_i > 0, \ i = 1, 2.$$

Solution

1. Existence of a global solution $\hat{x}$ follows from the Weierstrass theorem (again we allow x_1 and x_2 to be zero).

2. Lagrange function

$$\mathcal{L}(x, \lambda) = \lambda_0 x_1 x_2 + \lambda_1(x_1^6 + x_2^6 - a^6).$$

Lagrange: $\mathcal{L}_x(x,\lambda) = 0_2^T \Rightarrow$

$$\frac{\partial \mathcal{L}}{\partial x_1} = \lambda_0 x_2 + \lambda_1 6 x_1^5 = 0 \quad \text{and} \quad \frac{\partial \mathcal{L}}{\partial x_2} = \lambda_0 x_1 + \lambda_1 6 x_2^5 = 0.$$

Again we put $\lambda_0 = 1$, as we may.

3. Elimination of λ_1 gives $x_1^6 = x_2^6$. Use of the admissibility conditions gives $x_1 = x_2 = a/\sqrt[6]{2}$.

4. $\hat{x} = (a/\sqrt[6]{2}, a/\sqrt[6]{2})^T$.

Auxiliary role of multipliers. We want to emphasize one essential detail. In the calculation above, we did not determine the Lagrange multiplier λ_1. We only used it to relate x_1 and x_2 at the optimum. This is the typical situation. *In most applications of the*

Lagrange multiplier rule, you do not need the Lagrange multipliers themselves. Their role is just to relate the variables of optimization at the optimum.

Warning. A quick reading of the multiplier rule might give an incorrect impression: that a local extremum of the given constrained problem is always a local extremum of the Lagrange function. The multiplier rule only states that a local extremum of the given problem is a *stationary point* of the Lagrange function. The following example illustrates this.

Example 3.2.5 *Show that the problem*

$$f_0(x) = x_1 x_2 \to \max, \ f_1(x) = x_1 + x_2 - 2 = 0, \ x_1, x_2 > 0$$

has the unique global solution $\hat{x} = (1, 1)^T$, and that this is a stationary point—but not a point of local extremum—of the Lagrange function

$$\mathcal{L}(x, \lambda) = \lambda_0 x_1 x_2 + \lambda_1 (x_1 + x_2 - 2)$$

for the selection of Lagrange multipliers $(\lambda_0, \lambda_1) = (1, -1)$.

Solution.

- **Global maximum.** Take an *arbitrary admissible* vector x of the given problem, different from $\hat{x} = (1, 1)^T$, and write $h = x - \hat{x}$. Then

$$h_1 + h_2 = 0 \quad \text{and} \quad h_1 \neq 0;$$

therefore $f(x) - f(\hat{x})$ equals

$$(\hat{x}_1 + h_1)(\hat{x}_2 + h_2) - \hat{x}_1 \hat{x}_2 = (1 + h_1)(1 - h_1) - 1 = -h_1^2 < 0,$$

as required.

- **Stationary point of the Lagrange function.** However, for an *arbitrary* point $x \in \mathbb{R}^2$ we have, writing again $h = x - \hat{x}$, that

$$\mathcal{L}(x, (1, -1)) - \mathcal{L}(\hat{x}, (1, -1))$$

equals

$$((1 + h_1)(1 + h_2) - (h_1 + h_2)) - 1 = h_1 h_2.$$

As $h_1 h_2 = o(h)$, it follows that $\hat{x}$ is a stationary point of the function $\mathcal{L}(x, (1, -1))$.

- **Not a local extremum of the Lagrange function.** As $h_1 h_2$ takes both positive and negative values arbitrarily close to 0_2, it follows that $\widehat{x}$ is not a point of local extremum of $\mathcal{L}(x,(1,-1))$.

Conclusion. Even for *simple* problems with constraints, the Lagrange method is much more convenient than the Fermat method. In all applications we may set $\lambda_0 = 1$, after showing that $\lambda_0 = 0$ leads to a contradiction. This gives a system of $m+n$ equations in as many unknowns, $x_1, \ldots, x_n, \lambda_1, \ldots, \lambda_m$, which can be used to find $x_1, \ldots, x_n$. Often it is not necessary to determine the multipliers $\lambda_1, \ldots, \lambda_m$.

3.2.3 The bad case $\lambda_0 = 0$

The "bad case" $\lambda_0 = 0$ can occur, as the following artificial examples show. Then the vectors $f_1'(\widehat{x}), \ldots, f_m'(\widehat{x})$ are linearly dependent: this is all the information the multiplier rule gives in this case. In particular, note that this is only a condition on the system of constraints

$$f_i(x) = 0, \ 1 \leq i \leq m,$$

as the objective function f_0 does not occur in this condition.

Usually, the bad case can be traced back to one of the following two reasons:

- either the admissible set is, near the point $\widehat{x}$, "degenerate" in some sense,

- or the equations describing the admissible set have been written down in a "degenerate way."

The following example illustrates "degeneracy of the admissible set near the point $\widehat{x}$."

Example 3.2.6 *Apply the multiplier rule to a problem of the form*

$$f_0(x) \to \min, \quad x_1^2 - x_2^3 = 0,$$

and to $\widehat{x} = 0_2$ *where* $f_0 : \mathbb{R}^2 \to \mathbb{R}$ *is any function that is differentiable at* 0_2.

Solution. The Lagrange equations are readily seen to hold at $\widehat{x} = 0$ with $\lambda_0 = 0$ and $\lambda_1 = 1$. The type of degeneracy of the curved line given by the equation $x_1^2 - x_2^3 = 0$ near the origin is described by the word *cusp* (Fig. 3.2).

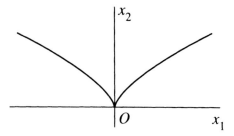

Figure 3.2 The curve $x_1^2 - x_2^3 = 0$ has a cusp at the origin.

The following example shows that the bad case $\lambda_0 = 0$ can occur for almost any problem if we write the constraints in a degenerate way.

Example 3.2.7 *Consider a problem of type $(P_{3.1})$ satisfying the assumptions of theorem 3.3 and let $\widehat{x}$ be a local solution for which $f_0'(\widehat{x})$ is not the zero vector. Apply the multiplier rule to the following, equivalent formulation of the problem*

$$f_0(x) \to \text{extr}, \quad (f_i(x))^2 = 0, \quad 1 \leq i \leq m.$$

Solution. The only information that the Lagrange equations give for this problem is just $\lambda_0 = 0$, as one readily checks.

Remark. Suppose that we consider problems of type $(P_{3.1})$ where the bad case does not occur and that we normalize λ_0 to be equal to 1. If we change the way we write down the constraints, then we get different Lagrange multipliers, of course. For example, if we multiply one of the constraints by a nonzero constant, then its Lagrange multiplier gets multiplied by the inverse of that constant.
Conclusion. The bad case $\lambda_0 = 0$ can occur if there is some degeneracy in the—formulation of the—problem.

3.2.4 Substantial illustration of the Lagrange multiplier rule

Comparison of two length concepts. The following example of the multiplier rule is more substantial. Moreover, it illustrates the difference between a pragmatic approach to practical problems and a desire to find out the essence of things. Sometimes, one has to take, instead of the ordinary length $\sqrt{x_1^2 + x_2^2}$ of a vector in the plane, another measure of the length, for example, $\sqrt[4]{x_1^4 + x_2^4}$, called the l_4-norm.

LAGRANGE: EQUALITY CONSTRAINTS

Pragmatic approach. Then it is useful to know that the expressions $\sqrt[4]{x_1^4 + x_2^4}$ and $\sqrt{x_1^2 + x_2^2}$ are of the same order of magnitude, in the sense that their quotient lies between two positive constants $C \leq D$ for all numbers x_1, x_2, that are not both zero. Usually, it is not even needed to have explicit values for C and D.

Desire to find out the essence. However, if you want to get to the bottom of the matter you will try to find the largest such C and the smallest such D.

Example 3.2.8

1. Show that there exist positive constants C and D such that

$$C\sqrt[4]{x_1^4 + x_2^4} \leq \sqrt{x_1^2 + x_2^2} \leq D\sqrt[4]{x_1^4 + x_2^4}$$

   for all numbers x_1, x_2.

2. Find explicit values for C and D.

3. Determine the largest such C and the smallest such D.

Solution. We observe that the quotient of $\sqrt{x_1^2 + x_2^2}$ and $\sqrt[4]{x_1^4 + x_2^4}$ is constant on all scalar multiples of any given nonzero vector. This leads us to consider the problem

$$f_0(x) = x_1^2 + x_2^2 \to \text{extr}, \quad f_1(x) = x_1^4 + x_2^4 - 1 = 0.$$

Our three questions can be reformulated in terms of this problem.

1. Show that the problem has finite positive minimal and maximal values. This follows from the Weierstrass theorem. To see this, we have to verify that the assumptions of the Weierstrass theorem hold. The objective function $x_1^2 + x_2^2$ is continuous, and the admissible set—the solution set of $x_1^4 + x_2^4 - 1 = 0$—satisfies the requirements:

   - it is nonempty as $(1, 0)$ is a solution;
   - it is closed as it is the zero set of the continuous function $x_1^4 + x_2^4 - 1$;
   - it is bounded as $-1 \leq x_1, x_2 \leq 1$ for all solutions of $x_1^4 + x_2^4 = 1$, using that $x_i^4 = (x_i^2)^2 \geq 0$, $i = 1, 2$.

2. Find a lower (upper) bound for this minimal (maximal) value. A possible answer is 1 (resp. $\sqrt{2}$), using the estimate

$$x_1^4 + x_2^4 \leq x_1^4 + 2x_1^2 x_2^2 + x_1^4 = (x_1^2 + x_2^2)^2$$

(resp.

$$(x_1^2 + x_2^2)^2 = x_1^4 + 2x_1^2 x_2^2 + x_2^4 \leq 2(x_1^4 + x_2^4),$$

using the inequality of the geometric-arithmetic means (remark 1.10)). This gives $C = 1$ and $D = \sqrt[4]{2}$.

3. Solve this optimization problem in order to determine its minimal (maximal) value.

Picture method. One does not need the multiplier rule to solve this optimization problem. The easiest way to solve it is geometrically, using a picture (Fig. 3.3). This figure gives the curves

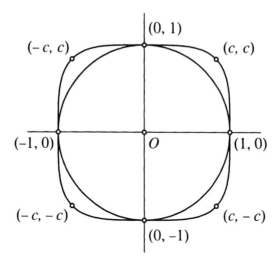

Figure 3.3 Comparing the modulus and the l_4-norm by means of their unit circles.

$$x_1^2 + x_2^2 = 1 \quad \text{and} \quad x_1^4 + x_2^4 = 1,$$

the "unit circles" of the modulus and the l_4-norm, and it shows that the points of global minimum (resp. maximum) are

$$(1, 0), (0, 1), (-1, 0), \text{ and } (0, -1)$$

LAGRANGE: EQUALITY CONSTRAINTS

(resp.
$$(c, c), (c, -c), (-c, c), \text{ and } (-c, -c),$$

where $c = 1/\sqrt[4]{2}$). Therefore, the minimal (resp. maximal) value is 1 (resp. $\sqrt{2}$). This leads to the constants $C = 1$ and $D = \sqrt[4]{2}$, and so the estimates given above happen to be optimal.

However, this picture method breaks down for problems with many variables. The analytical method using the multiplier rule, which we will demonstrate now, does not have this disadvantage.

Multiplier rule

1. Consider the problem
$$f_0(x) = x_1^2 + x_2^2 \to \text{extr}, \quad f_1(x) = x_1^4 + x_2^4 - 1 = 0.$$

   Existence of a global maximum $\widehat{x}_{\max}$ and a global minimum $\widehat{x}_{\min}$ follows from the Weierstrass theorem (the function $x_1^2 + x_2^2$ is continuous and the admissible set—the solution set of $x_1^4 + x_2^4 = 1$ —is closed, bounded, and nonempty).

2. Lagrange function
$$\mathcal{L}(x, \lambda) = \lambda_0(x_1^2 + x_2^2) + \lambda_1(x_1^4 + x_2^4 - 1).$$
   Lagrange: $\mathcal{L}_x(x, \lambda) = 0_2^T \Rightarrow$
$$2\lambda_0 x_1 + 4\lambda_1 x_1^3 = 0 \quad \text{and} \quad 2\lambda_0 x_2 + 4\lambda_1 x_2^3 = 0.$$

   We put $\lambda_0 = 1$, as we may ($\lambda_0 = 0$ would give $\lambda_1 \neq 0$, and then the Lagrange equations would give $x_1 = x_2 = 0$, contradicting the equality constraint).

3. Elimination of λ_1 gives $2x_1 x_2^3 = 2x_2 x_1^3$. We distinguish three cases:

   (a) $x_1 = 0 \Rightarrow x_2 = \pm 1$,
   (b) $x_2 = 0 \Rightarrow x_1 = \pm 1$,
   (c) $x_1^2 = x_2^2 \Rightarrow |x_1| = |x_2| = c$ with $c = 1/\sqrt[4]{2}$.

   That is, the candidate extrema are
   $(1, 0), (0, 1), (-1, 0), (0, -1), (c, c), (-c, c), (-c, -c), \text{ and } (c, -c)$.
   Comparison gives that the first four points have value 1 and the last four have value $\sqrt{2}$.

4. absmin = $\{(1,0),(0,1),(-1,0),(0,-1)\}$ and the minimal value is 1;

absmax = $\{(c,c),(-c,c),(-c,-c),(c,-c)\}$ and the maximum value is $\sqrt{2}$.

See exercise 3.6.16 for a generalization of this example.

Conclusion to section 3.2. Even for problems that can be solved easily geometrically, it is worthwhile to use the Lagrange method instead: often the resulting solution can be extended to more complicated problems.

3.3 APPLICATIONS TO CONCRETE PROBLEMS

3.3.1 Problem of Fermat (revisited)

Let us consider again the numerical example of the problem that Fermat used to illustrate his method (problem 1.4.1).

Problem 3.3.1 *Solve the problem of Fermat on the largest area of a right triangle with given sum $a = 10$ of the two sides that make a right angle.*

Solution

1. This problem can be modeled as follows:

$$f_0(x) = \frac{1}{2}x_1 x_2 \to \max, \quad f_1(x) = x_1 + x_2 - 10 = 0, \ x_1, x_2 > 0.$$

Existence of a global solution $\widehat{x}$ follows from the Weierstrass theorem (we allow x_1 and x_2 to be zero (this makes no difference here); then the admissible set is the closed interval with endpoints $(10, 0)$ and $(0, 10)$).

2. Lagrange function

$$\mathcal{L}(x, \lambda) = \lambda_0 x_1 x_2 + \lambda_1 (x_1 + x_2 - 10)$$

(we have omitted the factor $\frac{1}{2}$, as we may).

Lagrange: $\mathcal{L}_x(x, \lambda) = 0_2^T \Rightarrow$

$$\frac{\partial \mathcal{L}}{\partial x_1} = \lambda_0 x_2 + \lambda_1 = 0 \quad \text{and} \quad \frac{\partial \mathcal{L}}{\partial x_2} = \lambda_0 x_1 + \lambda_1 = 0.$$

We put $\lambda_0 = 1$, as we may.

3. $x_1 = x_2$. Using the equality constraint, we get $x_1 = x_2 = 5$.

4. $\widehat{x} = (5,5)^T$.

3.3.2 A problem of Kepler

We continue with another problem of Kepler (cf. exercise 1.6.29) from his work *A new stereometry of wine barrels*.

Problem 3.3.2 *Inscribe in a sphere the parallelepiped of maximal volume (Fig. 3.4).*

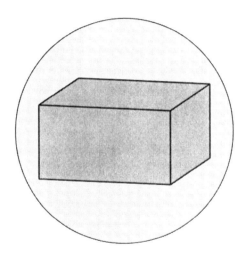

Figure 3.4 Kepler: inscribed parallelepiped of maximal volume.

Solution

1. The problem can be formalized as follows:

$$f_0(x) = 8x_1 x_2 x_3 \to \max, \quad f_1(x) = x_1^2 + x_2^2 + x_3^2 - 1 = 0,$$

$$x_i > 0, \ 1 \le i \le 3.$$

Existence of a global solution $\widehat{x}$ follows from the Weierstrass theorem (we allow x_1, x_2, and x_3 to be zero, as we may; this makes no difference here).

2. Lagrange function

$$\mathcal{L}(x,\lambda) = \lambda_0 x_1 x_2 x_3 + \lambda_1(x_1^2 + x_2^2 + x_3^2 - 1)$$

(we have omitted the factor 8, as we may).

Lagrange: $\mathcal{L}_x(x,\lambda) = 0_3^T \Rightarrow$

$\lambda_0 x_2 x_3 + 2\lambda_1 x_1 = 0, \ \lambda_0 x_1 x_3 + 2\lambda_1 x_2 = 0, \ \lambda_0 x_1 x_2 + 2\lambda_1 x_3 = 0.$

We put $\lambda_0 = 1$, as we may (if $\lambda_0 = 0$, then $\lambda_1 \neq 0$ and then the Lagrange equations would give $x_i = 0$, $1 \leq i \leq 3$, contradicting the equality constraint).

3. Elimination of λ_1 gives

$$x_1^2 x_2^2 = x_2^2 x_3^2 = x_1^2 x_3^2$$

and so, either one of the coordinates is zero—and then the f_0-value is zero—or we get, using the constraints of the problem, that $x_1 = x_2 = x_3 = 1/\sqrt{3}$, which has positive f_0-value.

4. $\widehat{x} = (1/\sqrt{3}, 1/\sqrt{3}, 1/\sqrt{3})^T$.

Inequality of the geometric-quadratic means. Note that in the same way one can solve the general problem

$$\prod_{i=1}^n x_i \to \max, \ \sum_{i=1}^n x_i^2 = 1, \ x_i \geq 0 \ (1 \leq i \leq n).$$

In example 3.2.3 (problem 3.3.2) we have $n = 2$ ($n = 3$). The answer is $\widehat{x}_i = 1/\sqrt{n}$. This leads to the following inequality between the geometric mean and the quadratic mean:

$$\left(\prod_{i=1}^n x_i\right)^{\frac{1}{n}} \leq \left(\frac{1}{n}\left(\sum_{i=1}^n x_i^2\right)\right)^{\frac{1}{2}}$$

3.3.3 Maximizing a linear function on a sphere

Now we apply the Lagrange multiplier rule to a problem with an equality constraint where application of the Fermat theorem would lead to an inconvenient solution, as reduction to an unconstrained problem would destroy part of the structure of the original problem. Moreover this application illustrates that when one has excluded the bad case $\lambda_0 = 0$, one need not put $\lambda_0 = 1$: sometimes it handier to put λ_0 equal to some other nonzero value. This does not make any difference: the Lagrange equations are homogeneous in the multipliers, and so each nonzero choice of λ_0 leads to the same solutions x.

Problem 3.3.3 *Solve the problem of maximizing a given linear function on the unit sphere.*

Solution. The solution of this problem is obvious from a geometrical point of view by consideration of the case $n = 2$ (Fig. 3.5). Now we

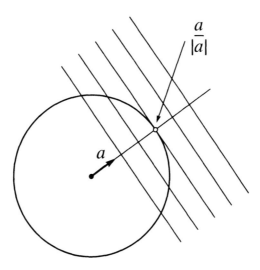

Figure 3.5 Maximizing a linear function on the unit circle using level curves.

give an analytical solution of the problem.

Choose a nonzero vector $a \in (\mathbb{R}^n)^T$.

1. $f_0(x) = a \cdot x \to \max$, $f_1(x) = |x|^2 - 1 = 0$, $x \in \mathbb{R}^n$.

   Existence of a global solution $\hat{x}$ follows from the Weierstrass theorem.

2. Lagrange function
$$\mathcal{L}(x, \lambda) = \lambda_0 a \cdot x + \lambda_1(|x|^2 - 1).$$

   Lagrange: $\mathcal{L}_x(x, \lambda) = 0_n^T \Rightarrow$

   $\lambda_0 a + 2\lambda_1 x^T = 0_n^T$, using problem 2.1.

   We put $\lambda_0 = -2$, as we may.

3. $\lambda_1 x = a^T$, and so $x = \pm |a|^{-1} a^T$, using $|x| = 1$.

Compare: $f_0(|a|^{-1} a^T) = |a|$ and $f_0(-|a|^{-1} a^T) = -|a|$, using $a \cdot a^T = |a|^2$.

4. $\hat{x} = |a|^{-1} a^T$ and $S_{\max} = |a|$.

3.3.4 The inequality of the arithmetic-quadratic means

The well-known Cauchy-Bunyakovskii-Schwarz inequality

$$a \cdot x \leq |a||x|$$

for vectors $a \in (\mathbb{R}^n)^T$ and $x \in \mathbb{R}^n$ (see theorem 2.1) is a corollary of the analysis of the previous problem. Many other inequalities can be proved in a standard way using the multiplier rule. Let us give one more example.

Problem 3.3.4 *Prove the following inequality between two types of average, the usual average or arithmetic mean and the quadratic mean,*

$$(z_1 + \cdots + z_n)/n \leq ((z_1^2 + \cdots + z_n^2)/n)^{1/2},$$

which holds for all $z_1, \ldots, z_n \in \mathbb{R}$.

Solution. We consider the problem

$$x_1 + \cdots + x_n \to \max. \quad x_1^2 + \cdots + x_n^2 = 1.$$

It is the special case of problem 3.3.3 that $a = (1, \ldots, 1)$, and so $|a| = \sqrt{n}$. It follows that it has the global solution

$$\hat{x} = (1/\sqrt{n}, \ldots, 1/\sqrt{n})^T,$$

and so it has optimal value $\sqrt{n}$. This implies the required inequality. Indeed, we get for all $z_1, \ldots, z_n \in \mathbb{R}$, not all equal to zero, writing

$$x_i = (z_1^2 + \cdots + z_n^2)^{-\frac{1}{2}} z_i, \ 1 \leq i \leq n,$$

that $x_1^2 + \cdots + x_n^2 = 1$ and so $x_1 + \cdots + x_n \leq \sqrt{n}$; rewriting this in terms of $z_1, \ldots, z_n$ gives the required inequality.

Justification of least squares approximation. This inequality—in combination with the easy inequality

$$z_1 + \cdots + z_n \geq (z_1^2 + \cdots + z_n^2)^{\frac{1}{2}}$$

LAGRANGE: EQUALITY CONSTRAINTS

for all nonnegative numbers $z_1, \ldots, z_n$ – gives a "justification" of the least squares approximation. We illustrate the meaning of this justification with the simplest example: the approximation of a cloud of points (x_i, y_i), $1 \leq i \leq k$, by a line $y = ax + b$. The most natural measure for this approximation would be the sum of the absolute values of the residues

$$r_i = y_i - (ax_i + b), \quad 1 \leq i \leq k.$$

However, for convenience one always uses the sum of the squares of the residues instead. The inequality of the arithmetic-quadratic means and the easy inequality above give

$$1 \leq \frac{(|r_1| + \cdots + |r_k|)^2}{r_1^2 + \cdots + r_k^2} \leq k.$$

Therefore, it does not make much difference which measure of approximation is chosen.

Inequalities between l_p-norms. In the same way as in the problem above, one can establish the more general inequality

$$(z_1 + \cdots + z_n)/n \leq ((z_1^q + \cdots + z_n^q)/n)^{1/q}$$

for all nonnegative $z_1, \ldots, z_n$ and all $q > 1$. As a corollary, one gets the inequality

$$((x_1^p + \cdots + x_n^p)/n)^{1/p} \leq ((x_1^q + \cdots + x_n^q)/n)^{1/q}$$

for $p < q$, replacing z_k by $x_k{}^p$ and q by q/p. In a similar way, one can establish the inequality

$$(x_1^p + \cdots + x_n^p)^{1/p} \leq (x_1^q + \cdots + x_n^q)^{1/q}$$

if $p > q$. These results are a simultaneous generalization of the result of example 3.2.8 in two different directions: this example is the special case that $n = 2$ and p and q are 2 and 4 (or 4 and 2). Exercise 3.6.16 is about settling these inequalities in the same way as in example 3.2.8 by using the multiplier rule.

These inequalities can be seen as the solution to a problem on the comparison of two different norms. For each number $p \geq 1$ one can define the length of a vector $x \in \mathbb{R}^n$ to be the expression

$$\|x\|_{l_p} = (x_1^p + \cdots + x_n^p)^{1/p},$$

called the l_p-norm of x. Moreover, the l_∞-norm can be defined either by a limit

$$\|x\|_{l_\infty} = \lim_{p \to +\infty} \|x\|_{l_p}$$

(this makes clear why ∞ is used in the notation) or, equivalently, by the explicit formula

$$\|x\|_{l_\infty} = \max_{1 \le i \le n} |x_i|.$$

The case $l = 2$ is the usual Euclidean norm, and the cases $l = 1$ and $l = \infty$ are also often used. It is useful to know that each two of these norms, l_p and l_q, are of the same order of magnitude. The inequalities above settle a more precise question: what are the sharpest constants c and d for which

$$c\|x\|_{l_p} \le \|x\|_{l_q} \le d\|x\|_{l_p}$$

for all $x \in \mathbb{R}^n$?

For an interesting application of these inequalities, we refer to the proof that the log-barrier of the semidefinite cone is self-concordant (see theorem 7.3).

3.3.5 Stakeholders versus shareholders

We give a model that tries to capture that there can sometimes be a difference of interest between stakeholders and shareholders of a firm, which might lead to problems.

Problem 3.3.5 *A firm has total revenue*

$$TR = 40Q - 4Q^2 + 2A,$$

where Q is its output and A is its advertising expenditure. Its total costs are

$$TC = 2Q^2 + 20Q + 1 + 4A.$$

To encourage the managers, that is, the stakeholders, to perform well, their salary is linked to how well the firm is doing. For practical reasons it is made to depend on the total revenue of the firm, but not on the total costs. However, the profit of the firm is of importance as well: to be more concrete, the shareholders will not accept a profit of less than 3.

LAGRANGE: EQUALITY CONSTRAINTS

What will be the best choice of output and advertising expenditure from the point of view of the managers? Is this choice also optimal from the point of view of the shareholders?

Solution. The problem can be modeled as follows:

$$TR(Q, A) \to \max, \quad TR(Q, A) - TC(Q, A) \geq 3, \quad Q \geq 0, \quad A \geq 0.$$

Now we replace the inequality constraint by the equality constraint

$$TR(Q, A) - TC(Q, A) = 3,$$

as it will not be optimal for the managers when the profit is higher than 3: then it would be better to have more costs, choosing higher levels of output and advertising expenditure, as this would increase total revenue. Note that this is far from optimal from the point of view of the shareholders.

Moreover, we omit the inequality constraints $Q, A \geq 0$, as we think that it will not be optimal for Q or A to be 0. If we will find later that the values for Q and A that are optimal for the modified problem are positive, then this will justify that these constraints have been omitted.

1. This gives the following problem:

$$f_0(Q, A) = 40Q - 4Q^2 + 2A \to \max,$$

$$f_1(Q, A) = 20Q - 6Q^2 - 2A - 4 = 0.$$

   Existence of a global solution $(\widehat{Q}, \widehat{A})$ follows from coercivity ($f_0(Q, A)$ can be written as $-10Q^2 + 60Q - 4$ on the admissible set).

2. Lagrange function

$$\mathcal{L}(x, \lambda) = \lambda_0(40Q - 4Q^2 + 2A) + \lambda_1(20Q - 6Q^2 - 2A - 4).$$

   Lagrange: $\mathcal{L}_x(x, \lambda) = 0_2^T \Rightarrow$

$$\lambda_0(40 - 8Q) + \lambda_1(20 - 12Q) = 0 \quad \text{and} \quad 2\lambda_0 - 2\lambda_1 = 0.$$

   We put $\lambda_0 = 1$, as we may.

3. Elimination of λ_1 gives $60 - 20Q = 0$. This leads to $Q = 3$ and $A = 1$. Note that the omitted constraints $Q, A \geq 0$ are satisfied.

160   CHAPTER 3

4. $(\widehat{Q}, \widehat{A}) = (3, 1)$.

This example illustrates that giving incentives that are successful is not so simple. In the example, the result will be a miserable profit of precisely 3, which is the minimum level that is acceptable to shareholders. In this example the managers, by the rules of the game, do not look upon profit as an objective of maximization. They view it as a *minimum constraint*.

The next two applications illustrate that there are optimization problems for which it would be either impossible or inconvenient to avoid the multiplier rule. Such applications represent the highest level of "multiplier craftsmanship." Usually, applications of the multiplier rule require less ingenuity.

3.3.6 The largest area of a hinged quadrangle

The art of geometry reached a high level in ancient Greece. There are very few examples of problems that the Greek geometers considered but were unable to solve. Here is one of these examples. It is a striking illustration of the power of the multiplier rule.

Problem 3.3.6 *The lengths of all sides of a quadrangle are fixed, but the sides are linked at vertices freely, so the angles between them can vary. Which position of the sides corresponds to the largest area of the quadrangle?*

Solution. The problem can be modeled as follows (Fig. 3.6). The area of the quadrangle equals

$$\frac{1}{2}(ad \sin \theta_1 + bc \sin \theta_2).$$

Applying the cosine rule to the triangle $\triangle ABD$ (resp. $\triangle BCD$) gives that

$$e^2 = a^2 + d^2 - 2ad \cos \theta_1$$

(resp. $e^2 = b^2 + c^2 - 2bc \cos \theta_2$).

Therefore, the following equality holds:

$$a^2 + d^2 - 2ad \cos \theta_1 = b^2 + c^2 - 2bc \cos \theta_2.$$

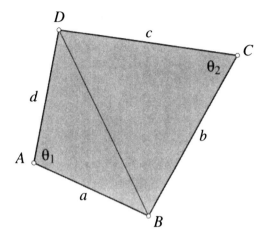

Figure 3.6 The hinged quadrangle.

1. Thus, we get the following problem:

$$f_0(\theta_1, \theta_2) = \frac{1}{2}(ad\sin\theta_1 + bc\sin\theta_2) \to \max,$$

$$f_1(\theta_1, \theta_2) = (a^2 + d^2 - 2ad\cos\theta_1) - (b^2 + c^2 - 2bc\cos\theta_2) = 0.$$

Existence of a solution follows from the Weierstrass theorem. Indeed, by assumption the admissible set is nonempty. Moreover, it is also closed, as f_1 is a continuous function. Furthermore, angles are determined up to constants which are multiples of 2π; therefore, we may add the constraints $\theta_1, \theta_2 \in [0, 2\pi]$, and this makes the admissible set bounded. Finally, the objective function f_0 is continuous.

2. Lagrange function

$$\mathcal{L}(\theta, \lambda) = \lambda_0 \tfrac{1}{2}(ad\sin\theta_1 + bc\sin\theta_2)$$
$$+ \lambda_1((a^2 + d^2 - 2ad\cos\theta_1) - (b^2 + c^2 - 2bc\cos\theta_2)).$$

Lagrange: $\mathcal{L}_\theta(\theta, \lambda) = 0_2^T \Rightarrow$

$$\frac{\partial \mathcal{L}}{\partial \theta_1} = \frac{1}{2}\lambda_0 ad\cos\theta_1 + 2\lambda_1 ad\sin\theta_1 = 0$$

and

$$\frac{\partial \mathcal{L}}{\partial \theta_2} = \frac{1}{2}\lambda_0 bc \cos\theta_2 - 2\lambda_1 bc \sin\theta_2 = 0.$$

We put $\lambda_0 = 1$, as we may.

3. Elimination of λ_1 gives

$$\tan\theta_1 = -\tan\theta_2$$

—using the formula $\tan\theta = \sin\theta/\cos\theta$—and hence—using that the sum of all angles of any quadrangle is 2π, and so $0 < \theta_1+\theta_2 < 2\pi$—we get

$$\theta_1 + \theta_2 = \pi.$$

This means that the four vertices of the quadrangle lie on one circle. Such a quadrangle is called a *circle quadrangle* (Fig. 3.7). There is of course only one such quadrangle for given a, b, c,

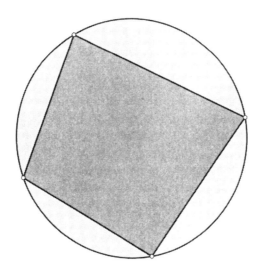

Figure 3.7 A quadrangle inscribed in a circle.

and d.

4. The unique global solution is the circle quadrangle with sides a, b, c and d.

LAGRANGE: EQUALITY CONSTRAINTS

We see that this problem, which withstood all attempts of geometers from antiquity till now to crack it by geometrical methods, is solved without any effort if the multiplier rule is used. Actually we do not know of any other method to solve it, so the Lagrange multiplier rule is the unique method to deal with this problem.

3.3.7 Basic property of symmetric matrices and the multiplier rule

We give a further illustration of the power of the Lagrange multiplier rule, by using it to establish the basic property of symmetric matrices. Another approach to this property has been given in section 2.4.6 and still another will be given in section 5.3.4. If $Av = \lambda v$ for an $n \times n$ matrix A, a nonzero vector $v \in \mathbb{R}^n$, and a number $\lambda \in \mathbb{R}$, then one says that λ is a real *eigenvalue* of A and v is a real *eigenvector* of A.

Problem 3.3.7 *Show that each symmetric $n \times n$ matrix A has a real eigenvalue.*

Solution

1. It turns out to be a fruitful idea to consider the problem
   $$f_0(x) = x^T A x \to \min, \quad f_1(x) = x^T \cdot x - 1 = 0.$$
   Existence of a global solution v follows from the Weierstrass theorem.

2. Lagrange function
   $$\mathcal{L}(x, \lambda) = \lambda_0 x^T A x + \lambda_1 (x^T x - 1).$$
   Lagrange: $\mathcal{L}_x(x, \lambda) = 0_n^T \Rightarrow$
   $$2\lambda_0 x^T A - 2\lambda_1 x^T = 0_n^T.$$
   We put $\lambda_0 = 1$, as we may.

3. $Ax = \lambda_1 x$.

4. $Av = \lambda_1 v$, as required.

This property of symmetric matrices is usually established at the cost of considerable effort. We have just seen that the Lagrange multiplier rule takes it in its stride and almost gives the impression that this property is a routine consequence of the definitions.

All stationary points are of interest. The stationary points of the optimization problem considered above correspond precisely to the eigenvectors of the matrix of length one. Therefore we have here the noteworthy phenomenon that *all stationary points of this optimization problem are of interest* ("eigenvectors"), not just the extrema. The solution of the problem is an eigenvector belonging to the smallest eigenvalue. If we would have maximized, rather than minimized, we would have obtained an eigenvector belonging to the largest eigenvalue.

This property of symmetric matrices has the following consequence, which is often used. An $n \times n$ matrix P is called *orthogonal* if $PP^T = I_n$ or, equivalently, if the columns of P are orthogonal to each other and have length one.

Theorem 3.4 *Orthogonal diagonalization of symmetric matrices.* A symmetric $n \times n$ matrix A can be written as a matrix product PDP^T, where D is a diagonal $n \times n$ matrix and P an orthogonal $n \times n$ matrix.

This theorem can be derived by induction with respect to n, the induction step being the statement of problem 3.3.7; we do not display this derivation. We only point out here that $A = PDP^T$ can be written as $AP = DP$; this means that the i-th column of P is an eigenvector of A with eigenvalue the i-th number on the diagonal of D for all i. The proof by induction works by producing the eigenvalues one by one, from small to large.

3.3.8 The problem of Apollonius

Finally, we will give a historical introduction to one of the greatest mathematical achievements from antiquity. In section 9.3 we will give a quick modern solution, using the multiplier rule.

Problems on extrema are found in the works of the greatest mathematicians of antiquity, Euclid, Archimedes and Apollonius (262BC–190BC). In his work *Conica* (*Conics*), the apex of ancient mathematics, Apollonius solves the problem of determining the distance between a point P and an ellipse in the two-dimensional plane.

At first sight, this might not look like a problem to get excited about. However, it is like an oyster, which only reveals the pearl inside on closer inspection. Here we have a minimum problem: to find the point Q on the ellipse which is closest to P. For this point, the line through P and Q intersects the given ellipse at the point Q at right angles (Fig. 3.8).

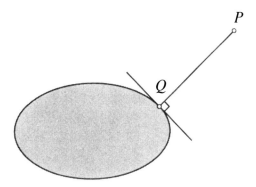

Figure 3.8 Geometric sense stationarity for a problem of Apollonius.

This observation suggested to Apollonius that he reduces his minimum problem to the problem of finding *all* the lines through a given point P that intersect the ellipse at right angles. Such a line is called a *normal*. This problem about the normals is of interest for its own sake. From our point of view, Apollonius applied the Fermat theorem to his minimum problem, and this suggested to him considering the problem of finding all stationary points.

We have here again—as in problem 3.3.7—the noteworthy phenomenon that *the stationary points* of an optimization problem—here the normals—are all of interest and not only the extremal points.

The astroid solves the mystery. What is the situation? Some experiments show that for some points P there are two normals, for other points three or even four. For example, if P is the center of the ellipse, then there are four normals. How does the number of normals depend on P?

Apollonius discovered the beautiful rule that determines this number. How he managed this is hard to imagine. He had to deal with the problem "barehanded," as the mathematical tools required to arrive at the answer—such as algebra—had not yet been developed. Moreover, the rule itself involves an intriguing star-shaped curve, now called an *astroid*, which is partly sticking out of the ellipse (Fig. 3.9).

This astroid holds the key to the answer of our problem. It is the pearl inside the oyster. Note that it is not at all obvious at first sight

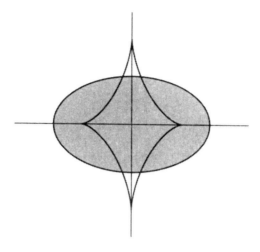

Figure 3.9 Solution involves astroid.

that this object has anything to do with the problem. Moreover, it does not belong to the "territory" of the mathematicians of antiquity. They considered lines, and the class of curves called *conics*, which includes circles, ellipses, parabolas, and hyperbolas.

An astroid belongs to an altogether different class of objects, which first turned up in the seventeenth century. However, unlikely as it may be, it is a fact that Apollonius somehow discovered the following rule:

Outside the astroid each point has two normals, inside the astroid four, and on the astroid itself three (except at the vertices, where there are two normals).

Moreover, he found the equation for the astroid. If the equation of the ellipse is

$$\left(\frac{x_1}{a_1}\right)^2 + \left(\frac{x_2}{a_2}\right)^2 = 1,$$

then the equation of the astroid is

$$(\xi_1 a_1)^{\frac{2}{3}} + (\xi_2 a_2)^{\frac{2}{3}} = (a_1^2 - a_2^2)^{\frac{2}{3}}.$$

In fact, he solved this problem not only for ellipses but more generally for conics; we will come back to this problem in exercises 9.4.3 and

LAGRANGE: EQUALITY CONSTRAINTS   167

9.4.4. Moreover, he showed how to find all the normals.

Later, the astroid will turn up in a completely different situation: problem 4.3.5 on a child drawing.

Conclusion to section 3.3. We have seen that it is possible to apply the method of Lagrange to problems of

- geometry (problems 3.3.1, 3.3.2, 3.3.3, 3.3.6, and section 3.3.8),

- algebra (the inequality of the geometric-quadratic means, problem 3.3.4, and inequalities between l_p-norms),

- economics (problem 3.3.5),

- matrix theory (problem 3.3.7).

For some problems, such as the problem of the hinged quadrangle (problem 3.3.6), it is the only known way to solve the problem.

Invitation to chapters eight and nine. At this point, you could continue with some problems in chapters eight and nine.

- Problem 8.4.1, "Prediction of flows of cargo," is a problem from maritime economics. The interest is that it allows reconstructing data in a seemingly hopeless situation. In addition, it shows how useful the Lagrange multiplier rule can be even for problems where the equality constraints are *linear*.

- Problem 9.2.3, "Certificate of redundancy for an equation from a linear system," represents one more illustration of how optimization methods can be used to prove results on matrices.

- In section 9.4, "The problem of Apollonius," we could not resist the temptation to discuss a related problem of Huygens on the "perfect pendulum," which led to one of the greatest science books ever published (cf. [31]).

3.4 PROOF OF THE LAGRANGE MULTIPLIER RULE

Why does the multiplier rule hold? This section will answer this question. The answer will provide more insight, but the contents of this section will not be used for solving concrete problems. The usual proof of the multiplier rule uses the implicit function theorem. However we will prove it using the *tangent space theorem* instead, as this leads to a more natural proof.

Advanced readers and experts are invited to compare the proof given here with the proof given in appendix G, "Conditions of Extremum from Fermat to Pontryagin." The latter proof is simpler, but given from a more advanced point of view: Brouwer's fixed point principle is used. This requires weaker smoothness assumptions on the functions defining the equality constraints than the present proof: only differentiability at the point of local extremum and continuity at some neighborhood of this point, instead of continuous differentiability at this point. Therefore, Brouwer's fixed point principle leads to a stronger result. Exercise 3.6.17 emphasizes the algorithmic character of the proof.

3.4.1 Motivation for the tangent space theorem

Plan. The aim of this section is to provide motivation for the tangent space theorem. To this end we recall the well-known implicit function theorem and sketch how the multiplier rule can be derived from it. Then we explain the disadvantages of this approach. The proof using the tangent space theorem will not have these disadvantages. The contents of this section, and in particular the implicit function theorem, will not be used anywhere outside this section. They only serve as a motivation for the relatively unknown tangent space theorem.

What is the problem? To write the necessary first order conditions for the problem $(P_{3.1})$, which is our task, we want to compare the value of f at $\widehat{x}$ to the value of f at solutions of $F(x) = 0_m$ near $\widehat{x}$. However, how can we be sure that such solutions exist at all? For example, the equation $x_1^2 + x_2^2 = 0$ has only one solution, $\widehat{x} = 0_2^T$. The following result, *the implicit function theorem*, clarifies the situation completely. It gives a regularity condition, which implies that the solution set of $F(x) = 0_m$ can be viewed locally at $\widehat{x}$ as the graph of a vector function; this vector function has the same differentiability properties as F.

Theorem 3.5 *Implicit function theorem. Let* $\widehat{x} \in \mathbb{R}^n$, U *a neighborhood of* $\widehat{x}$, *and*

$$F = (f_1, \ldots, f_m)^T : U \to \mathbb{R}^m$$

a continuously differentiable vector function for which $\operatorname{rank} F'(\widehat{x}) = m$, *and* $F(\widehat{x}) = 0_m$. *Choose* m *linearly independent columns of the matrix*

LAGRANGE: EQUALITY CONSTRAINTS

$F'(\hat{x})$, and let J be the resulting set of m indices from $\{1, \ldots, n\}$. Then the following statements hold true.

- In a suitable—sufficiently small—neighborhood of the point $\hat{x}$, the solution set of the vector equation

$$F(x) = 0_m (\Leftrightarrow f_1(x) = \cdots = f_m(x) = 0)$$

  determines the variables x_i $i \in \{1, \ldots, n\} \setminus J$, as functions of the variables x_j, $j \in J$.

- These functions are continuously differentiable, and the following formula of implicit differentiation holds:

$$\left(\frac{\partial x_i}{\partial x_j}\right)_{ij} = -\left(\frac{\partial f_k}{\partial x_i}\right)_{ki}^{-1} \left(\frac{\partial f_k}{\partial x_j}\right)_{kj}.$$

- If, moreover, F is r-times continuously differentiable, then these functions are r-times continuously differentiable as well.

The following numerical example illustrates this theorem.

Example 3.4.1 Show that the solution set of the equation

$$x_1^2 - x_2^3 = 1$$

determines at a suitable neighborhood of the point $(3, 2)^T$ the variable x_2 as a function of the variable x_1, say $x_2 = g(x_1)$. Calculate the first and second derivative of this function at $x_1 = 3$.

Solution. We could answer this question without using the implicit function theorem: we can give the explicit formula $x_2 = (x_1^2 + 1)^{1/3}$ for x_2 in terms of x_1, and then calculate the required derivatives. However, this calculation is not so pleasant. It is, even in this numerical example, more convenient to use the implicit function theorem. To begin with, we verify the regularity assumption, here $F(x) = x_1^2 - x_2^3 - 1$ and $\hat{x} = (3, 2)^T$. Therefore,

$$F'(x) = \left(\frac{\partial F}{\partial x_1}, \frac{\partial F}{\partial x_2}\right) = (2x_1, -3x_2^2),$$

and so $F'(\hat{x}) = (6, -12)$, which has rank one, as required. Therefore,

$$g'(x_1) = -(-3x_2^2)^{-1} 2x_1,$$

where $x_2 = g(x_1)$, and so $g'(3) = \frac{1}{2}$. One can continue in this way: g'' is by definition the derivative of g', and so

$$g'(x_1) = \frac{2}{3}\left(x_2^2 - x_1 \cdot 2x_2 \frac{dx_2}{dx_1}\right)/x_2^4,$$

where again $x_2 = g(x_1)$. Therefore, $g''(3) = -2/24 = -1/12$.

Note, in particular, that the implicit function theorem allows us to compute the *second* derivative of an implicit function.

Proof of the multiplier rule using the implicit function theorem. Now we can outline the usual way to prove the necessary conditions for equality constrained problems,

$$f(x) \to \text{extr}, \ F(x) = 0_m. \tag{$P_{3.1}$}$$

This makes use of the implicit function theorem. Use the constraints to eliminate variables: assume that the regularity condition holds and view $F(x) = 0_m$ as determining all variables x_i, $i \in \{1,\ldots,n\} \setminus J$ as functions of the variables x_j, $j \in J$ in a suitable neighborhood of $\widehat{x}$. Substitution in the objective function f gives an unconstrained optimization theorem. For this problem one can write the stationarity conditions, using Fermat. The resulting equations can be rewritten as Lagrange equations, using the explicit formula for the partial derivatives $\frac{\partial x_i}{\partial x_j}$ in the implicit function theorem.

Why the tangent space theorem is preferable. However, we will present the proof of the Lagrange multiplier rule in another way, which gives a better insight. The reason for this is as follows. The use of the implicit function theorem leads to a discrimination of the variables x_j, $j \in J$. However, there is no such discrimination in the formulation of the Lagrange equations. The *tangent space theorem*, to be given below, is essentially a variant of the implicit function theorem, but it is more suitable for our present purpose. It allows us to derive the multiplier rule without such an unwanted discrimination, and, moreover, without the not very illuminating calculations to which the use of the implicit function theorem leads.

3.4.2 Tangent space theorem: statement and geometric meaning

Plan. In this section we present the tangent space theorem. This is the main result of the differential calculus. Its sense is, roughly speaking, that it makes it possible to approximate the solution set of a system of nonlinear equations by the solution set of the "linearization" of this system.

LAGRANGE: EQUALITY CONSTRAINTS

The powerful Lagrange multiplier rule is just a consequence of the tangent space theorem, as we shall see in section 3.4.4. Therefore, the full power of the multiplier rule comes from the tangent space theorem 3.6. As one cannot expect to get anything of such great value for free, it should be no surprise that the proof of the tangent space requires some work. This is in contrast to all the other results we have met so far, which follow in a more or less routine way from the definitions.

A subset of $\mathbb{R}^n$ is called *an affine subspace* if it is the solution set of some system of linear equations in n variables. Geometrical examples of affine subspaces are lines in the plane and in space, and also planes in space. A vector r is called *orthogonal to an affine subspace A* if

$$r^T \cdot (a - b) = 0$$

for all $a, b \in A$.

Theorem 3.6 Tangent space theorem. *Let $\widehat{x} \in \mathbb{R}^n$, U a neighborhood of $\widehat{x}$, and $F = (f_1, \ldots, f_m)^T : U \to \mathbb{R}^m$ a vector function which is continuously differentiable at $\widehat{x}$, and for which $\mathrm{rank} F'(\widehat{x}) = m$ and $F(\widehat{x}) = 0_m$. We consider, along with the vector equation,*

$$F(x) = 0_m (\Leftrightarrow f_i(x) = 0,\ 1 \le i \le m), \qquad (*)$$

its linearized vector equation at $\widehat{x}$,

$$F'(\widehat{x})(x - \widehat{x}) = 0_m (\Leftrightarrow f_i'(\widehat{x}) \cdot (x - \widehat{x}) = 0,\ 1 \le i \le m). \qquad (**)$$

Then one can give for the set of solutions of $(*)$ that lie close to $\widehat{x}$ the following parametric description (Fig. 3.10):

$$x(\xi) = \xi + r(\xi),$$

where the parameter ξ runs over the solutions of $(**)$ that lie close to $\widehat{x}$ and the remainder $r(\xi)$ is orthogonal to the solution space of $(**)$ and is negligible in comparison to the distance between ξ and $\widehat{x}$, in the sense that

$$r(\xi) = o(\xi - \widehat{x}).$$

Precise version of the tangent space theorem. To be more precise, the conclusion of this theorem is as follows. For a suitable shrinking of U to a smaller neighborhood of $\widehat{x}$, which we will also call U, there exists for each solution $\xi \in \mathbb{R}^n$ of $(**)$ in U, a unique solution $x(\xi)$ of $(*)$ in U for which the difference vector $r(\xi) = x(\xi) - \xi$

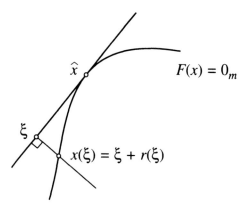

Figure 3.10 Tangent space theorem.

is orthogonal to the solution space of (∗∗). That is, the following implication holds:
$$F'(\widehat{x})v = 0 \Rightarrow (x(\xi) - \xi)^T \cdot v = 0.$$
Moreover, $x(\xi) = \xi + o(\xi - \widehat{x})$.

Tangent space. This theorem suggests calling the solution set of (∗∗) *the tangent space* at $\widehat{x}$ to the solution set of (∗). This extends the geometrical notions "tangent line to a curved line in the plane or in space," and "tangent plane to a surface in space." To be more precise, one can define the tangent space to any set in $\mathbb{R}^n$ at a point of this set; then the conclusion of the theorem above is that the tangent space to the solution set of (∗) at $\widehat{x}$ is precisely the solution set of (∗∗). We will give some numerical examples to illustrate how the theorem can be used to calculate tangent lines and tangent planes.

Why the tangent space theorem is a natural result. We emphasize that the statement of the theorem is very natural. Indeed, the linear approximation of the vector function $F : U \to \mathbb{R}^m$ at $\widehat{x}$ is given by
$$F(x) = F(\widehat{x}) + F'(\widehat{x})(x - \widehat{x}) + o(x - \widehat{x}),$$
by the definition of the derivative $F'(\widehat{x})$. This simplifies to
$$F(x) = F'(\widehat{x})(x - \widehat{x}) + o(x - \widehat{x}).$$

LAGRANGE: EQUALITY CONSTRAINTS

Therefore, it is natural to expect that the solution set of the vector equation $F'(\widehat{x})(x - \widehat{x}) = 0_m$ can be seen as the linear approximation of the solution set of $F(x) = 0_m$. This is indeed the case, and the tangent theorem gives a precise formulation of this fact.

Why the tangent space theorem is not obvious. It might be surprising at first sight that this natural result does not follow immediately from the definitions in some way. To see the core of the problem, note that the theorem implies that $\widehat{x}$ is not an *isolated solution* of $F(x) = 0_m$. That is, there exist solutions of $F(x) = 0_m$, other than $\widehat{x}$, in any ball with center $\widehat{x}$, however small its radius is. This existence statement might look self-evident, but proving it is about as hard as proving the tangent space theorem.

We illustrate the geometrical sense of the tangent space theorem for some numerical examples. To begin with, we consider a case where an alternative method is available.

Example 3.4.2 *Tangent line to a circle. Find the tangent line at the point $(3, 4)^T$ to the circle in the plane with center the origin and radius 5.*

Solution. It is easy to determine tangent lines to circles geometrically: the required line is orthogonal to the vector $(3, 4)$, clearly. Therefore, its equation is $3(x_1 - 3) + 4(x_2 - 4) = 0$.

However, we want to illustrate the tangent space theorem. We apply this theorem to

$$F(x) = x_1^2 + x_2^2 - 25 \text{ and } \widehat{x} = (3, 4)^T.$$

We have

$$F'(x) = (2x_1, 2x_2) \text{ and so } F'(\widehat{x}) = (6, 8).$$

Therefore, the tangent space theorem gives that

$$6(x_1 - 3) + 8(x_2 - 4) = 0$$

is the equation of the tangent line to the circle $x_1^2 + x_2^2 = 25$ at the point $(3, 4)$.

Now we consider some examples that are more interesting.

Example 3.4.3 *Tangent line to a plane curved line. Find the tangent line to the curved line $x_2^2 - x_1^3 = -2$ at the point $(3, 5)^T$.*

Solution. Here $n = 2$, $m = 1$, $F(x) = x_2^2 - x_1^3 + 2$, and $\hat{x} = (3,5)^T$. Then

$$F'(x) = (-3x_1^2, 2x_2) \text{ and so } F'(\hat{x}) = (-27, 10).$$

Therefore, the solution set of $F'(\hat{x})(x-\hat{x}) = 0$ is the line with equation

$$-27(x_1 - 3) + 10(x_2 - 5) = 0.$$

By the tangent space theorem, this is the required tangent line.

Example 3.4.4 *Tangent plane to a surface.* Find the tangent plane to the Fermat cubic surface $x_1^3 + x_2^3 + x^3 = 1$ at the point $(9, 10, -12)^T$.

Solution. Here $n = 3$, $m = 1$, $F(x) = x_1^3 + x_2^3 + x_3^3 - 1$, and $\hat{x} = (9, 10, -12)^T$. Then

$$F'(x) = (3x_1^2, 3x_2^2, 3x_3^2) \text{ and so } F'(\hat{x}) = (243, 300, 432).$$

Therefore, the solution set of $F'(\hat{x})(x - \hat{x}) = 0$ is the plane with equation

$$243(x_1 - 9) + 300(x_2 - 10) + 432(x_3 + 12) = 0.$$

By the tangent space theorem, this is the required tangent plane.

Example 3.4.5 *Tangent line to a curved line in space.* Find the tangent line to the intersection of the quadrics (=quadratic surfaces) $x_1 x_2 = x_3$ and $x_2^2 = x_1 x_3$ at the point $(1, 1, 1)^T$.

Solution. Here $n = 3$, $m = 2$, $F(x) = (x_1 x_2 - x_3, x_2^2 - x_1 x_3)^T$ and $\hat{x} = (1, 1, 1)^T$. Then

$$F'(x) = \begin{pmatrix} x_2 & x_1 & -1 \\ -x_3 & 2x_2 & -x_1 \end{pmatrix},$$

and so,

$$F'(\hat{x}) = \begin{pmatrix} 1 & 1 & -1 \\ -1 & 2 & -1 \end{pmatrix}.$$

Therefore, the solution set of $\ker F'(\hat{x})(x - \hat{x}) = 0_2$ is the set of solutions of the system of linear equations

$$(x_1 - 1) + (x_2 - 1) - (x_3 - 1) = 0,$$

$$-(x_1 - 1) + 2(x_2 - 1) - (x_3 - 1) = 0$$

("intersection of two planes"). Now we go over to a parameter representation, as this is a more convenient description for lines in space. Solving this system shows that the solution set is the line through the point $(1,1,1)^T$ with direction vector $(1,2,3)^T$. Therefore, the required tangent line has the parametric representation

$$(1,1,1)^T + \mu(1,2,3)^T = (1+\mu, 1+2\mu, 1+3\mu)^T,$$

where the parameter μ runs over the real numbers $\mathbb{R}$.

Conclusion. The tangent space theorem allows us to compute a linear approximation of the solution set of a system of nonlinear equations at a point. This linear approximation is called the tangent space. In particular, the tangent space theorem allows us to compute tangent lines to curves in the plane and in space, and also to compute tangent planes to surfaces.

3.4.3 Tangent space theorem: proof using Weierstrass and Fermat

Royal road to the proof. We recall that it took one hundred years after the discovery by Lagrange of the multiplier rule before a proof was found, and in particular, the mystery involving the "counter-examples" was solved by the introduction of the multiplier λ_0. The obstacle that took so long to overcome is essentially the task of proving the tangent space theorem (or, equivalently, the inverse function theorem or the implicit function theorem). The geometrical ideas of the various proofs that are now available for results such as the inverse function theorem are explained in section 1.5.1.

Now we are about to present a proof of the tangent space theorem. Given the fact that it took so long to prove this result, you might expect that the proof is rather complicated. In fact, it is now possible to give a reasonably straightforward proof, using the four-step method. It is true that the proof involves some technical details—some estimates that follow from the triangle inequality. However, the course of the proof is based on a straightforward geometric intuition.

In view of this, it is easier to follow the course of the proof below when you keep a concrete numerical example of a plane curve ($n = 2, m = 1$) in mind, maybe example 3.2.3. Then you can specify the proof to this numerical example, and by drawing some simple geometric figures, you will follow a "royal road" to a full understanding.

Plan of the proof. Choose a sufficiently small number $\varepsilon > 0$ and then an arbitrary solution ξ of (**) with $|\xi - \hat{x}| \leq \varepsilon$. Solve the following auxiliary problem with the four-step method (cf. Fig. 3.10):

$$|F(x)| \to \min, \quad x - \xi \text{ orthogonal to solution set } (**), \quad |x - \xi| \leq \varepsilon.$$

It turns out that this problem has a unique solution $x(\xi)$; moreover, $F(x(\xi)) = 0_m$ and $x(\xi) = o(\xi - \hat{x})$. This proves the conclusion of the tangent space theorem.

To begin with, we display two facts that will be used in the proof.

Matrix fact. The vectors that are orthogonal to the solution set of (**) are precisely the vectors of the form $F'(\hat{x})^T y$.

The precise formulation of the underlying matrix fact is as follows. Let A be an $m \times n$ matrix of rank m. Then the vectors ξ and $A^T y$ are orthogonal for each $\xi \in \mathbb{R}^n$ with $A\xi = 0_m$ and $y \in \mathbb{R}^m$. Moreover, each vector $x \in \mathbb{R}^n$ can be written in a unique way as

$$x = \xi + A^T y,$$

for all $\xi \in \mathbb{R}^n$ with $A\xi = 0_m$ and all $y \in \mathbb{R}^m$.

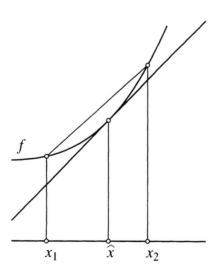

Figure 3.11 Illustrating the definition of strict differentiability.

LAGRANGE: EQUALITY CONSTRAINTS

Strict differentiability. The following inequality holds on a sufficiently small neighborhood U of $\hat{x}$:

$$|F(x_1) - F(x_2) - F'(\hat{x})(x_1 - x_2)| \leq 2s(\max(|x_1 - \hat{x}|, |x_2 - \hat{x}|))|x_1 - x_2| \quad (\dagger)$$

for all $x_1, x_2 \in U$, where

$$s(r) = \max_{|x - \hat{x}| \leq r} |F'(x) - F'(\hat{x})|$$

for all $r \geq 0$.

To verify this, one only has to choose U such that F is differentiable on U. We will not display the routine verification. This property of *strict differentiability* is illustrated in Figure 3.11.

Now we are ready to present the proof of the tangent space theorem.

Proof. We choose $\varepsilon > 0$ sufficiently small.

Moreover, we choose $c > 0$ such that

$$|F'(\hat{x})F'(\hat{x})^T y| \geq c|F'(\hat{x})^T y|, \quad (\ddagger)$$

as we may by corollary 2.10.

Take an arbitrary solution ξ of the linearized equation $(**)$ for which $|\xi - \hat{x}| \leq \varepsilon$.

- $F(x) = 0_m$ has a unique solution $x(\xi)$ for which $|x(\xi) - \xi| < \varepsilon$ and for which, moreover,

$$x(\xi) - \xi = F'(\hat{x})^T y(\xi)$$

for a—necessarily unique—$y(\xi) \in \mathbb{R}^m$.

1. Consider the problem

$$g(y) = |F(v(y))| \to \min, \ y \in \mathbb{R}^m, \ |v(y) - \xi| \leq \varepsilon.$$

   where $v(y) = \xi + F'(\hat{x})^T y$. Existence of a solution $\hat{y} = y(\xi)$ follows from Weierstrass.

2. Fermat: $g'(y) = 0_m^T \Rightarrow$

$$|F(v(y))|^{-1} F(v(y))^T F'(v(y)) F'(\hat{x})^T = 0_m^T,$$

   using the differential calculus (to be more specific, using that the derivative of $y \to |y|$ is $|y|^{-1} y^T$, and using the chain rule twice).

3. – **No stationary points.** The $m \times m$ matrix

$$F'(v(y))F'(\widehat{x})^T$$

is approximately equal to the—invertible—matrix $F'(\widehat{x})F'(\widehat{x})^T$ by the continuity of F' and the closeness of the points $v(y)$ and $\widehat{x}$ (this closeness follows from "$|\xi - \widehat{x}|$ is sufficiently small" and "$|v(y) - \xi| \leq \varepsilon$"). Therefore this matrix is invertible as well, and so the stationarity equations above can be simplified to

$$F(\xi + F'(\widehat{x})^T y) = 0_m.$$

However, g is not differentiable at solutions of this equation. Therefore there are no stationary points.

– **No boundary minima.** Let $\bar{y}$ be on the boundary of the admissible set, that is, $|F'(0_n)^T \bar{y}| = \varepsilon$. We will now show that $g(\bar{y}) > g(0_m)$, and so $\bar{y}$ is not a point of global minimum.

Applying (†) with $x_1 = \xi$ and $x_2 = \widehat{x}$, and using the triangle inequality, gives

$$g(0_n) \leq 2s(\varepsilon)|\xi - \widehat{x}|.$$

Applying (†) with $x_1 = v(y)$ and $x_2 = \xi$, and using the triangle inequality, gives the following lower estimate for $g(\bar{y})$,

$$|F'(\widehat{x})F'(\widehat{x})^T \bar{y}| - g(0_m) - 2s(|\xi + F'(\widehat{x})^T \bar{y}|)|F'(\widehat{x})^T \bar{y}|.$$

Therefore,

$$g(\bar{y}) - g(0_m) \geq c|\xi| - 4s(|\xi|)|\xi| - 2s(\sqrt{2}|\xi|)|\xi|.$$

This gives the promised inequality:

$$g(\bar{y}) > g(0_m) \text{ if } |\xi - \widehat{x}| \text{ is for sufficiently small,}$$

using that

$$\lim_{r \to 0} s(r) = 0$$

(this is precisely the condition $F \in C^1(0_n, \mathbb{R}^m)$, which holds by assumption).

– **Unique solution at point of nondifferentiability.**
We now prove that there is a unique solution of the problem and that this is a point of nondifferentiability. Given what we proved already, it suffices to verify that there can be at most one admissible point of nondifferentiability. Let y_i, $i = 1, 2$ be two such points. Then these point satisfy the equation $g(y) = 0$: indeed, g is differentiable at all points for which $g(y) \neq 0$, clearly. Applying (†) to $x_i = v(y_i)$, $i = 1, 2$, we get

$$|F'(\widehat{x})F'(\widehat{x})^T(y_1 - y_2)| \leq 2s(\sqrt{2}\varepsilon)|F'(\widehat{x})(y_1 - y_2)|,$$

and so, using (‡),

$$|F'(\widehat{x})F'(\widehat{x})^T(y_1 - y_2)|(2c^{-1}s(\sqrt{2}\varepsilon) - 1) \geq 0.$$

The second factor on the left-hand side is negative, as ε is sufficiently small, and therefore the first factor is zero. Hence, as the matrix $F'(\widehat{x})F'(\widehat{x})^T$ is invertible, it follows that $y_1 = y_2$.

4. For each solution ξ of (∗∗) with $|\xi - \widehat{x}| \leq \varepsilon$, there is a unique solution $x(\xi)$ of (∗) for which

$$(x(\xi) - \xi) = F'(0_n)^T y(\xi)$$

for a—necessarily unique—$y(\xi) \in \mathbb{R}^m$ and for which, moreover, $|x(\xi) - \xi| < \varepsilon$.

- $y(\xi) = o(\xi - \widehat{x})$. Applying (†) to $x_1 = \xi$ and $x_2 = \xi + F'(\widehat{x})^T y(\xi)$, and using the triangle inequality, we get

$$|F'(\widehat{x})F'(\widehat{x})^T y(\xi)| \leq 2s(\sqrt{2}\varepsilon)|F'(\widehat{x})^T y(\xi)| + |F(\xi)|.$$

This leads to

$$|F'(\widehat{x})F'(\widehat{x})^T y(\xi)|(1 - 2c^{-1}s(\sqrt{2}\varepsilon)) \leq 2s(\xi - \widehat{x})|\xi - \widehat{x}|.$$

This implies, as the matrix $F'(\widehat{x})F'(\widehat{x})^T$ is invertible, that $y(\xi) = o(\xi - \widehat{x})$, as required.

□

Remark. One advantage of the proof above is that it uses only our two standard tools, the Weierstrass theorem and the Fermat theorem. Moreover, it is shows clearly why the tangent space theorem holds, being based on a very simple geometric intuition. There is a slightly different proof, using the technique of Cauchy sequences and the modified Newton algorithm (or, equivalently, the contraction principle). This is the usual way to prove the tangent space theorem (or one of its variants such as the implicit function theorem). This resulting analytical proof is in fact shorter and moreover, it is an algorithm, as it does not require the Weierstrass theorem, which is an existence result. It is a matter of taste which proof is preferable.

Conclusion. The tangent space theorem has been proved by solving a suitable auxiliary optimization problem with the four-step method.

3.4.4 Proof of the multiplier rule

Finally, we prove the multiplier rule, using the tangent space theorem.

Proof. We stack the functions $f_1, \ldots, f_m : U_n(\widehat{x}, \varepsilon) \to \mathbb{R}$ into a column, and write

$$F = (f_1, \ldots, f_m)^T : U_n(\widehat{x}, \varepsilon) \to \mathbb{R}^m.$$

To simplify the formulas we assume that $\widehat{x} = 0_n$, as we may, without restricting the generality of the argument. Then $F(0_n) = 0_m$ and 0_n is a point of local minimum for the given problem.

The nonregular case. We consider the nonregular case: the rows of $F'(0_n)$ are linearly dependent. We choose $\lambda_0 = 0$ and a nontrivial linear relation between the rows of $F'(0_n)$, say,

$$\sum_{i=1}^{m} \lambda_i f_i'(\widehat{x}) = 0_n^T.$$

As $\lambda_0 = 0$, it follows that it makes no difference if we start the summation with $i = 0$. Therefore,

$$\sum_{i=0}^{m} \lambda_i f_i'(\widehat{x}) = 0_n^T,$$

as required.

The regular case. We consider the regular case: the rows of the $m \times n$ matrix $F'(0_n)$ are linearly independent. Choose an arbitrary element v of $\mathbb{R}^n$ for which $F'(0_n)v = 0_m$. We apply the tangent space

LAGRANGE: EQUALITY CONSTRAINTS 181

theorem. It gives that $tv + r(tv)$ is a solution of $F(x) = 0_m$ for $|t|$ sufficiently small, and that $r(tv) = o(t)$. Combining this with the local minimality of 0_n gives

$$0 \le f_0(tv + r(tv)) - f_0(0_n) = t f_0'(0_n) \cdot v + o(t).$$

It follows that $f_0'(0_n) \cdot v = 0$. As v is an arbitrary element of $\mathbb{R}^n$ with $F'(0_n)v = 0_m$, it follows that $f_0'(0_n)$ is a linear combination of the rows of $F'(0_n)$, by the theory of linear equations (corollary 2.11). This establishes the Lagrange equation with $\lambda_0 = 1$, as the rows of the $m \times n$ matrix $F'(0_n)$ are $f_1'(0_n), \dots, f_m'(0_n)$. □

Crash courses theory. This is a good moment to look up in appendices E and F, the crash courses on optimization theory, the parts concerning this chapter.

Proof for advanced readers. Advanced readers and experts might prefer the proof of the multiplier rule given in appendix G, 'Conditions of Extremum from Fermat to Pontryagin'.

Conclusion. The Lagrange multiplier rule follows readily from the tangent space theorem.

3.5 DISCUSSION AND COMMENTS

3.5.1 The secret of Lagrange's elimination method

We emphasize again—as in section 3.2.1—that the secret of Lagrange's elimination method is that by differentiating first and eliminating afterward, the hard task—elimination—is reduced from a nonlinear problem to a linear one. Now we will illustrate this with the numerical example 3.2.4. We will not use multipliers.

Fermat method
1. Eliminate x_2 from the constraint $x_1^6 + x_2^6 = a^6$, which is nonlinear. This leads to the following nonlinear expression for x_2 in terms of x_1:

$$x_2(x_1) = (a^6 - x_1^6)^{1/6}.$$

2. Differentiate

$$f(x_1, x_2(x_1)) = x_1(a^6 - x_1^6)^{1/6}.$$

This leads to the following formula:

$$(a^6 - x_1^6)^{1/6} + x_1(1/6)(a^6 - x_1^6)^{-5/6}(-6x_1^5).$$

3. Find stationary points: put the result of step 2 equal to zero, and solve the resulting equation. This leads to the following solutions $x_1 = \pm a/\sqrt[6]{2}$. Finally, taking into account the inequality constraint $x_1 > 0$, we are led to the solution $x_1 = x_2 = a/\sqrt[6]{2}$.

Lagrange method

1. Differentiate the constraint $x_1^6 + x_2(x_1)^6 = a^6$ and the objective function $f(x_1, x_2(x_1)) = x_1 x_2(x_1)$. This leads to the following *linear* equation in the derivative $\frac{dx_2}{dx_1}$,

$$6x_1^5 + 6x_2(x_1)^5 \frac{dx_2}{dx_1} = 0, \qquad (*)$$

and to the formula,

$$\frac{d}{dx_1} f(x_1, x_2(x_1)) = x_2(x_1) + x_1 \frac{dx_2}{dx_1}. \qquad (**)$$

2. Eliminate $\frac{dx_2}{dx_1}$ from the linear equation $(*)$. This leads to the following expression for $\frac{dx_2}{dx_1}$ in terms of x_1, x_2,

$$\frac{dx_2}{dx_1} = -x_1^5 x_2^{-5}.$$

3. Find stationary points: put $(**)$ equal to zero, substitute the result of step 2, and solve the resulting equation, in combination with the constraints.

We display the simple calculations:

$$x_2 + x_1(-x_1^5 x_2^{-5}) = 0 \Rightarrow x_2 = \pm x_1.$$

Using the constraints, we are led to the solution

$$x_1 = x_2 = a/\sqrt[6]{2}.$$

Conclusion. If you carry out the two tasks—elimination and differentiation—for a numerical example of an equality constrained problem, then you see that it is more convenient to differentiate first and to eliminate afterwards.

3.5.2 Lagrange's method with and without multipliers

Plan. We have already made the point that the role of the multipliers in Lagrange's elimination method is not an essential one. Now we will show how the description of Lagrange's method without multipliers, which was illustrated in section 3.5.1, can be turned into the usual description, which contains multipliers. For the sake of simplicity, we will only consider the following special case of problem ($P_{3.1}$),

$$f_0(x_1, x_2) \to \min, \quad f_1(x_1, x_2) = 0.$$

However, this time, we will do the differentiation step in terms of infinitesimal ($\approx$ "infinitely small") increments.

Lagrange's method without multipliers. We write down the stationarity conditions, to begin with. Let (x_1, x_2) be a solution of the problem. Each infinitesimal increment (dx_1, dx_2) that is admissible—that is, for which the admissibility equation

$$f_1(x_1 + dx_1, x_2 + dx_2) = 0$$

holds—gives a "negligible" increment of the objective function; that is, the following stationarity condition holds true:

$$f_0(x_1 + dx_1, x_2 + dx_2) = f_0(x_1, x_2).$$

Differentiation of the admissibility equation for $(x_1 + dx_1, x_2 + dx_2)$,

$$f(x_1 + dx_1, x_2 + dx_2) = 0,$$

leads to the following linear equation in dx_1 and dx_2:

$$\frac{\partial f_1}{\partial x_1} dx_1 + \frac{\partial f_1}{\partial x_2} dx_2 = 0, \tag{1}$$

and differentiation of the stationarity condition above leads to another linear equation in dx_1 and dx_2:

$$\frac{\partial f_0}{\partial x_1} dx_1 + \frac{\partial f_0}{\partial x_2} dx_2 = 0, \tag{2}$$

This shows that the following implication holds true for arbitrary infinitesimal increments dx_1, dx_2: equation (1) implies equation (2). As these equations are both linear in dx_1 and dx_2, this implication holds for arbitrary numbers dx_1, dx_2, not only for infinitesimals. This completes the description of the Lagrange method without the use of multipliers.

Enter the multipliers. Now we are ready to introduce the multipliers. By the theory of linear equations, the implication above can be reformulated as follows: equation (2) is a scalar multiple of equation (1). This gives

$$\frac{\partial f_1}{\partial x_i} = \mu \frac{\partial f_0}{\partial x_i}, \ i = 1, 2.$$

These are precisely the Lagrange equations, with $\lambda_0 = 1$ and $\lambda_1 = -\mu$.

Conclusion. The point of introducing multipliers is that this simplifies the necessary conditions.

3.5.3 Interpretations of the multiplier rule

Plan. We are going to focus on the multipliers themselves. This section gives additional insight; it is not used for solving concrete problems. The insight is that the Lagrange multipliers, which play an auxiliary role in the solution of concrete problems, are of independent interest as well, because of an interpretation as shadow prices. This section is a more precise version of the sketch in the introduction of this chapter. We let notation be as in theorem 3.3.

Formula for the Lagrange multipliers. We consider the regular case $\mathrm{rank} F'(\widehat{x}) = m$; assume the Lagrange equations hold with Lagrange multiplier

$$\lambda = (1, \lambda_1, \ldots, \lambda_m) = (1, \bar{\lambda}).$$

The following formula holds true:

$$\bar{\lambda} = -f_0'(\widehat{x})(F'(\widehat{x}))^T (F'(\widehat{x})(F'(\widehat{x}))^T)^{-1}. \qquad (*)$$

We display the derivation of this formula: we have

$$\mathcal{L}_x(\widehat{x}, \lambda) = 0_n^T \Leftrightarrow f_0'(\widehat{x}) + \bar{\lambda} F'(\widehat{x}) = 0_n^T.$$

Now we multiply from the right by the matrix $F'(\widehat{x})^T$:

$$f_0'(\widehat{x})(F'(\widehat{x}))^T = -\bar{\lambda}(F'(\widehat{x})(F'(\widehat{x}))^T) \Rightarrow$$

$$\bar{\lambda} = -f_0'(\widehat{x})(F'(\widehat{x}))^T (F'(\widehat{x})(F'(\widehat{x}))^T)^{-1}.$$

Interpretation of the multipliers (informal version). Let $y \in \mathbb{R}^m$ be close to 0_m. For each x close to $\widehat{x}$ that is admissible for the problem

$$f_0(x) \to \min, \ f_1(x) = y_1, \ldots, f_m(x) = y_m, \qquad (P_y)$$

one has
$$f_0(x) = f_0(\widehat{x}) - \lambda_1 y_1 - \cdots - \lambda_m y_m.$$
This follows from the following chains of—approximate—equalities, writing $h = x - \widehat{x}$, that is, $x = \widehat{x} + h$:
$$f_0(x) - f_0(\widehat{x}) \approx f_0'(\widehat{x})h = -\lambda_1 f_1'(\widehat{x})h - \cdots - \lambda_m f_m'(\widehat{x})h$$
and
$$f_i'(\widehat{x})h \approx f_i(x) = y_i \text{ for all } i \in \{1, \ldots, m\},$$
where we have used $f_i(\widehat{x}) = 0$ for all $i \in \{1, \ldots, m\}$ to establish the second chain.

Multipliers as a measure of the sensitivity of the optimal value. Consider the optimal value of the problem (P_y), that is, the value of f_0 at a point of global minimum of (P_y). This optimal value will in general depend on the choice of y, that is, it is a function of y. We denote this optimal value by $S(y)$. Using the approximation above, one can "derive" that "usually" the partial derivative $\frac{\partial S}{\partial y_i}(0)$ of S with respect to y_i in $y = 0_n$ exists and is equal to minus the Lagrange multiplier λ_i for $1 \leq i \leq m$, that is,
$$S'(0) = -(\lambda_1, \ldots, \lambda_m).$$
Thus we get the following interpretation.

The Lagrange multiplier λ_i is a measure for how sensitive the optimal value of the problem $(P_{3.1})$ is to small changes in the right-hand side of the constraint $f_i(x) = 0$.

Multipliers as shadow prices. Let the problem be one of minimizing costs. Which price are you prepared to pay to prevent a small change of the constraint $f_i(x) = 0$ into $f_i(x) = y_i$ for some $y_i \in \mathbb{R}$? Well, then the optimal value changes from $f_0(\widehat{x})$ to $f_0(\widehat{x}) + \lambda_i y_i$. Therefore, you are prepared to pay a price slightly less than λ_i. As this price is not determined openly, it is called a *shadow price*.

Economic interpretation of the multiplier rule. Furthermore, we give the economic interpretation of the multiplier rule. If you are allowed to choose the x freely provided you pay a fine λ_i for each unit that $f_i(x)$ exceeds zero, then the cost of choosing x is
$$f_0(x) + \lambda_1 f_1(x) + \cdots + \lambda_m f_m(x).$$

Here we use the convention that a negative fine means a reward. This is precisely the Lagrange function with $\lambda_0 = 1$. Therefore, by the multiplier rule this cost optimization problem has the same solution as the original problem with constraints. This can be viewed in the following way.

By the introduction of the right fines and rewards for not observing the constraints, the cost minimizer will be led to observe the constraints by his own desire to minimize costs, rather than being forced to it by some sort of law.

Interpretation of multipliers (formal version). Finally, we give a *precise* version of the intuitive results described in this section. This extends the envelope theorems of the previous chapters. We will use the *terminology* of first and second order conditions; we will not yet use the sufficiency of these conditions (this is proved in chapter five).

Theorem 3.7 *Envelope theorem for problems with equality constraints.* Let $\widehat{x}$ be a solution of $(P_{3.1})$, assume $\operatorname{rank} F'(\widehat{x}) = m$, and assume that the first and second order sufficient conditions hold. (To be explicit, f_i is twice continuously differentiable and $f_i(\widehat{x}) = 0$, for $1 \leq i \leq m$ and there exists $\widehat{\lambda} \in \mathbb{R}^m$ such that

$$\mathcal{L}_x(\widehat{x}, 1, \widehat{\lambda}) = 0_n^T, \text{ and } h^T \mathcal{L}_{xx}(\widehat{x}, 1, \widehat{\lambda}) h > 0$$

for all nonzero vector $h \in \ker F'(\widehat{x})$.)

We choose open balls $U = U_n(\widehat{x}, R)$ and $V = U_m(0_m, S)$ such that for all $y \in V$ the perturbed problem

$$f_0(x) \to \min, \ x \in U, \ f_i(x) = y_i, \ 1 \leq i \leq m, \qquad (P_y)$$

has a unique point of minimum $x(y)$, as we may.

Then the vector function $x(\cdot)$ on V is continuously differentiable and $x(0_m) = \widehat{x}$. Moreover, the Lagrange equations hold for the problem (P_y) and the point $x(y)$ with the choice of Lagrange multipliers $(1, \lambda(y))$, where

$$\lambda(y) = -\frac{d}{dy} f_0(x(y)).$$

This theorem can be derived, after local elimination of the equality constraints, from the envelope theorem 2.13.

LAGRANGE: EQUALITY CONSTRAINTS

Usual illustration of the multiplier rule. Figure 3.12 gives the usual illustration of the multiplier rule for the case of two variables and one constraint: the level curve of the objective function f_0 at the optimal point $\hat{x}$ is tangent to the curve determined by the equality constraint $f_1(x) = 0$. That this is indeed the geometric content of the

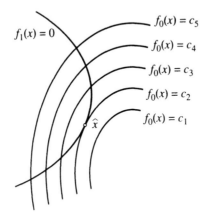

Figure 3.12 Illustration of the multiplier method using level curves.

multiplier rule in the present case can be seen as follows. To this end, we have to give the formal definition of the concepts tangent vector and normal vector.

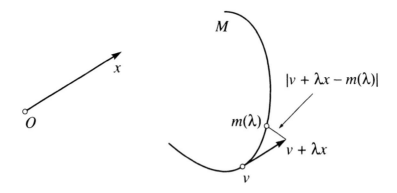

Figure 3.13 The definition of tangent space.

Tangent vector. Let a subset M of $\mathbb{R}^n$ and a point $v \in M$ be given. Then a *tangent vector* to M at v is a vector $x \in \mathbb{R}^n$ for which

$v + \lambda x + o(\lambda) \in \mathcal{M}$. Written out more fully, this condition means that there exists a vector-valued function $\lambda \to m(\lambda)$ from $[-1, 1]$ to $\mathcal{M}$ for which

$$v + \lambda x - m(\lambda) = o(\lambda)$$

This definition is illustrated in Figure 3.13.

The difference between the concepts "tangent line to a curve" and the concept "tangent vector" is illustrated in Figure 3.14.

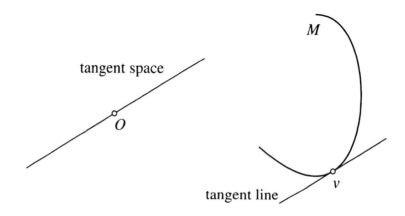

Figure 3.14 Tangent line and tangent space.

Tangent space. Let a subset $\mathcal{M}$ of $\mathbb{R}^n$ and a point $v \in \mathcal{M}$ be given. Then *the tangent space* to $\mathcal{M}$ at v is the set of all vectors $v + x$ where x is a tangent vector to $\mathcal{M}$ at v.

Normal vector. A *normal vector* to $\mathcal{M}$ at v is a vector in $\mathbb{R}^n$ that is orthogonal to all tangent vectors at $\mathcal{M}$ at v.

Usual geometrical illustration of the multiplier rule. Now we can give the promised relation between the Lagrange multiplier rule and Figure 3.12. The tangent space theorem shows that $f'_0(\widehat{x})^T$ (resp. $f'_1(\widehat{x})^T$) is a normal to the curved line with equation

$$f_0(x) = f_0(\widehat{x}) (\text{ resp. } f_1(x) = 0).$$

Therefore, the property that these two curves are tangent at the point $\widehat{x}$ corresponds to the property that these normal vectors are scalar

LAGRANGE: EQUALITY CONSTRAINTS

multiples of each other: this is precisely the multiplier rule.

Conclusion. We have focussed attention on the multipliers themselves. To begin with, we have given a formula for them. Then we have given an interpretation as shadow prices. This is an extension of the envelope theorem to problems with equality constraints. Finally, we have given the usual geometrical illustration of the multiplier rule.

3.5.4 Refined tangent space theorem

The tangent space theorem might leave a taste for more: why only a *linear* approximation? The following refinement gives quadratic and even higher order approximations. This result can be used for writing down the second and higher order necessary conditions (cf. section 5.4).

Theorem 3.8 Refined tangent space theorem. *Let $\widehat{x} \in \mathbb{R}^n$, U a neighborhood of $\widehat{x}$, and $F = (f_1, \ldots, f_m)^T : U \to \mathbb{R}^m$ a vector function that is r-times continuously differentiable at $\widehat{x}$, and for which $\mathrm{rank} F'(\widehat{x}) = m$ and $F(\widehat{x}) = 0_m$. Then, one can give for the set of solutions of $F(x) = 0_m$ that lie close to $\widehat{x}$ the following parametric description:*

$$x(v) = \widehat{x} + v + p(v) + r(v),$$

where the parameter v is a vector that runs over those solutions of $F'(\widehat{x})x = 0_m$ that lie close to $\widehat{x}$, and where $p(v)$ and $r(v)$ are characterized as follows: they are orthogonal to the solution space of

$$F'(\widehat{x})x = 0_m,$$

$p(v)$ depends polynomially on v of degree r with zero constant term and zero linear term, and $r(v) = o(v)$.

Remark. The polynomials $p(v)$ can in each case be calculated explicitly by solving the systems of linear equations to which their definition leads. The set of vectors

$$\widehat{x} + v + p(v),$$

where v runs over the solution set of $F'(\widehat{x})x = 0_m$, is called the r-th order approximation of the solution set of $F(x) = 0_m$ at the point $\widehat{x}$.

We illustrate this result and the calculation of $p(v)$ with the following numerical example.

Example 3.5.1 *Calculate the quadratic approximation to the curve $x_1^2 - x_2^3 - 1 = 0$ at the point $\widehat{x} = (3, 2)^T$.*

Solution. Then $F'(\widehat{x}) = (6, -12)$. Therefore, the solution space of $F'(\widehat{x})x = 0_m$ is the set of vectors $(2t, t)^T$, $t \in \mathbb{R}$, and its orthogonal complement is the set of vectors $(s, -2s)^T$, $s \in \mathbb{R}$. To find the quadratic approximation, we have to solve the following equation in the variable c:

$$F(3 + 2t + ct^2, 2 + t - 2ct^2) = o(t^2)$$

for all $t \in \mathbb{R}$.

This gives,

$$(3 + 2t + ct^2)^2 - (2 + t - 2ct^2)^3 - 1 = (4t^2 + 6ct^2) - (6t^2 - 8ct^2)$$

$$= (14c - 2)t^2 = o(t^2),$$

and hence $c = 1/7$. Therefore, the quadratic approximation of

$$x_1^2 - x_2^3 - 1 = 0$$

at the point $\widehat{x} = (3, 2)^T$ has the following parametric description

$$x(t) = (3 + 2t + \frac{1}{7}t^2, 2 + t - \frac{2}{7}t^2)^T, \ t \in \mathbb{R}.$$

3.6 EXERCISES

We begin with five numerical examples.

Exercise 3.6.1 $e^{x_1 x_2} \to \max$, $x_1^3 + x_2^3 = 1$.

Exercise 3.6.2 $x_1^2 + 12x_1x_2 + 2x_2^2 \to$ extr, $4x_1^2 + x_2^2 = 25$.

Exercise 3.6.3 $x_1 x_2^2 x_3^3 \to$ extr, $x_1^2 + x_2^2 + x_3^2 = 1$.

Exercise 3.6.4 *Find the extrema of the function of two variables $x_3 = f(x_1, x_2)$ that is implicitly defined by the following relation*

$$F(x_1, x_2, x_3) = x_1^2 + x_2^2 + x_3^2 - x_1x_3 - x_2x_3 + 2x_1 + 2x_2 + 2x_3 - 2 = 0.$$

At the end of this section we give more exercises of this type: exercises 3.6.18–3.6.21.

If you want a challenge, you can try the following intriguing exercise; at first sight it might look impossible to find the solution. The name of the game is to discover what information the Lagrange equations

LAGRANGE: EQUALITY CONSTRAINTS

give you here. In particular, we emphasize again that it is usually not necessary to compute the Lagrange multipliers: their role is "to relate the variables at the optimum."

Exercise 3.6.5 * $\sum_{i=1}^{5} x_i^4 \to$ extr, $\sum_{i=1}^{5} x_i = 0$, $\sum_{i=1}^{5} x_i^3 = 0$, $\sum_{i=1}^{5} x_i^2 = 4$.

Some more exercises of this type are given at the end of this section: 3.6.22–3.6.25. There we give a more explicit hint on how to solve such problems.

Basic problem of the consumer. Let us consider an economic example of a constrained problem. It concerns a consumer who chooses how much of his income to spend on one good and how much to leave over for expenditures on other goods. The individual preferences of the consumer can be modeled by a so-called utility function, which measures the pleasure his choice brings him. In the following exercise, a numerical example is given of the basic problem of consumer theory.

Exercise 3.6.6 *Playing games or eating ice creams. A family on holiday go for a day with their son to the churches and museums of Florence. The son gets 5 euros to spend in a game hall or on ice cream. Playing one game costs 50 euro cents and an ice cream costs 1 euro. His pleasure in playing x games and eating y ice creams is given by the following utility function*

$$U(x, y) = 3 \ln x + 2 \ln y$$

(you may replace this utility function by

$$\tilde{U}(x, y) = x^3 y^2$$

if you want to avoid logarithms).
How many games and ice creams will give him the greatest pleasure?

Exercise 3.6.7 *Consider the problem from the previous exercise. Determine how changes in the budget and in the prices of games and ice cream affect the optimal way to divide the budget over games and ice cream.*

Sometimes, we have to allow that the budget is not spent entirely. A numerical example of such a consumer problem is given in exercise 4.6.10; to solve it, the methods of chapter four are needed.

Basic problem of the producer. We can also look from the point of view of a producer, rather than that of a consumer. Suppose a producer chooses, among the possible ways of producing a given output, the one that minimizes her cost. These possibilities are modeled by a so-called production function. In the following exercise a numerical example is given of the basic problem of producer theory.

Exercise 3.6.8 *How much to produce. Assume that a firm can produce 90 units of output using 9 units of input X and 9 units of input Y. However, there is no need to use equal units of input X and Y. In fact the firm's technological possibilities can be represented by the production function*

$$Q = 10\sqrt{X}\sqrt{Y}.$$

1. *If the price of X is 8 dollars and the price of Y is 16 dollars, is the input of 9 units of X and 9 units of Y the most efficient way to produce 90 units of output?*

2. *What must be the ratio of input prices for this input combination to be efficient?*

3. *What is the least cost way to produce 400 units of output if the price of X is 1 dollar and the price of Y is 2 dollars?*

In praise of trade. We have given the basic problems of consumer and producer theory. Often, producers are also consumers. Then trade provides an essential incentive. If there were no trade, then a producer, like a farmer, would concentrate on producing food to satisfy the needs of his own family in an optimal way rather than on being as productive as possible.

The following example illustrates that both society and producers benefit from trade: a producer-consumer who tries to maximize her utility from consumption will automatically maximize her production—for the benefit of society—and she will achieve a higher level of utility than she would have achieved without the possibility of trade.

Exercise 3.6.9 **Net demander or net supplier. Assume that a worker has exactly 100 hours to allocate between wallpapering and painting. His output of wall-papering X and painting Y depends only on the hours of labor he spends, in the following way:*

$$X = \sqrt{L_X}, \quad Y = \sqrt{L_Y}.$$

The root function is chosen to capture the fact that when doing the same job for a long time the worker gets bored and slows down.

1. **Profit maximization.** *If the worker is offered fixed prices $P_X = 10$ and $P_Y = 5$, how should he allocate his time between wallpapering and painting, to maximize his profits?*

2. **Utility maximization.** *Now assume further that he decides instead to work 100 hours in his own house, wallpapering and painting, and that he has the following utility function:*

$$U = 10\sqrt{X}\sqrt{Y}.$$

    How should he allocate his time between wallpapering and painting to maximize his utility?

3. **Producing, trading, and consuming.** *If he can also let someone else wallpaper and paint in his house at the same fixed prices $P_X = 10$ and $P_Y = 5$, then he has the opportunity to spend part of the time that he allocates to wallpapering, working for someone else and paying someone else to paint in his house.*

    That is, he trades wallpapering for painting and becomes a net supplier of wallpapering and a net demander of painting. He can also do the converse.

    What should he do to maximize his utility? Is he a net demander or a net supplier of each of the two "goods" wallpapering and painting?

4. *Compare the outcomes of (i), (ii), and (iii) and show that they illustrate the following phenomena. Profit maximization leads to larger production (measured in money) than utility maximization without trade. Moreover, utility maximization with trade leads to the same allocation as profit maximization and to higher utility than utility maximization without trade.*

Now we consider the problem of distributing fixed endowments of two goods among two consumers in the best possible way for both of them; that is, the *social welfare*, to be defined in a suitable way, has to be maximized.

Exercise 3.6.10 Social welfare. *Someone is about to take his 22 X-Men comic books on a holiday. As he wants a bit of variation in his reading, he goes to visit a friend, who has 13 Spider-Man comic books and proposes to redistribute their comic books between them. Then each of them will end up with an interesting mix for his holiday.*

Their aim is to distribute the collection of 35 books in such a way that the social welfare is maximal. We take the social welfare here to be the sum of their utilities. Let the utility of the Spider-Man (resp. X-Men) fan be $4\ln(s) + \ln(x)$ (resp. $5\ln(s) + 20\ln(x)$) if he has s Spider-Man comic books and x X-Men comic books.

1. *Which distribution of the comic books gives maximal social welfare?*

   Comments on 1. Both like their own books four times as much as the other's. However, the X-Men fan likes reading in general five times as much as the Spider-Man fan. As a result, the X-Men fan has even more pleasure reading a Spider-Man comic book than the Spider-Man fan has.

   At first sight, this might suggest that you achieve maximal social welfare by giving all 35 books to the X-Men fan. However, this is not the case, because of the logarithmic utility functions U, which are monotonic increasing ($U' > 0$) with decreasing speed ($U'' < 0$) and so U is strictly concave. This concavity models that the pleasure from the second comic book is not as great as from the first one, and so on.

2. *Comment on the fairness of the outcome obtained in 1.*

3. *Estimate how much this maximal social welfare would increase if the Super-Man fan suddenly finds one more comic book in his room.*

Exercise 3.6.11 *Holding and ordering cost.* A firm's inventory of a certain homogeneous commodity is depleted at a constant rate per unit time, and the firm reorders an amount x of the commodity, which is delivered immediately, whenever the level of inventory is zero. The annual requirement for the commodity is A, and the firm orders the commodity n times each year, so

$$A = nx.$$

The firm incurs two types of inventory costs: a holding cost and an ordering cost. The average stock of inventory is $x/2$, and the cost of holding one unit of the commodity is C_h, so $C_h x/2$ is the holding cost.

The firm orders the commodity as stated above, n times a year, and the cost of placing one order is C_0, so $C_0 n$ is the ordering cost. The total cost is then

$$C = C_h x/2 + C_0 n.$$

LAGRANGE: EQUALITY CONSTRAINTS

1. Show in a picture how the inventory level varies over time.

2. Minimize the cost of inventory, C, by choice of x and n subject to the constraint $A = nx$. Find the optimal x as a function of the parameters C_0, C_h and A. Interpret the Lagrange multiplier.

Here is a warning against a popular pitfall.

Exercise 3.6.12 *The following text taken from a book on mathematics for management contains grave errors. Sort them out.*

"Consider the general problem of finding the extreme points of $z = f(x,y)$ subject to the constraint $g(x,y) = 0$. Clearly the extreme points must satisfy the pair of equations

$$f_x(x,y) = 0, \ f_y(x,y) = 0$$

in addition to the constraint $g(x,y) = 0$. Thus, there are three equations that must be satisfied by the pair of unknowns x, y. Because there are more equations than unknowns, the system is said to be overdetermined and, in general, it is difficult to solve. In order to facilitate computation..."

(A description of the Lagrange method follows.)

The next exercise illustrates that one can get the wrong solution if one applies the multiplier rule in a careless way.

Exercise 3.6.13 * *Show that the Lagrange multiplier rule suggests the wrong solution $(x_1, x_2) = (9, 4)$ for the problem*

$$2x_1 + 3x_2 \to \max, \ \sqrt{x_1} + \sqrt{x_2} = 5,$$

and explain this.

Paradox. Here is a paradox: we presented the hinged quadrangle problem 3.3.6 as a very tough geometrical problem, and used this fact to illustrate the power of the Lagrange multiplier rule. Now we present an elegant geometrical solution that is very simple.

Exercise 3.6.14 *The hinged quadrangle (revisited).* *Suppose you know that the closed curved line in the two-dimensional plane that encloses the largest area is a circle. How can you derive the solution of the hinged quadrangle problem from this fact?*

Solution. Take the quadrangle with given sides that is a circle quadrangle, draw the circle, and view the resulting disk as consisting of five pieces: the circle quadrangle and four circle sectors (Fig. 3.15). Now start to hinge the quadrangle, while leaving the four circle sec-

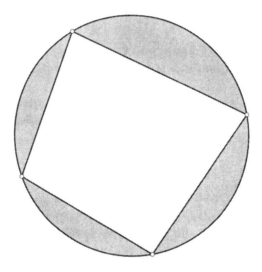

Figure 3.15 A deceptively simple solution.

tors attached to the sides, without altering their shape. Then the varying figure has the four circle sector pieces of the original circle as a boundary. When is the area of this figure in the plane maximal?

Well, use the familiar fact that among all figures in the plane with a given circumference, the one with the largest area is a circle. It follows that the initial position of our figure has maximal area: the position when it was a disk.

Now we observe that this new maximization problem is equivalent to the original problem of the hinged quadrangle, as the contribution of the four circle sectors to the total area remains constant throughout. We conclude that the solution of the hinged quadrangle problem is a circle quadrangle, as required.

Solution of the paradox. What is the solution of the paradox? Well, this way of solving the hinged quadrangle problem is elegant, but it is not entirely fair. It is a derivation from another fact ("the familiar fact"), which itself is much harder to establish.

Inequalities are always connected with the solution of optimization

LAGRANGE: EQUALITY CONSTRAINTS

problems. Therefore, they are a remarkable testing ground for the general theory. For example, we have already proved the inequality of Cauchy-Bunyakovskii-Schwarz and the inequalities of the arithmetic-geometric means and the geometric-quadratic means with the use of optimization methods. Here is another example.

Exercise 3.6.15 * *Inequality of Jacques Salomon Hadamard (1865–1963).*
Let $A = (a_{ij})_{i,j=1}^n$ be a square matrix of order n. Establish the following inequality:

$$(\det A)^2 \leq \prod_{i=1}^n \left(\sum_{j=1}^n a_{ij}^2 \right).$$

The following exercise gives an extension of the result of example 3.11; see also the remarks made after problem 3.4. For an interesting application of this extension we refer to the proof that the log-barrier of the semidefinite cone is self-concordant (see lemma 7.3).

Exercise 3.6.16 * *Comparing l_p-norms.* Establish the following inequalities for nonnegative numbers $x_1, \ldots, x_n$ and numbers $p, q \geq 1$ by means of the Lagrange multiplier rule:

1. if $p > q$ then $(x_1^p + \cdots + x_n^p)^{\frac{1}{p}} \leq (x_1^q + \cdots + x_n^q)^{\frac{1}{q}}$,

2. if $p < q$ then $(\frac{1}{n}(x_1^p + \cdots + x_n^p))^{\frac{1}{p}} \leq (\frac{1}{n}(x_1^q + \cdots + x_n^q))^{\frac{1}{q}}$.

In the next exercise we present the proof of the Lagrange multiplier rule in an algorithmic spirit.

Exercise 3.6.17 Let $\widehat{x}$ be an admissible point of a problem $(P_{3.1})$ for which the Lagrange equations do not hold for any nonzero Lagrange multiplier. Then there exists an admissible variation of descent

$$x_\alpha = \widehat{x} + \alpha \bar{x} + o(\alpha) \text{ with } \alpha \geq 0,$$

where $\bar{x} \in \ker F'(\widehat{x})$ and

$$f_0(x_\alpha) < f_0(\widehat{x}) \text{ for sufficiently small } \alpha > 0.$$

Finally, we give the promised additional exercises of the types of exercise 3.6.4 and exercise 3.6.5.

Exercise 3.6.18 *Find the extrema of the function of one variable $x_2 = f(x_1)$ that is implicitly defined by the relation*

$$F(x_1, x_2) = x_2^2 - x_1^3 - x_1^2 = 0.$$

Exercise 3.6.19 *Find the extrema of the function of one variable $x_2 = f(x_1)$ that is implicitly defined by the relation*

$$F(x_1, x_2) = x_2^2 - x_1^3 - x_1^2 = 0, \; x_2 \geq 0.$$

Exercise 3.6.20 *Find the extrema of the function of one variable $x_2 = f(x_1)$ that is implicitly defined by the relation*

$$F(x_1, x_2) = x_1 x_2 - \ln x_1 + \ln x_2 = 0 \quad (x_1 > 0, x_2 > 0).$$

Exercise 3.6.21 *Find the extrema of the function of two variables $x_2 = f(x_1, x_2)$ that is implicitly defined by the relation*

$$F(x_1, x_2, x_3) = x_3^5 + 2x_1^3 x_3^2 + x_2^4 x_3 + x_1 x_2 x_3 - 1 = 0.$$

Exercise 3.6.22 * **The problem of maximal entropy.** *For n positive numbers $x_1, \ldots, x_n$ such that $\sum_{k=1}^{n} x_k = 1$, find the minimal possible value of the following sum :*

$$\sum_{k=1}^{n} x_k \ln x_k.$$

(Statistical interpretation: if the x_i, $1 \leq i \leq n$ are viewed as a distribution, then minus this sum is called the entropy of this distribution.)

Exercise 3.6.23 * *What are the largest and the smallest possible values of the sum of squares of n numbers, if the sum of the fourth powers equals 1?*

Exercise 3.6.24 * *If the sum of five values (not necessarily positive) is 1, and the sum of the squares is 13, what is the smallest possible value of the sum of the cubes?*

Exercise 3.6.25 * *If the sum of five values is 1, and the sum of the squares is 11, what is the largest possible value of the sum of the cubes?*

Hint for the last four exercises and for exercise 3.6.5: keep in mind that usually it is not necessary to compute the Lagrange multipliers. In each of these exercises, the multiplier rule leads to the discovery of an equation that is satisfied for each coordinate of each local optimal point.

Chapter Four

Inequality Constraints and Convexity

> There are two ways of viewing a classical curve or surface like a conic, either as the locus of points or as an envelope of tangents.
>
> R.T. Rockafellar
>
> Supplement: the crucial property of the strictly convex case is that the mapping from points to tangents has an inverse.

- How can one find the minima of an n-dimensional convex problem with inequality constraints; that is, how can one solve the problem

$$f_0(x) \to \min, \quad f_i(x) \leq 0, \quad 1 \leq i \leq m,$$

where f_i, $0 \leq i \leq m$, are convex functions of $x_1, \ldots, x_n$?

4.0 SUMMARY

The need for inequality constraints. It is not always optimal to use up a budget completely. For example, in a problem of investment, it might be too risky to invest the entire capital. Such a situation can be modeled by an inequality constraint. Another example arises from the seemingly mysterious phenomenon of the discount on airline tickets with a Saturday stayover. This mystery has a simple explanation: price discrimination between the market for businesspeople and the market for tourists. Which discount is optimal for the airplane company? The discount should not be so big that it becomes attractive for businesspeople as well. This requirement can be modeled as an inequality constraint. For yet another example, consider investors. They want high return but at the same time low risk. Here, a trade-off has to be made. For example, you could maximize mean return with an upper bound on the risk, or minimize risk with a lower bound on the mean return. That is, again we are led to inequality constraints.

Why convexity? The advantage of convexity is a technical one: if an optimization problem is convex, then the necessary conditions for local optimality are in fact *criteria for global optimality*. As a first example of convexity, we mention the problem whether three lines in space have a unique "waist," (a triangle with a vertex on each line and sum of the sides a local minimum). This problem has a—hidden—convex structure and this structure leads to an easy solution. The authors know of no other way to settle this problem. This is an example of the usefulness of just the *concept* of convexity; the *methods* of convexity are not needed here. Optimization problems with inequality constraints often turn out to be convex. In economic examples, there is often a conceptual reason for this. We will give three examples.

1. The law of diminishing returns. "The more you have, the less additional pleasure you get from an increase." This leads to the concavity of utility functions (these are functions that measure the pleasure of a consumer).

2. Arbitrage. In all financial markets, arbitrageurs are active. If, for example, some share is much cheaper in Bangkok than in Chicago, then there is an arbitrage opportunity: you can quickly buy in Bangkok and sell in Chicago. The crucial question is always: does there exist an arbitrage opportunity? This depends on the set of all *exchange opportunities*. This set is convex, and this will allow us to settle the problem.

3. Second welfare theorem. A more advanced example is given by the second welfare theorem (see chapter eight). It shows that one can achieve any socially desirable distribution of goods, just by means of a suitable price system, and without any additional interference. Again, here it will be the convex structure that leads to this result.

There is another reason why convex problems occur frequently. It is natural to try to use the freedom of modeling, to choose a convex model! This explains to some extent the popularity of linear programming problems. In fact, when we consider (in chapters six and seven) the numerical solution of optimization problems, rather than the exact solution, then we will argue that the class of problems for which this can be done efficiently is essentially the class of convex problems.

Lagrange-Karush-Kuhn-Tucker-John conditions. For convex problems with inequality constraints, a variant of the multiplier method is available. It gives a criterion for global optimality, and it is usually called the KKT method. It has the following great advantage over the multiplier method. Its analysis leads to the distinction of many cases. Often, you have a hunch which case is the most promising one. If this case leads to a solution of the KKT conditions, then this must be a global solution of the given problem. Then you do not

INEQUALITY CONSTRAINTS AND CONVEXITY

have to consider all the other cases anymore. The proof of the KKT theorem will be seen to be a consequence of the following separation theorem, called the *supporting hyperplane theorem*, which is the central property of convex sets.

For each point on the boundary of a given convex set in n-dimensional space, there is a hyperplane that has the convex set completely on one of its two sides.

In dimension two (the "plane"), a hyperplane is a line, in dimension three ("space"), a hyperplane is a plane.

4.1 INTRODUCTION

Convexity. Convexity plays a central role in optimization. In this chapter we consider the *classical* convex problem (with equality and inequality constraints). In chapter ten we discuss briefly an example of a *modern* convex problem: semidefinite programming problems.

Three advantages of convexity. What is the reason for the great significance of *convexity* for solving optimization problems? It is that for convex problems the theory of first order necessary conditions gives all one could hope for:

- the first order necessary conditions are also sufficient,

- there is no difference between local and global solutions.

The reason for these excellent properties is that problems of this type are not only smooth (that is, all functions involved are differentiable), but also convex. Moreover, if all functions involved are *strictly* convex, one has the following additional property:

- there is at most one solution.

Essence of convexity. Why is it "convexity" that leads to these excellent properties? The core of the matter is as follows. The problem of minimizing a differentiable function f of n variables has the three properties above, provided the derivative $f'(x) : \mathbb{R}^n \to \mathbb{R}^n$, has an inverse for all x.

This invertibility is the essence of the "strict convexity" of a function f, as expressed in the supplement to the epigraph of this section.
For example, for the function f defined by $f(x) = x^2$ the derivative f' is given by $f'(x) = 2x$ and one can invert $y = 2x$, giving $x = \frac{1}{2}y$.
However, for the function g defined by $g(x) = x^3$, the derivative g' is given by $g'(x) = 3x^2$ and one cannot invert $y = 3x^2$; an attempt to invert gives the "multi-valued inverse" $x = \pm\sqrt{y/3}$. A further illustration of invertibility will be given in problem 4.5 on child drawing.

The dual view. You might have seen the following high school experiment with iron dust and a magnet. It is in the spirit of the words of Rockafellar from the epigraph chosen for this chapter, *"viewing a [...] curve as an envelope of tangents."* These two ways are called called *primal* and *dual*, and play an essential role in the numerical solution of convex optimization problems. The basic example of convex optimization problems is LP problems. These can be solved by two

INEQUALITY CONSTRAINTS AND CONVEXITY

types of algorithms: simplex methods and interior point methods. For both types the role of the dual view is indeed an essential one. For interior point methods, we will see this in section 6.9. For simplex methods, we refer to any textbook treatment of the simplex method.

Figure 4.1 Magnet arranges iron dust into curves.

Put iron-dust on the table and hold a magnet under the table; you will see that the dust forms curves as a result (Fig. 4.1).

The explanation is that each piece of dust is like a small needle that will follow the direction of the magnetic field. All these directed needles ("tangents") together suggest a collection of curves.

Conclusion to the section. Three advantages of convex problems are: "sufficiency," only *global* solutions exist, and—in the case of strict convexity—there exists at most one solution.

Royal road. If you want a shortcut to the applications, then it suffices to read the definitions of convex sets and functions, of the KKT conditions, as well as the statement of the main result, the KKT theorem 4.4, which is a variant of the Lagrange multiplier rule, and its supplement, theorem 4.5. Then you can read examples 4.2.5 and 4.2.7 and all applications in sections 4.3 and 4.6.

4.2 MAIN RESULT: KARUSH-KUHN-TUCKER THEOREM

Motivation for inequalities. Inequality constraints appeared only in the twentieth century. Why so late? The reason is that the laws of nature are expressed without inequality. Inequality constraints appear when one tries to formalize the actions of human beings. For example, a budget does not always have to be used up completely; an inequality constraint is the appropriate way to model a budget.

One specific motivation for inequality constraints comes from multiobjective situations: often there are several, sometimes conflicting objectives to be maximized (or minimized) simultaneously. Then one can make up one's mind which objective is the most important and decide upon acceptable minimal (resp. maximal) levels for the other objectives. This leads to modeling the situation as an optimization problem with a number of inequality constraints. Multiobjective problems are considered, for example, in [20].

4.2.1 Linear programming problems

The most popular problems with inequality constraints are *linear programming problems* (LP problems), also called, more appropriately, linear optimization problems. In these problems, the functions that describe the objective function and the (in)equality constraints are all affine (= linear + constant).

Example 4.2.1 *Serving two islands with one boat. Shipping company Doeksen wants to serve two Wadden Islands (islands at the north of the Netherlands) with one boat. For trip 1, Doeksen receives 30 euros/hour, with required fuel 1 liter/hour and required personnel 2 man-hours/hour. For trip 2, Doeksen receives 20 euros/hour, with required fuel 2 liters/hour and required personnel 1 man-hour/ hour. The available resources are 160 liters of fuel (restricted capacity fuel tank) and 140 man-hours, and the ship can be used for 90 hours.*

Model this problem as an LP problem.

Solution. The problem can be modeled as follows:

$$f_0(x) = 30x_1 + 20x_2 \to \max, \quad f_1(x) = x_1 + 2x_2 - 160 \leq 0,$$

$$f_2(x) = 2x_1 + x_2 - 140 \leq 0,$$

$$f_3(x) = x_1 + x_2 - 90 \leq 0, \quad f_4(x) = -x_1 \leq 0, \quad f_5(x) = -x_2 \leq 0,$$

where x_i is the number of hours the boat serves island i ($i = 1, 2$) and $30x_1 + 20x_2$ is the total amount that Doeksen receives. The first constraint represents the restricted capacity of the fuel tank, the second constraint the restricted number of available man-hours, and the third constraint the limited availability of the ship.

Historical comments. Linear programming problems were considered for the first time in 1939 by Leonid Vitalyevich Kantorovich (1912–1986). He modeled problems of economic planning as LP problems. This was not appreciated by the Soviet authorities and he was discouraged from continuing these investigations. Then, independently, George Dantzig (born 1914) considered LP problems and devised in 1947 an efficient algorithm to solve these problems, the *simplex method.*

To this day, the simplex method is one of the most popular mathematical techniques. It is used to save costs and increase profits. We will discuss this technique, and the more recent *interior point methods*, in chapters six and seven on algorithms.

Necessary conditions for LP problems. What about necessary conditions for LP problems? A remarkable feature of these problems is that these come in pairs, as we will see in section 4.5.1: to an LP minimization (resp. maximization) problem one can associate an LP maximization (resp. minimization) problem with the same optimal value but different variables. This *duality theory* of LP problems gives in particular necessary conditions for LP problems. These conditions can also be obtained by writing down for the case of LP problems the Karush-Kuhn-Tucker conditions, which are discussed below.

Need for an algorithm. Why do we need algorithms like the simplex method to solve LP problems, when we have the KKT conditions? The reason for this is that the following feature limits the usefulness of these conditions as a means for finding the optima. For a problem of n variables with m inequality constraints, the KKT conditions give a collection of 2^m systems of $m+n$ equations in $m+n$ unknowns (see the first remark after the formulation of the KKT conditions) and the information that each solution of the LP problem is a solution of one of these systems (moreover the KKT conditions give the nonnegativity of m of these unknowns). For an LP problem, these are systems of *linear* equations and therefore each system can be solved easily, but their number 2^m grows rapidly with increasing m. This explains the need to solve LP problems by an algorithm, unless m is very small.

Conclusion. The most popular problems are LP problems. It is not convenient to solve these problems using necessary conditions, as this leads to distinguishing many cases. Two convenient classes of *algorithms* to solve these problems are the simplex and interior point methods.

4.2.2 Nonlinear programming problems

Let $\widehat{x}$ be a point of $\mathbb{R}^n$ and $f_i : U_n(\widehat{x}, \varepsilon) \to \mathbb{R}$, $0 \leq i \leq m$, continuous functions of n variables, defined on an open ball with center $\widehat{x}$, such that $f_i(\widehat{x}) \leq 0$, $1 \leq i \leq m$.

Definition 4.1 *Main problem: inequality constrained optimization. The problem*

$$f_0(x) \to \min, \quad f_i(x) \leq 0, \ 1 \leq i \leq m, \qquad (P_{4.1})$$

is called a finite-dimensional problem with inequality constraints or a non-linear programming problem.

Remark. If you know in advance which inequality constraints are satisfied as equalities for a solution and which are not, then you can reduce to a problem with equality constraints by omitting the inequality constraints which are not satisfied as equalities and making the other ones into equality constraints. Then you are in a position to apply the Lagrange multiplier rule.

Example 4.2.2 *Model the problem to find the point closest to $(2,3)$ with sum of the coordinates not larger than 2 and with first coordinate not larger than 2 in absolute value (Fig. 4.2).*

Solution. The problem can be modeled as a problem with inequality constraints:

$$f_0(x) = (x_1 - 2)^2 + (x_2 - 3)^2 \to \min,$$

$$f_1(x) = x_1 + x_2 - 2 \leq 0, \quad f_2(x) = x_1^2 - 4 \leq 0.$$

Now we consider a problem where the nonlinearity is visible in the figure. In example 4.2.2 this is not the case: the nonlinear constraint

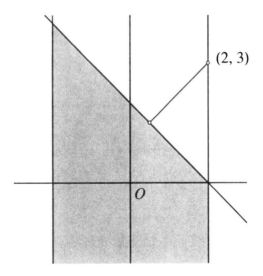

Figure 4.2 A nonlinear programming problem.

$x_1^2 - 4 \leq 0$ can be written as a system of two linear constraints, $x_1 - 2 \leq 0$, $-x_1 - 2 \leq 0$, and the only essential nonlinearity—of the objective function f_0—is not visible in the figure. However, in passing we mention that the example above represents the usual situation. Indeed, the most interesting nonlinear convex problems consist of the minimization of convex functions with linear constraints.

Example 4.2.3 *Model the problem to find the point that is closest to $(2,3)$ among the points with first coordinate at least 1 that lie on the disk with center the origin and radius 2 (Fig. 4.3).*

Solution. The problem can be modeled as a problem with inequality constraints:

$$f_0(x) = (x_1 - 2)^2 + (x_2 - 3)^2 \to \min,$$

$$f_1(x) = x_1^2 + x_2^2 - 4 \leq 0, \quad f_2(x) = 1 - x_1 \leq 0.$$

Example 4.2.4 *Model the problem to solve the inequality*

$$g(x_1, \ldots, x_n) \leq 0$$

with the additional requirement that you want the solution that is as close as possible to a given point $\bar{x}$.

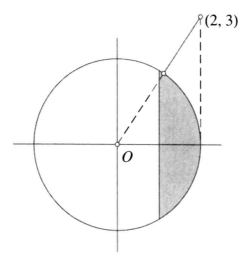

Figure 4.3 Another nonlinear programming problem..

Solution. This problem can also be modeled as a problem with inequality constraints:

$$f_0(x) = |x - \bar{x}| \to \min, \quad f_1(x) = g(x) \leq 0.$$

Problem type $(P_{4.1})$ is the traditional way to generalize the LP problem to nonlinear problems. A more modern problem type, $(P_{10.1})$, using convex cones, will be discussed in chapter ten.

Conclusion. The problem type considered in this section is the nonlinear extension of the celebrated LP problem.

4.2.3 The Karush-Kuhn-Tucker theorem

The function

$$\mathcal{L}(x, \lambda) = \sum_{i=0}^{m} \lambda_i f_i(x), \quad \lambda = (\lambda_0, \ldots, \lambda_m)$$

is called *the Lagrange function of* $(P_{4.1})$, and the variables λ_i, $0 \leq i \leq m$, are called the *Lagrange multipliers* of the problem; the row vector λ is called a selection of Lagrange multipliers. Assume that all functions f_i, $0 \leq i \leq m$, are differentiable at $\widehat{x}$.

Karush-Kuhn-Tucker conditions. The *Karush-Kuhn-Tucker conditions* (KKT conditions) are said to hold at $\widehat{x}$ for a nonzero selection of Lagrange multipliers λ with $\lambda_0 \geq 0$ if the following conditions

INEQUALITY CONSTRAINTS AND CONVEXITY

hold:

(α) $\mathcal{L}_x(\widehat{x},\lambda) = 0_n^T$ (*the stationarity condition*);
(β) $\lambda_i \geq 0$, $1 \leq i \leq m$ (*the nonnegativity conditions*);
(γ) $\lambda_i f_i(\widehat{x}) = 0$, $1 \leq i \leq m$ (*the conditions of complementary slackness*).

Remark. The only essential difference with the Lagrange conditions is the nonnegativity of the Lagrange multipliers. Note that the i-th complementary slackness condition leads to the distinction of two cases: $\lambda_i = 0$ or $f_i(\widehat{x}) = 0$. In all, this leads to a distinction of 2^m cases.

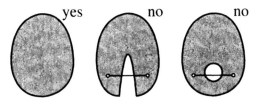

Figure 4.4 Illustrating the definition of a convex set.

The concepts *convex set* and *convex function* are illustrated by Figures 4.4 and 4.5. The precise definitions are as follows.

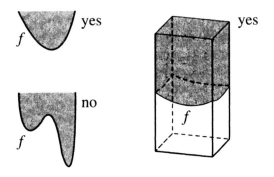

Figure 4.5 Illustrating the definition of convex function.

Definition 4.2 *Convex set. A set $A \subseteq \mathbb{R}^n$ is called convex if for each two points x and x' in A, all points*

$$x_\alpha = (1-\alpha)x + \alpha x'$$

with $0 \leq \alpha \leq 1$ belong to A.

Definition 4.3 *Convex function. A function $f : A \to \mathbb{R}$ on a convex set A is called convex if the epigraph of f*

$$\text{epi} f = \{(x,\rho) \mid x \in A, \ \rho \in \mathbb{R}, \ \rho \geq f(x)\}$$

is a convex set in $\mathbb{R}^{n+1}$.

Alternatively, the convexity of functions can be characterized by the convexity of the set A together with the inequality

$$f(x_\alpha) \leq (1-\alpha)f(x) + \alpha f(x'),$$

for all $x, x' \in A$ and all $\alpha \in [0,1]$.

Geometrical view on convex functions. For $n = 1$, the epigraph can be viewed as the set of points in the plane lying on or above the graph of f. For $n = 2$, one can think of a curved surface in three-dimensional space on which concrete or lava is poured; after awhile you get "stone," and this is the epigraph of the function that is described by the curved surface.

The basic facts on convex functions are given in appendix B. Here we only mention that a local minimum of a convex function is also a global minimum and that a strictly convex function has at most one minimum. Thus there is no need to distinguish between local and global minima.

Theorem 4.4 *The Karush-Kuhn-Tucker theorem—necessary conditions for smooth and convex programming problems. Consider a problem of type $(P_{4.1})$. Assume that the problem is smooth at $\widehat{x}$, and convex in the following sense. The functions*

$$f_i : U_n(\widehat{x}, \varepsilon) \to \mathbb{R}, \ 0 \leq i \leq m,$$

are differentiable at $\widehat{x}$, and they are convex.

Then, if $\widehat{x}$ is a point of minimum, there exists a nonzero selection of Lagrange multipliers λ such that the KKT conditions hold at $\widehat{x}$.

The following supplements to theorem 4.4 will be used in the solution of concrete problems. A point $\bar{x} \in \mathbb{R}^n$ is called a *Slater point* if $f_i(\bar{x}) < 0$, $1 \leq i \leq m$.

INEQUALITY CONSTRAINTS AND CONVEXITY

Theorem 4.5 *Sufficiency of the KKT conditions.*

1. *The KKT conditions are sufficient for optimality, provided $\lambda_0 = 1$. To be more precise, if the KKT conditions hold at $\widehat{x}$ for a selection of Lagrange multipliers λ with $\lambda_0 = 1$, then $\widehat{x}$ is a point of minimum.*

2. *Assume there exists a Slater point. If the KKT conditions hold for a selection of Lagrange multipliers λ, then λ_0 cannot be zero.*

On Slater points. Slater points appear here for the first time. This concept comes back later repeatedly, for example, in section 4.5.3 and in chapter ten. Their existence can often be demonstrated by a concrete example of such a point. It is of great interest to have Slater points. Indeed, the second statement of theorem 4.5 shows that this implies that the problem is regular, in the sense that the bad case $\lambda_0 = 0$ cannot occur. This implies in its turn, by the first statement of theorem 4.5, that each solution of the KKT conditions is a solution of the optimization problem. In other words, if a Slater point exists, then the necessary conditions are sufficient as well.

Conclusion. For convex problems with inequality constraints, necessary and sufficient conditions are given, the KKT conditions. These differ from the Lagrange conditions by the presence of nonnegativity conditions and conditions of complementary slackness.

4.2.4 The advantages of the KKT conditions

The KKT conditions have two advantages over the multiplier rule.

Great advantage. In the first place, we will describe a great advantage. For both methods we are led to something that is not very pleasant: as many as 2^m cases have to be distinguished and analyzed. Indeed, one has to distinguish for each inequality whether $f_i(\widehat{x})$ is negative (and so $\lambda_i = 0$) or zero, and this leads to 2^m cases, as there are m inequalities.

Note that 2^m grows rapidly with m. Often, one of the first cases we analyze gives us a point that satisfies all conditions of the multiplier rule (resp. the KKT theorem), either by our good fortune or because we had a hunch where to look for the solution. Then, in the case of the KKT theorem, we know that we have found a solution, as the KKT conditions are sufficient. Therefore, we can stop as soon as we hit upon just one solution of the KKT conditions. The fact that the KKT conditions hold is a "certificate of optimality."

In the case of the Lagrange multiplier rule, we are forced to continue with our longwinded analysis of all 2^m cases. We emphasize again that 2^m grows rapidly with m, so it is not practical to consider all cases, unless m is very small. Only at the end of this search, a comparison of all "candidate solutions" shows which one is the optimum, using the Weierstrass theorem. Note that there is no need the Weierstrass theorem if you have found a solution of the KKT conditions.

Small advantage. Moreover, the KKT equations have the small advantage that they allow us to exclude candidate solutions for which one of the Lagrange multipliers is negative. Sometimes this small advantage is of crucial importance. A striking illustration of this is given in the analysis of problem 9.4.1 of Apollonius.

Conclusion. The KKT conditions have a certain advantage over the multiplier rule: you do not always have to search through all the cases. Moreover, the nonnegativity conditions might eliminate some candidates.

4.2.5 Illustrations of the KKT theorem

We illustrate the KKT conditions with the first example given above.

Example 4.2.5 *Use the KKT theorem to solve the problem of example 4.2.2.*

Solution. One does not need the KKT conditions to solve this problem. The easiest way to solve it is geometrically, using a picture (Fig. 4.2). However, this method breaks down for problems with many variables. The analytical method using the KKT conditions, which we demonstrate now, does not have this disadvantage.

1. The problem can be modeled as follows, as we have seen:

$$f_0(x) = (x_1 - 2)^2 + (x_2 - 3)^2 \to \min, \quad f_1(x) = x_1 + x_2 - 2 \leq 0,$$

$$f_2(x) = x_1^2 - 4 \leq 0.$$

We recall that it is not necessary to apply the Weierstrass theorem here, because the problem is convex.

2. Lagrange function

$$\mathcal{L}(x, \lambda) = \lambda_0((x_1 - 2)^2 + (x_2 - 3)^2) + \lambda_1(x_1 + x_2 - 2) + \lambda_2(x_1^2 - 4),$$

put $\lambda_0 = 1$ ($(0, 0)$ is a Slater point).

KKT:

$$\frac{\partial \mathcal{L}}{\partial x_1} = 2(x_1 - 2) + \lambda_1 + 2\lambda_2 x_1 = 0,$$

$$\frac{\partial \mathcal{L}}{\partial x_2} = 2(x_2 - 3) + \lambda_1 = 0,$$

$$\lambda_1, \lambda_2 \geq 0,$$

$$\lambda_1(x_1 + x_2 - 2) = 0,$$

$$\lambda_2(x_1^2 - 4) = 0.$$

3. Distinguish cases.

   Case 1: no constraint is tight. We recall that this means that $x_1 + x_2 - 2 < 0$ and $x_1^2 - 4 < 0$. Then the KKT conditions can be simplified to

   $$\lambda_1 = \lambda_2 = 0, \ 2(x_1 - 2) = 0 \ \text{and} \ 2(x_2 - 3) = 0.$$

   However, the resulting point $(2, 3)$ is not admissible, as it does not satisfy $x_1 + x_2 - 2 \leq 0$.

   Case 2: only the second constraint is tight. Then the KKT conditions can be simplified to

   $$\lambda_1 = 0, \ x_1^2 = 4, \ 2(x_2 - 3) = 0,$$

   $$2(x_1 - 2) + 2\lambda_2 x_1 = 0, \ \lambda_2 \geq 0.$$

   However, the resulting point $(2, 3)$ is not admissible, as we have noted already.

   Case 3: only the first constraint is tight. Then the KKT conditions can be simplified to

   $$\lambda_2 = 0, \ x_1 + x_2 = 2, \ 2(x_1 - 2) + \lambda_1 = 0,$$

   $$2(x_2 - 3) + \lambda_1 = 0, \ \lambda_1 \geq 0.$$

   The resulting point is $(\frac{1}{2}, \frac{3}{2})$ with $\lambda_1 = 3$. This point is admissible, clearly. It follows that it is a global solution.

At this point we can stop as we have hit upon a solution of the KKT conditions; uniqueness of this solution follows from the strict convexity of the objective function f_0.

4. There is a unique closest admissible point to (2,3): it is the point $(\frac{1}{2}, \frac{3}{2})$. Its Lagrange multipliers are $\lambda_1 = 3$ and $\lambda_2 = 0$ (and $\lambda_0 = 1$). The fact that the KKT conditions hold constitutes a proof of global optimality.

Hunch. If we had a hunch that only the first constraint is tight (case 3) in the optimum, then we could have solved the problem even faster: it would have been enough to investigate this case only.

To illustrate the advantages of the KKT theorem over the multiplier rule, let us consider what would change if we solved the example above using the multiplier rule.

Example 4.2.6 *Use the multiplier rule to solve the problem of example 4.2.3.*

Solution. If we use the multiplier rule, then the calculations will be the same as above, but now we have to consider all cases and this leads to *four* candidate-solutions:

$$(\frac{1}{2}, \frac{3}{2}),\ (-2, 3),\ (2, 0),\ (-2, 4).$$

We can exclude $(-2, 3)$ because of a negative Lagrange multiplier, and the last two belong to a case that we did not have to consider above. A comparison gives

$$f_0(\frac{1}{2}, \frac{3}{2}) = \frac{9}{2},\ f_0(-2, 3) = 16,\ f_0(2, 0) = 9,\ f_0(-2, 4) = 17;$$

we see that of these four outcomes $\frac{9}{2}$ is the smallest one. To complete the analysis, we apply the Weierstrass theorem, as we may: the admissible set is closed and nonempty and it becomes bounded (without losing the first two properties) if we add the constraint $f_0(x) \leq M$, for a sufficiently large constant $M > 0$—observe that this constraint determines a disk with center $(2, 3)$ and radius $\sqrt{M}$.

As an additional illustration of the KKT conditions we consider the second example given above.

Example 4.2.7 *Use the KKT theorem to solve the problem of example 4.2.3.*

INEQUALITY CONSTRAINTS AND CONVEXITY

Solution. Again, one does not need the KKT conditions to solve this problem. The easiest way to solve it is geometrically, using a picture (Fig. 4.3). However, we want to illustrate the use of the KKT conditions.

1. The problem can be modeled as follows, as we have seen:

$$f_0(x) = (x_1 - 2)^2 + (x_2 - 3)^2 \to \min, \quad f_1(x) = x_1^2 + x_2^2 - 4 \leq 0,$$

$$f_2(x) = 1 - x_1 \leq 0.$$

2. Lagrange function

$$\mathcal{L} = \lambda_0((x_1 - 2)^2 + (x_2 - 3)^2) + \lambda_1(x_1^2 + x_2^2 - 4) + \lambda_2(1 - x_1),$$

put $\lambda_0 = 1$ ($(3/2, 0)$ is a Slater point).

KKT:

$$\frac{\partial \mathcal{L}}{\partial x_1} = 2(x_1 - 2) + 2\lambda_1 x_1 - \lambda_2 = 0,$$

$$\frac{\partial \mathcal{L}}{\partial x_2} = 2(x_2 - 3) + 2\lambda_1 x_2 = 0,$$

$$\lambda_1, \lambda_2 \geq 0,$$

$$\lambda_1(x_1^2 + x_2^2 - 4) = 0,$$

$$\lambda_2(1 - x_1) = 0.$$

3. Suppose we have a hunch that in the optimum only the first constraint is tight. We start by considering this case, that is, $x_1^2 + x_2^2 = 4$, $1 - x_1 \neq 0$. Then the KKT conditions lead to one solution: $x_1 = 4/\sqrt{13}$, $x_2 = 6/\sqrt{13}$ and $\lambda_1 = (\sqrt{13}-2)/2$, $\lambda_2 = 0$. By the sufficiency of the KKT conditions, it follows that this is a solution of the problem. By the strict convexity of f_0, it even follows that the solution is unique. Therefore, we do not have to consider other cases.

4. The problem has a unique solution, $\widehat{x} = (4/\sqrt{13}, 6/\sqrt{13})$.

Conclusion. For optimization problems with a very small number of inequality constraints, say m, the KKT method requires distinguishing a small number, 2^m, of cases. This method has some advantages over applying the multiplier rule to all 2^m cases.

4.2.6 Application of the KKT theorem to LP problems

Finally, we apply the KKT theorem to the special case of linear programming problems, and show that this leads to the *complementary slackness* conditions, a fundamental property of LP problems.

Example 4.2.8 *Consider the problem*

$$c^T \cdot x \to \min, \; Ax \leq b,$$

where $c \in \mathbb{R}^n$, A is an $m \times n$ matrix, and $b \in \mathbb{R}^m$, and assume that it has a Slater point $\bar{x}$, that is, $A\bar{x} < b$. Write the KKT conditions for this problem.

Solution. For each solution $\hat{x}$ of the given problem, there is a nonnegative row vector $\lambda \in (\mathbb{R}^m)^T$ for which

$$c^T + \lambda A = 0_n^T,$$

and the complementary slackness conditions hold: $\lambda_i = 0$ for each $i \in \{1,\ldots,m\}$ for which $(Ax)_i < b_i$.

A different view on the KKT theorem for LP problems will be given in section 4.5.1.

Historical remarks. In 1948, Albert W. Tucker (1906–1995) and his student Harold W. Kuhn (1925–) found necessary conditions in the case where the functions describing the objective and the inequality constraints are all *convex* (cf. theorem 4.11). Later it was discovered that a student, William Karush (1918–1997), had given these conditions already in 1939 in his—unpublished—master's thesis.

A crucial property of the necessary conditions of Karush-Kuhn-Tucker is that they are also *sufficient* for global optimality. Thus the Weierstrass theorem does not have to be used in the analysis of such problems.

In 1948 Fritz John (1910–1994) found necessary conditions in the case where the functions describing the objective and the inequality constraints are smooth enough (in the same sense as in the Lagrange

INEQUALITY CONSTRAINTS AND CONVEXITY

multiplier rule), instead of convex (cf. theorem 4.13). These conditions are easier to use, but they are not sufficient for optimality.

If the functions are both smooth and convex, then one gets necessary conditions that are both easy to use and sufficient for optimality. This is the situation considered in theorem 4.4 above: theorem of Lagrange-Karush-Kuhn-Tucker-John might be a more appropriate terminology for this result.

Conclusion to section 4.2. Applying the KKT theorem to LP problems, we get a fundamental property of LP problems, the complementary slackness conditions.

4.3 APPLICATIONS TO CONCRETE PROBLEMS

4.3.1 Maximizing a linear function on a sphere (revisited)

To begin with, we take another look at the problem of maximizing a linear function on a sphere. This problem has been solved in the previous section, using the Lagrange multiplier rule (problem 3.3.3). We have seen that the Lagrange conditions have two solutions: one is the global maximum and the other one is the global minimum. These had to be compared afterward. We will see that the KKT conditions have only one solution: the global minimum can be excluded because of a negative multiplier. Note an additional advantage over the treatment with the multiplier rule: one does not have to use the Weierstrass theorem, as the KKT conditions are *sufficient* for optimality.

Problem 4.3.1 *Let a nonzero vector $a = (a_1, \ldots, a_n)$ be given. Solve the problem*

$$a \cdot x = \sum_{i=1}^{n} a_i x_i \to \max, \quad |x|^2 = \sum_{i=1}^{n} x_i^2 = 1.$$

Solution. This problem gives the opportunity to illustrate a useful trick: we replace the equality constraint by the corresponding inequality constraint and go over at the same time to a minimization problem.

1. This gives the convex problem

$$f_0(x) = -a \cdot x \to \min, \quad f_1(x) = |x|^2 - 1 \leq 0. \qquad (Q)$$

2. Lagrange function

$$\mathcal{L}(x, \lambda) = \lambda_0(-a \cdot x) + \lambda_1(|x|^2 - 1),$$

put $\lambda_0 = 1$ (0_n is a Slater point).

KKT:

$$\mathcal{L}_x = -a + 2\lambda_1 x^T = 0_n^T,$$

$$\lambda_1 \geq 0,$$

$$\lambda_1(|x|^2 - 1) = 0.$$

3. Distinguish cases.

   Case 1: the constraint is not tight; that is, $|x|^2 - 1 < 0$. Then the KKT conditions can be simplified to $\lambda_1 = 0$ and $-a = 0_n^T$. Therefore, this case leads to contradiction, as a is a nonzero vector.

   Case 2: the constraint is tight; that is, $|x|^2 - 1 = 0$. Then $\lambda_1 \neq 0$, and so the KKT conditions can be simplified to x is a vector of length one, and is a positive scalar multiple of a^T. This determines x uniquely: $x = |a|^{-1} a^T$.

   As the KKT conditions with $\lambda_0 = 1$ are sufficient, it follows that $\widehat{x} = |a|^{-1} a^T$ is the unique point of global minimum of the problem (Q). Comparison of (Q) with the given problem shows that $\widehat{x} = |a|^{-1} a^T$ is the unique point of global minimum of the given problem as well. Indeed, the only essential change when going over to (Q) is that the admissible set is larger ("was sphere, becomes closed ball"), but we were fortunate that the solution of (Q) turned out to be also admissible for the given problem ("$\widehat{x}$ lies on the sphere").

4. The given problem has a unique global solution, $\widehat{x} = |a|^{-1} a^T$.

4.3.2 Discount on airline tickets

Why discount. We will look into a seemingly mysterious phenomenon. If you buy a plane ticket, then you get a considerable discount if your stay includes a Saturday night. What is the reason of this discount?

INEQUALITY CONSTRAINTS AND CONVEXITY

A dilemma. Here is the explanation. Airline companies face two market demands, by businesspeople and by "normal" people such as families going on holiday. Businesspeople are prepared to pay much more for a ticket. Therefore, it is tempting to charge these two categories of customers different prices for the same service. However, such open price discrimination is not allowed. Now it would seem that there are two choices: either a low price is chosen, and then all customers are prepared to pay, or a high price is chosen, and then only the businesspeople are prepared to pay. In the last case, many customers are lost; in the first case, there is a loss of profit from the business people.

Way out of the dilemma. One day someone hit upon an idea to avoid both losses: give discounts on tickets with a Saturday stayover. Why does this work? Well, it leads to self-selection by the customers. The businesspeople fly to go to a meeting for a day or two; they are not prepared to put up with a Saturday stayover. The normal people, on the other hand, gladly include a Saturday stayover in order to have the discount. So one gets the desired price discrimination after all, but this form of it is not forbidden.

This explanation leaves us with one question: what is the optimal discount?

Problem 4.3.2 *How do airline companies determine the size of the account on tickets with a Saturday stayover? Here you have to take into account the "danger" that businesspeople and not only tourists might be tempted to buy a ticket with a Saturday stayover if the discount is very attractive.*

This problem will be modeled as an optimization problem with $m = 4$ inequality constraints, and so the KKT conditions will lead to the distinction of $2^m = 16$ cases. The solution will illustrate that you do not always have to look at all the 2^m cases of the KKT-conditions.

Solution
Modelling utilities. Let the utility of traveling without a Saturday stayover be

$$u = z - \alpha p + t,$$

where $z \geq 0$ is the pleasure of spending Saturday night at home, p the price of a ticket, $\alpha \geq 0$ is a coefficient expressing how much it "hurts" to pay the price, and t is the pleasure of making the trip. In a similar way, let the utility of traveling with a Saturday stayover be given by

$$u = -\alpha p + t.$$

Difference between two types of travelers. Now we model the difference between normal travelers (tourists) and businesspeople: $z_n < z_b$ and $\alpha_n > \alpha_b \geq 0$. In fact, let us take $z_n = 0$.

Modeling individual rationality. First, the airline company wants to choose the prices p_n (resp. p_b) for a flight with (resp. without) a Saturday stayover in such a way that for the tourists (respectively businesspeople) the ticket is attractive enough to buy. This gives the constraints of individual rationality:

$$-\alpha_n p_n + t \geq 0, \qquad\qquad IR_n$$

$$z_b - \alpha_b p_b + t \geq 0. \qquad\qquad IR_b$$

Modeling incentive compatibility. Moreover, the more attractive of the two tickets for tourists (resp. businesspeople) should be the ticket with (resp. without) a Saturday stayover. This gives the constraints of incentive compatibility:

$$-\alpha_n p_n + t \geq z_n - \alpha_n p_b + t, \qquad\qquad IC_n$$

$$z_b - \alpha_b p_b + t \geq -\alpha_b p_n + t. \qquad\qquad IC_b$$

Modeling the objective function. Finally, the objective of the airplane company is to get the highest average price for a ticket. This is $p_n/2 + p_b/2$, if a person coming to buy a ticket is as likely to be a tourist as a businessperson.

Simplification. The prospect of applying the KKT conditions to this problem with four inequality constraints, and then distinguishing sixteen cases, might not be attractive. Therefore, we try to simplify the presentation of the problem. To begin with, two of the inequality constraints can be omitted: IR_b and IC_n. Indeed,

$$IC_b \text{ and } \alpha_n > \alpha_b \geq 0 \text{ give } z_b - \alpha_b p_b + t \geq -\alpha_n p_n + t,$$

and this is nonnegative by IR_n; this proves that IR_b can be omitted. We omit IC_n, as it amounts to $p_b \geq p_n$ (recall that $z_n = 0$), which we guess to hold strictly in the optimal situation. Moreover, the other two inequality constraints must hold as equalities in the optimal situation, as one readily sees.

In all, we get the following optimization problem:

$$f(p_n, p_b) = p_n/2 + p_b/2 \to \max,$$

$$-\alpha_n p_n + t = 0, \ z_b - \alpha_b p_b + t = -\alpha_b p_n + t.$$

The constraints of this problem form a system of two linear equations in two variables p_n and p_b. This system has a unique solution:

$$p_n = \alpha_n^{-1} t, \ p_b = \alpha_n^{-1} t + \alpha_b^{-1} z_b.$$

Therefore, we cannot be choosy: this point is optimal for this optimization problem. One sees immediately that $p_b > p_n$, and so we get that this solution is in fact the—unique—solution of the original problem. A different modeling of this problem will be discussed in problem 8.2.

4.3.3 Mean-variance investment

In this section, we consider methods to solve problems with inequality constraints. You might question the practical need for such methods. Indeed, is it not obvious in advance in each practical problem, whether a given inequality will be tight or slack in the optimal situation? Then, if it is slack, we can omit it, and if it is tight, we can replace it by the corresponding *equality* constraint. For example, it might seem farfetched to imagine that a budget constraint could ever be slack, as it might not seem optimal to leave part of a budget unused. The following problem gives a natural example where it is not optimal to use a budget completely.

Each investor is confronted with the need to make a trade off between two conflicting objectives, high expected return and low risk. We explore two possibilities to make this trade off:
- maximizing expected return for a given risk, the *acceptable risk level* (problem 4.3.3).

- minimizing risk for a given expected return, the *acceptable expected return* (exercise 4.6.7).

A professional investor will use tables of mean returns and estimates of the variance expected for combinations of stocks. These numbers are available by subscription from financial service companies. They are computed based on the history of the stocks.

Problem 4.3.3 *An agent can invest up to $1,000, distributing her investment between Heineken and Philips. Let x_1 (resp. x_2) denote the amount she invests in Heineken (resp. in Philips). The mean return on the investment of the portfolio is*

$$E[r_p] = .10x_1 + .12x_2,$$

and the mean variance is

$$\sigma_p^2 = .005x_1^2 - .006x_1x_2 + .008x_2^2$$

(E is expectation and σ is standard deviation). Our agent wishes to choose x_1 and x_2 subject to keeping the standard deviation at or below $25, or equivalently, keeping the variance at or below 625. A higher variance seems an unhealthy risk.

Which investment decision gives maximal mean return, subject to this constraint on the standard deviation?

Solution

1. This problem can be modeled as a convex problem:

$$f_0(x) = -.10x_1 - .12x_2 \to \min,$$

$$f_1(x) = .005x_1^2 - .006x_1x_2 + .008x_2^2 - 625 \leq 0,$$

$$f_2(x) = x_1 + x_2 - 1000 \leq 0.$$

The function defining the first inequality constraint is indeed convex (this can be seen by writing it in the form $x^T A x$ for a symmetric 2×2 matrix and checking that it is positive definite).

2. Lagrange function

$$\mathcal{L} = \lambda_0(-.10x_1 - .12x_2) + \lambda_1(.005x_1^2 - .006x_1x_2 + .008x_2^2 - 625)$$

$$+ \lambda_2(x_1 + x_2 - 1000),$$

put $\lambda_0 = 1$ ($(0,0)$ is a Slater point).

KKT:

$$\frac{\partial \mathcal{L}}{\partial x_1} = -.10 + .01x_1\lambda_1 - .006x_2\lambda_1 + \lambda_2 = 0,$$

$$\frac{\partial \mathcal{L}}{\partial x_2} = -.12 - .006x_1\lambda_1 + .016x_2\lambda_1 + \lambda_2 = 0, \ \lambda_1, \lambda_2 \geq 0,$$

$$\lambda_1(.005x_1^2 - .006x_1x_2 + .008x_2^2 - 625) = 0, \ \lambda_2(x_1 + x_2 - 1000) = 0.$$

3. This system has a unique solution $(348.01, 270.01)$; it comes from the case that the first constraint is tight and the second one is slack. We will omit the analysis that leads to this result.

4. The concrete interpretation of this result is that the following portfolio is optimal: she should invest \$348.01 in Heineken and \$270.01 in Philips. This leads to an expected return of \$67.20, a variance of 625, and a standard deviation of \$25. How much of the thousand dollars has been invested? Well, she has invested \$348.01+\$270.01=\$618.02.

Finally, suppose that, instead of optimizing, our agent had relied on two ideas: she wants to diversify and she wants to invest a bit more in Philips than in Heineken as Philips has a higher mean return than Heineken. For example, consider the choice $x_1 = 400$ and $x_2 = 600$. This gives expected return \$112, which is very attractive, but the standard deviation is \$47.33, which seems an unhealthy risk.

In the exercises 4.6.6 and 4.6.7 we will continue this topic.

4.3.4 Unique waist of three lines in space

The following problem will illustrate that just the definition of convexity—rather than the power of the KKT conditions—can be enough to solve a problem. Many years ago some Ph.D. students were challenged by John Tyrrell, lecturer at King's College London, with the following puzzle:

Problem 4.3.4 *Show that three lines in space in a sufficiently general position (at least two of them should be not parallel) have a unique waist.*

This can be visualized as follows. An elastic band is stretched around three iron wires in space that have a fixed position. By elasticity it will slip to a position where its total circumference is minimal. The challenge is to show that this final position does not depend on the initial position of the elastic band; it depends only on the position of the three wires.

Modeling the problem. A precise modeling of this problem is suggested by Figure 4.6. Let l_1, l_2, l_3 be three lines in three-dimensional space, pairwise disjoint and not all mutually parallel. Consider the minimization problem

$$f(p_1, p_2, p_3) = |p_1 - p_2| + |p_2 - p_3| + |p_3 - p_1| \to \min,$$

subject to $p_i \in l_i$ $(i = 1, 2, 3)$.

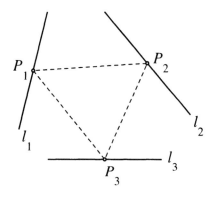

Figure 4.6 The waist of three lines.

The concept of local minimum of this problem makes precise the idea of "waist of three lines in space." The challenge is to show that this problem has a *unique* point of local minimum.

Frustration. Dr. Tyrrell told the students that, to the best of his knowledge, a solution of this simple-looking problem was not known. His words carried great weight: he was an expert in all sorts of problems. Everyone tried to find a solution, for example, by eliminating the constraints, applying Fermat's theorem, and carrying out all sorts of algebraic manipulations on the resulting equations in order to establish the desired uniqueness. Nothing worked.

Convexity comes to the rescue. Recently, we again came across the problem. This time it offered no resistance: the following elementary insight into optimization problems allows a straightforward solution of the puzzle. A successful analysis usually depends on the exploitation of not only the smoothness but also the (strict) *convexity* of the problem. The crucial observation is that the objective function f is convex: this can be verified using the triangle inequality. Then, all local solutions are global solutions, and so it remains to show that the set of global solutions of the problem consists of precisely one point.

Existence. To begin with, the Weierstrass theorem, in combination with the fact that f is coercive on the set of admissible points (p_1, p_2, p_3), shows that there exists at least one point of global minimum. Therefore, to complete our proof, it suffices to verify that each of the three terms in the expression defining f is convex and that at least one of these is strictly convex: then it follows that f is strictly convex, and the proof is completed. The convexity of the three terms is immediate.

Strict convexity. Note that a function is strictly convex precisely if its restriction to each line is strictly convex. Applying this observation to f gives the following criterion for strict convexity of f. For each triplet $(p_1(t), p_2(t), p_3(t))$ of parametric descriptions of the three lines l_i, $1 \leq i \leq 3$, of the form

$$p_i(t) = q_i + tr_i,$$

with t running over $\mathbb{R}$ and with $q_i, r_i \in \mathbb{R}^3$ for $1 \leq i \leq 3$, the function $f(p_1(t), p_2(t), p_3(t))$ of one variable t is strictly convex. Now we will check this.

To begin with, not all three difference functions

$$p_1(t) - p_2(t), \ p_2(t) - p_3(t), \ p_3(t) - p_1(t),$$

can be constant, as the lines l_1, l_2, l_3 are not all mutually parallel. Without restricting the generality of the argument we assume that $p_1(t) - p_2(t)$ is not constant. Then $p_1(t) - p_2(t)$ is a parametric description of a line in $\mathbb{R}^3$ not through the origin—as l_1 and l_2 have no common points. Therefore, the term $|p_1(t) - p_2(t)|$ is a strictly convex function of t, as desired. This can be seen by viewing this term as the restriction—to the line in question—of the norm function on $\mathbb{R}^3$, which is convex (a consequence of the triangle inequality $|x + y| \leq |x| + |y|$), and for which the only deviation from strict convexity is its behavior on lines through the origin.

Characterization. The problem of the unique waist of three lines is solved. However, now we become curious, what is this unique position of the elastic band? We apply the Fermat theorem. By what we have just proved, we know in advance that this will give stationarity conditions having a unique solution, and that this is the unique solution of the optimization problem. Let $(\widehat{p}_1, \widehat{p}_2, \widehat{p}_3)$ be the solution of the problem and let $\widehat{p}_1 + tr_1$ be a parametric description of line l_1.

We consider the one-dimensional convex problem

$$f(\widehat{p}_1 + r_1 t, \widehat{p}_2, \widehat{p}_3) \to \min.$$

Application of the Fermat theorem gives the equation

$$\frac{(\widehat{p}_1 - \widehat{p}_2)^T}{|\widehat{p}_1 - \widehat{p}_2|} \cdot r_1 + \frac{(\widehat{p}_1 - \widehat{p}_3)^T}{|\widehat{p}_1 - \widehat{p}_3|} \cdot r_1 = 0.$$

This means that the angles between the line l_1 and the sides $\widehat{p}_1\widehat{p}_2$ and $\widehat{p}_1\widehat{p}_3$ of the triangle $\widehat{p}_1\widehat{p}_2\widehat{p}_3$ are equal. One can proceed in the same way for the other two lines l_2 and l_3 and obtain the corresponding equality of angles for each of them.

Geometrical interpretation of the solution. It is possible to put this geometrical description of the solution of the problem in the following form. Let three lines in three-dimensional space, l_1, l_2, and l_3, be given that are pairwise disjoint and not all mutually parallel. Then there is a unique point in space with the following property.

It is the intersection of the bisectrices of the triangle which has as vertices the orthogonal projections of the point on the three given lines.

We have proved this statement, which we could not find in the literature, just by using the *concept* of convexity. It is doubtful whether this geometric fact could have been established in any different way. The concept of convexity can sometimes even be the key to find an explicit solution of an optimization problem, as we will illustrate in exercise 4.6.5.

In exercise 4.6.4 the problem of the waist of *two* lines is considered.

4.3.5 Child drawing

The following observation by one of the authors is an additional illustration of the supplement to the epigraph of this chapter.

Problem 4.3.5 *A child draws a line on a sheet of paper along a drawer. She holds the drawer in such a way that one end of it is somewhere at the bottom line of the sheet and the other one somewhere at the left side of the sheet. She continues drawing lines for all such positions of the drawer. The result surprises her. Drawing all these*

INEQUALITY CONSTRAINTS AND CONVEXITY

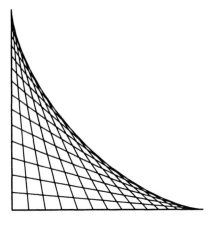

Figure 4.7 Child's drawing.

straight lines has led to something that is not straight: a curved line. The lines are tangents of this curve (Fig. 4.7).

What is the equation of this curve?

We see here invertibility "in action": a curved line is constructed from its tangents.

We will need the following well-known relation between the sine and cosine functions:
$$\sin^2 \varphi + \cos^2 \varphi = 1$$
for all $\varphi \in \mathbb{R}$.

Solution. We take the length of the drawer as the unit of length. We calculate the point of intersection of two neighboring lines, say,
$$\frac{x_1}{\cos \varphi_i} + \frac{x_2}{\sin \varphi_i} = 1, \ i = 1, 2,$$
with $\varphi_1 \approx \varphi_2$. This gives
$$x = \left(\frac{\sin \varphi_1 - \sin \varphi_2}{\tan \varphi_1 - \tan \varphi_2}, \frac{\cos \varphi_1 - \cos \varphi_2}{\cot \varphi_1 - \cot \varphi_2} \right),$$
using the formulas $\tan \varphi = \sin \varphi / \cos \varphi$ and $\cot \varphi = \cos \varphi / \sin \varphi$. We fix φ_2 and let φ_1 tend to φ_2. Then the point of intersection tends to the point
$$(\cos^3 \varphi_2, \sin^3 \varphi_2).$$

This follows from the formulas for the derivative of the functions sin, cos, tan, and cot and the definition of the derivative. The coordinates $x_1 = \cos^3 \varphi_2$ and $x_2 = \sin^3 \varphi_2$ of this point satisfy the equation

$$x_1^{2/3} + x_2^{2/3} = 1, \qquad (*)$$

using the formula

$$\sin^2 \varphi + \cos^2 \varphi = 1.$$

That is, the curve in the child's drawing consists of the nonnegative solutions of the equation $(*)$. The curve consisting of all solutions of this equation is called an *astroid* (Fig. 4.8). It also turns up in the

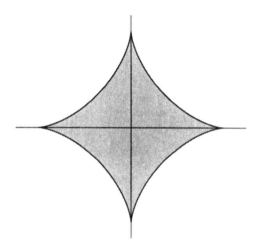

Figure 4.8 Astroid.

solution of the problem of Apollonius (section 3.3.8).

Conclusion to section 4.3. We have seen that it is possible to apply the KKT conditions to problems of

- geometry (problems 4.3.1 and 4.3.4),
- economics (problem 4.3.2),
- finance (problems 4.3.3),
- drawing (problem 4.3.5).

INEQUALITY CONSTRAINTS AND CONVEXITY

Invitation to chapters eight and nine. In section 8.5 the solution of Nash is given of the following problem: what is a fair outcome for two individuals who have a bargain opportunity? This problem is also considered in exercise 4.6.13.

In section 8.6, arbitrage-free bounds for the prices of assets are given. In section 8.7, the solution of Black and Scholes is given for the following problem: what is the fair price for a European call option? The equation of Black and Scholes determines a unique fair price. This result revolutionized the financial world. In section 8.8, a convenient probabilistic criterion for the absence of arbitrage opportunities is presented.

Problem 8.9.1, "How to assign a penalty," illustrates the celebrated *minimax theorem* of von Neumann.

Another application is the *second welfare theorem* (section 8.10). This is a classical theoretical result on the question whether a price system can lead to an allocation that is not only efficient but also fair. If the answer is negative, then this might suggest that it is necessary to interfere in the labor market to correct for unequal chances due to different backgrounds. The second welfare theorem supports the view that price mechanisms can achieve each reasonable allocation of goods without interference, provided one corrects the starting position; for example, in the case of the labor market, this can be achieved by means of education.

In problem 9.4.1 "Determine the distance to an ellipse from a point outside the ellipse," the nonnegativity of the Lagrange multipliers in the KKT conditions is used to single out the optimal point among the stationary points.

In section 9.3 we make the point that there exists an "algorithm" that produces proofs of all the results of the theory of systems of linear inequalities.

4.4 PROOF OF THE KARUSH-KUHN-TUCKER THEOREM

Why do the KKT conditions hold? This section will answer this question. The answer will provide more insight, but the contents of this section will not be used for solving concrete problems.

4.4.1 Proof of theorem 4.5

We begin with the proof of theorem 4.5. In the proof we will refer to the three KKT conditions $(\alpha), (\beta), (\gamma)$. Note that condition (α) implies that $\hat{x}$ is a global minimum of $\mathcal{L}(x, \lambda)$, as $\mathcal{L}(x, \lambda) = \sum_{i=0}^{m} \lambda_i f_i(x)$

is a convex function of x for each λ in the nonnegative orthant.

Proof.
1. Let x be an admissible point; then—normalizing λ_0 to be 1, as we may—we get

$$f_0(x) \overset{(\beta)}{\geq} \mathcal{L}(x,\lambda) \overset{(\alpha)}{\geq} \mathcal{L}(\widehat{x},\lambda) \overset{(\gamma)}{=} f_0(\widehat{x}).$$

2. Let $\bar{x}$ be a Slater point and assume that the KKT conditions hold at $\widehat{x}$ for a nonzero selection of Lagrange multipliers λ. To prove that $\lambda_0 \neq 0$, we will argue by contradiction: if $\lambda_0 = 0$, then

$$0 = \mathcal{L}(\widehat{x},\lambda) \overset{(\alpha)}{\leq} \mathcal{L}(\bar{x},\lambda) = \sum_{i=1}^{m} \lambda_i f_i(\bar{x}) < 0,$$

which is the desired contradiction.

$\square$

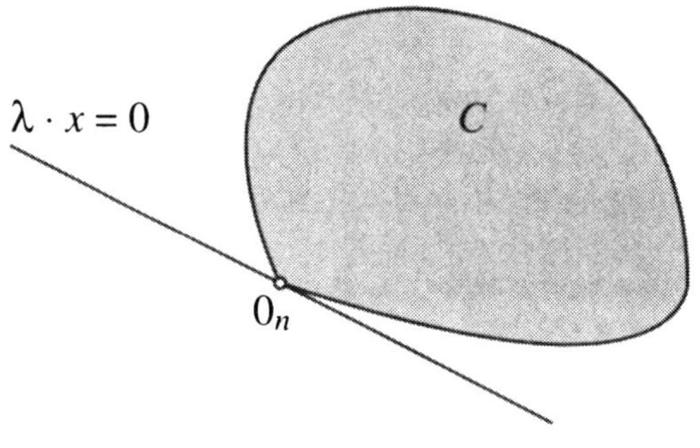

Figure 4.9 Supporting hyperplane theorem.

4.4.2 Supporting hyperplane theorem

Plan. The KKT theorem is just a consequence of the supporting hyperplane theorem, as we shall see in the next section. This is the main theorem of the convex calculus. It sense is, roughly speaking, that it gives the possibility to approximate convex nonlinear objects by linear ones. Therefore the full power of the KKT theorem comes from the supporting hyperplane theorem. Thus the situation runs parallel to that of the proof of the Lagrange multiplier rule. That is, the supporting hyperplane theorem plays a similar role in the proof of the KKT theorem to that of the tangent space theorem 3.6 in the proof of the Lagrange theorem.

Theorem 4.6 *Supporting hyperplane theorem—the main theorem of the convex calculus. If C is a convex set in $\mathbb{R}^n$ and 0_n is a point of C that is not an interior point of C, then there exists a nonzero $\lambda \in (\mathbb{R}^n)^T$ such that $\lambda \cdot x \geq 0$ for all $x \in C$ (Fig. 4.9).*

The geometrical meaning of this result is that the hyperplane with equation $\lambda \cdot x = 0$ supports the set C at the point 0_n (that is, $0_n \in C$, and C lies entirely on one of the two sides of the hyperplane).

In the proof, we will use the following concepts.

Closure. The *closure* of a set A in $\mathbb{R}^n$ is the smallest closed set in $\mathbb{R}^n$ containing A. In other words, it is the intersection of all closed sets in $\mathbb{R}^n$ containing A. It is denoted cl$A$.

Cone. The *cone spanned by a set A* in $\mathbb{R}^n$ is the smallest cone containing A. It is the set of all vectors ρa, with $\rho \in \mathbb{R}_+$, the nonnegative numbers, and $a \in A$. It is denoted as $\mathbb{R}_+ \cdot A$. Thus, it is the union of all *rays*

$$R_a = \{\rho a : \rho \in \mathbb{R}_+\},\ a \in A.$$

Proof.

Degenerate case. If int$C = \emptyset$, then C is contained in a hyperplane $\mathcal{H} = \{x \in \mathbb{R}^n \mid \lambda \cdot x = 0\}$ for a nonzero vector $\lambda \in (\mathbb{R}^n)^T$. Then λ satisfies the required conditions.

Regular case: geometrical idea of the proof. If $\bar{c} \in \text{int}C$, then we let $\widehat{x}$ be a point in the set

$$D = \text{cl}(\mathbb{R}^+ \cdot C)$$

of minimal distance to the point $-\bar{c}$. Such a point exists by the Weierstrass theorem, as can be seen by adding the constraint

$$|x + \bar{c}| \leq |\bar{c}|$$

to make the admissible set bounded, while keeping it closed and nonempty (it still contains 0_n). One can even prove the uniqueness of this point, but we do not need this. Then the choice $\lambda = (\widehat{x} + \bar{c})^T$ satisfies the requirements. Figure 4.10 illustrates that this is obvious from a geometrical point of view.

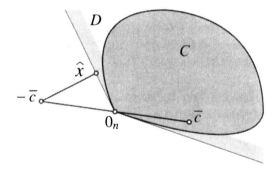

Figure 4.10 Geometrical proof of the supporting hyperplane theorem.

Analytical verification

- **Choice of hyperplane.** Let us verify this analytically as well. Choose $\varepsilon > 0$ such that the open ball $U_n(\bar{c}, \varepsilon)$ is contained in C. Then the set

$$\mathbb{R}_+ \cdot U_n(-\bar{c}, \varepsilon) \setminus \{0_n\}$$

has no points in common with C; otherwise we would get a contradiction with the convexity of C and the assumption that 0_n is a boundary point of C (Fig. 4.11). It follows that $-\bar{c}$ is

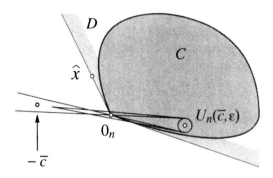

Figure 4.11 Proof of $-\bar{c} \notin D$.

not contained in the set D, and in particular $\widehat{x} \neq -\bar{c}$. Therefore the solution set of the equation

$$(\widehat{x} + \bar{c})^T \cdot x = 0$$

INEQUALITY CONSTRAINTS AND CONVEXITY 233

is a hyperplane, as required.

- **The hyperplane is supporting.** Finally we check that this hyperplane supports D at the point $\widehat{x}$, which implies that it supports C at the point $c = 0_n$. We will use that the set D is convex; this follows from the convexity of C. Choose an arbitrary point $a \in D$. Then one has

$$|ta + (1-t)\widehat{x} + \bar{c}| \geq |\widehat{x} + \bar{c}|$$

for all $t \in [0,1]$, by the minimality property of $\widehat{x}$ and the convexity of D. The left-hand side can be rewritten as

$$|t(a - \widehat{x}) + (\widehat{x} + \bar{c})|.$$

Now, take the square of both sides of the inequality, rewrite the left-hand side (use the formula $|v|^2 = v^T \cdot v$, expand, and apply on two of the terms of the resulting expression again the formula $|v|^2 = v^T \cdot v$, but now "from right to left"), and subtract $|\widehat{x} + \bar{c}|^2$ from both sides. The resulting inequality is

$$t(t|a - \widehat{x}|^2 + 2(\widehat{x} + \bar{c})^T \cdot (a - \widehat{x})) \geq 0.$$

As $t \in [0,1]$ is arbitrary, it follows that

$$(\widehat{x} + \bar{c})^T (a - \widehat{x}) \geq 0.$$

As $a \in D$ is arbitrary, this proves that the chosen hyperplane supports D at the point $\widehat{x}$, as required.

□

Conclusion. We have just given the second proof that requires a genuine idea (the first one was the proof of the tangent space theorem 3.6). It is the proof of the basic theorem from which we will derive the KKT theorem in the next section. The proof given above is the standard one: it makes use of the Weierstrass theorem. Just as in the case of the tangent space theorem, this proof is essentially an application of the four-step method for solving optimization problems; we will make this clear in section 4.5.2. There is a second—completely different—proof. This will be given in section 4.5.2.

4.4.3 Proof of the KKT theorem

Thus prepared, we can prove theorem 4.4.

Proof.
Without loss of generality, assume $f_0(\hat{x}) = 0$. Consider the set

$$C = \{y \in \mathbb{R}^{m+1} \mid \exists x \in \mathbb{R}^n \; y_i \geq f_i(x), \; 0 \leq i \leq m\}.$$

- C is a convex set (if x is "good" for y and x' is "good" for y', then

$$x_\alpha = (1-\alpha)x + \alpha x' \text{ is "good" for } y_\alpha = (1-\alpha)y + \alpha y'$$

for all α with $0 \leq \alpha \leq 1$).

- $0_{m+1} \notin \mathrm{int}\, C$ (if $0_{m+1} \in \mathrm{int}\, C$, then

$$(-\varepsilon, 0, \ldots, 0)^T \in C$$

for sufficiently small $\varepsilon > 0$; this contradicts the minimality property of $\hat{x}$).

From the supporting hyperplane theorem it follows that there exists a nonzero $\lambda \in (\mathbb{R}^{m+1})^T$ such that

$$\lambda \cdot y \geq 0 \text{ for all } y \in C. \tag{$*$}$$

Substituting in $(*)$ suitable elements of C gives the KKT conditions $(\alpha), (\beta), (\gamma)$, and $\lambda_0 \geq 0$:

1. $\mathbb{R}_+^{m+1} \subseteq C$ (as $\hat{x}$ is good for all $y \in \mathbb{R}_+^{m+1}$) $\overset{(*)}{\Rightarrow} (\beta)$ and $\lambda_0 \geq 0$.

2. If $f_i(\hat{x}) < 0$ for some $i \geq 1$, then

$$y = (0, \ldots, 0, f_i(\hat{x}), 0, \ldots, 0) \in C \overset{\beta,(*)}{\Rightarrow} \lambda_i \leq 0 \Rightarrow \lambda_i = 0 \Rightarrow (\gamma).$$

3. $(f_0(x), \ldots, f_m(x))^T \in C$ for all $x \in \mathbb{R}^n$ $\overset{(*)}{\Rightarrow}$

$$\sum_{i=0}^m \lambda_i f_i(x) \geq 0 = \sum_{i=0}^m \lambda_i f_i(\hat{x})$$

for all $x \in \mathbb{R}^n$. This implies (α) by the Fermat theorem.

Crash courses theory. This is a good moment to look up in appendices E and F, the crash courses on optimization theory, the parts concerning this chapter.

Conclusion to section 4.4. The KKT theorem follows readily from the supporting hyperplane theorem.

4.5 DISCUSSION AND COMMENTS

4.5.1 Dual problems

> You can see everything with one eye, but looking with two eyes is more convenient.   *anonymous expert on algorithms*

Plan and motivation. If you want to solve a convex optimization problem numerically, that is, by an algorithm, then one of the central ideas is duality theory. Its practical value is that it associates to any given convex problem—called *the primal problem*—a concave problem—called *the dual problem*—such that admissible elements of the dual problem give *lower* bounds for the value of the given problem. Note that admissible elements of the primal problem give *upper* bounds for this value, clearly. Often, for example for solvable LP problems, this leads to lower and upper bounds that are arbitrarily close. Therefore, if you want to solve a problem numerically, it is a good idea to solve its dual as well. The epigraph above expresses this experience of experts, in the form of a metaphor. Duality theory is essentially a reformulation of the KKT theorem. That is why we present it here. However, for the same reason, it gives nothing new if you want to solve a convex problem analytically.

A further use of duality theory is that it gives some insight; the solution of the dual problem allows a shadow price interpretation. Moreover, this solution is precisely the Lagrange multiplier from the KKT theorem.

The idea of duality theory was already given in the smooth case, in section 1.5.3.

In the same way one can consider a given concave problem as the primal problem. Then the dual problem is a convex problem.

Dual problems for LP problems in standard form. We begin with LP problems. Consider an LP problem (P) in standard form,

$$c^T \cdot x \to \min, \ Ax = b, \ x \geq 0_n.$$

That is, the problem is to minimize a linear function on the intersection of a *flat* (=affine subspace) and the first orthant.

Then its dual problem (D) is defined to be the LP problem

$$b^T \cdot y \to \max, \quad A^T y \leq c.$$

The given LP problem is called the primal problem.

Idea of the dual problem. If you have found an admissible element x of the primal problem and you are viewing it as an approximation of the solution, then it is natural to ask how far away you still are from the value $v(P)$ of the primal problem (P). Now suppose that you have found a solution y of the inequality $A^T y \leq c$. This gives you the information that you are not farther away from the optimal value than

$$c^T \cdot x - b^T \cdot y,$$

as $b^T \cdot y \leq v(P) \leq c^T \cdot x$. Indeed,

$$c^T \cdot x \geq (A^T y)^T x = y^T A x = y^T b = b^T y,$$

and as this holds for arbitrary admissible elements x of (P) and y of (D), it follows that

$$b^T \cdot y \leq v(P) \leq c^T \cdot x.$$

That is, each solution of the inequality $A^T y \leq c$ gives the following lower bound for $v(P)$: $b^T \cdot y$. Of course, you want to have the best possible upper bound. Thus you are led to the dual problem

$$b^T \cdot y \to \max, \quad A^T y \leq c.$$

In fact, the following more precise statement is proved by the argument above:

Weak duality. $b^T \cdot y \leq v(D) \leq v(P) \leq c^T \cdot x$ for all admissible elements x of (P) and y of (D).

In fact, a stronger result holds:

Strong duality. $v(D) = v(P)$, unless the problems (P) and (D) both have no admissible elements.

The following result gives the complete picture.

INEQUALITY CONSTRAINTS AND CONVEXITY

Theorem 4.7 *Main theorem on LP problems.* *Let a primal LP problem in standard form and its dual be given.*

1. *Duality theorem.* *The minimum value of the primal problem equals the maximal value of the dual problem, unless these problems are both infeasible.*

2. *Existence result.* *An LP problem has a solution if its value is finite.*

3. *Complementary slackness conditions.* *Let $\widehat{x}$ (resp. $\widehat{y}$) be an admissible element of the primal (resp. dual) problem. Then the complementarity conditions*

$$\widehat{x}_i (A^T \widehat{y} - c)_i = 0,\ 1 \leq i \leq n,$$

hold if and only if $\widehat{x}$ is a solution of the primal problem and $\widehat{y}$ is a solution of the dual problem.

We note that the complementary (slackness) conditions are the basis of the primal-dual version of the simplex method, the classical algorithm to solve LP problems. In particular, Dantzig's uv-method for the transport problem and the network-simplex method for the minimum flow problem are based on the complementary slackness conditions.

Dual problems for LP problems in symmetric form. One of the other equivalent ways to write LP problems is in symmetric form,

$$c^T \cdot x \to \min,\ Ax \geq b,\ x \geq 0_n.$$

We only give the description of the dual problem of such a problem:

$$b^T \cdot y \to \max,\ A^T y \leq c,\ y \geq 0_m.$$

Duality theory for arbitrary convex problems. The idea of the construction of the dual problem can be extended to arbitrary convex problems. Convex problems can always be written in the following primal conic form,

$$c^T \cdot x \to \min,\ Ax = b,\ x \succeq 0_n,$$

$\succeq$ is a vector ordering, defined by

$$x_1 \succeq x_2 \text{ precisely if } x_1 - x_2 \in C,$$

where C is a convex cone in $\mathbb{R}^n$. To be more precise, C is a closed pointed, solid, convex cone. That is, we write the given convex problem as a problem to minimize a linear function on the intersection of a flat and a suitable convex cone. For details, and the proof that arbitrary convex problems can be put in conic form, we refer to chapter ten. The analogy between a problem in conic form and the primal LP problem in standard form is striking. In fact, if in this type of problem we choose C to be the first orthant in $\mathbb{R}^n$, then the conic problem specializes precisely to the LP problem in standard form. Therefore, the problem in conic form is a very natural extension of the primal LP problem in standard form. Moreover, it is of great interest because of the result that all convex problems can be put in conic form.

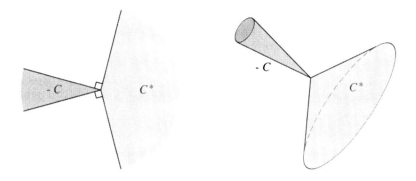

Figure 4.12 Dual cones.

Now we define the dual problem of the primal conic problem to be the problem

$$b^T \cdot y \to \max, \ A^T y \preceq^* c,$$

where $\preceq^*$ is a vector ordering, defined by

$$y_1 \preceq^* y_2 \text{ precisely if } y_2 - y_1 \in C^*,$$

where C^* is *the dual cone* of C. This dual cone is defined to be

$$C^* = \{y \in \mathbb{R}^n : y^T \cdot x \geq 0 \ \forall x \in C\}.$$

This is a closed convex cone. The concept of a dual cone is illustrated for $n = 2$ and $n = 3$ in Figure 4.12.

The reason for defining the dual problem in this way is that we can extend the method above to obtain lower bounds for the value of a

INEQUALITY CONSTRAINTS AND CONVEXITY

given convex problem (= the primal conic problem). Indeed,

$$c^T \cdot x \geq (A^T y)^T x = y^T A x = y^T b = b^T y$$

for arbitrary admissible elements x of the primal conic problem and arbitrary admissible elements y of the dual conic problem. Thus we get the following weak duality result for convex problems (strong duality does not extend to the general case).

Theorem 4.8 Main duality theorem for convex problems. *Let a convex problem be given in primal conic form. Then its value is at least equal to the value of its dual conic problem.*

Practical use of the duality theorem. For many convex cones of interest one has a explicit description for the dual cone, and then the duality theorem can be used to get explicit lower bounds for the value of a given convex problem.

Relation with KKT. Duality theory is a reformulation of the KKT theorem for the case of the convex problem with inequality constraints ($P_{4.1}$). We will not write out explicitly the formula manipulation that shows this. The KKT theorem can be extended to convex problems written in primal conic form; this result is essentially equivalent to duality theory for convex problems.

Shadow price interpretation of the solution of the dual problem. The shadow price interpretation of solutions of dual problems follows by combining the shadow price interpretation of the Lagrange multipliers in the KKT theorem with the relation between duality theory and the KKT theorem.

Relation of the explicit definition and the picture approach. The idea of the definition of the dual problem and its relation to the primal problem has been given in chapter one for smooth problems in terms of a picture (Fig. 1.10). The same idea can be used for arbitrary *convex* problems ($P_{4.1}$) (no assumptions of differentiability are needed), where we replace $f_i(x) \leq 0$ by $f_i(x) \leq y_i$ to define the required perturbation. It can be applied even more generally to convex problems written in primal conic form, where we replace the equation $Ax = b$ by $Ax = b + y$ to define the required perturbation. We do not display the details here, but only in chapter ten.

4.5.2 Geometric sense of the proofs of the supporting hyperplane theorem

Plan and motivation. The power of the Karush-Kuhn-Tucker theorem comes from the supporting hyperplane theorem. Indeed, we have seen that the KKT theorem is an immediate corollary. The supporting hyperplane theorem can be established in two completely different ways. We will present the geometrical idea of each of these proofs.

To reach the essence of the matter, we consider a variant of the supporting hyperplane theorem that is equivalent to it. In order to do this, we have to introduce a special type of convex set, *convex cones*. Investing some time in this concept will pay off: all convex sets can be turned into convex cones, and this greatly simplifies the analysis of convex sets. Both proofs presented in this section illustrate this general phenomenon. However, we have chosen not to make use of convex cones in the main text of this book (except for the presentation of unification in chapter ten).

Convex cone. A *convex cone* C in $\mathbb{R}^n$ is a subset of $\mathbb{R}^n$ that contains 0_n and that is closed under addition of vectors and under multiplication with nonnegative scalars:

$$v, w \in C \Rightarrow v + w \in C$$

and

$$\lambda \in \mathbb{R}_+, v \in C \Rightarrow \lambda v \in C.$$

Alternatively, a convex cone in $\mathbb{R}^n$ can be defined as a convex set in $\mathbb{R}^n$ that has the property that it is a union of the origin 0_n and a collection of open rays $\mathcal{S}_a = \{\rho a : \rho > 0\}$, where a runs over a suitable set of nonzero vectors of $\mathbb{R}^n$.

Dual of a convex cone. The proper way of defining the dual cone is as lying in the dual space $(\mathbb{R}^n)'$, rather than in the space $\mathbb{R}^n$ of the original cone, as we have done in section 4.5.1. That is, we define the dual C' of a convex cone C in $\mathbb{R}^n$ to be the set

$$C' = \{w \in (\mathbb{R}^n)^T : w \cdot c \geq 0 \ \forall c \in C\}.$$

Thus C^* and C' are each other's transposes; both versions are in use.

Thus prepared, we can formulate the promised variant of the supporting hyperplane theorem.

INEQUALITY CONSTRAINTS AND CONVEXITY

Theorem 4.9 *The dual D' of a convex cone D in $\mathbb{R}^n$ that is not the whole space, $D \neq \mathbb{R}^n$, contains nonzero elements, $D' \neq \{0_n\}$.*

Reduction from convex sets to convex cones. The concept of "convex cone" is a special case of the concept "convex set." It might be surprising at first sight that, conversely, "convex set" may be viewed as a special case of "convex cone." This is illustrated in Figure 4.13 for convex sets in the plane $\mathbb{R}^2$. In this figure, $\mathbb{R}^2$ is modeled as a

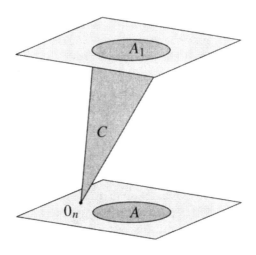

Figure 4.13 Associating a convex cone C to a convex set A.

horizontal plane, say the floor. The point 0_n, with $n = 2$, on the floor represents the origin $(0,0)^T \in \mathbb{R}^2$. A convex set $A \subseteq \mathbb{R}^2$ is represented as a convex plane figure lying on the floor. Take this figure and lift it up vertically to level 1. This brings it to position A_1. Then take the convex cone consisting of the union of all rays that start at 0_n and run through a point of A_1. The result of this construction is the convex cone C. Note that no information has been lost by the transition from A to C. Indeed, from C we can recover A: intersecting C with the horizontal plane at level 1, gives the set A_1; dropping it vertically on the floor gives A.

Now we turn to the case of general dimension n. We proceed in the same way as in the case $n = 2$. Thus we associate to a convex set A in $\mathbb{R}^n$ the following convex cone in $\mathbb{R}^{n+1}$:

$$C = \{\alpha(a^T, 1)^T : \alpha \geq 0,\ a \in A\}.$$

Note that $(a^T, 1)^T$ denotes the column vector in $\mathbb{R}^{n+1}$ that you get by adding 1 at the bottom of the column vector a in $\mathbb{R}^n$.

Relation with the supporting hyperplane. We sketch the derivation of the supporting hyperplane theorem 4.6 from this result. Apply this result to the smallest convex cone in $\mathbb{R}^{n+1}$ that contains the points $(c^T, 1)^T$, $c \in C$ and the point $(0_n^T, -1)^T$, after having checked that this is not the whole space $\mathbb{R}^{n+1}$. The resulting nonzero element of D' is of the form $(a, 0)^T$ with $a \in (\mathbb{R}^n)^T$, and this gives the required supporting hyperplane $a \cdot x = 0$ of C at 0_n.

We begin with the proof of theorem 4.9 that is in the same spirit as the proof of the supporting hyperplane theorem that we have chosen in the main text. This is the standard proof.

Proof 1: using the four step method (Weierstrass). We assume that D contains an interior point $\bar{d}$, without restriction of the generality of the argument. Indeed, the general case can be reduced to this case by replacing $\mathbb{R}^n$ by the smallest linear subspace containing D. Moreover, we assume that D is closed, as we may without restriction to the generality of the argument, replacing D by its closure and noting that $-\bar{d}$ cannot be contained in this closure.

We apply the four-step method to the problem of the distance from $-\bar{d}$ to D. We will show that this problem has a solution $\hat{d}$, and that $(\hat{d} + \bar{d})^T$ is a nonzero element of D'.

1. Consider the problem
$$f(d) = |d + \bar{d}| \to \min, \ d \in \mathrm{cl}D,$$
where $\mathrm{cl}D$ denotes the closure of D, the smallest closed set containing D. A solution $\hat{d}$ exists by coercivity, as $|d + \bar{d}| \to +\infty$ if $|d| \to +\infty$. Note that $\hat{d}$ is not equal to $-\bar{d}$ as $-\bar{d} \notin \mathrm{cl}D$.

2. Take an arbitrary $d \in D$. Write $r = d - \hat{d}$. Define $d_\alpha = \hat{d} + \alpha d$ for $\alpha > 0$ and define $g(\alpha) = f(d_\alpha)$. Then
$$g(\alpha) \geq g(0) \Rightarrow |v + \alpha d|^2 \geq |v|^2.$$

3. Expanding, simplifying and using that α can be chosen arbitrarily small, give $v^T \cdot d \geq 0$.

4. $v^T = (\hat{d} + \bar{d})^T$ is a nonzero element of D' as $d \in D$ is arbitrary.

Remark. Note that here working with convex cones leads to the following simplification: instead of having to separate by a hyperplane a closed convex set from a point on its boundary, we have to separate it by a closed convex set from a point outside this set, and this is easier.

Proof 2: by induction. The second proof is nonstandard in the present finite-dimensional context. However, it is the universal proof: it can be extended to an infinite-dimensional context, which is needed for the results from the calculus of variations and optimal control. This proof is essentially the proof of the Hahn-Banach theorem from functional analysis.

Rough outline. Note that we have obtained a nonzero element of D^* in one blow in the proof using Weierstrass. In the universal proof, this element is produced gradually by constructing a chain of linear subspaces $L_1 \subseteq \cdots \subseteq L_{n-1}$, where L_i has dimension i and does not contain any interior point of D, $1 \leq i \leq m$. This implies that L_{n-1} is a hyperplane that is the zero set of the required nonzero element of D'. Now we sketch the idea of this proof for $n = 2$ and $n = 3$.

Geometrical idea ($n = 2$). For $n = 2$, each convex cone D is an "angle" of at most π. Then, turning a line with endpoint 0_2, we can clearly arrive at a position L_1 of the line that has D completely on one side. This line is the zero set of a nonzero element of D'.

Geometrical idea ($n = 3$). For $n = 3$, we choose an interior point $\bar{d}$ of D and a plane through $\bar{d}$ and 0_n. This plane intersects D in a cone E, and we can choose a line L_1 in this plane that has this cone E on one side, by the case $n = 2$, which already has been settled. Then we turn a plane through this line till we arrive at a position L_2 of the plane that has D completely on one side. Indeed this is essentially the same problem that we considered in the case $n = 2$, when we turned a line. This plane is the zero set of the required nonzero element of D'.

In this way one can continue, and this leads to a proof for all n.

Remark. Note that here, working with convex cones leads to the following simplification: we can prove the theorem recursively, for higher and higher dimension ("complete induction with respect to the dimension"). Each step to a higher dimension is seen to be equivalent to having to settle the two-dimensional case. Thus the theorem is reduced to the case of dimension two, and this is easy.

Conclusion. We have seen that there are two proofs of the supporting hyperplane theorem, based on completely different ideas. One proof uses the Weierstrass theorem. The other one is the universal proof: it can be extended to the infinite-dimensional case; this is needed for the proofs of the infinite-dimensional theory, which is needed for the solution of problems of the calculus of variations and optimal control.

4.5.3 Shadow price interpretation of the KKT multipliers.

Plan. We give a shadow price interpretation of the KKT multipliers. We will see that this is more simple than for the Lagrange multipliers. The reason for this is that "linearization by means of supporting hyperplanes" is a simpler concept than "linearization by means of tangent spaces."

The Lagrange multipliers allow the following shadow price interpretation. For each $y \in \mathbb{R}^m$ we can consider the *perturbed* problem of $(P_{4.1})$,

$$f_0(x) \to \min, \ f_1(x) \leq y_1, \ldots, f_m(x) \leq y_m. \qquad (P_y)$$

We let $S(y)$ be the optimal value of (P_y). Let R be the set of $y \in \mathbb{R}^m$ for which $S(y)$ is not $+\infty$ or $-\infty$. Note that $S(y) = -\infty$ means that the problem (P_y) is unbounded below and $S(y) = \infty$ means that (P_y) has no admissible points. Thus $S(y)$ is a function on R.

Theorem 4.10 *Shadow price interpretation of the KKT multipliers. Consider the situation of theorem 4.4. Assume that the KKT conditions hold for the point $\hat{x} \in \mathbb{R}^n$ and the Lagrange multiplier $\lambda = (\lambda_0, \ldots, \lambda_m)$ with $\lambda_0 = 1$. Let the subset R of $\mathbb{R}^m$ and the function S on R be as defined above. Then R is a convex set and S is a convex function on R. Moreover,*

$$S(y) \geq S(0) - \sum_{i=1}^{m} \lambda_i y_i,$$

for all $y \in R$.

If the Slater condition holds, then S behaves nicely on a neighborhood of 0_m in the sense that there is a small positive number $\varepsilon > 0$ such that $U_m(0_m, \varepsilon) \subseteq R$.

If the problem is one of *cost minimization*, then this result can be stated as follows. It is worth paying a price of at most λ_i to prevent a change in the right-hand side of the i-th constraint. Indeed, then you have to pay $-\sum_{i=1}^{m} \lambda_i y_i$ to prevent that the cost minimization problem $(P) = (P_0)$ is replaced by (P_y), bringing your total cost to

$$S(0) - \sum_{i=1}^{m} \lambda_i y_i,$$

no more than $S(y)$, which is the value of the cost minimization problem (P_y).

Conclusion. We have given, without making any assumptions, a simple shadow price interpretation for the KKT multipliers. Note the contrast with the shadow price interpretation of the Lagrange multipliers given in theorem 3.7, which requires very strict assumptions.

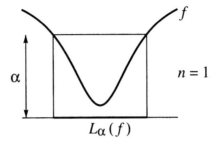

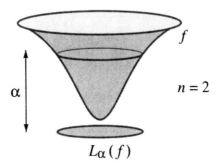

Figure 4.14 Illustration of quasi-convex functions.

4.5.4 Variants of the KKT theorem

Plan and motivation. It is natural to try giving variants of the KKT theorem. In this theorem assumptions of convexity and smoothness (that is, differentiability) are made. What happens if we assume only (quasi)convexity *or* smoothness?

Moreover, the KKT theorem is given for problems with inequalities only. It is of interest to include equality constraints as well.

Here we will indicate what sort of variants of the KKT theorem are possible. The proofs of these variants except one are similar to the proof of the KKT theorem given above. The exception is the variant where only smoothness assumptions are made. This is called the John theorem.

We will use in the applications the following variants of theorem 4.7. We need some definitions first. A function f on a convex set A is called *quasiconvex* if for each $\alpha \in \mathbb{R}$ the level set

$$\mathcal{L}_\alpha(f) = \{x \in A \mid f(x) \leq \alpha\}$$

is a convex set (Fig. 4.14). A function f is called *concave* (resp. *quasiconcave*) if $-f$ is strictly convex (resp. quasiconvex).

Variants

1. If the constraints also include equality constraints $f_i(x) = 0$, with f_i affine, that is, $f_i(x) = c_i \cdot x + \alpha_i$, then one has to make the following changes. In the definition of a Slater point: it should satisfy these equality constraints. In the definition of the KKT conditions: for the multipliers of the equality constraints one has neither the non-negativity conditions nor the conditions of complementary slackness.

2. The theorem holds as well if we require that the functions f_i, $0 \leq i \leq m$, are only quasiconvex instead of convex.

3. **The Karush-Kuhn-Tucker theorem.** The theorem holds as well if we drop the smoothness condition and add an inclusion constraint $x \in A$ for some convex set A, provided in the KKT conditions we replace the stationarity condition by the following *minimum principle*:

$$\min_{x \in A} \mathcal{L}(x, \lambda) = \mathcal{L}(\widehat{x}, \lambda).$$

In fact, this is the "official" KKT theorem. This minimum principle is not equivalent to the stationarity condition, as a convex function can be nondifferentiable: the function $\mathcal{L}(x, \lambda)$ might have a "kink" at $\widehat{x}$—then the stationarity condition is not defined—or $\widehat{x}$ might be a boundary point of A—then the stationarity condition might not hold. We note here that minimum principles play a central role in the necessary conditions for optimal control problems, which will be considered in chapter twelve. We now display this "official" KKT theorem, but not the proof, as it is essentially the same as the proof of theorem 4.4 that we have given.

Theorem 4.11 Karush-Kuhn-Tucker theorem—necessary conditions for convex programming problems. *Consider a problem of the type*

$$f_0(x) \to \min, \ f_i(x) \leq 0, \ 1 \leq i \leq k,$$

$$f_i(x) = 0, \ k+1 \leq i \leq m, \ x \in A,$$

where A is a convex set in $\mathbb{R}^n$, $f_i : A \to \mathbb{R}$, $0 \leq i \leq k$ (resp. $k+1 \leq i \leq m$), are convex (resp. affine) functions on A.

Then, if $\widehat{x}$ is a point of minimum, there exists a nonzero selection of Lagrange multipliers λ such that

- $\min_{x \in A} \mathcal{L}(x, \lambda) = \mathcal{L}(\widehat{x}, \lambda)$,
- $\lambda_i \geq 0, \ (0 \leq i \leq k)$,
- $\lambda_i f_i(\widehat{x}) = 0, \ (1 \leq i \leq k)$.

Moreover, the following supplement holds true.

Theorem 4.12 Sufficient conditions for convex programming problems. *Consider the same situation as in the theorem above. Then the necessary conditions are sufficient as well, if $\lambda_0 = 1$. If there exists a Slater point $\bar{x}$—that is, an admissible point for which the inequality constraints hold strictly—then the bad case $\lambda_0 = 0$ cannot arise.*

4. **The John theorem.** The conclusion of the KKT theorem 4.4 holds as well for problems $(P_{4.1})$ without the convexity assumptions, provided we strengthen the smoothness condition. This result is the John theorem, also called the KKT theorem for smooth problems.

Theorem 4.13 *John theorem—necessary conditions for smooth programming problems.* Consider a problem of the type $(P_{4.1})$. Assume that the problem is smooth at $\hat{x}$ in the following "strong" sense: $f_0 : U_n(\hat{x}, \varepsilon) \to \mathbb{R}$ is differentiable at $\hat{x}$ and the functions $f_i : U_n(\hat{x}, \varepsilon)$, $1 \leq i \leq m$, are continuously differentiable at $\hat{x}$.

Then, if $\hat{x}$ is a point of local minimum, there exists a nonzero selection of Lagrange multipliers λ such that the KKT conditions hold at $\hat{x}$.

However, in the present case theorem 4.5 does not hold, that is, *without the convexity conditions the KKT conditions are not sufficient*. We note that in the John theorem we can also allow *equality* constraints, $f_i(x) = 0$. Then these functions f_i do not have to be of the form $f_i(x) = c_i \cdot x + \alpha_i$. They can be any functions that are continuously differentiable at $\hat{x}$. Allowing equality constraints leads to the following changes in the KKT conditions: for the multipliers of the equality constraints one does not have the nonnegativity conditions and the conditions of complementary slackness. The proof of the John theorem is given in the analytical style crash course on optimization theory.

A mixed smooth-convex problem. The problem here is the first mixed smooth-convex problem we come across. The convexity comes from the presence of inequality constraints: these can be written as an inclusion constraint

$$(-f_1(x), \ldots, -f_m(x))^T \in \mathbb{R}_+^n,$$

and the first orthant $\mathbb{R}_+^n$ is a *convex* set. The proof of the John theorem requires tools of smooth analysis (the tangent space theorem), as well as tools of convex analysis (the supporting hyperplane theorem). It is given in appendix F, "Crash Course on Optimization Theory: Analytical Style."

5. If the function f_0 is strictly convex, then there can be at most one point of minimum $\hat{x}$.

6. One can turn the theorem and its variants into results on concave (resp. quasi-concave) functions "by means of minus signs."

Conclusion. In applications, the functions defining the objective and the inequalities are usually both convex and smooth. However, if these are only convex or only smooth, then the KKT theorem still holds, after a suitable modification of the conclusion, which we displayed above. We also have displayed the KKT theorem in the case that there are equality constraints and a convex inclusion constraint, besides the inequality constraints. For the John theorem, one needs to do more work: it is a result on a mixed smooth-convex problem, and in the proof the tools of smooth and convex analysis have to be combined.

4.5.5 Existence of solutions for convex problems

Plan and motivation. The most tricky point of the four-point method can be simplified if the problem is convex. This is the verification of boundedness or coercivity, necessary to establish existence by means of the Weierstrass theorem. Note that one needs the Weierstrass theorem for convex problems that cannot be solved analytically, but only by characterization of the solution by means of equations.

Boundedness: in general tricky. For convex problems, one has to prove existence of solutions—by the Weierstrass theorem—if the KKT conditions cannot be solved explicitly. This involves the verification of the boundedness of the admissible set (or, of the coercivity of the objective function). As we have seen, this is often the most subtle point of the whole solution of an optimization problem.

Boundedness for convex problems: easy. However, for *convex* problems it is relatively easy to settle the existence of solutions, by virtue of the following convenient criterion. A closed convex set containing the origin is bounded precisely if it does not contain any rays (a ray is a half-line with endpoint the origin).

Existence for convex problems. Now we will formulate a general existence result for solutions. It can be derived from this criterion.

Lower semicontinuity. A function is called *lower semicontinuous* if its epigraph is a closed set. For example, continuity implies lower semicontinuity. The sense of introducing the concept of lower semicontinuity is as follows: in the Weierstrass theorem the assumption of continuity can be replaced by that of lower semicontinuity. This is especially useful in the context of minimizing convex functions. Convex functions are continuous at all interior points of their domain. However, at boundary points they might exhibit "strange behavior." An example is given in appendix B. To exclude such behavior, the lower semicontinuity condition gives the natural condition of good behavior at boundary points for convex functions. It is slightly weaker than the condition that the convex function be continuous at all boundary points.

Descent along a ray. If a convex function is defined on a ray $\{tv \mid t \geq 0\}$ and if, moreover, $f(tv)$ is a monotonic decreasing function of t, then one says that f *descends along this ray*.

Theorem 4.14 *Weierstrass theorem for convex problems. Let $f : A \to \mathbb{R}$ be a lower semicontinuous convex function on a convex subset A of $\mathbb{R}^n$ that contains the origin. Assume that A contains no ray on which f descends. Then the problem $f(x) \to \min$, $x \in A$, has a global solution.*

However, we emphasize once again that for convex problems we only have to use the Weierstrass theorem if we cannot solve the necessary conditions explicitly, such as in problem 4.3.4 on the waist of three lines in space, and in exercise 4.6.11.

Conclusion. For convex problems, one does not have to carry out the most tricky verification of the Weierstrass theorem: that the admissible set is bounded. It suffices to check the absence of descent directions: as the admissible set is convex, this implies that the admissible set is bounded.

4.6 EXERCISES

We begin with three numerical exercises. We recommend solving the first one without using necessary conditions.

Exercise 4.6.1 $x_1^2 + x_2^2 + x_3^2 \to \text{extr}$, $x_1 + x_2 + x_3 \leq 12$, $x_1, x_2, x_3 \geq 0$.

Exercise 4.6.2 $x_1^2 - x_1 x_2 + x_2^2 + 3|x_1 - x_2 - 2| \to \min$.

Exercise 4.6.3 $e^{x_1 - x_2} - x_1 - x_2 \to \min$, $x_1 + x_2 \leq 1$, $x_1, x_2 \geq 0$.

In the next exercise we consider problem 4.3.4, but with two lines instead of three.

Exercise 4.6.4 *Waist of two lines.* Solve the problem of the shortest distance between two lines in three-dimensional space. That is, how does one find a point on each line such that the distance between the two points is minimal? Give a geometrical solution as well as an analytical one.

The following exercise is a striking illustration of the convenience of using (strict) convexity: it helps to solve the stationarity conditions analytically (that is, by an explicit formula); it might not be possible to do this in any other way.

Exercise 4.6.5 *Minimize the average distance of a point in $\mathbb{R}^n$ to the $n+1$ vectors consisting of the standard basis e_i, $1 \leq i \leq n$, and the origin.*

At first sight, it is not at all obvious that this problem can be solved analytically. For example, note that for $n = 2$ this is a special case of the problem of Fermat and Torricelli (problem 2.4.4), which can only be solved "by characterization", not "by a formula."

Exercise 4.6.6 *What will the agent in the mean variance investment problem 4.3.3 do with the remaining \$381.98? Presumably she will put it into a risk-free investment such as a savings account, say with an interest of 3.3%. Would she have arrived at another investment decision if she had taken this risk-free investment possibility into account right from the beginning?*

In problem 4.3.3 a trade-off was made between high expected returns and low risk by maximizing the expected return for given risk. Another possibility for the investor is to minimize risk for a given expected return, as in the following exercise.

Exercise 4.6.7 **Risk minimization. A person is planning to divide her savings among three mutual funds having expected returns of 10%, 10%, and 15%. Her goal is a return of at least 12%, while minimizing her risk. The risk function for an investment in this combination of funds is*

$$200x_1^2 + 400x_2^2 + 100x_1x_2 + 899x_3^2 + 200x_2x_3,$$

where x_i is the proportion of her savings in fund i. Determine the proportions that should be invested in each fund. Would it help if she could go short, that is, if the x_i are allowed to be negative?

On all financial markets, arbitrageurs are active. Like vultures looking for prey, they try to spot "arbitrage opportunities," which allow them to make a lot of money quickly. If, for example, some share is much cheaper in Bangkok than in Chicago, then you quickly buy in Bangkok and sell in Chicago. Arbitrageurs have a beneficial effect: their activities tend to diminish such differences, bringing the markets more in equilibrium. In the following two exercises we offer a method that is useful if you have been trying in vain to spot an arbitrage opportunity. The method can lead to absolute *certainty* that there is no arbitrage opportunity, and so that the market is in equilibrium.

Exercise 4.6.8 * *Exchange, arbitrage, and prices (1).*
Consider a market where people exchange goods, maybe without using money. Each offer holds in two directions, so that if you get a bottle of vodka for a Gouda cheese, then you can also come with a bottle of vodka in order to get a Gouda cheese. It would be interesting to have the possibility to start with some goods and then make various exchanges, at the end of which you have your original bundle plus some extra goods. This is called an arbitrage opportunity.

Show that, if there are no arbitrage opportunities, then there exist prices for all goods such that—for each offer of exchange on the markets—the two bundles of goods that can be exchanged have the same value (the converse implication holds as well of course). These prices are not always unique.

Modeling. Let us model this exercise. There are n goods on the market, say, $1, \ldots, n$. There are k *basic exchange opportunities*, that is, possibilities to exchange a bundle of goods for another one; this is represented by vectors $x_1, \ldots, x_k$ in $\mathbb{R}^n$. The signs of the coordinates indicate whether one should give or take the good in question: − means "give" and + means "take." For example, a vector with all coordinates zero except the coordinate of Gouda cheese (resp. vodka bottles), which is −1 (resp. +1) represents the basic exchange opportunity to give a Gouda cheese and take a bottle of vodka in return.

More generally, an *exchange opportunity* is defined to be a linear combination of the k basic exchange opportunities $x_1, \ldots, x_k$. Here we have tacitly made the assumptions that all opportunities to exchange two bundles of goods hold in both directions and that each positive multiple of an opportunity exchange also holds on this market. An "arbitrage opportunity" is defined to be a linear combination of the vectors $x_1, \ldots, x_k$ for which all n coordinates are nonnegative and at least one is positive.

The question is to show that there exists a nonzero nonnegative row vector $p \in \mathbb{R}^n$ with $p^T \cdot x_i = 0$ for $1 \leq i \leq k$ if there is no arbitrage opportunity.

Exercise 4.6.9 Exchange, arbitrage and prices (2).
This question is similar to the previous exercise with the following difference: exchange opportunities do not necessarily hold in both directions. For example, for the exchange of currency, the rate of exchange is not the same in both directions.

The question is here to show that there exists a nonzero nonnegative row vector $p \in \mathbb{R}^n$ with $p^T \cdot x_i \leq 0$ for $1 \leq i \leq k$ if there is no arbitrage opportunity.

Hint. The difference in comparison with the previous exercise is that one should define an *exchange opportunity* here as a linear combination of the basic exchange opportunities $x_1, \ldots, x_k$ *with nonnegative coefficients*. For the solution, one should use separation of convex sets, a variant of the supporting hyperplane theorem, given in appendix B.

We consider again the basic problem of consumer theory. This time, in contrast to the previous section, we allow that the budget not be spent entirely; moreover, there might be other, inequality constraints. Modeling this problem as a *convex* problem, requires a minor trick.

Exercise 4.6.10 *A consumer has income $I > 0$, and faces a positive price vector $p \in (\mathbb{R}^3)'$ for the three commodities she consumes. All commodities must be consumed in nonnegative amounts. Moreover, she must consume at least two units of commodity 2, and cannot consume more than one unit of commodity 1. Assuming $I = 4$ and $p = (1,1,1)$, calculate the optimal consumption bundle if the utility function is given by $u(x_1, x_2, x_3) = x_1 x_2 x_3$. What if $I = 6$ and $p = (1, 2, 3)$?*

Exotic parametrization of the positive quadrant in $\mathbb{R}^2$ and positive orthant in $\mathbb{R}^3$. The following geometrical fact might seem at first sight an isolated curiosity. A point x in the plane $\mathbb{R}^2$ is called *positive* if both coordinates are positive, $x_i > 0$, $i = 1, 2$.

If P and D are two orthogonal lines in the two-dimensional plane $\mathbb{R}^2$, both containing at least one positive point, then each positive point in $\mathbb{R}^2$ can be written in precisely one way as

$$(p_1 d_1, p_2 d_2)$$

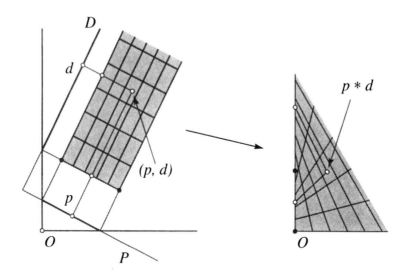

Figure 4.15 Exotic parametrization of the positive orthant.

where (p_1, p_2) (resp. (d_1, d_2)) is a positive point of P (resp. D).

This fact is illustrated in Fig 4.15. The shaded area represents the pairs of points

$$(p, d) \quad \text{with} \quad p \in P,\ p > 0_2,\ d \in D,\ d > 0_2$$

in the following way. It consists of all points in the plane for which the orthogonal projections on P and D are points with positive coordinates. In section 6.9.3, a detailed treatment will be given. Here you can try to investigate this phenomenon by yourself by doing the exercise below.

The fact we just described can be extended to a line and a plane in three-dimensional space $\mathbb{R}^3$ that are orthogonal to each other, both containing at least one positive point. If you try to establish these innocent looking statements in a straightforward way, it might be harder than you expect. Convexity can help here, as we will see below, where we will give hints on how to establish these facts in the n-dimensional case.

Exotic parametrization of the positive orthant in $\mathbb{R}^n$. These facts are more than curiosities. Their extension to $\mathbb{R}^n$ is the basis of one of the most efficient types of optimization algorithms, *the interior point methods for LP problems*, as we will see later, in section 6.9. We need some definitions.

Affine subspace. The set of solutions of a system of linear equations $Ax = b$ is called an *affine subspace* of $\mathbb{R}^n$. For example, lines in the two-dimensional plane are affine subspaces of $\mathbb{R}^2$ and lines or planes in three-dimensional space are affine subspaces of $\mathbb{R}^3$.

Direction vector. For an affine subspace P of $\mathbb{R}^n$, we call the difference of two of its vectors, that is, $p_1 - p_2$ with $p_1, p_2 \in P$, a *direction vector* of P.

Orthogonal complements. Two affine subspaces of $\mathbb{R}^n$ are called *orthogonal complements* if the direction vectors of each of them consist of all vectors of $\mathbb{R}^n$ which are orthogonal to all direction vectors of the other one. For example, two lines in the two-dimensional plane that are orthogonal in the usual sense ("they make a right angle") are orthogonal complements. To give another example, a line and a plane in three-dimensional space that are orthogonal in the usual sense are orthogonal complements.

Hadamard product. The *Hadamard product* $v * w$ of two vectors $v, w \in \mathbb{R}^n$ is defined coordinatewise by $v * w = (v_1 w_1, \ldots, v_n w_n)^T$. A point $v \in \mathbb{R}^n$ is called *positive* if all its coordinates are positive; then one writes $v > 0_n$.

Exercise 4.6.11 * *Exotic parametrization of the positive orthant in $\mathbb{R}^n$.* Let two affine subspaces P and D of $\mathbb{R}^n$ be given that are orthogonal complements. Assume that both contain positive points. Show that each positive point of $\mathbb{R}^n$ can be written in a unique way as the Hadamard product of two positive points, one of P and one of D (Fig. 4.15).

Hint. Analyze for each positive vector x of $\mathbb{R}^n$ the problem

$$f_x(p, d) = p^T \cdot d - x^T \cdot \ln(p * d) \to \min, \ p \in P, \ d \in D, \ p > 0_n, \ d > 0_n.$$

Show that f_x is strictly convex. Here the logarithm has to be taken coordinatewise; that is, we write $\ln v = (\ln v_1, \ldots, \ln v_n)^T$ for each positive point v of $\mathbb{R}^n$.

For the next exercise it is recommended to experiment with building towers from beer coasters (or cd boxes), before you try to solve it by doing calculations.

Exercise 4.6.12 *Tower of Pisa.* You have a large number of beer coasters (or cd boxes) and you want to build a tower on the boundary of a table such that the top coaster (cd-box) sticks out as far as possible over the border of the table (Fig. 4.16).

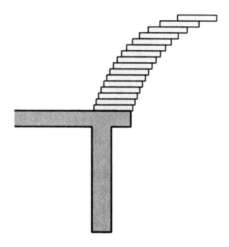

Figure 4.16 Optimal tower of Pisa.

How to do this? How far can you get away from the table if you have enough beer coasters (cd boxes)?

Exercise 4.6.13 **Nash bargaining.** *Suppose that C is a compact convex set in the plane $\mathbb{R}^2$ and that a point s in C is specified. Two individuals argue over which point in C should be adopted. If they agree on x, then the first individual receives a utility of x_1 units and the second a utility of x_2 units. If they fail to agree, it is understood that the result will be the status quo point s.*

The Nash bargaining solution for this problem is the solution of the problem

$$f(x) = (x_1 - s_1)(x_2 - s_2) \to \max, \ x \in C, \ x_i \geq s_i, \ i = 1, 2.$$

1. *Prove that this problem has a unique solution.*

2. *Determine the Nash bargaining solution if C is the set of solutions of $3x_1^2 + 4x_2^2 \leq 10$ and $s = (1, 1)^T$.*

Details about Nash bargaining are given in section 8.5.
We give a variant of problem 4.3.5 on child drawing.

Exercise 4.6.14 * Child drawing (2). *Consider again a child who draws straight lines from the left-hand side of her paper to the bottom line of her paper. This time she has arranged that the sum of the*

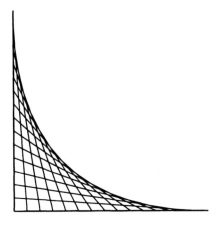

Figure 4.17 Child drawing 2.

distances of the endpoints of each line to the bottom-left corner of the paper is constant (Fig. 4.17).

All these lines are tangents (and supporting lines) of a curve. Find the equation of this curve.

Hidden convexity: nonconvex problems, modeled as convex problems. If one succeeds in modeling a problem as a convex problem, then this should be considered as a success. The material of this section already suggests this, such as the solution of the waist of three lines problem 4.3.4. Moreover, convex problems usually can be solved numerically in an efficient way (by the ellipsoid method or the barrier method, which are presented in chapters six and seven). We give two striking examples of this surprising phenomenon. We consider two popular problems that at first glance have no convexity properties whatsoever. We show how to model these as convex problems.

Exercise 4.6.15 *Minimizing a polynomial. Model the problem of finding the minimum of a polynomial function of one variable $p(x)$ of degree n as a convex problem.*

The usual formulation of this problem is $p(x) \to \min$, $x \in \mathbb{R}$. This is seldom a convex problem, as a polynomial function is "rarely" convex. At first sight, it might look impossible to give a reformulation of this problem as a convex problem. However, it can be done, as we will see now.

Solution. The problem can be modeled as follows:

$$-\rho \to \min, \ p - \rho \succeq 0,$$

where $\succeq$ is the inequality on the space of polynomials of x of degree n that is defined by

$$p_1 \succeq p_2 \text{ precisely if } p_1(x) \geq p_2(x) \text{ for all } x \in \mathbb{R}.$$

This is a convex problem, as the set of all polynomials $p(x)$ of degree n that are nonnegative—that is, with $p(x) \geq 0$ for all x—is a convex set.

The interest of this example is that the condition that a polynomial be nonnegative is equivalent to the condition that there exists a positive semidefinite matrix, the elements of which satisfy a number of linear equations. This leads to a reduction of the given optimization problem to a semidefinite optimization problem. This can be solved efficiently by barrier methods (cf. [44]).

Exercise 4.6.16 *Largest eigenvalue of a symmetric matrix. Model the nonconvex problem of finding the largest eigenvalue of a symmetric $n \times n$ matrix A as a convex problem.*

Solution. This problem can be formulated as follows:

$$\lambda \to \min, \ \lambda I_n - A \succeq 0_{n \times n},$$

where $\succeq$ is the ordering defined on the symmetric $n \times n$ matrices by

$$B \succeq C \text{ if } x^T B x \geq x^T C x \text{ for all } x \in \mathbb{R}^n.$$

This is readily seen to be a convex problem and it can be solved efficiently by *barrier methods*. We will come back to this example in chapter seven after we have defined the class of semidefinite problems.

Exercise 4.6.17 *Let $f(x)$ be a differentiable function of one variable and let $a, b \in \mathbb{R}$ be two numbers with $a < b$. Consider the problem*

$$f(x) \to \min, \ a \leq x \leq b,$$

and a number $\widehat{x} \in [a, b]$. Verify that the John theorem gives the following necessary condition for local minimality of $\widehat{x}$:

$f'(\widehat{x})$ is zero if $a < \widehat{x} < b$, it is nonnegative if $\widehat{x} = a$, and it is nonpositive if $\widehat{x} = b$.

INEQUALITY CONSTRAINTS AND CONVEXITY 259

Finally, we draw attention to the fact that the proof of the Karush-Kuhn-Tucker theorem can be presented in an algorithmic spirit.

Exercise 4.6.18 *Let $\widehat{x}$ be an admissible point of a problem $(P_{4.1})$ for which the KKT conditions do not hold for any nonzero Lagrange multiplier. Then there exists a direction vector of admissible descent $\bar{x} \in \mathbb{R}^n$. This means that there exists an admissible variation of descent

$$x_\alpha = \widehat{x} + \alpha\bar{x} + o(\alpha)$$

with $\alpha \geq 0$, for which $f_0(x_\alpha) < f_0(\widehat{x})$ for sufficiently small $\alpha > 0$. Show this.

Chapter Five

Second Order Conditions

N'y a-t-il pas là une lacune a remplir?

Isn't there a gap to be filled?

J. Liouville

- The second order conditions are essentially the only obstructions to local optimality besides first order conditions.

5.0 SUMMARY

Second order conditions distinguish whether a stationary point is a local maximum or a local minimum. However, usually they are not very helpful for solving concrete optimization problems. The following insight is of some interest, though: they are essentially the only obstructions to local minimality besides the first order conditions.

Smallest eigenvalue of a symmetric matrix. The main challenge to us was to test the second order condition for a minimization problem that has as stationary points the eigenvectors of a symmetric matrix and as global solution an eigenvector which belongs to the smallest eigenvalue. To our delight, the second order condition picked out the global solution among the stationary points.

5.1 INTRODUCTION

The material in this chapter is mainly of theoretical interest: to solve a concrete optimization problem, one only needs the first order conditions (Fermat, Lagrange, Karush-Kuhn-Tucker, John) together with the Weierstrass theorem. It gives insight into the gap between stationarity and local optimality and shows how this gap can be filled. The words of Liouville from the epigraph can be applied to any situation where you have not yet reached the essence of things (Liouville expressed his doubts on a "proof" of Fermat's Last Theorem that was presented at a meeting of the French Academy).

The first order conditions given so far are not the only obstructions to local optimality in smooth problems. For example, these conditions do not distinguish maxima from minima. Therefore, we complement the first order conditions with second order conditions. We will see that there are essentially no other obstructions to local optimality. To be more precise, we will see that the first order conditions together with a slight strengthening of the second order conditions are sufficient for local optimality. Thus, second order conditions often allow us to verify that a stationary point is a local minimum or maximum.

We will consider finite-dimensional problems with equality constraints ($P_{3.1}$) only. This includes the problem types ($P_{1.1}$) and ($P_{2.1}$). Second order conditions for problems with inequality constraints are discussed in the crash course on optimization theory.

Royal road. If you want a shortcut to the most striking application, problem 5.3.3, then it is sufficient preparation to read statement a) of theorem 5.3 and the definition of the quadratic form Q.

5.2 MAIN RESULT: SECOND ORDER CONDITIONS

Plan. We will give a precise form of the result that there are essentially no other obstructions to optimality than the first and second order necessary conditions. We will do this for problems with equality constraints (problems with inequality constraints are considered in the analytical style crash course on optimization theory). We will give the second order conditions in two forms: with and without Lagrange multipliers. Moreover, we will give a sketch of an algorithm to check whether the second order condition holds.

SECOND ORDER CONDITIONS

5.2.1 Formulation of the result

Definition 5.1 *Let $f : U_n(\hat{x}, \varepsilon) \to \mathbb{R}$ be a function of n variables, defined on an open ball with center $\hat{x}$. Then f is called twice differentiable at $\hat{x}$ if, to begin with, it is differentiable at $\hat{x}$, and moreover, there exists a symmetric $n \times n$ matrix $f''(\hat{x})$ such that*

$$f(\hat{x}+h) = f(\hat{x}) + f'(\hat{x}) \cdot h + \frac{1}{2} f''(\hat{x})[h, h] + o(|h|^2),$$

where we have written

$$f''(\hat{x})[h, h] = h^T f''(\hat{x}) h.$$

Then $f''(\hat{x})$ is called the *second derivative* (or *Hessian*) of f at $\hat{x}$. One calls f twice continuously differentiable at $\hat{x}$ if $f''(x)$ exists in a neighborhood of $\hat{x}$ and is continuous at $\hat{x}$. Then one writes $f \in C^2(\hat{x})$.

Theorem 5.2 *Second order conditions. Let $\hat{x} \in \mathbb{R}^n$ and let*

$$f_i \in C^2(\hat{x}), \ 0 \leq i \leq m.$$

Assume that $f_i'(\hat{x})$, $1 \leq i \leq m$, are linearly independent. Consider the problem

$$f_0(x) \to \min, \ f_i(x) = 0, \ 1 \leq i \leq m. \qquad (P_{3.1})$$

We write $F = (f_1, \ldots, f_m)^T$.

1. *If $\hat{x}$ is a local minimum of $(P_{3.1})$, then there exists a selection of Lagrange multipliers $\lambda = (1, \lambda_1, \ldots, \lambda_m)$ such that for the Lagrange function*

$$\mathcal{L}(x, \lambda) = f_0(x) + \sum_{i=1}^{m} \lambda_i f_i(x)$$

the following conditions hold true:

$$\mathcal{L}_x(\hat{x}, \lambda) = 0_n^T,$$

$$\mathcal{L}_{xx}(\hat{x}, \lambda)[h, h] \geq 0 \ \text{for all}, \ h \in \ker F'(\hat{x}).$$

2. *If there exists $\lambda = (1, \lambda_1, \ldots, \lambda_m)$ such that*

$$\mathcal{L}_x(\hat{x}, \lambda) = 0_n^T,$$

$$\mathcal{L}_{xx}(\hat{x}, \lambda)[h, h] > 0 \ \text{for all nonzero } h \in \ker F'(\hat{x}),$$

then $\hat{x}$ is a strict local minimum of $(P_{3.1})$.

The proof of this result is given in appendix D, "Crash Course on Optimization Theory: Analytical Style."

Second order conditions without multipliers. We give a formulation of the second order conditions without Lagrange multipliers as well. This is possible: formula (∗) in section 3.5.3 allows us to eliminate the Lagrange multipliers. We consider the situation of the theorem and assume that $\widehat{x}$ is a stationary point of $(P_{3.1})$. We recall that the matrix $F'(\widehat{x})F'(\widehat{x})^T$ is invertible, using the fact that the rows $f'_i(\widehat{x})$, $1 \leq i \leq m$, of the matrix $F'(\widehat{x})$ are linearly independent. We define

$$Q[h,h] = (f''_0(\widehat{x}) - f'_0(\widehat{x})F'(\widehat{x})^T(F'(\widehat{x})F'(\widehat{x})^T)^{-1}F''(\widehat{x}))[h,h].$$

Substitution of formula (∗) in section 3.5.3 into $\mathcal{L}_{xx}(\widehat{x},\lambda)[h,h]$ gives immediately the expression $Q[h,h]$. This leads to the following reformulation of the second order conditions.

Theorem 5.3 *Second order conditions without multipliers. We consider the situation of the theorem and assume that $\widehat{x}$ is a stationary point of $(P_{3.1})$.*

1. *If $\widehat{x}$ is a local minimum of $(P_{3.1})$, then $Q[h,h] \geq 0$ for all h for which $F'(\widehat{x})h = 0_m$.*

2. *If $Q[h,h] > 0$ for all nonzero h for which $F'(\widehat{x})h = 0_m$, then $\widehat{x}$ is a strict local minimum of $(P_{3.1})$.*

In particular, the second order necessary (resp. sufficient) condition for unconstrained problems $(P_{2.1})$—this is the case $m = 0$—is that the Hessian $f''(\widehat{x})$ is positive semidefinite (resp. definite).

5.2.2 Algorithm for determination of definiteness

To verify the definiteness conditions of the theorem, it is not necessary to calculate the eigenvalues, which is relatively complicated. There is an efficient algorithm, to be called *symmetric sweeping*, to verify the conditions of the theorem, a version of Gaussian elimination,

> where you carry out in each step the same operations on the rows and on the columns.

This does not change the "definiteness" of the matrix and keeps it symmetric.

SECOND ORDER CONDITIONS

In the special case of unconstrained problems, the algorithm ends either if the matrix is brought into diagonal form or if it gets stuck (this happens if you get a nonzero row for which its "diagonal element" is zero). In the first case, you can use that a diagonal matrix is positive definite (resp. semidefinite) precisely if all diagonal elements are positive (resp. nonnegative). In the second case, the matrix is indefinite (it takes positive values as well as negative ones).

The general case can be dealt with in a similar way, or it can be reduced to the special case by elimination, but we do not give the details here.

5.2.3 Illustrations of second order conditions

Example 5.2.1 *Apply the second order sufficient conditions to a twice differentiable function f of one variable and a stationary point $\widehat{x}$.*

Solution. A stationary point $\widehat{x}$ is a point of local minimum (respectively maximum) if the number $f''(\widehat{x})$ is positive (respectively negative).

Example 5.2.2 *Apply the second order conditions to the problem to minimize the function $f(x) = x_1^2 + 2ax_1x_2 + 5x_2^2$.*

Solution. It is easy to solve the problem. For example, we can write

$$f(x) = x_2^2 \left(\left(\frac{x_1}{x_2} \right)^2 + 2a \left(\frac{x_1}{x_2} \right) + 5 \right),$$

and then the second factor can be viewed as a quadratic polynomial in one variable $t = \frac{x_1}{x_2}$. However, it is our aim to illustrate the second order conditions.

If $a = \pm\sqrt{5}$, then $f(x) = (x_1 \pm \sqrt{5}x_2)^2$ and the stationary points are the scalar multiples of $(\sqrt{5}, \mp 1)$; all of these are points of global minimum.

If $a \neq \pm\sqrt{5}$, there is a unique stationary point $\widehat{x} = (0,0)^T$. The second order necessary (resp. sufficient) condition is that the Hessian matrix $\begin{pmatrix} 1 & a \\ a & 5 \end{pmatrix}$ is positive semidefinite (resp. definite). We apply the symmetric sweeping method sketched in section 5.2.2. We subtract a times the first row from the second one and apply the corresponding operation to the columns. The result is the diagonal matrix

$\begin{pmatrix} 1 & 0 \\ 0 & 5-a^2 \end{pmatrix}$. This matrix is positive definite if $-\sqrt{5} < a < \sqrt{5}$ and it is indefinite if $|a| > \sqrt{5}$.

It follows that the unique stationary point $\hat{x} = (0,0)^T$ is a point of strict local minimum if $-\sqrt{5} < a < \sqrt{5}$ and that it is not a point of local minimum if $|a| > \sqrt{5}$.

Example 5.2.3 *Apply the second order sufficient conditions to the problem*

$$f_0(x_1, x_2) = x_1^2 - x_2^2 \to \min, \quad F(x) = x_1^2 + x_2^2 - 1 = 0.$$

Solution. Using the multiplier rule, one finds that there are four stationary points $\hat{x}$:

$$(1,0), (-1,0), (0,1), (0,-1).$$

We calculate the quadratic form Q,

$$Q(h) = h^T \begin{pmatrix} 2 & 0 \\ 0 & -2 \end{pmatrix} h$$

$$-(2\hat{x}_1, -2\hat{x}_2)(2\hat{x}_1, 2\hat{x}_2)^T ((2\hat{x}_1, 2\hat{x}_2)(2\hat{x}_1, 2\hat{x}_2)^T)^{-1} h^T \begin{pmatrix} 2 & 0 \\ 0 & 2 \end{pmatrix} h.$$

This gives

$$Q(h) = 4\hat{x}_2^2 h_1^2 - 4\hat{x}_1^2 h_2^2.$$

We have to consider this quadratic form on the solution set of $2\hat{x}_1 h_1 + 2\hat{x}_2 h_2 = 0$.

If $\hat{x}$ is $(1,0)$ or $(-1,0)$, this means that we have to consider

$$Q(h) = -4h_2^2$$

for all nonzero h with $h_1 = 0$: this is always negative. It follows that $(1,0)$ and $(-1,0)$ are strict local maxima.

If $\hat{x}$ is $(0,1)$ or $(0,-1)$, we have to consider

$$Q(h) = 4h_1^2$$

for all nonzero h with $h_2 = 0$: this is always positive. It follows that $(0,1)$ and $(0,-1)$ are strict local minima.

Conclusion to section 5.2. Nonnegativity of the second derivative of the Lagrange function on a suitable subspace is necessary for

SECOND ORDER CONDITIONS

local optimization. Positivity is even sufficient. As the gap between nonnegativity and positivity is small, we get that the first and second order necessary conditions are essentially the only obstructions to local optimality.

5.3 APPLICATIONS TO CONCRETE PROBLEMS

5.3.1 Shortest distance from a point to a linear subspace

Problem 5.3.1 *Consider the problem of finding the shortest distance from a given point in $\mathbb{R}^n$ to a given linear subspace of $\mathbb{R}^n$. Write the second order conditions for this problem.*

Solution. Denote the given point by p and let the linear subspace be given as the solution set of the system of linear equations $Ax = 0_m$, where A is a given $m \times n$ matrix, the rows of which are linearly independent. The problem can be written as follows—after translating p to the origin—with $b = Ap$:

$$|x|^2 \to \min, \ Ax = b.$$

This problem has a unique stationary point,

$$\widehat{x} = A^T(AA^T)^{-1}b.$$

The second order condition is:

$$2h^T h \geq 0 \text{ for all } h \in \mathbb{R}^n.$$

This condition holds true, as well as the sufficient second order condition: $2h^T h > 0$ for all nonzero $h \in \mathbb{R}^n$.

5.3.2 Minimization of a linear function on a sphere

Problem 5.3.2 *Consider the problem to minimize a linear function on the unit sphere in $\mathbb{R}^n$. Write the second order conditions for this problem.*

Solution. The problem can be modeled as follows:

$$c^T \cdot x \to \min, \ x^T \cdot x = 1,$$

where c is a given nonzero vector from $\mathbb{R}^n$. The second order condition is

$$-c^T \cdot (2\widehat{x})((2\widehat{x}^T) \cdot (2\widehat{x}))^{-1} 2h^T \cdot h \geq 0$$

for all $h \in \mathbb{R}^n$. That is, $c^T \cdot \widehat{x} \leq 0$. We recall that there are two stationary points,

$$-|c|^{-1}c \quad \text{and} \quad |c|^{-1}c.$$

The first stationary point satisfies the second order necessary condition and the second stationary point does not satisfy it, as we see.

The sufficient second order condition is

$$c^T \cdot \widehat{x} < 0.$$

Again we see that the first stationary point satisfies this condition and the second stationarity point does not. It follows that the first point is a strict local minimum and the second point is not a local minimum. In a similar way one can check that this second point is a strict local maximum.

5.3.3 Minimization of a quadratic form on a sphere

Now we consider a more interesting test case for the first and second order conditions: an optimization problem that has as stationary points the eigenvectors of a given symmetric matrix, normalized to have length 1, and as solutions the ones that belong to the smallest eigenvalue. We could not find in the literature the analysis of this problem with second order conditions. We were curious to see whether the second order conditions manage to pick out the solutions among the stationary values and if so, how.

Problem 5.3.3 *Consider the problem of minimizing a given quadratic form on $\mathbb{R}^n$ on the unit sphere. Write the second order conditions.*

For two symmetric $n \times n$ matrices A and B we write $A \succeq B$ if $A - B$ is a positive semidefinite matrix, that is, if

$$x^T A x \geq x^T B x$$

for all $x \in \mathbb{R}^n$.

Solution. For the sake of simplicity we will assume that A has n different eigenvalues (the general case can be done in a similar way). The problem can be modeled as follows:

$$x^T A x \to \min, \ x^T \cdot x = 1,$$

SECOND ORDER CONDITIONS

where A is a given symmetric $n \times n$ matrix. We considered this problem before, in the solution of problem 3.3.7. The second order condition is

$$-(2\widehat{x}^T A) \cdot (2\widehat{x})((2\widehat{x}^T)(2\widehat{x}))^{-1}(2h^T h) + 2h^T A h \geq 0$$

for all $h \in \mathbb{R}^n$. That is,

$$2h^T(-\widehat{x}^T A \widehat{x} I_n + A)h \geq 0$$

for all $h \in \mathbb{R}^n$. This can be written as

$$A \succeq (\widehat{x}^T A \widehat{x})I_n.$$

We recall from the analysis of problem 3.3.7 that the problem has n pairs of stationary points: for each of the n eigenvalues of A, the corresponding two eigenvectors of length 1 are stationary points. Let $\widehat{x}$ be such an eigenvector and μ_i the corresponding eigenvalue. Then the second order condition boils down to

$$A \succeq \mu_i I_n.$$

This means that μ_i is the smallest eigenvalue of A.

Conclusion. The second order conditions manage to pick out the point of minimum among all the stationary points.

5.3.4 Basic property of symmetric matrices and second order conditions

Finally, we illustrate the method of the second order conditions by using them to establish the basic property of symmetric $n \times n$ matrices. We have already given two other proofs of this property, in order to illustrate other optimization methods (cf. sections 2.4.6 and 3.3.7). It is convenient to derive the first and second order conditions for this application from first principles. We have already shown how to reduce the theorem to the case $n = 2$ (cf. section 2.4.6). The case $n = 2$ can be settled by explicit formulas, of course, but it is of interest to do this instead by using the method of first and second order conditions.

Theorem 5.4 *Orthogonal diagonalization of symmetric matrices. For each symmetric $n \times n$ matrix A there exists an orthogonal $n \times n$ matrix P such that $P^T A P$ is a diagonal matrix.*

Proof. For each symmetric matrix B, we let $g(B)$ be the sum of the squares of the off-diagonal elements of B. We consider the following optimization problem:

1. $f(P) = g(P^T A P) \to \min$, $P^T P = I$. It has a solution $\widehat{P}$ by Weierstrass. To prove the theorem it suffices to show that $f(\widehat{P}) = 0$. We assume that $\widehat{P} = I$, as we may, after replacing A by $\widehat{P}^T A \widehat{P}$. Thus, to prove the theorem, it suffices to show that A is a diagonal matrix.

2. It suffices to consider the case $n = 2$, as we have explained above. Note that for $n = 2$, our task is to show that a_{12} is zero. The quadratic approximation to the solution set of the equation $P^T P = I$ has the following parametric description:

$$Q(t) = \begin{pmatrix} 1 - \frac{1}{2}t^2 & t \\ -t & 1 - \frac{1}{2}t^2 \end{pmatrix}.$$

This follows from theorem 3.8. We omit the derivation. Now we have to substitute $P = Q(t)$ into $f(P)$ and to put the coefficient of t equal to zero ("first order condition" (FOC)) and to write that the coefficient of t^2 is nonnegative ("second order condition" (SOC)). Well, the two—equal—off-diagonal elements of the symmetric 2×2 matrix $Q(t)^T A Q(t)$ are seen—after matrix multiplication—to be equal to

$$a_{12} + (a_{11} - a_{22})t - 2a_{12}t^2 + o(t^2).$$

Therefore its square equals

$$a_{12}^2 + 2a_{12}(a_{11} - a_{22})t + [(a_{11} - a_{22})^2 - 4a_{12}^2]t^2 + o(t^2).$$

Now we can write the necessary conditions for minimality:

- FOC: $a_{12}(a_{11} - a_{22}) = 0$,
- SOC: $(a_{11} - a_{22})^2 - 4a_{12}^2 \geq 0$.

3. The FOC gives either $a_{12} = 0$—and then we are through—or $a_{11} = a_{22}$. In the latter case, the SOC gives $-4a_{12}^2 \geq 0$, and so $a_{12} = 0$—and then we are through as well.

4. Conclusion: the theorem is proved.

SECOND ORDER CONDITIONS

Examination of stationary points. In the proof above we have essentially solved the following optimization problem:

$$f(P) = g(P^T A P) \to \text{extr}, \ P^T P = I,$$

where A is a symmetric 2×2 matrix, not a scalar multiple of the identity matrix I. The results are as follows. The stationary points can be split by the second order condition into two nonempty sets, the global minima and the global maxima. An admissible P is a global minimum if $P^T A P$ is a diagonal matrix; this diagonal matrix does not depend on P; its diagonal elements are the eigenvalues of A. An admissible P is a global maximum if the diagonal elements of $P^T A P$ are equal and if, moreover, it is not a diagonal matrix; this matrix $P^T A P$ does not depend on P. □

5.4 DISCUSSION AND COMMENTS

Plan. Now we go on and write down higher order necessary and sufficient conditions for optimality for all types of problems.

Theorem 5.5 *Higher order necessary and sufficient conditions for one-variable problems.* Consider a problem of type $(P_{1.1})$. Assume that the function $f : (\widehat{x} - \varepsilon, \widehat{x} + \varepsilon) \to \mathbb{R}$ is r times differentiable at $\widehat{x}$ and that $f^{(k)}(\widehat{x}) = 0$ for $1 \leq k < r$.

- **Necessary conditions.** If the point $\widehat{x}$ is a local minimum (resp. maximum) of $(P_{1.1})$, then $f^{(r)}(\widehat{x})$ is nonnegative (resp. nonpositive) if r is even, and $f^{(r)}(\widehat{x})$ is zero if r is odd.

- **Sufficient conditions.** If r is even and $f^{(r)}(\widehat{x})$ is positive (resp. negative), then $\widehat{x}$ is a local minimum (resp. maximum) of $(P_{1.1})$.

Higher order conditions for unconstrained n-variable problems. The task of writing higher order conditions for unconstrained n-variable problems $(P_{2.1})$ can be reduced to the same task for one-variable problems, by using the observation that if $\widehat{x} \in \mathbb{R}^n$ is a local minimum (resp. maximum) of a function $f : U_n(\widehat{x}, \varepsilon) \to \mathbb{R}$, then 0 is a local minimum (resp. maximum) of the problem $g_{\bar{x}}(t) = f(\widehat{x} + t\bar{x}) \to$ extr for each $\bar{x} \in \mathbb{R}^n$ of length one.

Higher order conditions for equality-constrained n-variable problems. The task of writing higher order conditions for equality-constrained n-variable problems $(P_{3.1})$ can be reduced to the same task for unconstrained problems, in the following way. Assume that f and F are r times differentiable. Use the refined tangent space theorem 3.8 (or the implicit function theorem) to write the r-th order approximation at $\hat{x}$ of the solution set of the equation $F(x) = 0_m$. Substitute the result into the objective function f. Write the higher order conditions up to order r of the resulting unconstrained problem.

Example 5.4.1 *The approach in this section has been used in section 5.3.4 to write down the first and second order conditions for a concrete equality constrained problem.*

Example 5.4.2 *Use the higher order conditions to analyze the problem $f_k(x) = x^k \to$ min for $k = 3, 4$.*

Solution.

- $k = 3$. $f_3'(0) = f_3''(0) = 0$, $f_3^{(3)}(0) = 6 \neq 0 \Rightarrow$ the point $\hat{x} = 0$ is a stationary point, but not a point of local minimum.

- $k = 4$. $f_4'(0) = f_4''(0) = f_4^{(3)}(0) = 0$, $f_4^{(4)}(0) = 24 >> 0 \Rightarrow$ the point $\hat{x} = 0$ is a local minimum.

In the same way one can use the higher order conditions to analyze the problem for $k > 4$.

Application. The only application of interest of higher order conditions known to us is in the proof of the fundamental theorem of algebra, given in chapter two.

Conclusion to section 5.4. One can write down in a straightforward way the higher order conditions of all unconstrained and equality-constrained n-variable problems.

5.5 EXERCISES

In the first exercise, we reconsider exercise 2.6.3.

Exercise 5.5.1 $2x_1^4 + x_2^4 - x_1^2 - 2x_2^2 \to$ extr. *Find the stationary point that is a local maximum.*

Exercise 5.5.2 $3x_1 + 4x_2 \to$ extr, $x_1^3 x_2^4 = 128$. *Find the stationary point that is a local minimum.*

Chapter Six

Basic Algorithms

> The tremendous power of the simplex method is a constant surprise to me.
>
> *G. Dantzig*

- How does one compute the solution of an optimization problem numerically by an iterative procedure?

6.0 SUMMARY

Nonlinear optimization is difficult. The most important thing to realize is that it is in principle impossible to have an algorithm that can solve efficiently all nonlinear optimization problems. This is illustrated, on the one hand, by reformulating one of the greatest mathematical problems of all times, Fermat's Last Theorem, as a nonlinear optimization problem. On the other hand, we use ball-coverings to *prove* that even the best algorithm to optimize a well-behaved function on a block in n-dimensional space is inadequate.

Linear optimization. This state of affairs is in sharp contrast with the situation for linear optimization problems, also called linear programming problems (LP). Here, two efficient classes of algorithms are available, simplex methods and interior point methods (also called barrier methods). We will describe the idea of the simplex algorithm in terms of an insect climbing a topaz, a semi-precious stone with many facets. Similarly, we could view interior point methods in terms of an insect that gnaws a path to the top through the interior of the topaz.

Descent direction. If you want to improve an approximation of a point of minimum of a function of n variables, then one problem is which direction you should go. This is the problem of the choice of the descent direction. The best-known descent directions are minus the gradient vector and the Newton direction, a modification of it. The Newton direction is more expensive to compute, but it is of better quality.

Line search. The second problem is how far you should go. This is the problem of "line search." We will show that each one of the basic methods—the Fibonacci method, the golden section method, the bisection method, and the Newton method—is optimal in some sense. In the case of the bisection method, you might have discovered this yourself if you have ever played the game "Guess who?"

Quality of the approximation. The idea of each algorithm is to produce better and better approximations. What does "better" mean precisely? There is more than one way to answer this question, and not all measures of quality are equivalent. Here is a result on a relation between two such measures. It might be surprising at first sight. The bisection method not only decreases by a factor two the worst case distance of x_N to the optimum $\hat{x}$, but also decreases the worst case distance of $f(x_N)$ to the optimal value $f(\hat{x})$.

Quality of an algorithm. When are we satisfied with an algorithm? Roughly speaking, if each step of an algorithm gives at least a constant number of additional correct decimals, then we are happy. This is called *linear convergence*. Sometimes we get better convergence: in the Newton method, the convergence is spectacular, once we come close enough to the minimum. Each step the number of accurate decimals doubles at least. This is called *quadratic convergence*.

Center of gravity method. It is tempting to try to extend the bisection method to two or more variables. For example, if you want to minimize a convex function f on a bounded, closed, convex set G, then an idea would be to compute the gradient of f at a point c somewhere in the middle of G; the hyperplane through the point orthogonal to this gradient vector cuts G in two parts and one of them can be thrown away, the part in which direction this gradient vector points. There is an obvious choice for the point c in the middle: the center of gravity of G. This "algorithm" has one fatal weakness: it is not practical to compute the center of gravity.

Ellipsoid method. The idea of the ellipsoid method is as follows. Start with a convex function defined on an ellipsoid. Do one step of the gravity method. The result of this is a convex function defined on a "half-ellipsoid." Find an ellipsoid of small volume that contains this half-ellipsoid. Then consider the problem of minimizing the function on this new ellipsoid instead of on the smaller half-ellipsoid. The resulting step turns the problem of minimizing a convex function on an ellipsoid into a problem of the same type, where the progress is that the volume of the ellipsoid is reduced by a guaranteed factor.

The result on the bisection method about approximation of the optimal value extends to the ellipsoid value.

6.1 INTRODUCTION

The joy of optimization software. Many applications ask for the solution of linear or nonlinear optimization problems. Often one cannot solve the equations of the necessary conditions analytically (that is, by a formula) in an efficient way. Then one has to be satisfied with a numerical solution. That is, one looks for an algorithm that leads in an efficient way to a good approximation of the solution. Modern optimization software is impressive, and it is a great experience to use it, or at least to see it used! It is strongly recommended to try out and to visualize with a computer the algorithms in this and the next chapter, for some simple examples to begin with. In this way these algorithms come to life.

Packages give wrong answers. At first sight there seems to be no problem. It should suffice to choose one of the many excellent optimization packages that are available. Well, they are there in order to be used, so why not just use them? There does not seem to be a compelling need to be aware of precisely how they work. Few people know precisely how a dvd player or a pc works, but we all make good use of these inventions. However, if you "just push the buttons" of optimization packages, you might be in for unpleasant surprises. Superficially all is well: you ask the software to solve a problem, and then it tells you what the optimal solution is. Unfortunately, the shocking reality is that this so-called solution is often far from optimal.

Packages will always give wrong answers. This is no mistake in the software that can be corrected. There is a fundamental reason for this: it will never be possible to devise all-purpose algorithms for nonlinear optimization that always work. We will give two reasons for this bad news.

- On the one hand, we take one of the most famous mathematical problems of all times, Fermat's Last Theorem, and reformulate it as a nonlinear optimization problem. An algorithm that could solve all nonlinear optimization problems, could therefore solve this problem in a routine way. However, this problem withstood for hundreds of years the attempts of the greatest scientists to solve it. In fact, almost all mathematical problems can be formulated as nonlinear optimization problems. It follows that it is highly unlikely that an all purpose algorithm for nonlinear optimization problems exists that always works.

- On the other hand, we will identify a "reasonable looking" class of optimization problems and prove a lower bound for its "complexity" that makes clear that an algorithm to deal with problems of this class cannot exist.

Need for insight in algorithms. What to do? In order to use optimization packages, you need to have some knowledge of the algorithms on which they are based. Then you will understand why some of them are guaranteed to work, whereas for others such guarantees are impossible. Moreover, for each problem you have to choose which optimization package to use, and again this requires some insight in the underlying algorithms. In fact often, when you start looking for the right algorithm for a given optimization problem, it is already too late. Often it is better to think about algorithms in advance, in the modeling stage of your problem. Then you can try to head toward a type of problem for which efficient and reliable algorithms exist. Also, it is useful to realize that failings of all-purpose algorithms are there forever: if you use these algorithms—in order to save costs, for example—a realistic benchmark should be how much you can save, and not how well the optimal situation has been approximated.

Efficient and reliable algorithms. The development of algorithms is in full swing. When to use which algorithm is a matter of taste and experience. The subject of numerical solution of optimization problems is vast. The aim of this section is modest: to give a flavor of the subject that leaves a taste for more. Subtleties being of later worry, the main story is as follows. We aim for "ideal" algorithms: we want them to be both "efficient" and "reliable." What do we mean by these two words? The following simpleminded interpretation turns out to be fruitful. We call an algorithm "efficient" if each step of the algorithm leads to a guaranteed number of additional decimals. We call it "reliable" if it provides a "certificate of quality" along with each approximate solution. These strict requirements restrict the vast field of optimization algorithms to one for which we can sketch the state of the art in two chapters.

Role of convex problems. The state of the art is that such "ideal algorithms" are only available for *convex* optimization problems. To wit, one has *interior point methods* and *cutting plane methods* (for example, the ellipsoid method). Our aim is to give an introduction to these methods. Then you can try to write implementations of these methods, to use professional software with confidence, and/or to begin reading the scientific literature on this subject.

Basic algorithmic ideas. As preparation, it is useful to become familiar with the two basic algorithmic ideas. We now explain why this is advisable.

- Everyone who has ever shared a cake with someone, has probably discovered the basic idea of the *bisection method*. This basic optimization method is the best preparation for the cutting plane methods. Once you have decided to look at the bisection method, it is worth the effort to consider two related optimization methods as well: the golden search method and the Fibonacci method.

- One of the greatest algorithmic ideas is the Newton method. It can be used for every unconstrained optimization problem to turn a reasonable approximation into an excellent one with amazing speed: the number of correct decimals is *doubled* at each step of the algorithm! Also, the Newton method is a central ingredient of the interior point methods. Once you have decided to look at the Newton method, it is worth the effort to consider the *modified Newton method* as well: it is the universal tool in proofs of optimization methods, as we have emphasized already (chapter three).

LP problems. Finally, we say a few words about the most popular class of problems, linear optimization problems, usually called linear programming (LP) problems. There are two types of algorithms to solve LP problems, simplex methods and interior point methods. You might know the celebrated simplex method, the older of the two, from high school. It may be the most used mathematical technique in the world (usually to save costs). We will give a geometrical description of its main idea, but in order to save space, we have chosen not to display the bookkeeping devices to carry out the algorithm. The interior point methods, which are more recent, will never replace it, but have several advantages—guaranteed efficiency, higher speed for large problems, and greater flexibility.

L'art pour l'art. We have emphasized so far that it is useful to have some insight into optimization algorithms. There is more: each algorithm represents a great victory of the mind. If you want to get to the bottom of some algorithm because of scientific curiosity into the reason for its tremendous power, then you will be amply rewarded for your efforts. Most optimization algorithms have a great intrinsic beauty. In particular, it is of interest to note that an optimization algorithm is usually itself optimal in some sense.

Aim of this chapter. In this chapter we try to give a first impression of the wonderful world of optimization algorithms.

Royal road. Read the solution of example 6.2.1, taking note of the lower bound for the number of test points in section 6.2.2: after this you will realize that an algorithm that solves all nonlinear optimization problems efficiently is impossible. Read in sections 6.3.1 and 6.3.2 the geometrical sketches of the two main methods to solve LP problems, the simplex methods and the interior point methods. Read section 6.4.1 to see how the problem of designing optimization algorithms splits up into the problem of line search and the problem of direction of descent search. Read example 6.4.1 to understand what is the problem of line search. Read, in sections 6.4.3, 6.4.4, 6.4.5, and 6.4.6, the descriptions of the main line search methods, the Fibonacci method, the golden section method, the bisection method, and the Newton method. Take note of the exceptional speed of the Newton method. Read in section 6.5 the descriptions of the two main methods of descent direction search, the gradient method and the Newton method. Read in section 6.7 the description of the—theoretical but not implementable—center of gravity method, with the following two aims: to see that it is an attempt to extend the bisection method to two or more variables, and as a preparation for the ellipsoid method. Then read in section 6.8 the description of the ellipsoid method. Finally, read in section 6.9 the description of interior point methods. Take note of the fact that the class of optimization problems that one can expect to solve efficiently with an algorithm is the class of convex optimization problems, and that the two main methods to solve optimization methods are the ellipsoid method and the interior point methods.

6.2 NONLINEAR OPTIMIZATION IS DIFFICULT

Plan and motivation. For linear optimization problems there are two successful all-purpose algorithms, the simplex method and interior point or barrier methods.

It might be tempting to look for an extension to an algorithm to solve all *nonlinear* optimization problems numerically in an efficient way. Unfortunately such an algorithm does not exist, and in fact it *cannot* exist. In this section we will demonstrate this in two ways.

BASIC ALGORITHMS

6.2.1 All mathematical problems are optimization problems

Plan. To begin with, we make the point that it is extremely unlikely that such an algorithm exists, as almost all mathematical problems can be formulated as the problem of solving a nonlinear optimization problem numerically. We give an example to illustrate this.

Example 6.2.1 *Fermat's Last Theorem. Suppose you had the opportunity to send a civilization somewhere out in space a short message, and you wish to use this to help the development of this civilization, and, at the same time, to show how far our own civilization has progressed.*

Then you could pose the challenge to show that there are no solutions in positive integers of the equation

$$x^n + y^n = z^n$$

for $n \geq 3$. This statement is known as Fermat's Last Theorem. For centuries it was one of the greatest challenges of mathematics.

Model this problem of Fermat as a nonlinear optimization problem.

Solution. Fermat's Last Theorem can be formulated as the statement that the optimal value of the following nonlinear optimization problem is positive:

$$f(x, y, z, n) = (x^n + y^n - z^n)^2 \to \min, \ x, y, z, n \in \mathbb{R},$$

$$\sin \pi x = \sin \pi y = \sin \pi z = \sin \pi n = 0, \ xyz \neq 0, \ n \geq 3.$$

Note that it suffices to solve this problem numerically, as you know that the optimal x, y, z, and n and the optimal value are integers, so you can round off.

Historical comments. The story of this problem of Fermat is full of wonderful anecdotes (cf. [69]). In 1994 it was finally solved by Andrew Wiles as a culmination of many of the greatest developments of mathematics in the last two centuries, many of which resulted from attempts to prove Fermat's Last Theorem. This solution involves so many different ideas and methods that it would not—yet—be possible to give a series of courses for Ph.D. students leading up to a complete proof.

Conclusion. We have reformulated Fermat's Last Theorem, until recently one of the greatest mathematical challenges, as an optimization problem. This strongly suggests that it is extremely unlikely that there exists an algorithm that can solve all nonlinear optimization problems in an efficient way.

6.2.2 Proof of the impossibility of a universal algorithm

Plan. Now we will demonstrate in a second way that a universal algorithm to solve all nonlinear optimization problems efficiently cannot exist. We formulate a class of nice optimization problems and *prove* that, even for relatively small problems, even the best algorithm would require a forbidding amount of computation.

Suppose you want to find the minimum of a function f of n variables $x_1, \ldots, x_n$, which are allowed to run over certain ranges, say,

$$x_i \in [M_i, N_i]$$

for all i.

Lipschitz condition. Assume that we know that f does not vary too much in the following sense: we have a positive number K such that

$$|f(x) - f(y)| \leq K|x - y|$$

for all admissible vectors x and y. This is called a K-Lipschitz function. This concept lies in strength in between the properties continuity and continuous differentiability. The following example gives a precise version.

Example 6.2.2
1. *Let f be a continuously differentiable function f of one variable x that is defined on the interval $[M, N]$. Write*

$$K = \max_{M \leq x \leq N} |f'(x)|.$$

   Show that f is a K-Lipschitz function.

2. *Show that each K-Lipschitz function is continuous.*

Solution

1. This follows from the identity

$$f(y) - f(x) = \int_x^y f'(t) dt$$

in combination with the inequalities

$$\left|\int_x^y f'(t)dt\right| \leq \int_x^y |f'(t)|dt \leq (\max_{t\in[x,y]} |f'(t)|)|y-x|.$$

2. Immediate from the definition.

This example and its solution can be extended to the n-variable case.

We want to calculate the value of a given function f at a number of test points with the aim of finding a sufficiently good approximation of a point of minimum of f. To be more precise, we are given a small positive number $\varepsilon > 0$ and we want to find an admissible point $\bar{x}$ for which $f(\bar{x}) - f(\hat{x})$ is smaller than ε, to be called an ε-solution.

Grid method. The simplest possible algorithm for this type of problem is the *grid method*. Choose the *grid* on the domain of f that is determined by dividing the range of the i-th coordinate into a number, say r_i, of equal parts for $1 \leq i \leq n$. Then calculate the value of f at all the $\prod_{i=1}^n (r_i + 1)$ points of the grid and take the minimum of these values. Note that for each point in the domain the distance to the nearest grid point is at most the square root of the sum of the squares $(\frac{N_i - M_i}{2r_i})^2$. The grid method leads to an ε-solution if the r_i, $1 \leq i \leq n$ have been chosen in such a way that K times the maximal distance from a point in the domain of f to its nearest grid-point is smaller than ε. The grid method is not very efficient: usually it requires 'many function evaluations' in order to achieve the required precision. Therefore it is natural to ask whether there do exist efficient methods of the type under consideration.

We are going to see that the answer to this question is negative. In order to formulate this question and the answer to it in a more precise way, we define for a given algorithm and a given precision ε the number $N(\varepsilon)$ to be the worst case number of test points needed to obtain the required precision. Here 'worst case' means that this number of test points suffices for each K-Lipschitz function on the domain $\{x : M_i \leq x_i \leq N_i, 1 \leq i \leq n\}$. We can use this definition to measure the efficiency of an algorithm. For example, the smaller the number $N(\varepsilon)$ is, for a given ε, the more efficient the algorithm can be considered to be for this given ε. A different way to compare algorithms is to consider, roughly speaking, how fast $N(\varepsilon)$ grows if ε tends to zero. The slower this growth, the faster the algorithm is then considered to be.

Now we can give a more precise formulation of the answer we are going to give to the question above: we will give a lower bound for $N(\varepsilon)$ that holds for *each algorithm*, and we will see that this lower bound is, in a sense, very large.

How to outwit an algorithm. The following thought experiment gives a lower bound for the number of test points $N(\varepsilon)$ that is required in order to be sure that we will find an approximation of the required precision ε. It is valid for *all* algorithms of the type under consideration. Suppose somebody wants to test an algorithm and asks you to give him an example of a function. Now you want to have some fun at his expense and, pretending to be helpful, you offer to calculate for him the value of your function at all the test points he gives you. Therefore, it is not relevant to him to know during the execution of the algorithm what the function is.

Now he starts giving you test points. In reality you have not chosen any function, and for all test points he comes up with, you tell him, after the pretense of a calculation, that the value at that point is zero. After a number of test points, say T, your friend will stop and will declare that the optimal value of the problem is zero.

Then you come into action and look whether there exists an admissible point P which has distance more than ε/K from all the test points. If you can find such a point, then it will not be hard for you to construct a K-Lipschitz function which has value smaller than $-\varepsilon$ at P and value zero at all the test points. You show this function to your friend, and he has to admit that his algorithm did not achieve the desired precision. If you cannot find such a point P, this means a victory for the algorithm.

Ball-coverings do the job. Now we come to the central question: does there exist a clever strategy of choosing the test points that always leads to victory in an efficient way, that is, for T not too large. Such a strategy should have led in the experiment related above to the conclusion that there exists no admissible point P which has distance more than ε/K from all the test points.

This can be reformulated in terms of coverings of the admissible region by balls. It means that the admissible region is completely covered by the collection of balls with radius ε/K and centers the test points. This implies that the n-dimensional volume of the admissible region is smaller than or equal to the sum of the n-dimensional volumes of the balls of this collection. That is, the following inequality holds true:

$$\prod_{i=1}^{n}(N_i - M_i) \leq T(\varepsilon/K)^n V_n,$$

where V_n is the n-dimensional volume of the unit ball in $\mathbb{R}^n$. For "reasonable" numerical instances this leads to a lower bound for the number of test points T—and so for $N(\varepsilon)$—of astronomical size.

Conclusion. We have "proved" that there cannot exist an algorithm that solves all nonlinear optimization problems in an efficient way. We have done this by considering a nice looking class of optimization problems, and we have seen that even the best possible algorithm would lead to a forbidding amount of computation.

Conclusion to section 6.2. It is not reasonable to expect that there will ever be an all-purpose algorithm that will solve all optimization methods numerically in an efficient way. Therefore we present in this chapter and the next one a collection of the most successful optimization algorithms. Most of these algorithms are optimal algorithms in some sense, as we will see. That is, each of them can be viewed as the solution of an optimization problem.

6.3 MAIN METHODS OF LINEAR OPTIMIZATION

Plan. We present the two—totally different—algorithmic ideas to solve linear optimization problems, the simplex methods and the interior point methods. For the latter we only give a first glimpse of the idea; more details are given in section 6.9.

LP problems. Many pragmatic problems can be modeled as a problem with equality and inequality constraints, where the functions that describe the objective and the constraints are linear. We recall from section 4.2.1 that such problems are called *linear optimization problems* or, more commonly, *linear programming problems (LP)*. One of the standard ways to write such a problem is as follows:

$$c \cdot x \to \min, \; Ax = b, \; x \geq 0_n,$$

where c is an n-dimensional row vector, A an $m \times n$ matrix, and b an m-dimensional column vector. There are two efficient types of algorithms to solve such problems, the *simplex method* and *interior point methods* (also called *barrier methods*).

6.3.1 Idea of the simplex method: insect climbing a topaz

Plan. Now we will present the simplex method in geometrical language.

Example of an LP problem: insect and topaz. Let us consider a tiny insect that wants to climb to the top of a topaz, a semiprecious stone with many facets, which is lying on a table. The insect follows the edges, as this gives a better grip on the slippery topaz. At each vertex it has to make a choice how to continue.

Geometrical description of the simplex method: insect following edges. All the simplex method tells about this choice is not to choose an edge that goes down. This clearly agrees with common sense. There are several variants of the simplex method, corresponding to rules about which of the edges that go up one should choose, called *pivoting rules*. We mention the two best-known rules. One can choose the edge that goes up most steeply. Alternatively, one can choose the edge for which the other vertex is highest. The insect has to continue in this way till it reaches the top.

Extension of the simplex method to arbitrary LP problems. What is the relation of this "insect climbing a topaz"-problem and LP problems? Well, if we take the LP problem above with $n = 3$ and $c^T \cdot x = -x_3$, then the admissible set is a sort of topaz with m facets (provided the admissible set is bounded). That is, the problem of the insect wanting to reach the highest point of the topaz is an LP problem. The method given above is precisely the simplex method for this problem. If you translate the geometrical description of the algorithm for the "insect climbing a topaz"-problem into an analytical one in terms of c, A and b, then you will find that this can be extended to arbitrary LP problems; here we only note that one can rewrite each LP problem by means of an additional variable x_{n+1} in such a way that the objective function is x_{n+1}. The resulting extension is precisely the simplex method.

Complexity of the simplex method. The simplex method is not always efficient. For all versions of the simplex method, examples have been constructed—with great difficulty—where the simplex method visits all the vertices of the admissible set before it reaches a point of extremum. However, if you solve problems arising from pragmatic needs rather than specially constructed problems, you will find that the LP algorithm is very efficient.

How to put the simplex method to work. To carry out the simplex method, you need to describe it in an analytical style. This

BASIC ALGORITHMS

can be found in almost all textbooks on optimization. The most popular such description is the one in terms of *tableaux*.

Conclusion. The most popular class of optimization problem is that of *LP problems*, and the most popular method for solving it is *the simplex method*. The geometrical idea of this algorithm can be understood by an allegory of an insect climbing along the edges of a topaz to its top.

6.3.2 Idea of interior point methods: insect gnawing a path through a topaz

Plan. Now we offer an introduction to the geometrical idea of interior point methods.

Historical comments. Since the discovery of the simplex method, it has been a challenge to find an algorithm for LP problems that is efficient for *all* instances. It is possible to formulate this question in a precise way. This can be sketched as follows: is there an algorithm for LP problems that is *polynomial* in the sense that the length of the calculations can be bounded by a polynomial in the total length of the numerical data of the problem? Loosely speaking, one can "hope" that a polynomial algorithm is also efficient in practice.

Khachyan. In 1979 Leonid Khachyan discovered a polynomial algorithm for LP problems that is totally different from the simplex method. This raised great expectations, but this algorithm turned out not to be efficient in practice, but only in theory.

Karmarkar. Then in 1983 Narendra K. Karmarkar hit the headlines of major world newspapers. He had discovered another polynomial algorithm, which he claimed to be very efficient in practice. This type of algorithm was later called an *interior point method*. He worked for a private company, though, and the source code of the algorithm was kept secret for commercial reasons.

Impact of Karmarkar's discovery. This energized the scientific community, and the next years saw tremendous efforts of many researchers to "rediscover" the ideas of Karmarkar, to develop them, and to beat his implementation. On the other hand, experts of the simplex method fought for all they were worth and achieved enormous improvements in the performance of the simplex method. As a result, now LP problems can be solved a million times faster than before 1983. A factor of a thousand is due to faster computers and another factor of a thousand is due to better theory.

Gnawing insect. To explain the idea of the interior point methods in vivid terms, let us consider again our insect climbing a topaz problem. Now the insect somehow gnaws a path to the top through the interior of the topaz. The basic idea of this is sketched at the end of chapter one using the concept *barrier*.

How to put interior point methods to work. We will give *a coordinate-free* description of the v-space presentation of interior point methods in section 6.9.4. Another, more far reaching approach is by using barriers; this is explained in chapter seven.

Conclusion. Interior point methods are an attractive alternative for the simplex method, especially for large problems.

6.4 LINE SEARCH

6.4.1 Why "line search" and "direction of descent search"

All optimization algorithms proceed in steps. At the beginning of each step you possess a point $\bar{x}$ that is your current approximation of a point of minimum. The aim of the step is to move to a nearby point $\bar{\bar{x}}$ which is a better approximation in some sense. Finding $\bar{\bar{x}}$ is the same thing as finding the difference vector $\bar{\bar{x}} - \bar{x}$. This small vector has, like every vector, a length and a direction. Accordingly, in each step of the algorithm the two main questions are, in which direction to move ("direction of descent") and how far ("line search"). Therefore, the problem of designing an algorithm for n-dimensional minimization problems is often split up into two problems: "descent direction search" and "line search." In this section we consider the second question only, and in the next section we will consider the first problem.

Conclusion. The problem of designing optimization algorithms can be split up into the problem of line search and the problem of descent direction search.

6.4.2 Type of functions for which line search is desirable and possible.

Plan. We present the class of unimodular functions. We make clear that this class is the right one for line search, for two reasons.
   1. One-dimensional problems often arise because of some trade-off, and this leads to unimodularity.

BASIC ALGORITHMS

2. For unimodular functions there is a basic algorithmic idea that makes line search possible.

Suppose that we have made up our mind in which direction to move and it remains only to decide how far we will move along this line. This is called a *line search*. As we are moving along a given line, we might as well restrict our discussion in this section to the optimization of functions of one variable. We will describe various possible ways to do a line search.

You cannot say which one is the best one: this depends on the situation. However, we will see that each of the possible ways to do a line search is optimal for some situation. To be more precise, it will turn out that each of the optimization algorithms presented below is itself a solution of some optimization problem!

Unimodular functions. Optimization problems of one variable that we have to solve for pragmatic reasons are often in the following spirit.

Example 6.4.1 *Soup without salt is tasteless, but if a lot of salt has been added, then you do not want it either. If you start to add salt gradually, the soup will become better and better, up to a point. After this point the quality of the soup gets worse. This point gives the optimal soup.*

Suppose we want to discover how much salt will give the most tasty soup. Each day we decide how much salt to put in the soup and then we taste it and compare it in our mind with the soups of the previous days.

What is the best algorithm to do this, that is, how much salt should we put in each day, given our experiences so far?

Now we formalize this type of optimization problem. We consider the case of minimization. A function $f : [a, b] \to \mathbb{R}$ on an interval $[a, b]$ is called *unimodular* if there is a point $\widehat{x} \in [a, b]$ such that the function decreases to the left of $\widehat{x}$ and increases to the right of it. For example, each strictly convex, differentiable function on $f : [a, b] \to \mathbb{R}$ for which $f'(a) \leq 0$ and $f'(b) \geq 0$ is unimodular. We note that the point $\widehat{x}$ is the unique point of global minimum and that there are no other points of local extremum.

The following easy property of unimodular functions $f : [a, b] \to \mathbb{R}$ is the basis of most line search algorithms. If we have evaluated the function at two points, say, $c < d$ and $f(c) < f(d)$ (resp. $f(c) > f(d)$), then the point of minimum of $f : [a, b] \to \mathbb{R}$ cannot lie to the right (left) of d (see Fig. 6.1).

A first attempt to do line search: the greedy approach. At first sight it might seem a good idea to choose c and d near the middle of the interval $[a, b]$. Then the comparison of f at c and d leads to a doubling of the precision with which we know the location of the optimum: we knew that it is contained in $[a, b]$, and after the comparison we either know that it is contained in $[a, d]$, or we know that it is contained in $[c, b]$, and the lengths of the intervals $[a, d]$ and $[c, b]$ are roughly half the length of $[a, b]$, as c and d lie close to the middle of the interval $[a, b]$.

How to avoid waste: do not be greedy. However, this idea is wasteful. This doubling of the accuracy costs two function evaluations. If we are less greedy and take the points c and d not so close to the middle, then we can use one of the two again in the next step. Thus, one needs for the next step only one new function evaluation. Going on in this way, one needs for each step, except the first one, only one new function evaluation. It will turn out that this gives a better improvement per function evaluation than when we double the accuracy at each step by doing two function evaluations near the middle of the current interval. The Fibonacci method and the golden search method give optimal ways to implement this idea.

Now we consider some methods to minimize a unimodular function f on an interval $[a, b]$.

Conclusion. The reasonable class of functions for which to develop a line search is the class of unimodular functions, for two reasons. On the one hand, often one-dimensional problems model a trade-off between two effects: therefore increasing the value of the variable gives initially an improvement, but after awhile a deterioration. On the other hand, there is an idea available that makes line search algorithms possible for unimodular functions.

6.4.3 Fibonacci method

Plan. To begin with, we will present a line search algorithm for the case when you have decided in advance how many function evaluations you are prepared to do.

Fibonacci sequence. The Fibonacci numbers form an infinite sequence $1, 2, 3, 5, 8, 13, 21, \ldots$ that is characterized by the first two terms and the property that each other term is the sum of its two predecessors in the sequence (this is more convenient for our purpose than the usual convention to let the Fibonacci sequence start with two ones).

First step. Take a Fibonacci number, say 13. Divide the interval $[a, b]$ into 13 equal parts. For convenience in describing the algorithm we assume that $a = 0$ and $b = 13$. We compute the value of f at 5 and 8, the two predecessors of 13 in the Fibonacci sequence, and see which of the two values is smallest. Suppose that we find $f(5) \leq f(8)$. Then we know for sure that the optimal point cannot lie beyond 8: this is an immediate consequence of the unimodularity of f (Fig. 6.1). If we find $f(5) \geq f(8)$, then we get that the point of minimum cannot

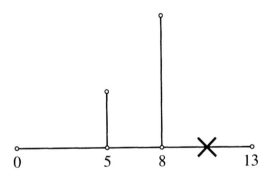

Figure 6.1 Fibonacci method.

lie before 5. This is the first step of the algorithm.

What does this step achieve? Before we started, our uncertainty about the position of the optimal point was great. It could lie anywhere on the interval $[0, 13]$, which has length 13. After this step the uncertainty is smaller; we have located the optimal point within an interval of length 8 (and if we were so lucky that $f(5) = f(8)$ of length 3). Note that 8 is the predecessor of 13 in the Fibonacci sequence.

Second step. How to continue? Suppose that we know that the optimal point lies on $[0, 8]$ (the case that it is $[5, 13]$ is similar). Note that already we have calculated the value of f at 5, one of the two predecessors of 8. Now we calculate the value at 3, the other predecessor. We observe which of these two values is larger. Then we proceed in a similar way as above and reduce the "interval of uncertainty" from length 8 to length 5.

Next steps. We continue to make such steps till the length of the interval of certainty is 2, and then we take the middle of this interval as our final approximation.

Complexity of the Fibonacci method

Special case: four steps. Let us look at the costs and benefits of this algorithm. For each step we had to do one function evaluation, for the first step two. The length of the uncertainty interval is reduced as follows: 13, 8, 5, 3, 2, that is, it is reduced in 4 steps—costing 5 function evaluations—from 13 to 2. The distance of the midpoint of the last uncertainty interval to the true point of minimum is therefore at most 1.

General case: $n - 2$ **steps.** Now we consider the general case. If we start with the n-th Fibonacci number F_n, instead of with the sixth Fibonacci number 13, then the algorithm would take $n - 2$ steps and cost $n - 1$ function evaluations. This would reduce the length of the interval of uncertainty $[a, b]$ by a factor $F_n^{-1} F_2$ and it would lead to an approximation for which the distance to the true point of minimum is not larger than $F_n^{-1}(b - a)$. The quotients of consecutive Fibonacci numbers tend to a limit

$$\lim_{n \to \infty} F_{n+1}^{-1} F_n = \frac{1}{2}\sqrt{5} - \frac{1}{2},$$

and so we see that, roughly speaking, each reduction of the uncertainty interval by a factor $\frac{1}{2}\sqrt{5} - \frac{1}{2}$ costs one function evaluation.

Remark. In particular, let us compare with the first "greedy" attempt, that is, to take each step two points near the middle. This gives a reduction of the uncertainty interval by a factor 2 for each two function evaluations, that is, a reduction by a factor $\frac{1}{2}\sqrt{2}$ for each function evaluation. We have $\frac{1}{2}\sqrt{2} \approx 0.7071$ and $\frac{1}{2}\sqrt{5} - \frac{1}{2} \approx 0.6180$. This shows what we have gained by being not too greedy.

Optimality of the Fibonacci method. If we want to minimize a unimodular function on an interval $[a, b]$, having fixed in advance the number of function evaluations we are prepared to carry out, and if we want to find an approximation "of highest guaranteed quality," then the Fibonacci method is the optimal algorithm. By "quality of an approximation" $\tilde{x}$ we mean—for the time being—that we always possess some upper bound for the distance from $\tilde{x}$ to the point of minimum $\hat{x}$. By "guaranteed" we mean that the upper bound should hold true every time the algorithm is applied, in particular in the "worst case."

Conclusion. If you want to do a line search for a unimodular func-

BASIC ALGORITHMS

tion, and you have decided in advance how many function evaluations you want to do, then the Fibonacci method is the best one, as far as worst-case performance is concerned.

6.4.4 Golden section method

Plan. Now we will present an algorithm for the case when you have not decided in advance how many function evaluations you are prepared to do.

Golden section point. Divide the interval $[a, b]$ into two parts such that the length of the larger part equals $\frac{1}{2}\sqrt{5} - \frac{1}{2}$ times the length of the whole interval. Then the following property holds: the quotient of the lengths of the smaller and larger part is also $\frac{1}{2}\sqrt{5} - \frac{1}{2}$. The interval is said to be divided according to the *golden section* and the point is called a *golden section point*.

First step. Now take the two golden section points of the interval $[a, b]$, calculate the value of the function f in each of them, and look at which point the value is minimal. Then, just as in the Fibonacci method, in either case we get a reduction of the uncertainty interval. If the two golden section points are x' and x'' with $a < x' < x'' < b$ and $f(x') < f(x'')$ (the other case is similar), then the uncertainty interval is reduced from $[a, b]$ to $[a, x'']$. Note that we already know the value of f in x' and that *this is one of the two golden section points of the new uncertainty interval* $[a, x'']$, by the property of the golden section given above. This finishes the first step of the algorithm.

Second step. The next step consists of taking the other golden section point of $[a, x'']$, computing the value of f at it, comparing the values at these two golden section points, and proceeding as in the first step.

Next steps. One continues with these steps as long as one wishes.

Complexity of the golden section method. Each step of the golden section method costs one function evaluation, except the first one, which costs two function evaluations; on the benefit side it leads to a reduction of the uncertainty interval by a factor $\frac{1}{2}\sqrt{5} - \frac{1}{2}$. This reminds us of the Fibonacci method. In fact, there is a close relation between the two methods. If we take the Fibonacci method and let the number of prescribed function evaluations tend to infinity, then this method "tends" to the golden section method.

Optimality of the golden section method. If we want to minimize a unimodular function on an interval $[a, b]$, without wanting to fix the number of function evaluations we are prepared to carry out in advance, and if we want to find an approximation "of highest guaranteed quality in the long run," then the golden section method is the optimal algorithm.

Conclusion. If you want to do a line search for a unimodular function, and you have not decided in advance how many function evaluations you want to do, then the golden search method is the best one, as far as worst-case performance is concerned.

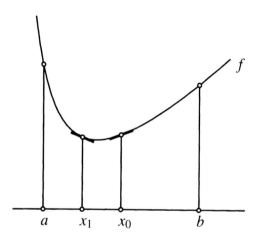

Figure 6.2 Illustrating the bisection method.

6.4.5 Bisection method

Plan. Now we will present an algorithm for the case when we can and want to evaluate derivatives at points (in particular, the function is differentiable). Note that this is the limiting case of function evaluation at two nearby points.

First step of the bisection method. We compute the derivative of $f : [a, b] \to \mathbb{R}$ at the midpoint $x' = \frac{1}{2}(a + b)$ of $[a, b]$ and look at its sign. If this is positive (resp. negative), then the minimum must lie on the interval $[a, x']$ (resp. $[x', b]$) (if we are so lucky that it is zero, then we are done: x' is the point of minimum) (Fig. 6.2).

This is the first step of the algorithm. One continues with these steps as long as one wishes.

BASIC ALGORITHMS

Complexity of the bisection algorithm. Each step of the bisection method costs one evaluation of the derivative of $f : [a,b] \to \mathbb{R}$ at a point and gives a reduction of the uncertainty interval by a factor $\frac{1}{2}$. This reduction is more favorable than in the case of the golden section and Fibonacci methods. However, it is usually more "costly" to evaluate the derivative of f at a point than to evaluate f itself at a point.

Optimality of the bisection method. If we want to minimize a differentiable unimodular function $f : [a,b] \to \mathbb{R}$, having fixed in advance the number of evaluations of the derivative of f we are prepared to carry out, and if we want to find an approximation "of highest guaranteed quality," then the bisection method is the optimal algorithm. A result about the efficiency of the way the bisection method approximates *the optimal value* of the problem, rather than the solution, is given in theorem 6.3.

We give two illustrations of—part of—the idea of this algorithm.

Example 6.4.2 *How to cut the cake? Two persons want to share a cake in a fair way. How to arrange the cutting of the cake?*

Solution. The solution is well known. One person will cut the cake into two pieces and the other one will choose first. This encourages the cutter to cut the cake as much as possible into two equal parts.

Example 6.4.3 *Guess who? In the game "Guess who?" you have color pictures of fifteen persons in front of you. Your opponent has one of them in mind, the "culprit," and you have to guess who. Each turn you can ask a question of the type: does the person wear glasses, yes or no? You have to find out in as few turns as possible (this is a simplified presentation of the game).*
What is the optimal strategy?

Solution. You should find a characteristic that half of the persons have. Then you can ask a question that will for each of the two answers reduce the number of suspects to at most eight. Then you should find a characteristic that half of the remaining persons have. This allows you to ask a question that will certainly reduce the number of suspects to at most four. Continuing in this way, you know after four turns who is the culprit.

Conclusion. If you want to do a line search for a unimodular differentiable function, and you can and want to evaluate its derivative at points, then the bisection method is the best one, as far as worst-case performance is concerned.

6.4.6 Newton method

Motivation and plan. Now we come to the most spectacular of all algorithms, the Newton method. The bisection method does not take into account the curvature of the graph of the function. This curvature is encoded in the second derivative (if this exists). Now we will present an algorithm for the case that we can and want to evaluate second derivatives at points (in particular, the function is twice differentiable). This is the Newton algorithm.

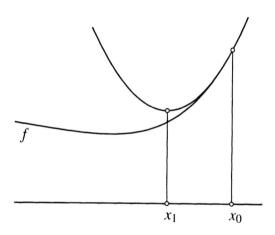

Figure 6.3 Illustrating the Newton method for the minimization of f.

Newton as a minimization method. We assume that a function $f : [a, b] \to \mathbb{R}$ is twice differentiable, with $f'' > 0$, $f'(a) \leq 0$, and $f'(b) \geq 0$. We compute the quadratic approximation of f at some point x_0 of the interval $[a, b]$, that is,

$$f(x_0) + f'(x_0)(x - x_0) + \frac{1}{2}f''(x_0)(x - x_0)^2,$$

and determine its point of minimum x_1 (Fig. 6.3).
This gives

$$x_1 = x_0 - f''(x_0)^{-1}f'(x_0).$$

That is, the first step takes the point x_0 as an initial approximation

BASIC ALGORITHMS

and goes over to the new approximation

$$x_1 = x_0 - f''(x_0)^{-1} f'(x_0).$$

The second step takes x_1 as a starting point and then proceeds in the same way as the first step. One continues with these steps as long as one wishes.

Newton as a root finding algorithm. Alternatively, one can present the Newton method as an algorithm for solving the stationarity equation $f'(x) = 0$. We compute the linear approximation of f' at x_0, that is,

$$f'(x_0) + f''(x_0)(x - x_0),$$

put it equal to zero, and take the unique solution (Fig. 6.4), which is

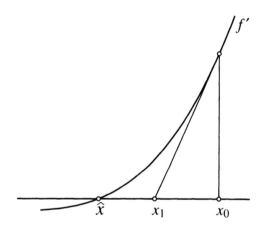

Figure 6.4 Illustrating the Newton method for solving $f'(x) = 0$.

$$x_1 = x_0 - f''(x_0)^{-1} f'(x_0),$$

and so we see that we get the same recursion.

Complexity of the Newton method. At each step of the Newton method the distance to the true minimum is squared, roughly speaking, as soon as you are close enough to the minimum. This can be expressed in a more spectacular way: at each step the number of reliable decimals doubles. The following result gives a precise formulation.

Theorem 6.1 *Quadratic convergence of the Newton method.*
Let $f : (\widehat{x} - \varepsilon, \widehat{x} + \varepsilon) \to \mathbb{R}$ be a function for which

$$f \in C^3(\widehat{x}), \quad f'(\widehat{x}) = 0 \text{ and } f''(\widehat{x}) > 0.$$

Choose a constant

$$K > \frac{1}{2}|f^{(3)}(\widehat{x})|f''(\widehat{x})^{-1}.$$

Then, for each number x_0 which is sufficiently close to $\widehat{x}$, one step of the Newton method leads to a new approximation x_1 with

$$|x_1 - \widehat{x}| < K|x_0 - \widehat{x}|^2.$$

Proof. Choose

$$K_1 > \frac{1}{2}|f^{(3)}(\widehat{x})| \quad \text{and} \quad K_2 > (f''(\widehat{x}))^{-1}$$

such that $K_1 K_2 = K$. By the method of approximation by Taylor polynomials one has the inequality

$$|f'(\widehat{x}) - (f'(x_0) + f''(x_0)(\widehat{x} - x_0))| < K_1|\widehat{x} - x_0|^2$$

for all x_0 which are so close to $\widehat{x}$ that

$$\frac{1}{2}|f^{(3)}(x_0)| < K_1.$$

If moreover x_0 is so close to $\widehat{x}$ that

$$(f''(x_0))^{-1} < K_2,$$

then dividing both sides of the inequality by $f''(x_0)$—and using the equality $f'(\widehat{x}) = 0$—leads to

$$\left|(x_0 - \frac{f'(x_0)}{f''(x_0)}) - x_0\right| < K_1 K_2 |x_0 - \widehat{x}|^2 = K|x_0 - \widehat{x}|^2,$$

as required. □

Optimality of the Newton method. In which sense is the Newton method the best one? Well, we can view the "smallness" of $|f'(x_0)|$ as a measure of the quality of the current approximate solution x_0. Now suppose we want to replace x_0 by a new approximate solution of optimal guaranteed quality, given the available information. Here the available information is that we know the values $f'(x_0)$ and $f''(x_0)$ and moreover, we know that the curvature of the graph of f does not

BASIC ALGORITHMS

vary too much, in the precise sense that we have a positive constant c for which $|f'''(x)| < c$ for all x. Then x_1 is the best choice of approximate solution (provided it is contained in the interval (a, b), of course) in the sense that it gives the best guaranteed quality.

Conclusion. If you want to do a line search for a unimodular differentiable function, and you can and want to evaluate its second (and first) derivative at points, then the Newton method is the best one, as far as worst case performance is concerned. Its performance is spectacular, once you are close enough to the minimum: at each step, the number of guaranteed correct decimals roughly doubles!

6.4.7 Modified Newton method

Motivation and plan. Now we will describe a variant of the Newton method. It is slower, but more stable and "cheaper"—it does not require the calculation of second derivatives—than the Newton method. Moreover, it plays the following central role in the theory of necessary conditions: it can be used to give an alternative proof of the tangent space theorem. This proof has the advantage that it can be extended to an infinite-dimensional setting. This extension is needed to establish the necessary conditions of the calculus of variations and optimal control.

Modified Newton method as a minimization method. We assume that the unimodular function $f : [a, b] \to \mathbb{R}$ we want to minimize is differentiable. Let a constant $C > 0$ be given, which represents an estimate of $f''(\widehat{x})$. We consider the recursion defined by the transition from a point x_0 to the unique point of minimum of the quadratic function

$$f(x_0) + f'(x_0)(x - x_0) + \frac{1}{2}C(x - x_0)^2.$$

This gives the recursion $x_1 = x_0 - C^{-1}f'(x_0)$.

Modified Newton as a root finding method. Alternatively, the recursion can be presented as a method to solve the stationarity equation $f'(x) = 0$: by the transition from a point x_0 to the unique solution of the linear equation

$$f'(x_0) + C(x - x_0) = 0.$$

This recursive method is called the *modified Newton method* with slope C (Fig. 6.5). The difference with the Newton method is that now the slanting lines are all parallel, as we use the same constant C in each recursion step.

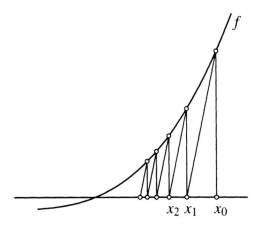

Figure 6.5 Modified Newton method for solving $f'(x) = 0$.

Complexity of the modified Newton method. At each step, the distance to the minimum is at most multiplied by a fixed fraction, provided we are close enough to the minimum and the constant C is suitably chosen. In terms of decimals, at each step we gain at least a constant number of reliable decimals. We give a precise statement of this result. Note that the Newton method is much faster.

Theorem 6.2 *Linear convergence of the modified Newton method.* Let $f : (\widehat{x} - \varepsilon, \widehat{x} + \varepsilon)$ be a function of one variable for which

$$f \in C^2(\widehat{x}), \ f''(\widehat{x}) > 0, \ f'(\widehat{x}) = 0 \text{ and } D < f''(\widehat{x}) < E.$$

Write

$$C = \frac{1}{2}(E + D) \quad \text{and} \quad \theta = (E - D)(E + D)^{-1}.$$

Then, for each starting point x_0 sufficiently close to $\widehat{x}$, the modified Newton method with slope C leads to a new approximation x_1 with

$$|x_1 - \widehat{x}| < \theta |x_0 - \widehat{x}|.$$

Proof. We only consider the case $x_0 > \widehat{x}$; the other case is similar. We have $D < f''(x_0) < E$ for all x_0 that are sufficiently close to $\widehat{x}$, using the assumption $D < f''(\widehat{x}) < E$ in combination with the continuity

BASIC ALGORITHMS

of f'' at $\widehat{x}$. On integrating from $\widehat{x}$ to x_0, this gives

$$D(x_0 - \widehat{x}) < f'(x_0) - f'(\widehat{x}) < E(x_0 - \widehat{x}).$$

This can be rewritten as the required inequality

$$\left|(x_0 - \frac{f'(x_0)}{C}) - \widehat{x}\right| < \theta|x_0 - \widehat{x}|,$$

as

$$E = (1 + \theta)C \quad \text{and} \quad D = (1 - \theta)C.$$

$\square$

Conclusion. The modified Newton method is one of the two pillars of the necessary conditions for optimization problems (the other one is "separation of convex sets"). Moreover, this algorithm is of practical value, giving a cheaper variant of the Newton method.

Conclusion to section 6.4. If you want to do a line search for a unimodular function, then it depends on the situation which one is the best. If you want to do only function evaluations, then it is the Fibonacci method if you know in advance how many function evaluations you are prepared to carry out; otherwise it is the golden section method. If you can evaluate the derivative of the objective function efficiently, then the best one is the bisection method. If you can even evaluate the second derivative of the objective function, then the best one is the Newton method. The modified Newton method is presented because of its theoretical importance: it is the core of the universal proof of the tangent space theorem.

6.5 DIRECTION OF DESCENT

Plan. We have seen that the problem of designing optimization algorithms can be split into the problems of "line search" and of "descent direction" search. We have done line search in the previous section. In this section we consider the other problem, how to find descent directions.

Direction of reasonable descent. We consider the question of which direction to move. We present three methods to do this; in each case the method gives direction vectors and not only directions. However, often one only uses the direction of this vector.

Idea of the gradient method (also called method of steepest descent). Suppose that you have lost your way in the mountains, and night has fallen. You know you must be close to a valley, and at the bottom of the valley is a hut where you can spend the night. Then it is a reasonable strategy to determine the direction of steepest descent, follow it for awhile, determine at your new position the direction of steepest descent, and follow this. Go on in this way till you have reached the bottom of the valley.

Now we describe this method of choosing a direction in a more formal way. If f is a function of n variables that you want to minimize and your current approximation of the minimum is x_0, then the vector

$$-f'(x_0)^T,$$

minus the gradient, points in the direction of *steepest descent*. The gradient method suggests looking for a next approximation of the form

$$x_0 - \alpha f'(x_0)^T$$

for some positive number α. How to choose α is a line search problem (see section 6.4).

This method is wasteful in the sense that it only uses the last computed gradient. The conjugate gradient method takes all computed gradients into account. This method will be given in chapter seven.

Newton method. The Newton method given in the previous section can be extended to the case of a twice continuously differentiable function f of n variables for which the second derivative is positive definite. This extension is straightforward by virtue of vector notation. If your current approximation of the minimum is x_0, then you take the quadratic approximation of f at x_0, that is,

$$f(x_0) + f'(x_0) \cdot (x - x_0) + \frac{1}{2}(x - x_0)^T f''(x_0)(x - x_0)$$

and choose as the new approximation x_1, the minimum of this quadratic function. That is,

$$x_1 = x_0 - f''(x_0)^{-1} f'(x_0)^T.$$

Note again that this is the unique solution of the vector equation

$$f'(x_0)^T + f''(x_0)(x - x_0) = 0_n.$$

BASIC ALGORITHMS

We see that the Newton method gives both direction and length of the direction vector. It gives the following formula for the direction vector:

$$-f''(x_0)^{-1} f'(x_0)^T.$$

However, often one only uses the direction of the vector. A comparison with the gradient method shows that the Newton method allows you to take into account the "curvature," and therefore it suggests modifying the direction of the gradient method, which is $-f'(x_0)$, by multiplication by the inverse of the symmetric matrix $f''(x_0)$. This is a much better direction, but it can involve great costs as the matrix $f''(x_0)$ has to be calculated and inverted. The theorem on the quadratic convergence—given for the one-variable case—and its proof can be extended to the n-variable case.

Modified Newton method. In the same way, the modified Newton method—and the analysis of its convergence—can be extended to the n-variable case.

Conclusion to section 6.5. The basic descent vector is the direction vector *minus the gradient*. This is the direction of steepest descent (locally). A better direction—taking into account curvature—is given by the Newton method.

6.6 QUALITY OF APPROXIMATION

Proposals of measures for quality of approximation. There are various ways to measure the quality of an approximation x_0 of the point of global minimum $\widehat{x}$ of a function $f : G \to \mathbb{R}$ of n variables defined on a subset G of $\mathbb{R}^n$. The main measures for the quality are

$$x_0 - \widehat{x},\ f(x_0) - f(\widehat{x}),\ \frac{f(x_0) - f(\widehat{x})}{f(\widetilde{x}) - f(\widehat{x})},\ |f'(x_0)|,$$

where $\widetilde{x}$ is the point of global *maximum* of f on G.

The following example illustrates the meaning of the first two of these measures.

Example 6.6.1 *Suppose we want to find an approximation to the richest person in the world. Will it make a difference whether we take the first or the second measure?*

Solution. If we take the first measure, then we will end up with someone living in the neighborhood of Bill Gates. If we take the second measure, then we might end up with the Sultan of Brunei, who is one of the richest people in the world.

Which measure is best? Let us make some remarks on these measures.
- The first one might seem to be the most natural one.

- On second thought, the second one is more appropriate, as the name of the game is to minimize the value of f, so this is what should be judged.

- However, when one has to compare the quality of the approximations for two different optimization problems, then it is clear that the third measure is better. An approximation for which this measure is smaller than a given positive number ε is called a ε-solution.

- Finally, the fourth measure, which you can use for unconstrained problems, has one practical advantage over the other measures: you can calculate it. Therefore you can use it to make a stopping rule for the algorithm. Here we note that the other ones involve the solution $\hat{x}$, which you do not know: it is the thing that you are desperate to approximate.

Sometimes these different measures are related. Here we give one result of interest about such a relation. The bisection method leads, by construction, to approximations that are very close to the true point of global minimum. However, what is not obvious at all is that it gives good approximations of the *value* of the problem as well. This is the sense of the following result.

Theorem 6.3 *On the performance of the bisection method.*
Let $f : [a, b] \to \mathbb{R}$ be a convex function and let x^* be a point of minimum of f. After N steps of the bisection method, the endpoint x_N of the uncertainty interval that has the smallest f-value is an ε-solution with $\varepsilon = 2^{-N}$. That is,

$$\frac{f(x_N) - \inf_{x \in [a,b]} f(x)}{\sup_{x \in [a,b]} f(x) - \inf_{x \in [a,b]} f(x)} \leq 2^{-N}.$$

Proof. The proof is based on Figure 6.6.
Here we have assumed, without restriction of the generality of the proof, that $x_N = a_N$, that is, $f(a_N) \leq f(b_N)$; we have drawn a

BASIC ALGORITHMS

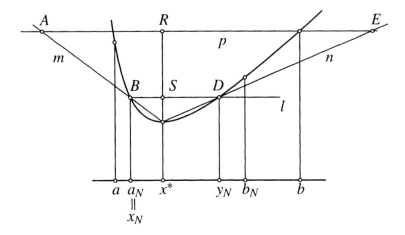

Figure 6.6 Performance of the bisection method.

horizontal line l through the point on the graph of f above x_N. Let $(y_N, f(y_N))$ be the other point of intersection of l with the graph of the function f.

Moreover, we have drawn two straight lines;

- one, m, through the points $(x^*, f(x^*))$ and $(x_N, f(x_N))$,

- and one, n, through the points $(x^*, f(x^*))$ and $(y_N, f(y_N))$.

Finally, we have drawn a horizontal line p on level

$$\sup_{x \in [a,b]} f(x);$$

in the picture we consider the case that this supremum is assumed at $x = b$. Let A, B, C, D, E, R, and S be the indicated points of intersection.

Now we make the central observation: the triangles $\triangle ACE$ and $\triangle BCD$ are similar. Therefore

$$\frac{|SC|}{|RC|} = \frac{|BD|}{|AE|}.$$

Now observe the following facts:

$$|SC| = f(x_N) - \inf_{x \in [a,b]} f(x),$$

$$|RC| = \sup_{x \in [a,b]} f(x) - \inf_{x \in [a,b]} f(x),$$

$$|BD| \leq b_N - a_N,$$

$$|AE| \geq b - a.$$

As $\frac{b_N - a_N}{b-a} = 2^{-N}$, it follows that $\bar{x}$ is an ε-solution with $\varepsilon = 2^{-N}$. □

Remark. We have based the proof on a picture. It is a routine exercise to write out this proof without appealing to a picture.

Conclusion to section 6.6. For measuring the quality of approximations, various possibilities have been discussed. For the bisection method, there is a surprising result: it performs well with respect to *approximating the value of the problem*, and not only with respect to approximating the point of optimum.

6.7 CENTER OF GRAVITY METHOD

Plan. The state of the art of solving optimization problems numerically is, roughly speaking, as follows: the class of optimization problems for this can be done with satisfactory speed and with a guarantee for the quality of the obtained "solution" is precisely the class of convex optimization problems. Two types of algorithms for convex problems have been discovered and developed in the last decades. We present introductions to these two types of algorithms in this section and the next ones.

What is better: ellipsoid method or interior point methods? The two types of algorithms for solving convex problems are *cutting plane methods*—and more in particular the *ellipsoid method*—and the *interior point methods*—also called *barrier methods*. Ellipsoid methods were the first of the two to be discovered. After an initial enthusiasm there was some disappointment: it turned out to be difficult to construct stable and satisfactory implementations. Then the exciting area of the interior point methods began, and the ellipsoid method was somewhat neglected. However the ellipsoid methods might be about to make a comeback. Recent attempts to implement the ellipsoid method have shown that the current generation of computers are not susceptible to the difficulties that were experienced in the past.

BASIC ALGORITHMS

Cutting plane algorithms. In this section and the next one, we present an introduction to *cutting plane algorithms*. This is an extension of the bisection method to convex problems of two and more variables.

Why not use KKT. Many convex problems can be written as an optimization problem with equalities and inequalities. For these problems the KKT conditions are available; however this does not help if we want to solve the problem: it leads to the distinction of too many cases to be of practical use if there are many inequalities.

Description of the optimization problem. We consider a smooth convex optimization problem (P) in $\mathbb{R}^n$. To be more precise, (P) is of the form

$$f(x) \to \min, \quad x \in G,$$

where G is a closed, bounded convex set in $\mathbb{R}^n$ (for example, G is a polyhedral set) and $f : G \to \mathbb{R}$ is a nonconstant, differentiable, convex function on G. Note that this includes LP problems. By the theorem of Weierstrass, (P) has a solution. However, we cannot expect to obtain an exact solution in general.

Choice of measure of the quality of approximation. We consider the following measure of quality for an element $x_0 \in G$, when viewed as an approximate solution. We take for this the fraction of the range of f that the value of $f : G \to \mathbb{R}$ in x_0 exceeds the value of the problem (P):

$$\frac{f(x_0) - \inf_{x \in G} f(x)}{\sup_{x \in G} f(x) - \inf_{x \in G} f(x)}.$$

This measure is a number in $[0,1]$; it equals 0 precisely if x_0 is a solution of (P). We choose as our aim to find for given small $\varepsilon > 0$ an element of G for which this measure of quality is $< \varepsilon$ (called an "ε-solution").

Cutting plane idea. The first step toward a generalization of the bisection method to the case of arbitrary n is given by the following straightforward result:

Theorem 6.4 *For each $c \in G$ for which $f'(c)$, the derivative of f at c, is a nonzero vector, all solutions of (P) are contained in the half-space H consisting of the solutions of the inequality*

$$f'(c) \cdot (x - c) \leq 0.$$

Proof. By the convexity of f, one has the inequality

$$f(x) \geq f(c) + f'(c) \cdot (x - c) \qquad \text{for all } x \in G.$$

Therefore, for each $x \notin H$, one has $f(x) > f(c)$ and so x is no solution of (P). □

Problem of choosing a point in the middle. Now the trouble starts: wishful thinking might suggest choosing $c \in G$ such that each hyperplane through c divides G into two equal halves, say, of equal volume. However, such a point does not exist in general, not even for a triangle in $\mathbb{R}^2$, as one readily checks (exercise 1.6.27).

Choice of a point in the middle: center of gravity. The best positive result in this direction is the following one. There is a systematic way to associate to G a point g that lies in some sense as much as possible "in the middle" of the set G: *its center of gravity*. Its precise definition is not so important for our purpose, but let us give the defining formula for completeness' sake: the i-th coordinate of g is given by

$$g_i = (\text{volume } G)^{-1} \int_G x_i dx_1 \ldots dx_n.$$

Center of gravity method. The first step of the center of gravity method is to compute the gradient of $f : G \to \mathbb{R}$ in the center of gravity g of G. Then all solutions of the problem (P) are contained in the intersection $G(1)$ of G and the half-space

$$f'(g) \cdot (x - g) \leq 0$$

(by theorem 6.4 above). Now repeat this step, with G replaced by $G(1)$. This leads to a subset $G(2)$ of $G(1)$ containing all solutions of the problem. Continuing in this way one gets a sequence $G(N)$, $N = 1, 2, \ldots$. These sets are called *uncertainty sets*, as they represent the uncertainty over the position of the solutions at some moment in carrying out the algorithm.

Theorem of Grünbaum. To analyze the performance of the center of gravity method, the following result can be used.

Theorem 6.5 *Grünbaum.* *Each hyperplane in $\mathbb{R}^n$ through the center of gravity g of a bounded convex set G in $\mathbb{R}^n$ divides G into two pieces, each of which has volume at most $1 - 1/e$ times the volume of G.*

The proof is relatively complicated; we omit it.

BASIC ALGORITHMS

Performance of the center of gravity method. The theorem of Grünbaum shows that there is a constant $c \in (0,1)$ such that the volume of the "uncertainty set"—that is, the set in which we know the solutions to lie—is at each step of the algorithm reduced by at least a factor c. To be more precise, this holds for $c = 1 - 1/e$. Let us display this result more formally.

Theorem 6.6 *On the performance of the center of gravity method. After N steps of the center of gravity method, the volume of the uncertainty set has shrunk to a size of at most $(1 - 1/e)^N$ times the volume of the size of G.*

Fatal flaw. Thus the performance of the center of gravity method is good. However, from a practical point of view it has a fatal flaw:

it is not possible to compute the center of gravity of the $G(k)$ in an efficient way.

In fact, even if G is very simple, say, the unit ball, then with increasing k the shape of the $G(k)$ becomes too complicated to compute the center of gravity.

Repairing the flaw. In the next section we will offer a more sophisticated but practical scheme (based on the idea of the center of gravity method), which avoids the problem of computing centers of gravity of complicated sets. For this scheme we will not need Grünbaum's theorem. This justifies that we have skipped the proof of the theorem.

Conclusion to section 6.7. The center of gravity method extends the bisection method and has good convergence properties. It has one weakness: in practice it is very hard to determine or even to approximate the center of gravity. This makes this method useless for practical purposes. However, the discussion above is a good preparation for the ellipsoid method, to be given below.

6.8 ELLIPSOID METHOD

The account of the ellipsoid method given below is based on a lecture by Vladimir Protasov.

Plan. The rough idea of the ellipsoid method is to replace the sequence of complicated convex sets $G(k)$ ($k \in \mathbb{N}$) constructed in the center of gravity method by a sequence of ellipsoids (linear transformations of unit balls). Two properties of ellipsoids combine to make this rough idea into a successful algorithm.

1. It is very easy to determine the center of gravity of an ellipsoid: it is the center of the ellipsoid.

2. A half-ellipsoid (= one of the two pieces into which an ellipsoid is divided by a hyperplane through its center) is contained in a "small" ellipsoid. By "small" we mean here that the quotient of the volumes of the new ellipsoid and the old ellipsoid does not exceed some constant $c(n)$ in the open interval (0,1) that only depends on the dimension n.

Note that, for example, the collection of balls does not have the second property.

Now we are going to present the ellipsoid method in a more systematic way.

Ellipsoids. An *ellipsoid* in $\mathbb{R}^n$ is defined to be a subset of $\mathbb{R}^n$ of the form

$$E(M, c) = \{c + Mz : |z| \leq 1\}$$

with $c \in \mathbb{R}^n$ and M an invertible $n \times n$ matrix. That is, an ellipsoid in $\mathbb{R}^n$ is the image of the unit ball $\{z : |z| \leq 1\}$ in $\mathbb{R}^n$ under an invertible affine map $z \to c + Mz$. The vector c is called the *center* of the ellipsoid. If $n = 2$ one calls $E(M, c)$ an *ellipse*.

Example 6.8.1 *The set of solutions of the inequality*

$$\frac{x_1^2}{b_1^2} + \frac{x_2^2}{b_2^2} \leq 1$$

(for given $b_1, b_2 > 0$) forms an ellipse in $\mathbb{R}^2$.

Example 6.8.2 *The previous example can be generalized to $\mathbb{R}^n$. If we take $c = 0$ and $M = \begin{pmatrix} b_1 & & 0 \\ & \ddots & \\ 0 & & b_n \end{pmatrix}$ with $b_1, \ldots, b_n > 0$, then the ellipsoid $E(M, c)$ is the set of solutions of the inequality*

$$\frac{x_1^2}{b_1^2} + \cdots + \frac{x_n^2}{b_n^2} \leq 1.$$

BASIC ALGORITHMS

The following result shows that the ellipsoids in example 6.8.2 are in a sense the most general ones.

Theorem 6.7 *Each ellipsoid in $\mathbb{R}^n$ can be transformed by an affine orthogonal transformation $y \to d + Py$, with P an orthogonal matrix (that is, $PP^T = I_n$), into the set of solutions of the inequality of the form*

$$\frac{x_1^2}{b_1^2} + \cdots + \frac{x_n^2}{b_n^2} \leq 1$$

for suitable positive numbers $b_1, \ldots, b_n$. The set—with multiplicity—$\{b_1, \ldots, b_n\}$ is uniquely determined by the ellipsoid.

Proof. We are going to derive the first statement of the theorem from the fact that each positive definite matrix can be transformed into a positive diagonal matrix by conjugation with an orthogonal matrix (theorem 3.4).

Let an ellipsoid $E(M, c)$ be given. Then the product MM^T is a positive definite matrix, as

$$x^T M M^T x = (M^T x)^T (M^T x) = |M^T x|^2 > 0$$

for each nonzero vector $x \in \mathbb{R}^n$. Now write

$$MM^T = P^T D P$$

with P an orthogonal $n \times n$-matrix and D a diagonal matrix with positive diagonal entries. Write

$$D = \begin{pmatrix} b_1^2 & & 0 \\ & \ddots & \\ 0 & & b_n^2 \end{pmatrix} \text{ and } B = \begin{pmatrix} b_1 & & 0 \\ & \ddots & \\ 0 & & b_n \end{pmatrix}$$

for suitable positive numbers $b_1, \ldots, b_n$.

To prove the first statement of the theorem it suffices to verify that the affine orthogonal transformation $y \to P(y - c)$ satisfies the requirements. Well, "$y \in E(M, c)$" means that

$$(y - c)^T (M^{-1})^T M^{-1} (y - c) \leq 1.$$

By the transformation

$$y \to P(y - c) = x$$

and the relation
$$MM^T = P^T B^2 P,$$
this can be rewritten as $x^T B^{-2} x \leq 1$, that is, as the inequality
$$\frac{x_1^2}{b_1^2} + \cdots + \frac{x_n^2}{b_n^2} \leq 1.$$
To prove the last statement of the theorem, it suffices to observe that $E(M,c)$ determines the matrix $(M^{-1})^T M^{-1}$ and that $b_1^{-2}, \ldots, b_n^{-2}$ are the eigenvalues of this matrix. □

Corollary 6.8 *Let notation be as in theorem 6.7. The volume of $E(M,c)$ equals the product of $b_1, \ldots, b_n$ and the volume of the unit ball in $\mathbb{R}^n$.*

Proof. We have vol $E(M, I) = \det M \cdot$ vol $E(I, 0)$, by the definitions of $E(M, I)$ and of determinant. By
$$MM^T = P^T B^2 P,$$
we get $\det M = b_1 \cdots b_n$ (replacing M by $-M$ if necessary). □

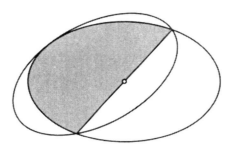

Figure 6.7 Ellipsoid method.

Including half-ellipsoids in ellipsoids of small volume. A *half-ellipsoid* of an ellipsoid E, also called a half of the ellipsoid, is defined to be one of the two parts into which the ellipsoid is divided by a hyperplane through its center. That is, for each nonzero vector w in $\mathbb{R}^n$ and each ellipsoid $E = E(M, c)$, the set $HE(M, c; w)$ of all $x \in E(M, c)$ for which
$$(x - c)^T \cdot w \leq 0$$

BASIC ALGORITHMS

is called a *half-ellipsoid* of E, where M is an invertible $n \times n$ matrix and $c \in \mathbb{R}^n$. Now we come to the key result. It generalizes the result of exercise 1.6.28 from $n = 2$ to arbitrary n.

Theorem 6.9 *Let an ellipsoid E in $\mathbb{R}^n$ be given, and choose a half-ellipsoid of E. Among the ellipsoids that contain the chosen half-ellipsoid of E (cf. Fig. 6.7), there is a unique one, $\bar{E}$, of minimal volume. The volume of this new ellipsoid $\bar{E}$ is smaller than the volume of the given ellipsoid E in the following sense:*

$$\frac{\text{vol}\bar{E}}{\text{vol}E} \leq e^{-1/(2n+2)}.$$

Remarks

- The estimate depends only on n, not on the chosen half-ellipsoid of E, and not even on the given ellipsoid E.

- $e^{-1/(2n+2)}$ is a positive constant < 1.

- The proof below will make clear that it is possible to give an explicit description of $\bar{E}$ in terms of the given half-ellipsoid. This fact will be used, as the transition from a given half-ellipsoid to the ellipsoid $\bar{E}$ will be part of the ellipsoid algorithm.

The proof, to be given below, might look technical, but essentially it is a high school exercise for the one-variable Fermat theorem.

Proof.
Reduction of a given half-ellipsoid to half of the unit ball. We assume that the half-ellipsoid is

$$B_n^- = \{x \in \mathbb{R}^n : |x| \leq 1,\ x_1 \leq 0\},$$

as we may by applying the affine map $x \mapsto M^{-1}(x - c)$: this maps the given ellipsoid $E(M, c)$ bijectively onto the unit ball.
Parametrization of all minimal ellipsoids containing half of the unit ball. The ellipsoids that contain the given half of the unit ball B_n^- and that are minimal with this property (that is, they do not contain any smaller ellipsoid with this property) can be described as follows. These are the ellipsoids that contain B_n^- and touch B_n^- at the point $(-1, 0, \ldots, 0)$ and at the points $\{x \in B_n^- : x_1 = 0\}$ (forming a sphere in the hyperplane $x_1 = 0$). A calculation shows that the

family of these ellipsoids can be parametrized as follows. For each $c \in (-\frac{1}{2}, 0)$ we consider the ellipsoid E_c given by the inequality

$$(c+1)^{-2}[(x_1 - c)^2 + \frac{2c+1}{(c+1)^2} \sum_{i=2}^{n} x_i^2] \leq 1.$$

Then E_c, $c \in (-\frac{1}{2}, 0)$ is the promised parametrization. We omit the verification.

Choice of the best ellipsoid E_c. Now we want to find the ellipsoid E_c with the smallest volume. To this end, we note to begin with that

$$\frac{\text{volume } E_c}{\text{volume unit ball}} = b_1 \cdots b_n = (c+1)\left[\frac{(c+1)^2}{2c+1}\right]^{\frac{n-1}{2}}.$$

This equals

$$\frac{(c+1)^n}{(2c+1)^{\frac{n-1}{2}}}.$$

Thus we get the following optimization problem:

$$h(c) = \frac{(c+1)^n}{(2c+1)^{\frac{n-1}{2}}} \to \min; \qquad c \in \left(-\frac{1}{2}, 1\right).$$

We solve this problem, using the four-step method.

1. The theorem of Weierstrass gives existence of a solution $\hat{c}$ of this problem, as $\lim_{c \downarrow -\frac{1}{2}} h(c) = +\infty$ and $\lim_{c \to +\infty} h(c) = +\infty$.

2. Fermat: $h'(c) = 0 \Rightarrow$

$$\frac{n(c+1)^{n-1}(2c+1)^{\frac{n-1}{2}} - (c+1)^n \frac{n-1}{2}(2c+1)^{\frac{n-3}{2}} 2}{(2c+1)^{n-1}} = 0.$$

3. $n(2c+1) - (c+1)(n-1) = 0$. Therefore, $c = -\frac{1}{n+1}$.

4. $\hat{c} = -\frac{1}{n+1}$.

Now

$$\frac{\text{volume } E_{\hat{c}}}{\text{volume unit ball}} = \frac{(\frac{n}{n+1})^n}{(\frac{n-1}{n+1})^{\frac{n-1}{2}}}.$$

BASIC ALGORITHMS

The right-hand side can be rewritten as

$$\left(1 - \frac{1}{n+1}\right)\left(1 + \frac{1}{n^2-1}\right)^{\frac{n-1}{2}}. \qquad (*)$$

Now use the inequality

$$1 + x \leq e^x.$$

Then one gets the following upper bound for $(*)$:

$$e^{-\frac{1}{n+1}}\left(e^{\frac{1}{n^2-1}}\right)^{\frac{n-1}{2}} = e^{-\frac{1}{2(n+1)}}.$$

$\square$

Great achievement. A great thing has been achieved by these high school calculations. This is the solution of a once famous problem. It establishes a fundamental fact on the complexity of convex optimization problems. It shows that these problems can be solved efficiently in the following sense: with each step of the algorithm we gain a guaranteed fixed number of decimals of accuracy (where we measure accuracy by the volume of the current ellipsoid).

The nightmare of the cigars. This result leaves one thing to be desired: the volumes of the successive ellipsoids might converge to zero, but this does not exclude that these ellipsoids might have the shape of thin, long "cigars." This would not give much information about the performance of the ellipsoid method. However, one can conclude good performance with respect to the following measure of quality: the value of the objective function in the point of approximation, just as for the bisection method. This only requires that we be a bit more careful in our choice of point of approximation at the end of the algorithm: for this we should not take automatically the center of the final ellipsoid.

Description of the ellipsoid algorithm. Let a convex function f on $\mathbb{R}^n$ and an ellipsoid E_0 be given. Suppose we want to solve the following problem:

$$f(x) \to \min, \ x \in E_0.$$

Compute the derivative of f at the center c_0 of E_0, and consider the half-ellipsoid

$$\{x \in E_0 : f'(c_0) \cdot (x - c_0) \leq 0\}.$$

Take the ellipsoid of smallest volume E_1, say, with center c_1, that contains this half-ellipsoid. Then repeat this step, but now starting with E_1 instead of with E_0. Continuing in this way we get after N steps a sequence of ellipsoids E_k, $1 \leq k \leq N$ with centers c_k. Finally, we determine among these centers the one that has the smallest f-value. This center, for which we write x_N, is chosen as the approximate solution after N steps.

Theorem 6.10 *On the performance of the ellipsoid method. After N steps of the ellipsoid method the point x_N is an ε-solution of the given problem, with $\varepsilon = e^{-\frac{N}{2n(n+1)}}$.*
That is,

$$\frac{f(x_N) - \inf_{x \in E_0} f(x)}{\sup_{x \in E_0} f(x) - \inf_{x \in E_0} f(x)} \leq \varepsilon.$$

Proof. We will sketch the proof. After N steps one has

$$\frac{\text{volume } E_N}{\text{volume } E_0} \leq \left(e^{-\frac{1}{2(n+1)}}\right)^N,$$

clearly. By a similar argument as was used in the proof of theorem 6.3 on the performance of the bisection method (cf. Fig. 6.6) one can show that this implies that x_N is an ε-solution of the given problem with $\varepsilon = e^{-\frac{N}{2n(n+1)}}$. $\square$

Khachiyan. This result represents an important breakthrough. The story is as follows. Since the discovery of the simplex method to solve LP problems, this method has been continuously improved. The resulting simplex methods are very fast and robust and became one of the most used mathematical techniques. They are used routinely everywhere in the world to save costs, for example. The surprising speed of the method raised the question whether a theoretical result could be proved on the complexity of the method that explained this efficiency. To be precise, the right way to formulate this question is whether the computational effort needed to solve an LP problem depends polynomially on its size. This fundamental problem is still unsolved, but in 1979 Leonid Khachiyan showed that the ellipsoid method invented by Shor can be used to solve LP problems and that this method has the required complexity. This is the meaning of the result above. We mention in passing that the good theoretical complexity bound initially raised expectations that the method could give a fast practical algorithm to solve LP problems. However, this turned out not to be the case. Only in 1984 the first algorithm was proposed, by Narendra

BASIC ALGORITHMS 315

Karmarkar, that was very efficient in practice and for which, moreover, a good theoretical complexity could be proved. This was the first interior point method.

Explicit formulas for writing codes. In order to write a code implementing this algorithm, we need explicit formulas for the ellipsoids $E(M_k, c_k)$ occurring in it. We display these formulas below for problems where $c_k \in \text{interior}(G_k)$ for all k with the suggestion to use them to write a code for the ellipsoid method. For each positive semidefinite matrix N we define $\sqrt{N}$, the square root of N, to be the positive semidefinite matrix M for which $M^2 = N$.

Let $E(M_0, c_0)$ be the starting ellipsoid, so $E_0 = E(M_0, c_0)$. So c_0 is the starting point of gravity and $N_0 = M_0^2$ is a positive definite matrix. To compute the ellipsoid in step k given the ellipsoid in step $k - 1$, the following formulas can be used. Let $f'(c_k)$ denote the derivative of f at the point c_k, which is the center of gravity in step k. Compute then in step k:

$$b_{k-1} = \frac{N_{k-1} f'(c_{k-1})}{\sqrt{(f'(c_{k-1}))^T N_{k-1} f'(c_{k-1})}},$$

$$c_k = c_{k-1} - \frac{1}{n+1} b_{k-1},$$

$$N_k = \frac{n^2}{n^2 - 1}(N_{k-1} - \frac{2}{n+1} b_{k-1} b_{k-1}^T),$$

$$M_k = \sqrt{N_k}.$$

Ellipsoid problem for general convex problems. In addition we mention one detail. Often the function f is not defined on the whole space $\mathbb{R}^n$ but only on some given convex set G in $\mathbb{R}^n$, and the minimum of f on G is sought. Then we should make a small change at the beginning of the algorithm: one should embed G into an ellipsoid E_0.

Another matter seems more problematic. The ellipsoid method, as described above, might lead to an ellipsoid with center c not in G. This could already happen at the beginning of the algorithm, after G has been embedded in an ellipsoid E_0. What to do? At such a difficult moment, the separation theorem for convex sets (see Appendix

B) comes to our rescue. It gives a half-space containing G with the center on its boundary (to this end we apply the separation theorem to the convex set G and the point c). Intersecting this half-space with the last ellipsoid gives a half-ellipsoid containing all solutions of our problem. Therefore, we can continue the algorithm in the usual way, embedding this half-ellipsoid into the ellipsoid of smallest volume.

Conclusion to section 6.8. The ellipsoid method is an adaptation of the center of gravity method. It makes use of the fact that the class of ellipsoids has so much flexibility that each half-ellipsoid is contained in an ellipsoid of smaller volume than the original ellipsoid. The ellipsoid algorithm is reliable and efficient, and it is easy to implement.

6.9 INTERIOR POINT METHODS

Plan. In this final section we give an introduction to the other type of algorithms to solve convex optimization problems: interior point methods. We give an elementary presentation of the simplest case, that of primal-dual short-step algorithms for *LP problems*. A special feature of this presentation in compared to standard treatments is that it is coordinate-free.

Two presentations: calculation of limit and barriers. In 1984, when Karmarkar published his epoch-making paper, interior point methods for solving linear programming problems seemed rather mysterious. By now, the basic idea can be explained in a relatively straightforward way: as *a method to calculate a limit numerically*, as we are going to see. In section 7.3 we will give a different presentation of interior point methods, using *barriers* for arbitrary convex problems. Here, we will not include an analysis of the complexity of this method. A discussion of the complexity will be considered in section 7.3. The approach to interior point methods given here—called the *v-space approach*—can be extended to semidefinite programming problems (cf. [73]).

6.9.1 Linear complementarity problems

Plan and motivation. The best versions of the interior point method solve the primal and dual problems at the same time. This is

BASIC ALGORITHMS 317

in the spirit of the epigraph of section 4.5.1: *you can see everything with one eye, but looking with two eyes is more convenient.* Primal and dual LP problems have been considered in section 4.5.1. It is convenient to present such algorithms in terms of *linear complementarity problems* (LCP). An LCP problem is one problem that is equivalent to a pair of primal-dual LP problems. In this section we explain what an LCP problem is.

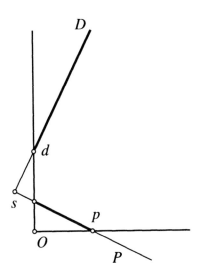

Figure 6.8 Intersection of orthogonal lines with positive quadrant.

Definition of an LCP problem. For each linear subspace $\tilde{P}$ of $\mathbb{R}^n$ and each vector s in $\mathbb{R}^n$ we let

$$P = \tilde{P} + s = \{\tilde{p} + s : \tilde{p} \in \tilde{P}\}$$

and

$$D = \tilde{D} + s = \{\tilde{d} + s : \tilde{d} \in \tilde{D}\},$$

where $\tilde{D}$ is the *orthogonal complement* of $\tilde{P}$ (that is, $\tilde{D}$ consists of all $d \in \mathbb{R}^n$ that are orthogonal to all $p \in \tilde{P}$). One can extend the terminology of orthogonal complements from linear subspaces to affine subspaces. Then the affine subspaces P and D are called a pair of orthogonal complementary affine subspaces. The choice of symbols P and D refers to the words primal and dual. We assume that there exist $p \in P$ with $p > 0_n$ and $d \in D$ with $d > 0_n$. We recall that we write $v > w$ for $v, w \in \mathbb{R}^n$ if $v_i > w_i$, $1 \leq i \leq n$. We consider the following problem (Q):

find two nonnegative vectors $p \in P$ and $d \in D$ that are orthogonal.

A problem of type (Q) is called a *linear complementarity problem* (LCP). This terminology is motivated by the observation that the orthogonality condition for two nonnegative vectors p and d is equivalent to the complementarity conditions $p_i d_i = 0$, $1 \leq i \leq n$.

Two-dimensional LCP. As a first illustration, let $n = 2$ and let P and D be two orthogonal lines in the plane $\mathbb{R}^2$, as in Figure 6.8.

Assume that both lines contain positive vectors and do not contain the origin. Then the problem (Q) has a unique solution $(\widehat{p}, \widehat{d})$; one glance at the picture suffices to spot it.

Three-dimensional LCP. The next case is already slightly more interesting: take $n = 3$ and choose P to be a line in $\mathbb{R}^3$ and D a plane in $\mathbb{R}^3$, orthogonal to the line P. Then the problem (Q) asks us to find a point p in P and a point d in D, both in the first orthant, such that the vectors p and d are orthogonal. We are going to give a geometrical description of the unique solution of this problem in the following special case. The line P intersects the x_1, x_2-plane (resp. the x_2, x_3-plane) in a point $\widehat{p}_1$ (resp. $\widehat{p}_2$) that lies in the interior of the first quadrant of this plane, as is shown in Figure 6.9.

Moreover, we assume that the point $\widehat{d}$ of intersection of D with the union of the positive x_1-axis and the positive x_3-axis is not the origin. Then the problem (Q) has a unique solution. It is $(\widehat{p}_2, \widehat{d})$ if $\widehat{d}$ lies on the x_1-axis: then $\widehat{p}_2$ and $\widehat{d}$ are orthogonal. It is $(\widehat{p}_1, \widehat{d})$ if $\widehat{d}$ lies on the x_3-axis: then $\widehat{p}_1$ and $\widehat{d}$ are orthogonal.

General LCP. If n is large, then the combinatorics of the situation are sufficiently rich to make the problem (Q) really interesting.

Conclusion. To each pair of affine subspaces in $\mathbb{R}^n$ that are orthogonal complements of each other, one can associate a linear complementarity problem: find a pair of nonnegative points, one in each of the two affine subspaces, with inner product zero.

6.9.2 Relation between LCP and LP

Plan. In this section we will give the promised relation between LCP and linear programming. The LCP problem (Q) above will be

BASIC ALGORITHMS

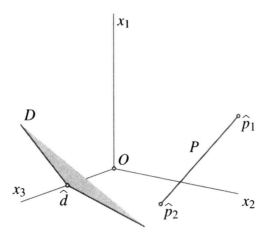

Figure 6.9 Intersection of orthogonal line and plane with the positive orthant.

related to the following pair of primal-dual LP problems

$$f(p) = s^T p \to \min, \; p \in P, \; p \geq 0_n, \qquad (P)$$

$$g(d) = s^T(s - d) \to \max, \; d \in D, \; d \geq 0_n. \qquad (D)$$

The promised relation of (Q) with (P) and (D) is as follows.

Theorem 6.11 *Let P and D be a pair of orthogonal complementary affine subspaces. Let (Q) be the associated LCP problem and let (P) and (D) be the associated LP problems. Then for each $\widehat{p} \in P$ and $\widehat{d} \in D$ the following two conditions are equivalent:*

- $(\widehat{p}, \widehat{d})$ *is a solution of* (Q),

- $\widehat{p}$ *is a solution of* (P) *and* $\widehat{d}$ *is a solution of* (D).

The statement of this theorem can be readily derived from the following equality:

$$s^T \cdot p - s^T \cdot (s - d) = p^T \cdot d$$

for all $p \in P$ and $d \in D$. This equality follows by rewriting the equality

$$(p-s)^T \cdot (d-s) = 0,$$

which holds as $p - s \in \tilde{P}$ and $d - s \in \tilde{D}$.

Conclusion. We have seen that we can reformulate the problem to find solutions for a pair of primal-dual LP problems (P) and (D), as the problem to find a solution for the LCP problem (Q).

6.9.3 Hadamard product bijection

Plan. We consider a mapping from the Cartesian product $P_+ \times D_+$ to $\mathbb{R}^n_+$ and establish the following two properties:

1. The inverse image of 0_n is the set of solutions of the LCP problem (Q).

2. The restriction of the mapping to $P_{++} \times D_{++} \to \mathbb{R}^n_{++}$ is a bijection.

Here P_+ (resp. D_+) is the set of nonnegative points in P (resp. D). Moreover, P_{++} (resp. D_{++}) is the set of positive points in P (resp. D).

This will suggest solving (Q) by computing a limit, as we will see in section 6.9.4.

Notation. We need some notation:

- the Cartesian product

$$P_+ \times D_+ = \{(p,d) : p \in P, \ p > 0_n, \ d \in D, \ d > 0_n\},$$

- the positive orthant in n-space $\mathbb{R}^n_+$,

- the Hadamard product $v * w = (v_1 w_1, \ldots, v_n w_n)^T \in \mathbb{R}^n$ of two vectors $v, w \in \mathbb{R}^n$.

Theorem 6.12 *Consider the mapping from the Cartesian product $P_+ \times D_+$ to $\mathbb{R}^n_+$, given by the Hadamard product, $(p,d) \mapsto p * d$. This mapping has the following two properties:*

1. *The inverse image of 0_n is the set of solutions of (Q).*

2. *The restriction to $P_{++} \times D_{++} \to \mathbb{R}^n_{++}$ is a bijection.*

BASIC ALGORITHMS

For each positive vector in n-space x we will write $(p(x), d(x))$ for the inverse image of x under the bijection from theorem 6.12. Then

$$p(x) * d(x) = x \text{ and } p(x) \in P,\ p(x) > 0_n,\ d(x) \in D,\ d(x) > 0_n.$$

Proof. The first statement is obvious from the definitions. We will formulate and solve an optimization problem (Q_x) for each $x \in \mathbb{R}^n_+$; the solution will give the statement of theorem 6.12. We introduce the following notation.

$\ln v = (\ln v_1, \ldots, \ln v_n)$ for all positive vectors $v \in \mathbb{R}^n$,

$v^r = (v_1^r, \ldots, v_n^r)$ for all positive vectors $v \in \mathbb{R}^n$ and all $r \in \mathbb{R}$.

This notation allows a convenient way of defining (Q_x) for each positive vector $x \in \mathbb{R}^n$:

1. Consider the problem

$$h_x(p, d) = p^T \cdot d - x^T \cdot \ln(p * d) \to \min,\ p \in P,\ p > 0_n,$$

$$d \in D,\ d > 0_n. \qquad (Q_x)$$

Existence and uniqueness of a global solution $(p(x), d(x))$ follows from the coerciveness and strict convexity of h_x.

2. Fermat:
   - $(d - x * p^{-1})^T \cdot \tilde{p} = 0$ for all $\tilde{p} \in \tilde{P}$,
   - $(p - x * d^{-1})^T \cdot \tilde{d} = 0$ for all $\tilde{d} \in \tilde{D}$.

3. $(p^{\frac{1}{2}} * d^{\frac{1}{2}} - x * p^{-\frac{1}{2}} * d^{-\frac{1}{2}})^T \cdot (p^{-\frac{1}{2}} * d^{\frac{1}{2}} * \tilde{p}) = 0\ \forall \tilde{p} \in \tilde{P}$,
   $(p^{\frac{1}{2}} * d^{\frac{1}{2}} - x * p^{-\frac{1}{2}} * d^{-\frac{1}{2}})^T \cdot (p^{\frac{1}{2}} * d^{-\frac{1}{2}} * \tilde{d}) = 0\ \forall \tilde{d} \in \tilde{D}$.

   Observe that the affine subspaces

   $$p^{-\frac{1}{2}} * d^{\frac{1}{2}} * \tilde{P} \text{ and } p^{\frac{1}{2}} * d^{-\frac{1}{2}} * \tilde{D}$$

   are orthogonal complements as $\tilde{P}$ and $\tilde{D}$ are orthogonal complements. Therefore, the conditions above are equivalent to

   $$p^{\frac{1}{2}} * d^{\frac{1}{2}} - x * p^{-\frac{1}{2}} * d^{-\frac{1}{2}} = 0_n.$$

   That is, $p * d = x$.

4. It follows that there are unique $p \in P$, $p > 0_n$, and $d \in D$, $d > 0_n$, for which $p * d = x$.

□

In Figure 6.10 the bijection of this theorem is illustrated for $n = 2$. If we view the lines P and D in $\mathbb{R}^2$ as coordinate axes, then the pairs

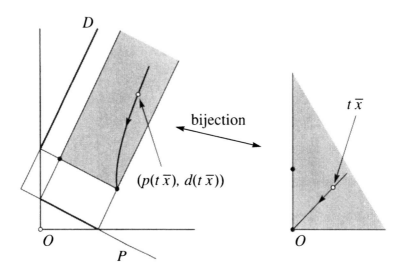

Figure 6.10 The central path.

of positive vectors (p, d) with $p \in P$ and $d \in D$ form a region in the P, D plane. The Hadamard product maps this region bijectively to the positive quadrant of $\mathbb{R}^2$.

6.9.4 Description of the interior point method

Plan. We will explain how to solve a linear programming problem by computing a limit.

Interior point method as computation limit. The idea of the interior point method is to find the solution of the LCP problem (Q)—and so of the primal and dual LP problems (P) and (D)—by computing the limit

$$(\widehat{p}, \widehat{d}) = \lim_{x \to 0_n} (p(x), d(x)),$$

where x tends to 0_n along the ray consisting of the vectors in the positive orthant with all coefficients equal to each other (to be called *the central ray*). The path that is then followed by the points $(p(x), d(x))$ is called the *central path* (Fig. 6.10). *It is a royal road to the solution.* It can be proved that this limit exists. It is readily seen that this limit must be a solution of the LCP problem (Q), as required.

How to calculate the limit. This limit is calculated algorithmically in the following way. For the sake of simplicity, we assume that we can somehow find positive vectors $\bar{p} \in P$ and $\bar{d} \in D$. Write $\bar{x} = \bar{p} * \bar{d}$. Now we take an infinite sequence of points in the positive orthant of n-space, $\bar{x} = x_0, x_1, \ldots$ in such a way that two consecutive terms are close and, moreover, the points of the sequence first move from $\bar{x}$ to the central ray, and then tend along the central ray to the origin 0_n. Then the required limit equals the following limit:

$$\lim_{k \to \infty} (p(x_k), d(x_k)).$$

Now we are ready to describe the algorithm to calculate this limit.

First step of the algorithm: from $(\bar{p}, \bar{d}) = (p(x_0), d(x_0))$ to $(p(x_1), d(x_1))$. We apply one step of the Newton method, with starting point $(\bar{p}, \bar{d})$, to the system

$$\begin{cases} p \in P, \\ d \in D, \\ p * d = x_1. \end{cases}$$

Note that the solution of this system is $(p(x_1), d(x_1))$. As $\bar{x} = x_0$ and x_1 are close, the starting point $(\bar{p}, \bar{d}) = (p(x_0), d(x_0))$ and the solution of the system $(p(x_1), d(x_1))$ are close. Therefore this step of the Newton method leads to a very good approximation of $(p(x_1), d(x_1))$.

We continue in this way. Thus the k-th step of the algorithm turns a very good approximation of $(p(x_{k-1}), d(x_{k-1}))$ into a very good approximation of $(p(x_k), d(x_k))$.

This finishes the explanation how the interior point method works. By being more precise, one can show that the sequence $x_1, x_2, \ldots$ can be chosen in such a way that the algorithm converges *linearly*, that is, at each step of the algorithm we gain at least a fixed number of digits of the decimal expansions of the coordinates of the limit $(\hat{p}, \hat{d})$.

Conclusion. The interior point method, one of the greatest discoveries of optimization theory, can be explained in a simple way: *as a method of solving LP problems by calculating a limit.*

Chapter Seven

Advanced Algorithms

Elegance should be left to tailors and shoemakers.

L. Boltzmann

- In this chapter we will give treatments of two advanced algorithms: the conjugate gradient method and the unified theory of interior point methods of self-concordant barriers.

7.1 INTRODUCTION

Einstein quoted the words of Ludwig Boltzmann in the preface of his famous popular book on relativity theory. However, in reality both relativity theory and Einstein's style of writing are very elegant. Maybe he interpreted these words as saying that elegance should not stand in the way of user-friendliness. This is how we read them. We present two beautiful and very successful methods, conjugate gradient methods and self-concordant barrier methods. The accounts can be read independently. We have tried to explain them in an analytical style, as simple and straightforward as we could. However, we do not provide elegant geometrical explanations of the essence of these two methods,.

7.2 CONJUGATE GRADIENT METHOD

This section is based on Lecture 21, Krylow Sequence Methods, of G. W. Stewart, *Afternotes goes to Graduate School* [72].

Motivation. Sometimes, an algorithm is preferable to a formula. For example, a system of n linear equations in n unknowns can be solved by a formula—provided the coefficient matrix is invertible—such as Cramer's rule or, equivalently, a formula involving the inverse of the coefficient matrix. However, it is much easier to solve such a system by Gaussian elimination. In the case of a positive definite, symmetric coefficient matrix that is very large and sparse, it can be done much faster *using an optimization algorithm, the conjugate gradient method*. For symmetric coefficient matrices, the system of equations can be seen as the stationarity equations of a quadratic function. This function is strictly convex (and coercive) if the coefficient matrix is positive definite. The conjugate gradient method is a refinement of the gradient method. The gradient method does not destroy the sparsity of the coefficient matrix, and so "keeps the problem small"—the Gaussian elimination does destroy the sparsity—and the conjugate gradient method shares this good property of the gradient method. The idea of the refinement comes out of the observation that the gradient method is wasteful in the sense that it uses only the gradient of the objective function in the current approximation to determine the direction of descent, which is taken to be minus this gradient. It does not use the gradients of the objective function at the previous approximations. These gradients give additional information about the best direction to move in. Moreover, they have already been computed, and so they are freely available. This observation suggests taking as direction of descent a suitable linear combination of all gradients computed—the one in the current approximation and the ones in the previous approximations. Of course, one expects that the last computed gradient will have the largest coefficient or weight. It turns out that the resulting algorithm can be written down, for the case of strictly convex quadratic functions, in such a way that

- sparsity is preserved,

- it is simple to implement,

- it does not lead to a significant loss of precision.

We emphasize that the conjugate gradient method can be seen either as a method to solve linear systems with a positive definite symmetric coefficient matrix, or as a method to minimize a strictly convex quadratic function. Here, we will take the latter point of view.

Minimizing large quadratic functions. We continue the consideration of the minimization of quadratic functions of n variables, problem 2.4.1,

$$\varphi(x) = \frac{1}{2}x^T A x - b^T \cdot x + \gamma \to \min, \qquad (P_{7.1})$$

but now from a different point of view: how can it be solved numerically? Here A is an $n \times n$ matrix, $b \in \mathbb{R}^n$, and $\gamma \in \mathbb{R}$. We assume that

A is symmetric and positive definite; for all practical purposes this is no restriction. The factor $\frac{1}{2}$ and the minus sign are there to simplify some formulas later on. This problem has a unique solution given by

$$x_* = A^{-1}b,$$

as we have seen in problem 2.2. At this point we might run into an unexpected numerical problem. Often the matrix A is very large, but sparse; that is, only a very small proportion of its entries are nonzero. Then our problem should be considered as being of reasonable size and so one might hope for an efficient solution method. However, Gaussian elimination, the method of choice for solving a system of linear equations, is not practical. Its iterations ruin the sparsity property gradually and so it runs about as slowly as it would on a dense matrix of the same size. This situation might seem hopeless. Surprisingly, there is a way out. The following algorithm gives all one could wish for. It starts with a given vector x_1, an initial approximation of the solution x_*, and produces a sequence of approximations

$$x_1, x_2, x_3, \ldots, x_{m+1}$$

of increasing quality. In fact, x_{m+1} is the exact solution of the problem. The sequence of approximations is constructed by recursion. To be more precise, the algorithm simultaneously constructs sequences of vectors r_k and s_k in $\mathbb{R}^m$, a sequence of vectors x_k in $\mathbb{R}^n$, and two sequences of numbers α_k and β_k, $k = 1, 2, \ldots$. Here is the detailed description of the intertwining recursions defining these sequences.

Conjugate gradient algorithm

1. $r_1 = s_1 = b - Ax_1$

2. for $k = 1, 2, \ldots$

3. $\alpha_k = |r_k|^2 / s_k^T A s_k$

4. $x_{k+1} = x_k + \alpha_k s_k$

5. $r_{k+1} = r_k - \alpha_k A s_k$

6. $\beta_k = -|r_{k+1}|^2 / |r_k|^2$

7. $s_{k+1} = r_{k+1} - \beta_k s_k$

8. end for $k = m + 1$.

This algorithm might seem a mystery at first sight! Why are these recurrence formulas convenient to use, what are the ideas behind them, what is the meaning of the quantities occurring in this algorithm, and what are the convergence properties of the algorithm? In this section we will answer these questions. The answers will provide a full insight in the algorithm. The properties of this algorithm are given in theorem 7.1 and interpretations and properties of the sequences r_k and s_k, $1, 2, \ldots$ are given in theorem 7.2. However, if you are just interested in applying the algorithm, the description above is all you need.

Some good features of the algorithm. To begin with, we point out some good features of these recurrence formulas. They are simple to implement and the calculations can be carried out easily and without loss of precision. Indeed, by the sparsity of A, it is practical to calculate the product of A with a vector, even though A is very large. Therefore, it is practical to compute Ax_1, As_k, and $s_k^T As_k$. Moreover, norms of vectors, such as $|r_k|$ and $|r_{k+1}|$, can be calculated accurately. Inner products $v^T w$ of vectors $v, w \in \mathbb{R}^n$ might lead to undesirable cancellation: the recursion above avoids inner products.

Sketch of the idea of the algorithm. Now we give a preliminary sketch of the idea of this algorithm. We need some notation. Let m be the rank of the sequence of vectors

$$(A^k(x_* - x_1))_{k \geq 0}$$

in $\mathbb{R}^n$. That is, the first m vectors of this sequence are linearly independent and all other vectors of it can be expressed as a linear combination of these vectors in a unique way. We observe in passing that we can compute these vectors easily, except the one with $k = 0$: for $k = 1$ one has

$$A(x_* - x_1) = b - Ax_1;$$

then repeatedly multiplying with A one gets

$$A^2(x_* - x_1), \ A^3(x_* - x_1), \ldots.$$

Calculating the excepted vector $x_* - x_1$ is the name of the game: then we would also have x_*, as it is the sum of $x_* - x_1$ and x_1. Well, the vector $x_* - x_1$ is a linear combination of the vectors

$$A^k(x_* - x_1), \quad 1 \leq k \leq m,$$

by the definition of m, using the invertibility of A. We would be through if we could get the coefficients of this linear combination. Unfortunately, this is not easy, but there is an ingenious change of basis such that the coefficients with respect of the new basis can be calculated. We will see this below; this rough sketch has in any case already brought the vectors $A^k(x_* - x_1)$ into the game.

Properties of the algorithm. In the following result we have collected the useful properties of the conjugate gradient algorithm. We need some more notation. We let K_k, $0 \le k \le m$, denote the set of all linear combinations of the vectors

$$A^i(x_* - x_1),\ 1 \le i \le k.$$

In particular, $K_0 = 0$. We let $\kappa = \kappa(A)$ denote *the condition number of A*, defined by

$$\kappa = \|A\|\|A^{-1}\|,$$

here $\|A\|$ is the matrix norm defined—in problem 2.7—by

$$\|A\| = \max_{|x|=1} |Ax|.$$

This is a quantity that usually turns up in the analysis of matrix algorithms. It equals the quotient of the largest and the smallest eigenvalue of A. In the special case that A is a diagonal matrix this is obvious, and this special case implies the general case as each symmetric matrix can be diagonalized by a suitable orthogonal matrix (theorem 3.4).

Theorem 7.1 *We consider the conjugate gradient algorithm for the problem $(P_{7.1})$ with given initial point x_1.*

1. *The k-th approximation of the algorithm x_k is the unique solution of the problem*

$$\varphi(x) \to \min,\quad x - x_1 \in K_{k-1},$$

   for $1 \le k \le m+1$.

2. *Each approximation has lower φ-value than its predecessor, up to x_{m+1}, which is the solution of the given problem:*

$$\varphi(x_1) > \varphi(x_2) > \cdots > \varphi(x_{m+1})$$

   and $x_{m+1} = x_*$.

3. The algorithm performs at least linearly if we measure the quality of the approximations by the error in the φ-value,

$$\varphi(x_{k-1}) - \varphi(x_*) \leq 2\left(\frac{\sqrt{\kappa} - 1}{\sqrt{\kappa} + 1}\right)^{2k} (\varphi(x_1) - \varphi(x_*)).$$

Interpretation and properties of the vectors r_k and s_k. The recurrence formulas defining the algorithm beg for a meaningful characterization of the vectors x_k, r_k, $1 \leq k \leq m+1$, and s_k, $1 \leq k \leq m$. For the x_k this is given in statement 1. of the theorem. The following result completes the picture.

Theorem 7.2 *We consider the conjugate gradient algorithm for the problem $(P_{7.1})$ with given initial point x_1.*

1. *The vectors r_k can be characterized as follows:*

    $$r_k = -\varphi'(x_k)^T = b - Ax_k, \ 1 \leq k \leq m+1;$$

    in particular r_k is a steepest descent vector for the function φ in the point x_k.

2. *The vectors $r_1, \ldots, r_m$ are nonzero and mutually orthogonal; that is, $r_k^T r_l = 0$ for $k \neq l$. Moreover $r_{m+1} = 0$.*

3. *The finite sequence of vectors $s_1, \ldots, s_m$ is obtained from the sequence $r_1, \ldots, r_m$ by orthogonalization with respect to the A-inner product, defined by $\langle v, w \rangle_A = v^T A w$:*

    $$s_k - r_k \in \mathrm{span}\{r_1, \ldots, r_{k-1}\}, \ 1 \leq k \leq m,$$

    and

    $$\langle s_k, s_l \rangle_A = 0 \text{ for } k \neq l.$$

4. *The vectors $r_1, \ldots, r_k$ form an orthogonal basis of the space*

    $$K_k = \mathrm{span}(A(x_* - x_1), \ldots, A^k(x_* - x_1))$$

    and the vectors $s_1, \ldots, s_k$ form an A-orthogonal basis of this space, $1 \leq k \leq m$.

ADVANCED ALGORITHMS

This is a good moment to say that the ingenious change of basis announced above will be the change from

$$A(x_* - x_1), \ldots, A^m(x_* - x_1)$$

to $s_1, \ldots, s_m$. It will be useful to consider an intermediate basis as well: this will be $r_1, \ldots, r_m$. The remainder of this section provides a deeper insight: theorems 7.1 and 7.2 will be proved. We will establish these theorems simultaneously.

Proof. Our strategy for proving the two theorems will be as follows. We *define* the vectors x_k and r_k, $1 \leq k \leq m+1$, and s_k, $1 \leq k \leq m$, by the characterizations from theorems 7.1 and 7.2, so we ignore the recursive definition given by the algorithm above. We will verify all other statements, including the recursions of the algorithm.

To begin with, we will show that $x_{m+1} = x_*$. This is a consequence of the observation that multiplication by the invertible matrix A gives a bijection between the linear relations between the vectors

$$x_* - x_1, \; A(x_* - x_1), \ldots, A^m(x_* - x_1)$$

and the linear relations between the vectors

$$A(x_* - x_1), \ldots, A^{m+1}(x_* - x_1),$$

as we will see. Choose a nontrivial linear relation

$$\sum_{i=1}^{m+1} \gamma_i A^i(x_* - x_1) = 0,$$

multiply by A^{-1}, and rewrite it as

$$\gamma_1(x_* - x_1) = \sum_{j=1}^{m}(-\gamma_{j+1}) A^j(x_* - x_1). \qquad (*)$$

The coefficient γ_1 cannot be zero; otherwise we would get a nontrivial linear relation between the vectors

$$A(x_* - x_1), \ldots, A^m(x_* - x_1),$$

contradicting the definition of m. Therefore, equality $(*)$ shows that $x_* - x_1 \in K_m$. This means that $x_{m+1} = x_*$ by the minimality properties defining x_{m+1} and x_*.

In the remainder of the proof, the following notation will be useful. For each $k \leq m$ and each vector $v \in K_k$, there exists a unique polynomial $p(t)$ of degree at most k with $p(0) = 0$ such that

$$v = p(A)(x_* - x_1):$$

indeed, for

$$v = \sum_{i=1}^{k} \gamma_i A^i (x_* - x_1)$$

this polynomial is $p(t) = \sum_{i=1}^{k} \gamma_i t^i$.

Our next aim is to show that $r_k \neq 0$ for all $k \leq m$. We write

$$x_k - x_1 = p_k(A)(x_* - x_1),$$

with $p_k(t)$ a polynomial of degree $\leq k - 1$ with $p(0) = 0$, using $x_k - x_1 = K_{k-1}$. Then

$$r_k = b - Ax_k = A(x_* - x_k)$$

$$= A(x_* - x_1) - A(x_k - x_1) = (A - Ap_k(A))(x_* - x_1).$$

This shows that $r_k \in K_k$ and that $r_k \neq 0$ as the polynomial $t - tp_k(t)$ is a nontrivial polynomial of degree $1 + \deg(p_k) \leq k \leq m$ with value 0 in $t = 0$.

We are going to show that r_k is orthogonal to all vectors in K_{k-1} for all $k \leq m$. We apply the Fermat theorem to the optimality property defining x_k. A calculation—using the chain rule and the definition of $r_k = -\varphi'(x_k)^T$—shows that the partial derivative

$$\frac{\partial}{\partial \gamma_i} \varphi(x_1 + \gamma_1 A(x_* - x_1) + \cdots + \gamma_{k-1} A^{k-1}(x_* - x_1))$$

equals $r_k^T A(x_* - x_1)$. Thus we get that the vector r_k is orthogonal to all vectors

$$A(x_* - x_1), \ldots, A^{k-1}(x_* - x_1),$$

and so it is orthogonal to all vectors in K_{k-1}, as desired. It follows in particular that $r_k \notin K_{k-1}$; otherwise, r_k would be orthogonal to itself, which is impossible as $r_k \neq 0$. Now we have proved that

$$r_k \in K_k \setminus K_{k-1}, \ 1 \leq k \leq m.$$

This implies, as $\dim K_k - \dim K_{k-1} = 1$ for all $k \in \{1,\ldots,m\}$, that $r_1,\ldots,r_k$ form a basis of K_k. In particular we also get

$$\varphi(x_1) > \varphi(x_2) > \cdots > \varphi(x_{m+1});$$

indeed if one had

$$\varphi(x_l) = \varphi(x_{l+1})$$

for some l, then $x_l = x_{l+1}$, by the uniqueness of the minimum property defining x_{l+1}, and so $r_l = r_{l+1}$, contradicting what we have just proved.

Now we turn to the vectors $s_1,\ldots,s_m$. The vectors $s_1,\ldots,s_k$ also form an A-orthogonal basis of K_k for all $k \leq m$, by the corresponding property of the vectors $r_1,\ldots,r_k$ and the definition of the vectors $s_1,\ldots,s_m$. The definition of the vectors $s_1,\ldots,s_m$ can be rewritten as follows, using $K_k = \mathrm{span}(r_1,\ldots,r_k)$, $1 \leq k \leq m$:

$$s_k - r_k \in K_{k-1},$$

$$s_k \perp_A K_{k-1}, \quad 1 \leq k \leq m.$$

Here $\perp_A$ denotes orthogonality with respect to the inner product $\langle v, w \rangle_A = v^T A w$.

Now choose $k \leq m$ and write $s_k - r_k$ out on the basis $s_1,\ldots,s_{k-1}$ of K_{k-1}. Let σ_i denote the coefficient of s_i. Then, using the A-orthogonality of the s_j, one gets

$$\sigma_i = \langle s_k - r_k, s_i \rangle_A = -\langle r_k, s_i \rangle_A = -\langle r_k, As_i \rangle.$$

For $i \leq k - 2$ one has

$$As_i \in K_{k-1} \text{ and so } \langle r_k, As_i \rangle = 0.$$

This proves that $s_k - r_k$ is a scalar multiple of s_{k-1}. That is,

$$s_k - r_k = -\beta_{k-1} s_{k-1}$$

for some β_{k-1}.

Now we consider $x_k - x_{k-1}$. This vector equals

$$(x_k - x_1) - (x_{k-1} - x_1) = (p_k(A) - p_{k-1}(A))(x_* - x_1),$$

and so it is an element of K_{k-1}. Let us write it as a linear combination of the vectors $s_1, \ldots, s_{k-1}$, say,

$$\sum_{i=1}^{k-1} \rho_i s_i.$$

Then, using the A-orthogonality of the s_j, we get

$$\rho_i = \langle x_k - x_{k-1}, s_i \rangle_A.$$

This equals $\langle r_{k-1} - r_k, s_i \rangle$, as

$$A(x_k - x_{k-1}) = r_{k-1} - r_k.$$

If $i < k-1$, then this equals zero as $s_i \in K_{k-2}$. This proves that

$$x_k - x_{k-1} = \alpha_{k-1} s_{k-1}$$

for some α_{k-1}. Moreover, multiplying this equality by $-A$ gives

$$r_k - r_{k-1} = -\alpha_{k-1} A s_{k-1}.$$

It remains to verify the expressions for α_k and β_k in the recursions defining the algorithm. We have

$$r_{k+1} = r_k - \alpha_k A s_k;$$

multiplying with s_k^T from the left, gives the formula

$$\alpha_k = \frac{|r_k|^2}{s_k^T A s_k},$$

using $r_{k+1} \perp s_k$ and using the equality

$$s_k^T r_k = r_k^T r_k,$$

which holds as $s_k - r_k \in K_{k-1}$ and $r_k \in K_{k-1}$.

We have

$$s_{k+1} = r_{k+1} - \beta_k s_k;$$

multiplying by r_k^T from the left and using $r_k^T r_{k+1} = 0$ gives

$$r_k^T s_{k+1} = -\beta_k r_k^T s_k.$$

The left-hand side equals

$$(r_{k+1} + \alpha_k A s_k)^T s_{k+1} = r_{k+1}^T s_{k+1},$$

and this equals $|r_{k+1}|^2$, as

$$s_{k+1} - r_{k+1} \in K_k \quad \text{and} \quad r_{k+1} \perp K_k.$$

The right-hand side equals $-\beta_k |r_k|^2$, as

$$s_k - r_k \in K_{k-1} \quad \text{and} \quad r_k \perp K_{k-1}.$$

On comparing, we get

$$\beta_k = -\frac{|r_{k+1}|^2}{|r_k|^2}$$

as desired. $\square$

7.3 SELF-CONCORDANT BARRIER METHODS

This section is based on Y. Nesterov, *Introductory Lectures on Convex Programming* [57]. The expository novelty in the account below is the systematic use that is made of "restriction to a line." This allows a reduction of all calculations in the proofs to the case of functions of one variable.

7.3.1 Introduction

Plan and motivation. We make clear what is the impact of the theory of self-concordant barriers on the theory and practice of optimization algorithms.

Historical comments. The idea of barrier methods is old and simple (see the description at the end of section 1.5.1), but for a long time good barriers and efficient ways to use them were not known. Then, efficient barrier methods, also called interior point methods, were developed for linear programming problems. We recalled the initial development briefly in section 6.3.2. After this, such methods were gradually discovered for other types of convex problems as well, for example, for quadratic and for semidefinite programming problems. Then a brilliant new theory was created by Yurii Nesterov and Arkady Nemirovskii. This clarified the situation completely. They showed that all barriers satisfying two new, so-called self-concordancy axioms lead to efficient algorithms. Here "efficient" is meant in a theoretical sense, but these algorithms turned out to be efficient in practice as well. All known efficient barriers turned out to satisfy the self-concordancy axioms. Moreover, Nesterov and Nemirovskii showed that each closed convex set with nonempty interior has a self-concordant barrier (theorem 2.5.1 in [56]). All that remains in each concrete case is to find an explicit self-concordant barrier for the convex set of admissible points. This "usually" turns out to be possible.

Abilities of minimization schemes. We give some indication of the progress made by barrier methods. This gives only a very rough idea; there is no simple way to describe this progress, as it depends on the application and on how the "speed" of algorithms is measured. Moreover, speed is not the only criterion, reliability is important as well. Having said all this, let us quote the opinion of an expert, Yurii Nesterov, in [57]:

"General minimization schemes, such as the gradient method and the Newton method, work well for up to four variables. Convex black-box methods, such as the ellipsoid method, "work well" for up to 1,000 variables. Self-concordant barrier methods work well for up to 10,000 variables. A special class of self-concordant barrier methods, which is available for linear, quadratic and semidefinite programming, the so-called primal-dual methods, works even well for up to 1,000,000 variables."

State of the art implementations of most algorithms are freely available on the Internet.

Conclusion. Self-concordant barrier methods unify the interior point methods and point the way toward finding efficient interior point algorithms for convex problems.

7.3.2 Barrier methods and Hadamard products

Plan. We explain the basic idea of barrier methods and relate it to the interior point method presented in section 6.9.

Idea of barrier methods. Consider a convex problem

$$f_0(x) \to \min, \ x \in Q,$$

where Q is a convex set in $\mathbb{R}^n$ and $f_0 : Q \to \mathbb{R}$ a convex function. We assume that it has a boundary point as a solution and that we want to find it by a barrier method. This means that we want to construct a function $F : \text{int} Q$ on the interior of Q—called a barrier function— such that $F(x) \to +\infty$ when x tends to ∂Q, the boundary of Q, and then we want to solve a family of problems without constraints,

$$t f_0(x) + F(x) \to \min. \tag{P_t}$$

The idea of this is that, by continuity, the solution $x^*(t)$ of this problem will change gradually if t changes, and tend to the solution of the given problem if $t \to \infty$. This is plausible as (P_t) is equivalent to the problem

$$f_0(x) + \frac{1}{t}F(x) \to \min,$$

and as this problem "tends to" the problem

$$f_0(x) \to \min$$

if $t \to +\infty$. It turns out that the solution for $t = 0$ can be calculated. Then one can choose an increasing infinite sequence of t's, $0 = t_1, \ldots, t_k, \ldots$, with limit $+\infty$, and solve the corresponding problems one after the other: an approximation of $x^*(t_k)$ is found by means of—one step of—the Newton method with starting point an approximation of $x^*(t_{k-1})$. Here we recall that the Newton algorithm works well if the starting point is close to the solution.

Connection between the barrier presentation and the Hadamard product presentation

The connection between the barrier presentation of interior point methods and the presentation using the Hadamard product, which was given in chapter six for LP problems, follows from the following facts. For the pair of primal-dual LP problems (P) and (D) considered in chapter six, the expression $\ln p * d$ is as a function of $p \in P$ and $d \in D$ a self-concordant barrier, and the stationarity equations of the function

$$h_x(p, d) = p^T \cdot d - x^T \cdot \ln(p * d),$$

which was considered in section 6.9.3, are equivalent to the equation $p * d = x$. This has been verified in the proof of theorem 6.12; moreover, the equation $p * d = x$ is the basis of the presentation of the interior point methods using the Hadamard method. We emphasize one difference between the presentation of interior point methods in this section and the one in chapter six: there, we considered a primal-dual method; in this section we consider a primal method. The reason for this is, in terms of barrier functions, as follows. In order to solve a convex optimization problem efficiently with interior point methods, one needs a "good" self-concordant barrier function for this problem. It can happen that such a barrier can be found for the primal problem but not for the dual problem. Therefore, primal interior point methods are more general than primal-dual ones. However, when both primal and dual problem have a "good" self-concordant barrier, then primal-dual interior point methods are more efficient than primal ones. Therefore, we have chosen to present a primal-dual interior point method for LP problems.

Conclusion. We have presented the simple idea behind barrier methods. Moreover, we have shown that the primal dual interior point method for LP problems, presented in section 6.9, can be described as a barrier method.

7.3.3 The three golden barriers and self-concordancy

Plan. We will give the three main examples of the self-concordant theory. Moreover we will formulate the theoretical result that for all convex sets there exist self-concordant barriers.

Reduction to barriers on convex cones. All convex problems can be reformulated as a problem

$$f_0(x) \to \min, \; x \in K \cap A,$$

where K is a cone (to be precise, a convex cone), A is an affine manifold in $\mathbb{R}^n$ (a parallel translation of a linear subspace), K is a convex cone in $\mathbb{R}^n$, and $f_0 : \mathbb{R}^n \to \mathbb{R}$ is a linear function. Thus the problem of finding barriers on convex sets is reduced to the problem for cones.

We give a sketch of this reformulation. Let a convex problem be given, say,

$$f_0(x) \to \min, \; x \in C,$$

with $C \subseteq \mathbb{R}^n$ a convex set and $f_0 : C \to \mathbb{R}$ a convex function. We assume that the objective function f_0 is linear, as we may without restriction of generality of the argument. Indeed, we may introduce an additional variable x_{n+1}, and write the problem in the form

$$x_{n+1} \to \min, \; x = (x_1, \ldots, x_n)^T \in C, \; x_{n+1} \geq f_0(x).$$

Then the objective function $(x_1, \ldots, x_{n+1})^T \to x_{n+1}$ is linear, as promised. Now we "homogenize" the admissible set C by introducing the convex cone

$$K = \{(x_0, \ldots, x_n) : x_0 > 0, \; (x_1/x_0, \ldots, x_n/x_0) \in C\} \cup \{0_{n+1}\}.$$

The given problem can be reformulated as

$$f_0(x_1, \ldots, x_n) \to \min, \; (x_0, \ldots, x_n) \in K, \quad x_0 = 1.$$

ADVANCED ALGORITHMS

Three golden barriers. There are three "golden cones" and each has a "golden barrier."
- *The nonnegative orthant* $\mathbb{R}^n_+$, with barrier $-\sum_{i=1}^n \ln x_i$.
- *The ice cream (or Lorenz) cone* Lor_n, the solution set of

$$x_{n+1} \geq (\sum_{k=1}^n x_k^2)^{1/2},$$

with barrier $-\ln(x_{n+1} - (\sum_{k=1}^n x_k^2)^{1/2})$.
- *The cone of positive semidefinite $n \times n$ matrices* $S_n \subseteq \mathcal{M}^{n,n}$ with barrier $-\ln \det X$.

Self-concordant barriers. A function $F : U \to \mathbb{R}$ is called a ν-*self-concordant barrier* (SCB) if it is a three times continuously differentiable closed convex function defined on an open convex set $U = \text{dom}F \subseteq \mathbb{R}^n$, such that $\forall x \in \text{dom}F$ and $\forall y \in \mathbb{R}^n \setminus \{0\}$ the function

$$f_{xy}(t) = f(t) = F(x + ty)$$

satisfies the following two inequalities:

$$|f'''(0)| \leq 2f''(0)^{\frac{3}{2}} \quad \text{and} \quad (f'(0))^2 \leq \nu f''(0). \tag{1}$$

If we drop the last inequality, then we get the concept of *self-concordant function*. This concept is useful as we have to consider functions $tf_0 + F$, where F is a self-concordant barrier and f_0 is linear: these are self-concordant functions, clearly.

One calls ν the *parameter* of the barrier. Note that, just as a function is convex precisely if its restriction to each line is convex, a function is a self-concordant barrier precisely if its restriction to each line is a self-concordant barrier.

Theorem 7.3 *The three golden barriers are SCB.*

Proof. We display the calculations only for the most interesting case S_n. We have, writing

$$Z = (\sqrt{X})^{-1}Y(\sqrt{X})^{-1},$$

and letting λ_i, $1 \leq i \leq n$, be the eigenvalues of X, that

$$f_{XY}(t) = f(t) = -\ln\det(X+tY)$$
$$= -\ln\det(\sqrt{X}(I+tZ)\sqrt{X}) = -\ln\det X - \sum_{i=1}^{n}\ln(1+\lambda_i t).$$

Therefore,

- $f'(0) = -\sum_{i=1}^{n}\lambda_i$,
- $f''(0) = \sum_{i=1}^{n}\lambda_i^2$,
- $f'''(0) = -2\sum_{i=1}^{n}\lambda_i^3$,

and so (1) follows from the classical inequalities

$$\left(\sum_{i=1}^{n}\lambda_i^3\right)^{\frac{1}{3}} \leq \left(\sum_{i=1}^{n}\lambda_i^2\right)^{\frac{1}{2}} \quad \text{and} \quad \left(\frac{1}{n}\sum_{i=1}^{n}\lambda_i\right) \leq \left(\frac{1}{n}\sum_{i=1}^{n}\lambda_i^2\right)^{\frac{1}{2}}.$$

We refer to section 3.3.4 (or to exercise 3.6.16) for derivations of these inequalities between the l_i-norms, $i = 1, 2, 3$. □

This lemma shows that the theory of self-concordant barriers applies to linear programming, quadratic programming and semidefinite programming.

For many other open convex sets, SCB barriers have been constructed that lead to efficient algorithms. A partial explanation for this pleasant phenomenon is given by the following fundamental and deep result of Nesterov and Nemirovskii (theorem 2.5.1 from [56]):

there is an absolute constant $C > 0$ such that each open convex set in $\mathbb{R}^n$ admits a Cn-self-concordant barrier.

In fact, the theorem gives an explicit self-concordant barrier, provided the convex set does not contain any line, but this barrier is not suitable for use in algorithms. Therefore, in practice, the task remains to find, for each given convex set, a self-concordant barrier that is suitable for use in algorithms. "Usually" this turns out to be possible.

Conclusion. The logarithmic barriers of the nonnegative orthant, the ice cream (or Lorenz) cone, and the cone of positive semidefinite matrices are self-concordant. Essentially for all convex sets there exists a self-concordant barrier.

7.3.4 Simplified presentation of the theory of self-concordant barriers: restriction to a line and integration of inequalities

Plan. We will explain and illustrate two simplifications of the presentation of the proofs of the self-concordant barrier theory. Their effect is that they reduce seemingly complicated proofs to standard high school exercises.

First trick: restriction to a line. The defining properties of a self-concordant barrier have the following pleasant property. They hold precisely if they hold for the restriction of the barrier to any line. Moreover, if we need to prove something about self-concordant barriers, then it always turns out that "all the action takes place on one straight line." Therefore, it suffices to give the proof in the one-dimensional case ("one variable") only.

Second trick: integration of inequalities. The defining properties of one-dimensional self-concordant barriers have the form of inequalities in the first, second, and third derivatives of the barrier. One can integrate these inequalities repeatedly in a straightforward way. As a result one gets certain other inequalities. It turns out that this extremely simple procedure gives routine proofs of all properties of self-concordant barriers that are used in the construction and analysis of barrier methods.

Illustration tricks: nonnegative definiteness of the Hessian matrix and convexity function. These two tricks will be used in each proof in the remainder of this section. Now we give a first illustration: we present the proof of a well-known result on convex functions in terms of these two tricks.

Example 7.3.1 *Show that a function $f \in D^2(\mathbb{R}^n)$ with $f''(x)$ nonnegative definite for all $x \in \mathbb{R}^n$ is convex.*

Solution. We will proceed in steps.
1. **Reduction of convexity to a line.** Note that a function $f : \mathbb{R}^n \to \mathbb{R}$ is convex precisely if its restriction to any line in $\mathbb{R}^n$ is convex (that is, if the function of one variable $\alpha \mapsto f(v + \alpha r)$ is convex $\forall v \in \mathbb{R}^n$, $\forall r \in \mathbb{R}^n \setminus \{0_n\}$).

2. **Reduction of nonnegative definiteness to a line.** Note that a function $f \in D^2(\mathbb{R}^n)$ has the property that $f''(x)$ is nonnegative definite for all $x \in \mathbb{R}^n$ precisely if this is true for the restriction to any line in $\mathbb{R}^n$ (that is, the second derivative of

the function g of one variable $\alpha \mapsto f(v + \alpha r)$ at $\alpha = 0$ is nonnegative $\forall v \in \mathbb{R}^n$, $\forall r \in \mathbb{R}^n \setminus \{0_n\}$). To see this, one has to note that $r^T f''(v) r = g''(0)$.

3. **Application of the first trick.** It follows that we can apply the first trick. As a consequence it is sufficient to give the proof for the case $n = 1$. Therefore, from now on, $n = 1$. We have to prove the defining property of convex functions,

$$f((1-\rho)a + \rho b) \leq (1-\rho)f(a) + \rho f(b), \ \forall a, b \in \mathbb{R}, \ \forall \rho \in [0,1].$$

We choose arbitrary $a, b \in \mathbb{R}$ with $a \leq b$, and $\rho \in [0, 1]$. We write $w_\rho = (1-\rho)a + \rho b$.

4. **First integration.** We start with the inequality

$$f''(x) \geq 0 \ \forall x \in \mathbb{R}^n.$$

Integrating this inequality from $x_1 \in \mathbb{R}$ to $x_2 \in \mathbb{R}$ with $x_1 \leq x_2$ gives the following inequality:

$$f'(x_2) - f'(x_1) = \int_{x_1}^{x_2} f''(x) dx \geq 0.$$

5. **Second integration.** Integrating the resulting inequality

$$f'(x) - f'(w_\rho) \geq 0$$

again, from w_ρ to b gives the following inequality:

$$f(b) - f(w_\rho) - (b - w_\rho) f'(w_\rho) = \int_{w_\rho}^{b} (f'(x) - f'(w_\rho)) dx \geq 0.$$

In the same way, integrating the resulting inequality

$$f'(w_\rho) - f(x) \geq 0$$

from a to w_ρ gives the following inequality:

$$(w_\rho - a) f'(w_\rho) - (f(w_\rho) - f(a)) = \int_{a}^{w_\rho} (f'(w_\rho) - f(x)) dx \geq 0.$$

Now we take the following nonnegative linear combination of these two inequalities: $1 - \rho$ times the first one plus ρ times the second one. The result of this is the following inequality:

$$(1 - \rho) f(a) + \rho f(b) - f(w_\rho) + (w_\rho - (1-\rho)a - \rho b) f'(w_\rho) \geq 0.$$

The last term of the left-hand side equals zero by the definition of w_ρ. Therefore, we get the required defining inequality of convex functions.

In the remainder of this chapter we will give an introduction to the art of using a self-concordant algorithm to design an algorithm. Some of the calculations are relatively technical, but the idea is always the same straightforward approach: restrict to a line and integrate—repeatedly—the two inequalities defining the concept of self-concordant barrier. Thus, if we accept the great idea of the self-concordant method, which is to consider barriers that satisfy these two inequalities, then all else flows more or less automatically. Moreover the achievement of this self concordant approach is impressive: *it allows us to solve numerically in an efficient way each nonlinear optimization problem that is convex and for which a convenient self-concordant barrier is available.*

Plan for the design of a self-concordant algorithm. We consider all issues that arise naturally in the design of an algorithm that aims to follow the central path.

- Definition of the central path.
- How to achieve stability.
- How to achieve speed.
- How to make a trade-off between stability and speed.
- The main path-following scheme.
- How to end.
- How to begin.

Conclusion. We have presented two "tricks" to simplify the presentation of the proofs of the theorems on self-concordant barriers: reduction to a line and repeated integration of inequalities. As an illustration we have presented the proof of a well-known result on convex functions in terms of these tricks.

7.3.5 Following the central path

Plan. We start the explanation of the self-concordant algorithm by defining the central path. The idea of the algorithm is to follow the central path. "At the end of the central path" lies a solution of the problem.

Central path. We consider a conic problem,

$$f_0(x) \to \min, \ x \in K \cap A,$$

for which we possess an SCB F on K. It turns out that the most convenient way to approximate the solution is by following—more or less—the so-called *central path*. This is defined to be the trajectory $x^*(t)$, $t \in [0, \infty)$, where $x^*(t)$ is the solution of the auxiliary problem

$$g(x) = tf_0(x) + F(x) \to \min, \ x \in \text{interior}(K \cap A),$$

Therefore, at each step (= replacement of an approximation by a new one), one has to make a trade-off between two conflicting aims:
- to move closer to the central path ("stability"),

- to move closer to the solution of the given problem ("speed").

Conclusion. The idea of self-concordant algorithms is to follow the central path. This involves a trade-off between stability and speed.

In the next three sections we will present techniques to move closer to the central path, to move closer to the solution, and to make a trade-off.

7.3.6 How to achieve stability: fixed penalty coefficient

Plan. We will show how to move closer to the central path.

Again we consider a conic problem,

$$f_0(x) \to \min, \ x \in K \cap A,$$

for which we possess an SCB F. Assume that $t \in [0, \infty)$ is given and also a vector $\bar{x} \in \text{interior}(K \cap A)$. We view $\bar{x}$ as an approximation of $x^*(t)$. We choose to measure the quality of the approximation of $x^*(t)$ by $\bar{x}$, by the difference

$$f_0(\bar{x}) - f_0(x^*(t)).$$

We want to choose the best possible approximation $\bar{x}$ of $x^*(t)$ given that we allow ourselves only the following information:

$$g(x) = tf_0(x) + F(x)$$

ADVANCED ALGORITHMS

is a self-concordant function on $K \cap A$ and the function value, gradient, and Hessian of the restriction of g to A are known at the point $\bar{x}$.

One-dimensional case. As a preparation we will make an exhaustive search of the one-dimensional case. Suppose we have the following information on a function of one variable g: it is a self-concordant function, its domain includes 0, and $g(0)$, $g'(0)$, and $g''(0)$ are given. We want to determine upper and lower bounds for the functions g'', g', and g on $[0, \infty)$. We note that the point $\bar{\bar{x}}$ where the upper bound is minimal is the solution of the problem above in the one-dimensional case (here $\bar{\bar{x}} = 0$).

The crucial observation is that the first self-concordant inequality can be written as follows:

$$\left|\frac{d}{ds}(g''(s)^{-\frac{1}{2}})\right| \leq 1 \quad \forall s.$$

This gives the possibility of integrating three times. We display the results; the details are straightforward.

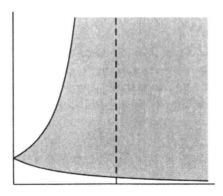

Figure 7.1 The graph of g'' lies in the shaded area.

First integration

$$(g''(0)^{-\frac{1}{2}} + s)^{-2} \leq g''(s) \leq (g''(0)^{-\frac{1}{2}} - s)^{-2},$$

where the left (resp. right) inequality holds for all s with $0 \leq s$ (resp. $0 \leq s < g''(0)^{-\frac{1}{2}}$). This is illustrated in Figure 7.1.

Second integration

$$g'(0) + g''(0)^{\frac{1}{2}}(1 - (1 + g''(0)^{\frac{1}{2}}s)^{-1}) \leq g'(s)$$

$$\leq g'(0) - g''(0)^{\frac{1}{2}}(1 - (1 + g''(0)^{\frac{1}{2}}s)^{-1})$$

where the left (resp. right) inequality holds for all s with $s \geq 0$ (resp. $0 \leq s < g''(0)^{-\frac{1}{2}}$. This is illustrated in Figure 7.2.

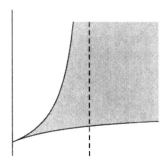

Figure 7.2 The graph of g' lies in the shaded area.

This makes clear what are the necessary and sufficient conditions that force g to have a minimum on $(0, \infty)$: the horizontal axis intersects the graphs of the lower and upper bounds for g' given above at two distinct points. To see this, note that a minimum of the convex function g on $(0, \infty)$ corresponds to a positive solution of $g' = 0$. This

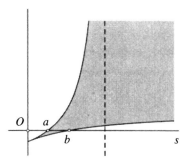

Figure 7.3 The graph of g' lies in the shaded area.

geometric condition can be expressed analytically as follows: $g'(0) < 0$

and $0 < \lambda < 1$, where

$$\lambda = \lambda_g(0) = -g''(0)^{-\frac{1}{2}} g'(0).$$

We calculate the point of intersection $A = (a, 0)$ (resp. $B = (b, 0)$) of the upper bound (resp. lower bound) with the horizontal axis (Fig. 7.3). This gives, writing $\Delta = -g''(0)^{-1} g'(0)$,

$$a = (1 + \lambda)^{-1} \Delta,$$

$$b = (1 - \lambda)^{-1} \Delta.$$

Note that Δ is a well-known expression: it is the direction vector of

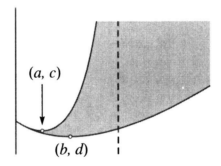

Figure 7.4 The graph of g lies in the shaded area.

the Newton method. In the present one-dimensional case this vector is just a number.

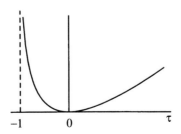

Figure 7.5 The graph of the function $\omega(\tau) = \tau - \ln(1 + \tau)$.

Third integration

$$g(0) + (g'(0) + g''(0)^{\frac{1}{2}})s - \ln(1 + g''(0)^{\frac{1}{2}}s)$$
$$\leq g(s) \leq g(0) + (g'(0) - g''(0)^{\frac{1}{2}})s - \ln(1 - g''(0)^{\frac{1}{2}}s) \quad \forall s < g''(0)^{-\frac{1}{2}},$$

where the left (resp. right) inequality holds for all s with $s \geq 0$ (resp. $0 \leq s < g''(0)^{-\frac{1}{2}}$). This is illustrated in Figure 7.4.

We will need the function

$$\omega(\tau) = \tau - \ln(1 + \tau);$$

its graph is drawn in Figure 7.5.

We calculate the point of minimum (a, c) (resp. (b, d)) of the upper bound (resp. lower bound) of g. This gives

$$c = g(0) - \omega(\lambda),$$

$$d = g(0) - \omega(-\lambda).$$

From Figures 7.2 and 7.4 one immediately reads off the solution of our main problem.

Let $\widehat{x}$ be the point of minimum of g. One should replace the approximation $\bar{x} = 0$ of $\widehat{x}$ by $\bar{\bar{x}} = (1 + \lambda)^{-1}\Delta$. The improvement of the value of the objective function, $g(\bar{x}) - g(\bar{\bar{x}})$ is, in the worst case $\omega(\lambda)$; the worst case is here that $\widehat{x} = (1 + \lambda)^{-1}\Delta$. We take the expression

$$\frac{\ln(g(\bar{\bar{x}}) - g(\widehat{x}))}{\ln(g(\bar{x}) - g(\widehat{x}))}$$

as a measurement of the relative improvement. We assume here that $g(\bar{x}) - g(\widehat{x}) < 1$, so if the expression takes a large value, then there is a big improvement. This is in the worst case

$$\frac{\ln(\omega(-\lambda) - \omega(\lambda))}{\ln \omega(-\lambda)};$$

the worst case is here $\widehat{x} = (1 - \lambda)^{-1}\Delta$. This worst case expression, seen as a function of λ, converges to $\frac{3}{2}$ if $\lambda \downarrow 0$, as one can check. This fact can be interpreted as follows, noting that $\lambda_g(0) = 0$ precisely if the old approximation $\bar{x} = 0$ is equal to the point of minimum $\widehat{x}$:

if the old approximation $\bar{x} = 0$ of $\widehat{x}$ is close enough, then the new approximation $\bar{\bar{x}}$ gives at least about $1\frac{1}{2}$ times as many reliable decimals of $f(\widehat{x})$ as the old one.

This guarantees that the number of reliable decimals grows at least exponentially as a function of the number of iterations. This is fast; in fact we would be happy with at least linear growth.

The general case. The derivation of the general case from the one-dimensional case is routine: it is given by restriction to lines. We now display the details. Take any vector $x \neq \bar{x}$ and define

$$v = \bar{x} - x \text{ and } \tilde{g}(s) = g(\bar{x} + sv).$$

Then we can apply the bounds for the one-dimensional case to the line through $\bar{x}$ and x. We calculate the following formulas relating $\tilde{g}$ and its first two derivatives to g and its first two derivatives:

$$\tilde{g}(0) = g(\bar{x}) \text{ and } \tilde{g}(1) = g(x),$$

$$\tilde{g}'(s) = g'(\bar{x} + sv) \cdot v \text{ and } \tilde{g}'(0) = g'(\bar{x}) \cdot v,$$

$$\tilde{g}''(s) = v^T g''(\bar{x} + sv)v \text{ and } \tilde{g}''(0) = v^T g''(\bar{x})v.$$

This leads to the following bounds for g.

Upper bound for g:

$$g(x) \leq g(\bar{x}) + g'(\bar{x}) \cdot v + \omega(-(r^T g''(\bar{x})r)^{\frac{1}{2}})$$

if $r^T g''(\bar{x})r < 1$. This upper bound for g is minimal for

$$\bar{\bar{x}} = x + (1+\lambda)^{-1}\Delta,$$

where we write

$$\Delta = -g''(\bar{x})^{-1}g'(\bar{x}) \text{ and } \lambda = (g'(\bar{x})g''(\bar{x})g'(\bar{x})^T)^{\frac{1}{2}}.$$

Lower bound for g:

$$g(x) \geq g(\bar{x}) + g'(\bar{x}) \cdot v + \omega((v^T g''(\bar{x})v)^{\frac{1}{2}}).$$

This lower bound for g is minimal for $\bar{x} + (1-\lambda)^{-1}\Delta$, provided $0 < \lambda < 1$.

We will display the derivation of the upper bound:

$$g(x) = \tilde{g}(1) \leq \tilde{g}(0) + \tilde{g}'(0) + \omega(-\tilde{g}''(0)^{\frac{1}{2}})$$

$$= g(\bar{x}) + g'(\bar{x}) \cdot v + \omega(-(v^T g''(\bar{x})v)^{\frac{1}{2}}).$$

Moreover, we sketch how to find the point of minimum of the upper bound: put the derivative of the function $g(\bar{x}+r)$ of the variable vector r equal to zero, and determine the direction of the optimal r and then its length.

Both statements on the lower bound are obtained in the same way as the corresponding statements on the upper bounds.

Conclusion (simplified version). We have shown how to move closer to the central path.

Conclusion (detailed version). Consider a self-concordant function $g(x)$—for example, $tf_0(x) + F(x)$—and a vector $\bar{x}$. Assume that the only thing we know about g is that it is a self-concordant function and that

$$g(\bar{x}),\ g'(\bar{x}),\ g''(\bar{x})$$

are known, with $g''(\bar{x})$ positive definite. Then we know for sure that g has a point of minimum $\hat{x}$ precisely if $0 < \lambda < 1$, where we write

$$\lambda = \lambda_g(\bar{x}) = (g'(\bar{x})g''(\bar{x})^{-1}g'(\bar{x})^T)^{\frac{1}{2}}.$$

Assume that $0 < \lambda < 1$. Then the vector $\bar{\bar{x}}$ with the lowest worst case g-value is

$$\bar{\bar{x}} = \bar{x} + (1+\lambda)^{-1}\Delta,$$

where we write

$$\Delta = -f''(\bar{x})^{-1}f'(\bar{x})$$

("damped Newton"). Moreover,

$$g(\bar{x}) - g(\bar{\bar{x}}) \geq \omega(\lambda)$$

and

$$\frac{\ln(g(\bar{\bar{x}}) - g(\hat{x}))}{\ln(g(\bar{x}) - g(\hat{x}))} \geq \frac{\ln(\omega(-\lambda) - \omega(\lambda))}{\ln \omega(\lambda)}.$$

This upper bound is $\frac{3}{2}+O(\lambda)$. Finally we give an explicit upper bound for the complexity of the damped Newton algorithm given above. Let $\bar{x}$ be given such that

$$\lambda_g(\bar{x}) < 1.$$

Then do damped Newton, starting with $\bar{x}$. This leads to a sequence of approximations $\bar{x} = x^{(0)}, x^{(1)}, \ldots$. Choose $\tau_0 > 0$ such that

$$\frac{\ln(\omega(-\tau) - \omega(\tau))}{\ln \omega(\tau)} \geq \frac{4}{3}$$

for all $\tau \in (0, \tau_0)$. Let N be the largest natural number less than or equal to

$$\frac{\omega(-\lambda_g(\bar{x}))}{\omega(\tau_0)}.$$

Then

$$\frac{\ln(g(x^{(N+k)}) - g(\hat{x}))}{\ln(g(x^{(N)}) - g(\hat{x}))} \geq \left(\frac{4}{3}\right)^k \quad \forall k \in \mathbb{N}.$$

7.3.7 How to achieve speed: raising the penalty coefficient

Plan. We will show how to move from a point "on" the central path to a point "on" the central path that lies closer to the solution (with minimal concern for stability).

We consider the situation where we have maximal stability, that is, $\bar{x}$ is on the central path, say $\bar{x} = x^*(\bar{t})$ for some $\bar{t}$. We want to increase the penalty coefficient as much as possible, say to $\bar{\bar{t}}$, with minimal concern for stability: we only want that $\bar{x}$ is a possible starting point for approximating $x^*(\bar{\bar{t}})$ with damped Newton.

Theorem 7.4 *Raising the penalty coefficient.* If $\bar{\bar{t}} - \bar{t} < \bar{t}/\nu$, then $\bar{x}$ is a possible starting point for approximating $x^*(\bar{\bar{t}})$, in the sense that

$$0 < \lambda_{\bar{\bar{t}}f_0+F}(\bar{x}) < 1.$$

Proof. Write $\Delta t = \bar{\bar{t}} - \bar{t}$. We want

$$(g'(\bar{x})g''(\bar{x})^{-1}g'(\bar{x})^T)^{\frac{1}{2}} < 1, \qquad (*)$$

with

$$g(x) = (t + \Delta t)f_0(x) + F(x).$$

As f_0 is linear, $f_0(x) = c \cdot x$ for some $c \in (\mathbb{R}^n)^T$. So, as $\bar{x} = x^*(\bar{t})$, we have

$$\bar{t}c + F'(\bar{x}) = 0.$$

Therefore,
$$g'(\bar{x}) = (\bar{t} + \Delta t)c + F'(\bar{x}) = \Delta t\, c = -\Delta t \frac{F'(\bar{x})}{\bar{t}}$$
and $g''(\bar{x}) = F''(\bar{x})$.

Therefore (∗) amounts to
$$\Delta t < \bar{t}(F'(\bar{x})F''(\bar{x})^{-1}F'(\bar{x}))^{-\frac{1}{2}}.$$

By the second self-concordant inequality, this will certainly hold if $\Delta t < \bar{t}/\sqrt{\nu}$. □

Conclusion. We have shown how far we can move in one step along the central path if we have only minimal concern for stability.

7.3.8 How to make a trade off

Plan. We have to find a right balance between our wish to move on along the central path ("speed") and our wish not to wander too far from this path ("stability").

Let $g : Q \to \mathbb{R}$ be a self-concordant function on a convex set $Q \in \mathbb{R}^n$ which we want to minimize, and let $\bar{x} \in Q$ be an approximate solution. The most natural measure of the quality of $\bar{x}$ as an approximate solution is the number
$$g(\bar{x}) - \inf_{x \in Q} g(x).$$

In particular, this number is always nonnegative, and it is zero precisely if $\bar{x}$ is a solution. In section 7.3.6 we analyzed the complexity of the problem to minimize a self-concordant function g, using this measure. This analysis forced another measure upon us:
$$\lambda_g(\bar{x}) = g'(\bar{x})g''(\bar{x})^{-1}g'(\bar{x})^T.$$

This number is nonnegative as well and it is zero precisely if $\bar{x}$ is a solution. We take this as a hint of nature and so we do the analysis of the trade-off between speed and reliability in terms of this alternative measure. We begin by observing that locally these measures are equivalent (this fact will not be used in what follows).

Theorem 7.5 *Let g be a self-concordant function and let $\bar{x} \in \mathrm{dom}(g)$ with $\lambda_g(\bar{x}) < 1$. Then the problem to minimize $g(x)$ has a unique*

solution x_g^*. Moreover,
$$\omega(\lambda_g(\bar{x})) \le g(\bar{x}) - g(x_g^*) \le \omega(-\lambda_g(\bar{x})).$$

All statements of this theorem have been proved in section 7.3.6. We recall that
$$\omega(\tau) = \frac{1}{2}\tau^2 + O(\tau^3), \ \tau \to +\infty.$$

Here we make use of the following notation
$$g(\tau) = O(h(\tau)) \ (\tau \to \infty)$$
if there is a constant $C > 0$ such that
$$g(\tau) \le Ch(\tau) \text{ for } \tau \text{ sufficiently large}$$
(*big Landau-O*).

Again we consider a given conic problem,
$$f_0(x) = c \cdot x \to \min, \ x \in K \cap A,$$
and a self-concordant barrier F on K. For each $t \in \mathbb{R}$ and each $x \in \text{int}(K \cap A)$, we define
$$\lambda(t, x, F) = \lambda_g(x),$$
where $g = tf_0 + F$. Let $\bar{t} \in \mathbb{R} \setminus \{0\}$ and $\bar{x} \in \text{interior}(K \cap A)$ be given such that $\lambda(\bar{t}, \bar{x}, F) < 1$. We begin by investigating how $\lambda(\bar{t}, \bar{x}, F)$ changes if we replace $\bar{t}$ by $\bar{\bar{t}} = \bar{t} + \Delta t$, keeping $\bar{x}$ fixed. For each $x \in \text{interior}(K \cap A)$, we consider the norm
$$\|y\|_x^* = (yF''(x)^{-1}y^T)^{\frac{1}{2}} \ \forall y.$$

Then
$$\lambda(\bar{\bar{t}}, \bar{x}, F) = \|\bar{\bar{t}}c + F'(\bar{x})\|_{\bar{x}}^*.$$

By the triangle inequality this is at most
$$|\Delta t|\|c\|_{\bar{x}}^* + \lambda(\bar{t}, \bar{x}, F).$$

Moreover,
$$\bar{t}\|c\|_{\bar{x}}^* = \|(\bar{t}f_0'(\bar{x}) + F'(\bar{x})) - F'(\bar{x})\|_{\bar{x}}^*,$$

which is, by the triangle inequality, at most

$$\|\bar{t}f_0'(\bar{x}) + F'(\bar{x})\|_{\bar{x}}^* + \|F'(\bar{x})\|_{\bar{x}}^* \leq \lambda(\bar{t}, \bar{x}, F) + \sqrt{\nu}.$$

Therefore, we conclude that

$$\lambda(\bar{\bar{t}}, \bar{x}, F) \leq \lambda(\bar{t}, \bar{x}, F) + |\Delta t| \left(\frac{\lambda(\bar{t}, \bar{x}, F) + \sqrt{\nu}}{\bar{t}} \right). \quad (**)$$

Now we investigate how $\lambda(\bar{\bar{t}}, \bar{x}, F)$ changes if we replace $\bar{x}$ by $\bar{\bar{x}}$, leaving $\bar{\bar{t}}$ fixed; here $\bar{\bar{x}}$ is defined as the result of a damped Newton step for the problem of minimizing $\bar{\bar{t}}f_0(x) + F(x)$ with starting point $\bar{x}$. That is,

$$\bar{\bar{x}} = \bar{x} + (1 + \lambda)^{-1}\Delta,$$

with $\lambda = \lambda(\bar{\bar{t}}, \bar{x}, F)$ and $\Delta = -g''(\bar{x})^{-1}g'(\bar{x})$.

Theorem 7.6 *Let g be a self-concordant function for which $g''(x)$ is positive definite for all x. Let $\bar{x} \in$ interior(dom g) with $\lambda_g(\bar{x}) < 1$. Let $\bar{\bar{x}}$ be the result of one step of damped Newton. Then*

$$\lambda_g(\bar{\bar{x}}) \leq \frac{2\lambda_g(\bar{x})^2}{1 + \lambda_g(\bar{x})}.$$

Proof. We may assume without restriction of the generality of the argument that dom$g \subseteq \mathbb{R}$, that $\bar{x} = 0$ and that $g'(0) < 0$ ("by restriction to the line through $\bar{x}$ and $\bar{\bar{x}}$"). Now we estimate

$$\lambda_g(\bar{\bar{x}}) = -\frac{g'(\bar{\bar{x}})}{g''(\bar{\bar{x}})^{\frac{1}{2}}}$$

by using the bounds for g and g'' which were obtained in section 7.3.6. The result turns out to be the statement of the theorem. We now display the calculations, which are all straightforward. Write

$$A = -g'(0), \quad B = g''(0).$$

Then $\bar{\bar{x}} = (1 + \lambda)^{-1}\Delta$ with

$$\lambda = \lambda_g(\bar{x}) = \frac{A}{B^{\frac{1}{2}}} \text{ and } \Delta = -\frac{g'(0)}{g''(0)} = \frac{A}{B}.$$

By section 7.3.2 we have

$$0 \leq -g'(\bar{\bar{x}}) \leq -g'(0) - g''(0)^{\frac{1}{2}}(1 - (1 + g''(0)^{\frac{1}{2}}\bar{\bar{x}})^{-1})$$

$$= A - B^{\frac{1}{2}}\left(1 - \left(1 + B^{\frac{1}{2}}\frac{A}{B(1+\lambda)}\right)^{-1}\right)$$

$$= A - B^{\frac{1}{2}}\left(1 - \left(1 + \frac{\lambda}{1+\lambda}\right)^{-1}\right) = A - B^{\frac{1}{2}}\frac{\lambda}{1+2\lambda} = B^{\frac{1}{2}}.$$

Moreover, again by section 7.3.2, we have

$$g''(\bar{\bar{x}}) \leq (g''(0)^{-\frac{1}{2}} + \bar{\bar{x}})^{-2} = \left(B^{-\frac{1}{2}} + \frac{A}{B(1+\lambda)}\right)^{-2}$$

$$= B\left(1 + \frac{\lambda}{1+\lambda}\right)^{-2} = B\left(\frac{1+\lambda}{1+2\lambda}\right)^2.$$

Therefore,

$$\lambda_g(\bar{\bar{x}}) = -\frac{g'(\bar{\bar{x}})}{g''(\bar{\bar{x}})^{\frac{1}{2}}} \leq \frac{B^{\frac{1}{2}}(2\lambda^2/(1+2\lambda))}{B^{\frac{1}{2}}(1+\lambda)/(1+2\lambda)} = \frac{2\lambda^2}{1+\lambda},$$

as required. □

Now we proceed as follows. We assume that

$$\lambda(\bar{t}, \bar{x}, F) \leq \beta$$

for some $\beta \in (0,1)$. We want to pick $\bar{\bar{t}}$ such that

$$\lambda(\bar{\bar{t}}, \bar{\bar{x}}, F) \leq \beta.$$

By theorem 2 it suffices to ensure that

$$\frac{2\lambda(\bar{\bar{t}}, \bar{x}, F)^2}{1 + \lambda(\bar{\bar{t}}, \bar{x}, F)} \leq \beta \qquad (***)$$

One readily checks that the function $h(\lambda) = 2\lambda^2/(1+\lambda)$ satisfies

$$h(0) = 0, \ h(1) = 1, \ h'(\lambda) = 2 - \frac{2}{(1+\lambda)^2};$$

this is positive for all positive λ. Therefore, h is strictly monotonic on $(0, \infty)$ and so for all $\beta \in (0,1)$ the equation $h(\lambda) = \beta$ has a unique root $\lambda_\beta \in (0,1)$; explicitly,

$$\lambda_\beta = \frac{\beta + \sqrt{\beta^2 + 8\beta}}{4}.$$

Moreover, for all $\beta \in (0,1)$ a number $\lambda \in [0, \infty)$ satisfies
$$2\lambda^2/(1+\lambda) \le \beta$$
precisely if $\lambda \le \lambda_\beta$.

Therefore (*) can be rewritten as $\lambda(\bar{\bar{t}}, \bar{x}, F) \le \lambda_\beta$. By inequality (**) it therefore suffices to ensure that
$$\beta + |\Delta t| \frac{\beta + \sqrt{\nu}}{\bar{t}} \le \lambda_\beta.$$

That is,
$$|\Delta t| \le \frac{-3\beta + \sqrt{\beta^2 + 8\beta}}{4(\beta + \sqrt{\nu})} \bar{t}.$$

Now we compute the value of β for which the numerator
$$-3\beta + \sqrt{\beta^2 + 8\beta}$$
is maximal—we note that the term $\beta + \sqrt{\nu}$ is dominated by the second term; this is readily verified to be
$$\beta = -4 + 3\sqrt{2} \approx 0.2426.$$

For this value of β we can choose $|\Delta t|$ such that
$$|\Delta t| = \frac{0.318}{4(0.243 + \sqrt{\nu})} \bar{t}.$$

Conclusion. We have given the best trade-off between speed and stability.

7.3.9 The main path-following scheme

Plan. By making repeatedly a step along the central path that strikes a balance between speed and stability, we get an algorithm for following the central path.

Now we are in a position to describe how to follow the central path. This description will not yet be a complete algorithm: the begin and the end of the algorithm will still be missing. For the time being, we assume that we are somewhere on the central path to begin with. Moreover we will not yet say when to stop the path-following scheme.

Given a conic problem,
$$f_0(x) = c \cdot x \to \min, \quad x \in K \cap A,$$
and a self-concordant barrier on K with parameter ν, choose
$$\beta = -4 + 3\sqrt{2} \approx 0.2426$$
("the centering parameter") and
$$\rho = \frac{-3\beta + \sqrt{\beta^2 + 8\beta}}{4(\beta + \sqrt{\nu})} \approx \frac{0.318}{0.972 + 4\sqrt{\nu}}.$$
Let $x_0 \in \text{dom}F$ and $t_0 > 0$ be given such that
$$\lambda(t_0, x_0, F) \le \beta.$$
Choose an accuracy $\varepsilon > 0$.

1. k-th iteration ($k \ge 0$). Set
$$t_{k+1} = (1 + \rho)t_k,$$
$$x_{k+1} = x_k - (1 + \lambda(t_k, x_k, F))^{-1} F^{(2)}(x_k)^{-1}(t_{k+1}c + F'(x_k)).$$

2. Stop the process if
$$\nu + \frac{(\beta + \sqrt{\nu})\beta}{1 - \beta} \le \varepsilon t_k.$$

We present the complexity result on this scheme.

Theorem 7.7 *The scheme terminates after no more than N steps, where*
$$N \le O\left(\sqrt{\nu} \ln\left(\frac{\nu(\beta + \sqrt{\nu})}{t_0 \varepsilon}\right)\right).$$
Moreover, at the moment of termination we have
$$f_0(x_N) - \inf_{x \in K \cap A} f_0(x) \le \varepsilon.$$

Conclusion. We have shown how to follow the central path and we have given the complexity of the resulting algorithm.

7.3.10 How to end

Plan. We will give a stopping criterion for a path-following algorithm that guarantees that we have found an approximation of the solution of required precision.

In this section and the next one we will discuss the remaining issues: how to end and how to begin. This requires the analysis of the second self-concordant inequality. This analysis, to be given now, runs entirely parallel to that of the first one given in section 7.3.6.

Assume that $\tilde{s}$ is a positive number and f is a self-concordant barrier with

$$\mathrm{dom} f \subseteq \mathbb{R}, \ [0, \tilde{s}) \subseteq \mathrm{dom} f \text{ and } f'(0) > 0.$$

We are going to determine the consequences of the second self-concordant property for the restrictions of f and f' to $[0, \tilde{s})$. The property can be written as follows:

$$\frac{d}{ds}(-f'(s)^{-1}) \geq \nu^{-1} \quad \forall s \in [0, \tilde{s}).$$

We are going to integrate this inequality twice. We display the results.

Information on the graph of $-f'$

$$-f'(0)^{-1} + \nu^{-1} s \leq -f'(s)^{-1} \leq 0.$$

This can be reformulated in terms of f' (Fig. 7.6):

$$f'(s) \geq f'(0) \left(1 - \frac{f'(0)}{\nu} s\right)^{-1}.$$

In particular,

$$\tilde{s} \leq \frac{\nu}{f'(0)}.$$

Information on the graph of f

$$f(s) \geq f(0) - \nu \ln\left(1 - \frac{f'(0)}{\nu} s\right).$$

Corollary 7.8 *Let F be a ν-self-concordant barrier. Then for any $x \in \mathrm{dom} F$ and $y \in \mathrm{closure}(\mathrm{dom} F)$ we have:*

$$F'(x)(y - x) \leq \nu.$$

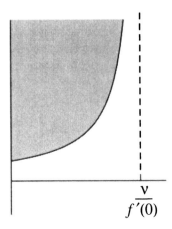

Figure 7.6 The graph of f lies in the shaded area.

Proof. Write
$$f(s) = F(x + s(y-x)).$$
Then $f'(0) = F'(x)(y-x)$ and so, as f is a ν-self-concordant barrier, the inequality
$$\tilde{s} \leq \frac{\nu}{f'(0)}$$
gives (with $\tilde{s} = 1$)
$$F'(x)(y-x) \leq \nu,$$
provided $f'(0) > 0$; if $f'(0) \leq 0$, then the inequality is obvious. □

Moreover, we need the following consequence of the first property of self-concordancy. Here $\|\cdot\|_x$ is defined by
$$\|y\|_x = (y^T F''(x) y)^{\frac{1}{2}}.$$

Theorem 7.9 *Let g be a self-concordant function and let $\bar{x} \in \mathrm{dom} g$ with $\lambda_g(\bar{x}) < 1$.*
Then the problem of minimizing $g(x)$ has a solution x_g^* and this is unique. Moreover,
$$\omega'(\lambda_g(\bar{x})) \leq \|x - x_g^*\|_x \leq -\omega'(-\lambda_g(\bar{x})).$$

All statements of this theorem have been proved in section 7.3.6 (as can be seen by restricting to the line through $\bar{x}$ and x_g^*).

Now we come to the main result of this section. Assume that $K \cap A$ is bounded. Let x^* be the solution of the given conic problem. The point $x^*(t)$ on the central path with parameter t satisfies

$$tc + F'(x^*(t)) = 0. \qquad (****)$$

Theorem 7.10 *For any $t > 0$ we have*

$$f_0(x^*(t)) - f_0(x^*) \leq \frac{\nu}{t}.$$

Let $\beta \in (0,1)$ and $x \in \text{interior}(K \cap A)$ with $\lambda(t, x, F) \leq \beta$. Then

$$f_0(x) - f_0(x^*) \leq \frac{1}{t}\left(\nu + \frac{(\beta + \sqrt{\nu})\beta}{1 - \beta}\right).$$

Proof. To begin with,

$$f_0(x^*(t)) - f(x^*) = c(x^*(t) - x^*)$$

$$= (\text{by } (****) \text{ above}) \frac{1}{t} F'(x^*(t))(x^* - x^*(t)) \leq (\text{by the corollary}) \frac{\nu}{t}.$$

Now we need the following inequality:

$$y^T z \leq \|y\|_x^* \|z\|_x \quad \forall x \in \text{dom} f, \ \forall y, z.$$

This follows immediately from the definitions and the inequality of Bunyakowskii-Cauchy-Schwarz. Let $\beta \in (0,1)$ and $x \in \text{interior}(K \cap A)$ with $\lambda(t, x, F) \leq \beta$. Then

$$t(f_0(x) - f_0(x^*(t))) = tc(x - x^*(t)) = ((tf_0'(x) + F'(x)) - F'(x))(x - x^*(t)).$$

By the version above of the Bunyakowskii-Cauchy-Schwarz inequality and by the triangle inequality, this is at most

$$(\|tf_0'(x) + F'(x)\|_x^* + \|F'(x)\|_x^*)\|x - c^*(t)\|_x.$$

Now

$$\|tf_0'(x) + F'(x)\|_x^* = \lambda(t, x, F) \leq \beta,$$

$$\|F'(x)\|_x^* = (F'(x)F''(x)^{-1}F'(x)^T)^{\frac{1}{2}} \leq \sqrt{\nu}$$

(using the second self-concordant inequality), and

$$\|x - x^*(t)\|_x \le -\omega'(-\lambda(t,x,F)) = \frac{\lambda(t,x,F)}{1-\lambda(t,x,F)} \le \frac{\beta}{1-\beta}.$$

We conclude that

$$t(f_0(x) - f_0(x^*(t))) \le (\beta + \sqrt{\nu})\frac{\beta}{1-\beta}.$$

Therefore,

$$f_0(x) - f_0(x^*) = (f_0(x) - f_0(x^*(t))) + (f_0(x^*(t)) - f_0(x^*))$$

and this is not more than

$$\le \frac{1}{t}(\beta + \sqrt{\nu})\frac{\beta}{1-\beta} + \frac{\nu}{t}.$$

This equals

$$\frac{1}{t}\left(\nu + \frac{(\beta+\sqrt{\nu})\beta}{1-\beta}\right),$$

as required. □

Conclusion. We have given an adequate stopping rule for the algorithm.

7.3.11 How to begin

Plan. To begin the path-following algorithm we need to possess a point on this path. Here we show how to find such a point under the mild assumption that we do possess an interior admissible point. The idea is to move gradually from this point to a very special point on the central path, *the analytical center*.

Traveling in France and the interior point method. For the sake of simplicity we restrict attention to the case that an interior point is available. We can explain the essence of the method by making a comparison with traveling in France. If you want to go from one city to another, the most convenient route is by way of Paris. In the same spirit, the most convenient route from a given interior point to the solution of a given conic problem is by means of some special point in the admissible set of the given problem, the *analytical center*, to be defined below.

Set-up. Let a conic problem

$$f_0(x) = cx \to \min, \ x \in K \cap A,$$

a self-concordant barrier F, and an interior point x_0 of $K \cap A$ be given. We assume that $K \cap A$ is bounded. Then F has a unique minimum x_F^*, as F is a strictly convex function on a bounded, open convex set in A and as $F(x) \to \infty$ if $x \to \partial(K \cap A)$. One calls x_F^* *the analytical center* of $K \cap A$ =closure (dom$F$).

Observation. Each y_0 in the interior of $K \cap A$ in A, different from x_F^*, lies on a unique central path. To be more precise, $d = -F'(y_0)$ is the unique row vector for which the central path

$$x^*(t) = x_d^*(t) = \mathrm{argmin}_x(tdx + F(x))$$

satisfies $x^*(1) = y_0$.

Sketch of the proof. As F is a strictly convex function on a bounded, open, convex subset of $\mathbb{R}^n$, F' is a bijection from $\mathbb{R}^n$ to $(\mathbb{R}^n)^T$. This implies the observation.

General scheme. The general scheme for solving our problem is as follows: choose a centering parameter $\beta \in (0,1)$ and a precision $\varepsilon > 0$; follow the central path that runs through y_0, from y_0 to x_F^* with damped Newton steps, *diminishing* each time the penalty coefficient t as much as possible without violating the centering condition; then follow the central path associated to our given problem, from x_F^* to its limit $\widehat{x}$, which is the solution of our problem; do this again with damped Newton steps, *increasing* each time the penalty coefficient t as much as possible without violating the centering condition. Stop if the required precision is attained.

We will not display the formulas that have to be used in the implementation of this algorithm.

Conclusion. We have shown how to begin the algorithm, starting with an arbitrary admissible point. The idea is to move gradually to the analytical center.

Chapter Eight

Economic Applications

There are many applications of optimization theory to economics. For mathematical applications, we hold those problems in the highest esteem that are the hardest to solve. For economic applications, the name of the game is different. To clarify the matter, we recall once more the words of Marshall: *"A good mathematical theorem dealing with economic hypotheses is very unlikely to be good economics."* The success of an economic application is measured by the light it sheds on examples that are important in real life.

We begin with some applications of the one-variable Fermat theorem.

8.1 WHY YOU SHOULD NOT SELL YOUR HOUSE TO THE HIGHEST BIDDER

Auctions. Auctions lead to many optimization problems of interest. Auctioneers want to organize the auction "in the best way" and sellers try to select the best bidding strategy. The impact of auction theory has become very substantial (cf. [51]). A article in 1995 in the *New York Times* hailed one of the first U.S. spectrum auctions—the design of which was based on suggestions by the academic designers Preston McAfee, Paul Milgrom, and Robert Wilson—as "The Greatest Auction Ever." The British spectrum auction of 2000, which raised about 34 billion pounds, the principal designers of which were professors Ken Binmore and Paul Klemperer, earned the former a commendation from the Queen and the title "Commander of the British Empire."

Selling a house or car. Suppose you have a house or a car that you want to sell. Your objective in selling the house is to maximize the expected revenue from the sale. Your own valuation of the house is 0. Now there are different ways in which you can sell the house. In a first price auction, bidders make a bid. The bidder with the highest bid gets the house and pays his own bid. In a second price auction, the bidder with the highest bid gets the house and pays the second highest bid for the house. In both cases you sell the house to the highest bidder, but is this actually optimal?

To analyze this question, let us put some structure on the problem. Assume there are n risk neutral bidders for the house. Let s_i denote bidder i's valuation for the house (that is, the maximum amount he is willing to pay for the house). We assume that the valuation s_i is drawn from the interval $[\underline{s}_i, \bar{s}_i]$ with density function $f_i(\cdot)$ and distribution function $F_i(\cdot)$. The distribution of the valuations may differ between buyers, as some buyers may be professional buyers who want to resell, while other buyers want to buy the house to live there. We assume that the draws from the valuation distributions are independent. Now instead of analyzing a game (like a first or second price auction) we consider the problem in a more abstract way. We consider a mechanism where each bidder reveals his valuation s_i. In the example below, it can be verified that you cannot gain anything by an untruthful revelation: this would not be optimal. The revelation determines the probability that bidder i gets the house and the amount of money he has to pay for it.

What does the optimal mechanism look like in the general case with n players? Let $x_i(s_1, \ldots, s_n)$ denote the probability that player i gets the good as a function of all the revealed valuations $s_1, \ldots, s_n$. Then the (marginal) probability that i gets the good can be derived as

$$x_i(s_i) = \int_{S_{-i}} x_i(s_1, \ldots, s_n) dF_{-i}(S_{-i}).$$

Here the subscript $-i$ denotes that we omit the i-variable. Further, let $t_i(s_i)$ denote the transfer from player i with valuation s_i to the seller.

Now we want to choose the functions $x_i(\cdot)$ and $t_i(\cdot)$ in such a way that the expected revenues from the house are maximal.

A bidder i with valuation s_i gets a payoff $u_i(s_i)$ which is determined by

$$u_i(s_i) = \max_{s'_i}[s_i x_i(s'_i) - t_i(s'_i)],$$

where s'_i is the announcement made by player i about his valuation. Using the envelope theorem we see that truthful revelation leads to

$$u'_i(s_i) = x_i(s_i).$$

One can show that the second order condition for this maximization problem is satisfied if and only if $x_i'(s_i) \geq 0$ for all s_i.

Hence, we can write the payoff of type s_i as

$$u_i(s_i) = u_i(\underline{s}_i) + \int_{\underline{s}_i}^{s_i} x_i(y) dy,$$

where type $\underline{s}_i$ will only participate if $u_i(\underline{s}_i) \geq 0$. The transfer from player i to the seller can be written as $t_i(s_i) = s_i x_i(s_i) - u_i(s_i)$.

This allows us to write the expected revenue R for the seller as follows:

$$R = \sum_{i=1}^{n} E_{s_i}[t_i(s_i)]$$

$$= \sum_{i=1}^{n} E_{s_i}\left[s_i x_i(s_i) - u_i(\underline{s}_i) - \int_{\underline{s}_i}^{s_i} x_i(y) dy\right]$$

$$= \sum_{i=1}^{n}\left[\int_{\underline{s}_i}^{\overline{s}_i} \left(s_i - \frac{1 - F_i(s_i)}{f_i(s_i)}\right) x_i(s_i) f_i(s_i) ds_i\right] - \sum_{i=1}^{n} u_i(\underline{s}_i).$$

Hence, to maximize expected revenue, you have sell to the bidder with the highest value for

$$\left(s_i - \frac{1 - F_i(s_i)}{f_i(s_i)}\right).$$

In particular, you choose $x_i(s_i) = 1$ if and only if

$$\left(s_i - \frac{1 - F_i(s_i)}{f_i(s_i)}\right) = \max_{s_j}\left(s_j - \frac{1 - F_j(s_j)}{f_j(s_j)}\right)$$

and $\left(s_i - \dfrac{1 - F_i(s_i)}{f_i(s_i)}\right) \geq 0,$

and you choose $x_i(s_i) = 0$ otherwise.

To see what this means, consider the following example with $n = 2$. Assume that s_1, the valuation of bidder 1, is uniformly distributed on $[0, 1]$, while s_2 is uniformly distributed on $[0, 2]$. Then if $s_1 = 0.75$, $s_2 = 0.8$, you know that bidder 2 has the highest valuation. However, it is optimal to sell to player 1 because $0.75 - \frac{1-0.75}{1}$ is more than $0.8 - \frac{2-0.8}{2}$. Moreover, if $s_1 < \frac{1}{2}$, $s_2 < \frac{2}{3}$, then although the valuations are positive (i.e., exceed your valuation of the house, which equals zero) it is optimal not to sell the house at all!

The intuition for this result is that by threatening not to sell the house, you can get a higher price from bidders with high valuations. Similarly, in the first example with $s_1 = 0.75, s_2 = 0.8$, you do not sell to player 2 because you want to get more money from the higher types of players 2 ($s_i > 1$). To get this additional money from the high 2 types, you threaten not to sell to them even if their valuation exceeds the valuation of the other buyer.

8.2 OPTIMAL SPEED OF SHIPS AND THE CUBE LAW

The next application is from maritime economics. It is a problem of making an optimal trade-off. Noteworthy in this example is *the cube law* for the dependence of the use of fuel on the speed.

Time charter contract. A time charter contract is a contract for a certain period, say half a year, between an operator (the owner of a ship) and a charterer (the owner of cargo). During this period, the operator will make trips for the charterer, say between Europe, Africa, and the Middle East. The time charter contract has a speed clause. This fixes the speed of the ship and the settlement of costs due to deviations from this speed demanded by the charterer. The need for such settlements is as follows. Sometimes, it can be very profitable to the charterer for the ship to increase its speed considerably to be somewhere in time to deliver or pick up cargo. This leads to a steep increase in fuel costs. The operator has to advance these costs, but he has no difficulty with this, as the settlement is described in the contract.

Cube law. This is done by using the cube law. This law assumes that fuel consumption per day is proportional to the cube of the speed. As a result, a charterer who wishes to order the operator to speed up faces a trade-off between the profits she expects to make and the additional costs she is facing. Let us explore such a trade-off.

Problem 8.2.1 *Optimal speed of ships and the cube law*

A charterer aims to maximize her daily profits by choosing an optimal speed. There are three factors that determine this speed: price of fuel, freight rates dictated by the market, and voyage distance. The following simple model gives some insight into how this problem can be solved. Time in port is ignored, so that the daily gross profit is given by

$$GS = \frac{RW}{d/(24s)} - C_R - pks^3,$$

ECONOMIC APPLICATIONS

where

GS = gross profit or surplus per day,
R = freight rate per ton of cargo,
W = deadweight available for cargo in tons,
d = distance steamed in nautical miles,
   (nautical mile=meridian minute=1851.85 m=1.16 miles),
s = speed in knots (nautical miles/hour),
C_R = running costs per day,
p = price of bunker fuel in dollars per ton,
k = constant of proportionality.

For a given freight rate and voyage the only variable is speed s. The charterer is confronted with an optimization problem: to maximize daily gross profits GS by choosing the speed s. Solve this problem.

Solution. To begin with, we show that this problem is indeed one about a trade-off; by the way, this is the case for every practical optimization problem of interest. We refer to the trade-off between the effect of s on the terms

$$\frac{RW}{(d/24s)} = 24\frac{RWs}{d} \text{ and } -pks^3.$$

Indeed, let us do a preliminary qualitative analysis of the problem. We want $24\frac{RWs}{d}$ to be big; to this end, we should choose s large; but then pks^3 is very large, which is bad as this term represents costs ("the minus sign"). But trying to keep fuel costs pks^3 small is not a good idea either: then we should choose s small; but then the term $24\frac{RWs}{d}$ is very small ("the voyage takes too many days"), which is bad.

Now we turn to a quantitative analysis. We use the result of the generalized problem of Tartaglia (exercise 1.6.25). This gives that the daily gross profits GS are maximal if the speed is chosen to be

$$\widehat{s} = \sqrt{8\frac{RW}{pkd}}.$$

We continue with an application of the multi-dimensional case of the Fermat theorem.

8.3 OPTIMAL DISCOUNTS ON AIRLINE TICKETS WITH A SATURDAY STAYOVER

We reconsider the matter of discounts on airline tickets with a Saturday stayover, discussed in problem 4.3.2. This time we model it in a different way. We will not model the behavior of the passengers in detail. Instead, we model it by means of their demand functions.

Problem 8.3.1 *Why do there exist discounts on airline tickets with a Saturday stayover? If you buy a plane ticket, then you get a considerable discount if your stay includes a Saturday night. How does the airline company determine the size of the discount?*

Solution. Let us recall the definition of "discount": if the price for a tourist (resp. a businessperson) is p_1 (resp. p_2), then the discount is $\frac{p_2 - p_1}{p_2}$.

Modeling. This optimization problem can be modeled as follows:

$$\pi(x_1, x_2) = p_1(x_1)x_1 + p_2(x_2)x_2 - C(x_1 + x_2) \to \max, \; x_1, x_2 > 0.$$

Here x_1 (resp. x_2) is the number of tickets sold to tourists (resp. to businesspeople), $C(x_1 + x_2)$ is the cost when total sales are $x_1 + x_2$, and the functions $p_1(\cdot)$ and $p_2(\cdot)$ describe the two market demands. We assume that $p_1(\cdot), p_2(\cdot)$ and $C(\cdot)$ are C^1-functions with $p_1'(\cdot), p_2'(\cdot)$ negative valued; these inequalities represent the reasonable assumption that higher prices means that fewer tickets are sold, for each of the two categories.

We announced above that p_1, p_2 are the variables which have to be chosen optimally. However, in this model we have followed the convention of choosing the demands x_1, x_2 as the variables which have to be chosen optimally. This seems less natural, but the two choices are mathematically equivalent. Indeed, as $p_1'(x_1) < 0$ for all x_1 and $p_2'(x_2) < 0$ for all x_2, the functions $p_1 = p_1(x_1)$ and $p_2 = p_2(x_2)$ can be inverted: $x_1 = x_1(p_1)$ and $x_2 = x_2(p_2)$. Therefore the model above is equivalent to the problem

$$\tilde{\pi}(p_1, p_2) = p_1 x_1(p_1) + p_2 x_2(p_2) - C(x_1(p_1) + x_2(p_2)) \to \max, \; p_1, p_2 > 0.$$

Seeing the unpleasant expression $C(x_1(p_1) + x_2(p_2))$, we realize that the convention referred to above is not so bad after all.

Economic interpretation of the solution. Now we will give an explicit description of how the elasticities of the demand by families and by business-people determine the optimal discount percentage. We do not display here the application of the Fermat theorem, the algebraic manipulations which lead to this description, and the conditions under which there is a solution. Introduce the revenue functions

$$R_1(x_1) = p_1(x_1)x_1 \text{ and } R_2(x_2) = p_2(x_2)x_2.$$

We recall that the marginal revenue functions $MR_1(x_1)$ and $MR_2(x_2)$ are defined to be the derivatives of the total revenue functions. Now we will express this in terms of elasticities. Let us recall this concept. The elasticity of demand x with respect to price p is defined by

$$\epsilon = \epsilon(x) = \frac{dx}{x} / \frac{dp}{p} = \frac{p}{x}\frac{dx}{dp}.$$

This measures which *relative* change in x is caused by a *relative* change in p. By relative change we mean "change measured in percentage." For example, if the absolute value of the elasticity is larger than 1, then each small percentage change in the variable x is magnified when you go over to the variable p. Such properties of elasticity mean that the elasticity $\frac{p}{x}\frac{dx}{dp}$ is often a more appropriate measure of sensitivity than the derivative $\frac{dx}{dp}$ itself. Economists find it convenient to work with elasticities, as one can often make an estimate for them. Often one calls the absolute value of $\frac{p}{x}\frac{dx}{dp}$ the elasticity, as the sign of $\frac{p}{x}\frac{dx}{dp}$ is often known and so not interesting. The formula

$$MR_i(\widehat{x}_i) = p_i(\widehat{x}_i)\left(1 + \frac{1}{\epsilon_i}\right), \quad i = 1, 2,$$

follows from $R_i(x_i) = p_i(x_i)x_i$ and the product rule for differentiation. Let

$$\varepsilon_i = \frac{p_i(\widehat{x}_i)}{\widehat{x}_i p'_i(\hat{x}_i)}$$

be the elasticity of the demand x_i with respect to the price p_i in the optimum, for $i = 1, 2$. Now we can present the promised expression of the optimal discount in terms of elasticities

$$\frac{p_2(\widehat{x}_2) - p_1(\widehat{x}_1)}{p_2(\widehat{x}_2)} = \frac{\frac{1}{\epsilon_1} - \frac{1}{\epsilon_2}}{1 + \frac{1}{\epsilon_1}}.$$

For example, if the elasticity in market 1 (families) is 4 while the elasticity in market 2 (the business travelers) is 2, then the optimal discount is 33,3 percent.

For many—but not all—choices of p_1, p_2 and C, this problem has a unique solution.

Criticism of the model. This model does not take into account that businesspeople might decide to accept a Saturday stayover if the difference in price is too big. This consumer behavior can also be modeled, leading to a problem with inequality constraints, as we have seen in problem 4.3.2.

Now we give some applications of the Lagrange multiplier rule.

8.4 PREDICTION OF FLOWS OF CARGO

Finally we give an advanced application of the Lagrange multiplier rule, this time to transportation. This leads to an optimization problem that has only *linear* constraints. Nevertheless it would be inconvenient to solve it without the Lagrange multiplier rule.

Information about transport flows is of great commercial interest. For example, one London-based company has representatives in all major ports in the world whose main job it is to register, sometimes using megaphones, which ships with which cargoes arrive and leave. However, no information is collected about where the ships come from and for which ports they are heading. Now the problem is to make an estimate of all the transport flows, having only this information. If we number the ports from 1 to N and let $T_{i,j}$ be the flow of transport from port i to port j, then the problem is to estimate the $N \times N$ distribution matrix $T = (T_{i,j})$ if all its row sums and column sums are given.

Variants. Sometimes one has some additional information: there are some routes of which one knows that they do not have any flow as they are too expensive. Then this additional information has to be taken into account as well. The so-called *method of maximal entropy* or *method of maximum likelihood*, in this context usually called the *RAS method*, to be explained below, gives satisfactory estimates for both variants of the problem.

Idea of RAS. It is based on the assumption that all decisions to transport cargo are taken independently of each other. This is modeled as follows. All units of cargo, say containers, are distributed with equal probability over the N^2 possible combinations of origin

ECONOMIC APPLICATIONS

and destination. Then it is natural to propose as an estimate the distribution matrix T that has maximal probability. Let us demonstrate the method of solution using a numerical example with $N = 2$. In real-world applications N is much larger, but the analysis of the problem is the same.

Problem 8.4.1 *Prediction of flows of cargo.* *An investor wants to have information about the four flows of cargo within an area consisting of two zones, 1 and 2, including the flows within each zone. For both zones data are available to him, not only for the total flow originating in this zone, $O_1 = 511$ and $O_2 = 1451$, but also for the total flow with destination in this zone, $D_1 = 1733$ and $D_2 = 229$. Observe that*

$$S = O_1 + O_2 = D_1 + D_2 = \text{total flow}.$$

However, the investor is not satisfied with this; he wants to have an estimation for T_{ij}, the flow from i to j, measured in containers, for all $i, j \in \{1, 2\}$. What is the most probable distribution matrix T_{ij} given the available data and assuming that all units of cargo are distributed over the four possibilities with equal probability?

Solution. We will use without proof the following fact. The following approximate formula for the natural logarithm of the probability of a distribution matrix T holds:

$$C - \sum_{i,j} [T_{ij}(\ln T_{ij}) - T_{ij}],$$

where C is a constant which does not depend on the choice of T. It follows that the problem can be modeled as follows:

$$\sum_{i,j} [T_{ij}(\ln T_{ij}) - T_{ij}] \to \min, \quad T_{ij} > 0 \quad \forall i, j,$$

subject to

$$\begin{cases} T_{11} + T_{12} = O_1, \\ T_{21} + T_{22} = O_2, \\ T_{11} + T_{21} = D_1, \\ T_{12} + T_{22} = D_2. \end{cases}$$

We have obtained the optimization problem above by going over from this probability to its logarithm, using the approximate formula, removing the constant C and the minus sign, and replacing maximization by minimization (because of the change of sign). Now we apply the Lagrange multiplier rule to this optimization problem. We form the Lagrange function

$$\mathcal{L} = \lambda_0 \sum_{i,j} [T_{ij}(\ln T_{ij}) - T_{ij}]$$

$$+ \sum_i \lambda_i \left(O_i - \sum_j T_{ij} \right) + \sum_j \lambda'_j \left(D_j - \sum_i T_{ij} \right)$$

with Lagrange multipliers $\lambda_0, \lambda_1, \lambda_2, \lambda'_1, \lambda'_2$ (one for the objective function and one for each equality constraint). Then we write the Lagrange equations

$$\frac{\partial \mathcal{L}}{\partial T_{ij}} = 0 \Leftrightarrow \lambda_0 \ln T_{ij} - \lambda_i - \lambda'_j = 0. \qquad (*)$$

We may assume without restriction that $\lambda_0 = 1$. To see this, observe first that we can omit the last equality constraint from the problem, as the sum of the first two equality constraints is the same as the sum of the last two ones and so one of the constraints is redundant. The resulting new Lagrange equations can be obtained from the ones above by putting the Lagrange multiplier of the last constraint equal to zero: $\lambda'_2 = 0$. Now assume, contrary to what we want to prove, that $\lambda_0 = 0$. Then the Lagrange equations give that the other multipliers $\lambda_1, \lambda_2, \lambda'_1$ would also be zero. This finishes the proof by contradiction that, in the new Lagrange equations, one has $\lambda_0 \neq 0$ and so, by scaling, we may take $\lambda_0 = 1$. It follows that in the original Lagrange equations we may take $\lambda_0 = 1$ (and $\lambda'_2 = 0$).

Now we can rewrite $(*)$ as $\ln T_{ij} = \lambda_i + \lambda'_j$ and so as $T_{ij} = e^{\lambda_i} e^{\lambda'_j}$. Substituting this in the four equality constraints of our problem we get the equations

$$\begin{aligned} e^{\lambda_1} e^{\lambda'_1} + e^{\lambda_1} e^{\lambda'_2} &= O_1, \\ e^{\lambda_2} e^{\lambda'_1} + e^{\lambda_2} e^{\lambda'_2} &= O_2, \\ e^{\lambda_1} e^{\lambda'_1} + e^{\lambda_2} e^{\lambda'_1} &= D_1, \\ e^{\lambda_1} e^{\lambda'_2} + e^{\lambda_2} e^{\lambda'_2} &= D_2. \end{aligned}$$

This gives $e^{\lambda_i} = \frac{O_i}{e^{\lambda'_1} + e^{\lambda'_2}}$ and $e^{\lambda'_j} = \frac{D_j}{e^{\lambda_1} + e^{\lambda_2}}$.

Adding the first two equations gives

$$(e^{\lambda_1} + e^{\lambda_2})(e^{\lambda'_1} + e^{\lambda'_2}) = S.$$

It follows that $T_{ij} = \frac{O_i D_j}{S}$.
Substituting the given data gives the required estimates.

Completion of the analysis. To complete the analysis, it remains to prove the existence of a solution and the verification that a solution has all its entries strictly larger than zero. The function $g(t) = t \ln t - t$ has the following properties: $g(t)$ tends to $+\infty$ if $t \to +\infty$ and to 0 if $t \downarrow 0$; moreover, the derivative $g'(t)$ tends to $-\infty$ if $t \downarrow 0$. Taking this into account and applying the Weierstrass theorem gives the desired existence result and shows that a solution has all its entries strictly larger than zero.

The method we have just demonstrated is very flexible. For example, let us discuss the variant of the problem, for general N, where we have the following additional information in advance: for a number of combinations of origin and destination we know that they have flow zero. Then we can proceed in the same way as above to write down an optimization problem, but now we add a constraint $T_{ij} = 0$ for each combination of origin i and destination j for which we know in advance that the flow will be zero. Then we apply the Lagrange multiplier rule and solve the resulting system of equations numerically.

Finally we present some applications of the Karush-Kuhn-Tucker theorem and of convexity.

8.5 NASH BARGAINING

In the seminal paper [55], John Nash considered the problem of what would be a fair outcome if two people have an opportunity to bargain. We are going to describe his result, following the approach in the original paper of Nash. This paper originally for classroom use, but then Nash became aware that the result was worth publishing. A *bargaining opportunity* is modeled as a convex set F in the plane $\mathbb{R}^2$ together with a point $v \in F$, such that the set of all $x \in F$ with $x \geq v$ is closed and bounded. It is assumed that there exists an element $w \in F$ for which $w > v$. The two people can either come to an agreement—the points of F represent the possible agreements—or not; if not, then they will have to accept as outcome the point v.

The point v is called *the disagreement point*. For each outcome, the utility for person i is the i-th coordinate of the point corresponding to the outcome ($i = 1, 2$). Nash illustrated the concept of bargaining opportunity with the following numerical example.

Example 8.5.1 *Bill and Jack are in a position to barter goods but have no money with which to facilitate exchange.*

Bill's goods	Utility to Bill	Utility to Jack
book	2	4
whip	2	2
ball	2	1
bat	2	2
box	4	1

Jack's goods	Utility to Bill	Utility to Jack
pen	10	1
toy	4	1
knife	6	2
hat	2	2

1. How can this bargaining opportunity be modeled by a convex set F and a disagreement point v?

2. What should they do?

Solution

1. The bargaining set F is the set of linear combinations (here we will write elements of $\mathbb{R}^2$ as rows)

$$p_1(-2, 4) + p_2(-2, 2) + p_3(-2, 1) + p_4(-2, 2) + p_5(-4, 1)$$

$$+p_6(10, -1) + p_7(4, -1) + p_8(6, -2) + p_9(2, -2),$$

with $0 \leq p_i \leq 1$ ($1 \leq i \leq 9$), and the disagreement point v is the origin. This statement requires some explanation. Let a *basic* bargaining outcome represent the changes in utilities if one of the two individuals gives one of the objects in his possession to the other one; for example, the basic bargaining outcome that Bill gives his book to Jack is represented by the point $(-2, 4)$ ("the pay-off"), as Bill loses two units of utility and Jack gains four units of utility. The following example illustrates the idea

ECONOMIC APPLICATIONS

of taking a scalar multiple of a basic bargaining outcome with scalar in the interval $[0, 1]$. Bill and Jack might agree to throw a die and then Bill will give his book to Jack if five comes up (and otherwise he will keep it). This bargaining opportunity is represented by the point $\frac{1}{6}(-2, 4) = (-\frac{1}{3}, \frac{2}{3})$; in probabilistic terms, its first (resp. second) coordinate is the expectation of the utility of Bill (resp. Jack). Continuing in this way, one gets an interpretation of each of the linear combinations above as a bargaining outcome.

2. We will see below that the *fair* outcome is as follows:

   Bill gives Jack: book, whip, ball, and bat;

   Jack gives Bill: pen, toy, and knife.

Fair outcome rule. To address the question what is a fair outcome of the bargain, we have to consider the more general problem of finding a fair rule ϕ to assign to an arbitrary bargaining opportunity (F, ϕ) an outcome $\phi(F, v)$ in F. This rule should only depend on the underlying preferences of the two individuals and not on the utility functions that are used to describe these preferences (if a utility function u is used to describe a preference, then the utility functions that describe the same preference are the affine transformations of u: $cu + d$ with $c > 0$). It is natural to impose the following axioms on this rule. We need the following concept. A pair (F, v) is called *symmetric* if there exist affine transformations of the two utility functions such that after transformation the following implication holds: if $(a, b) \in F$, then $(b, a) \in F$, that is, the set F is symmetric with respect to the line $x_1 = x_2$; this means that both persons are in a symmetrical situation, as they are confronted with exactly the same bargaining opportunity.

1. There are no $x \in F$ for which $x > \phi(F, v)$.

   This represents the idea that each individual wishes to maximize the utility to himself of the ultimate bargain.

2. If (E, v) is another bargaining opportunity with $E \subseteq F$, and $\phi(F, v) \in E$, then $\phi(E, v) = \phi(F, v)$.

   This represents the idea that eliminating from F feasible points (other than the disagreement point) that would not have been chosen should not affect the solution.

3. If (F, v) is symmetric and utility functions are chosen such that this is displayed, then $\phi(F, v)$ is a point of the form (a, a).

This represents equality of bargaining skill: if the two people have the same bargaining position, then they will get the same utility out of it.

Theorem 8.1 Nash. *There is precisely one rule ϕ that satisfies the three axioms above: $\phi(F, v)$ is the unique solution of the following optimization problem:*

$$f(x) = (x_1 - v_1)(x_2 - v_2) \to \max, \ x \in F, \ x \geq v.$$

Proof. For each pair (F, v) the problem has a solution by Weierstrass; this solution is unique, as, taking minus the logarithm of the objective function, we can transform the problem into an equivalent convex problem with a strictly convex objective function

$$-\ln(x_1 - v_1) - \ln(x_2 - v_2).$$

Now we can assume, by an appropriate choice of the utility functions of the two individuals, that the two coordinates of this solution are equal to 1 and that $v_1 = v_2 = 0$. Then $(1, 1)$ is the solution of the problem $x_1 x_2 \to \max, \ x \in F, \ x \geq 0_2$. The set F is contained in the half-plane G given by $x_1 + x_2 \leq 2$; let us argue by contradiction. Assume that there exists a point $x \in F$ outside this half-plane; then the segment with endpoints v and x is contained in F, by the convexity of F, and this segment contains points near v with $x_1 x_2 > 1$, which contradicts the maximality of $(1, 1)$. The axioms above give immediately that $\phi(G, 0_2) = (1, 1)$ as G is symmetric, and so that $\phi(F, 0_2) = (1, 1)$ as $\phi(G, 0_2) \in F$. □

Verification of the Nash solution. Now we are ready to verify that the solution proposed above is indeed the Nash bargaining solution. The necessary and sufficient conditions for the convex optimization problem that defines the Nash solution lead to a convenient finite algorithm to calculate the solution. Now we will display the calculations, without making any comments on the algorithm and its derivation. Calculate, for the objects in the possession of Bill, the quotients of the utility of Bill and that of Jack:

book $\frac{2}{4} = \frac{1}{2}$, whip $\frac{2}{2} = 1$, bat $\frac{2}{2} = 1$, ball $\frac{2}{1} = 2$, box $\frac{4}{1} = 4$.

Then calculate, for the objects in the possession of Jack, the quotients of the utility of Jack and that of Bill:

pen $\frac{1}{10}$, toy $\frac{1}{4}$, knife $\frac{2}{6} = \frac{1}{3}$, hat $\frac{2}{2} = 1$.

ECONOMIC APPLICATIONS

Then find a positive number r such that the sum of all basic bargaining opportunities for which the quotients corresponding to objects in possession of Bill (resp. Jack) is not larger than r (resp. $1/r$), plus some linear combination with coefficients in $[0, 1]$, of any basic bargaining opportunities for which the quotient corresponding to objects in possession of Bill (resp. Jack) is equal to r (resp. $1/r$) is a positive vector with quotient of first and second coordinate equal to r. We can take $r = 2$ and take

$$(-2, 4)+(-2, 2)+(-2, 1)+(-2, 2)+(10, -1)+(4, -1)+(6, -1) = (12, 6).$$

This gives automatically the Nash solution.

We give another example, in order to illustrate that solutions can be probabilistic.

Example 8.5.2 *Jennifer and Britney are in a position to barter goods but have no money with which to facilitate exchange.*

Jennifer's goods	Utility to Jennifer	Utility to Britney
ball	4	1
Britney's goods		
bat	1	2
box	2	2

1. How can this bargaining opportunity be modeled by a convex set F and a disagreement point v?

2. What should they do?

Solution

1. The bargaining opportunity can be modeled by the set of linear combinations

$$p_1(4, -1) + p_2(-1, 2) + p_3(-2, 2)$$

with coefficients in $[0, 1]$ and with disagreement point the origin.

2. The Nash solution turns out to be $(4, -1) + (-1, 2) + \frac{1}{2}(-2, 2)$. This represents the following outcome: Jennifer gives her ball to Britney, Britney gives her bat to Jennifer, and a coin is tossed: if it is heads, then Britney gives her box to Jennifer (otherwise she keeps it).

8.6 ARBITRAGE-FREE BOUNDS FOR PRICES

In the next three sections we consider applications to finance. Consider m assets such as stocks, bonds, and options, with prices at the beginning of an investment period $p_1, \ldots, p_m$, and with values at the end of the investment period $v_1, \ldots, v_m$. Let $x_1, \ldots, x_m$ be the investments in the assets at the beginning of the period, and let short positions be allowed: $x_i < 0$ means a short position in asset i. Then the cost of the initial investment is

$$p^T \cdot x = p_1 x_1 + \cdots + p_m x_m$$

and the final value of the investment is

$$v^T \cdot x = v_1 x_1 + \cdots + v_m x_m.$$

The values of the assets at the end of the investment period, $v_1, \ldots, v_m$, are uncertain. We will assume that only n possible scenarios are possible, and that the final value of the asset i for scenario j is $v_i^{(j)}$.

At first sight, it might be a surprise that you can give bounds for the initial price of one of the assets, say, p_m, if you only know the initial prices of the other assets, $p_1, \ldots, p_{m-1}$, and, moreover, the values $v_i^{(j)}$ of all the assets after the investment period in all the possible scenarios. In particular, you do not have to know how likely each one of the scenarios is, and the attitude of investors toward risk does not play any role.

The reason for this is that it is generally assumed that there are no *arbitrage opportunities*, that is, guaranteed money-making investment strategies. Indeed, arbitrageurs are constantly on the look out for such opportunities. Their activities lead to changes in the initial prices, making these opportunities disappear as soon as they arise. One type of arbitrage opportunity is the existence of an investment vector x with $p^T \cdot x < 0$, and with, in all possible scenarios, a nonnegative final value, that is,

$$\left(v^{(j)}\right)^T \cdot x \geq 0,$$

for $j = 1, \ldots, n$. The condition $p^T \cdot x < 0$ means that you are paid to accept the investment mix, and the condition $\left(v^{(j)}\right)^T \cdot x \geq 0$, for $j = 1, \ldots, n$, means that, in all scenarios, the value of the investment will be nonnegative at the end of the period. This is said to be *type A arbitrage*; here you obtain money immediately and never have to pay anything.

ECONOMIC APPLICATIONS

This leads to the following bounds.

Problem 8.6.1 *Arbitrage-free bounds on prices.* Show that the following arbitrage A-free bounds for the price of one of the assets, say, p_m, hold true if you know the initial prices of the other assets, $p_1, \ldots, p_{m-1}$, as well as the values $v_i^{(j)}$ of all the assets after the investment period for all the possible scenarios. This price p_m lies between the extremal values of the following LP problem:

$$f(p_m) = p_m \to \text{extr}, \quad Vy = p, \quad y \geq 0_n,$$

where V is the $m \times n$ matrix of values $v_i^{(j)}$ of the m assets under all n possible scenarios.

In particular, if the assets contain an asset with price equal to the value in each possible scenario, then the system $Vy = p$ gives that $\sum_{i=1}^n y_i = 1$, and then y can be interpreted as a probability distribution on the scenario's. Moreover, then the system $Vy = p$ can be interpreted as follows: the price of each asset equals the expectation of its value after the investment period with respect to the distribution y. Then one says that y is a *martingale*.

In the solution of this problem we will use Farkas's lemma. This result is useful in situations where it is necessary to show that a system of linear inequalities has no solution. Farkas's lemma helps to reduce this task of showing nonexistence to the easier task of showing the existence of a solution for another system of linear inequalities.

Theorem 8.2 *Farkas's lemma.* The system of inequalities

$$Ax \leq 0_m, \quad c^T \cdot x < 0,$$

where A is an $m \times n$ matrix and $c \in \mathbb{R}^n$, and the system of inequalities

$$A^T y + c = 0_n, \quad y \geq 0_m,$$

are strong alternatives. That is, precisely one of the two is solvable.

Proof. We outline the proof. Apply theorem 4.7 to the LP problem

$$f(x) = c^T \cdot x \to \min, \quad Ax \leq 0_m,$$

and its dual

$$g(y) = 0 \to \max, \quad A^T y + c = 0_n, \quad y \geq 0_m.$$

This leads to the statement of the theorem. □

A different proof is given in section 9.3. Now we are ready to give the solution of problem 8.6.1.

Solution. The assumption that type A arbitrage does not exist gives that the inequality system

$$p^T \cdot x < 0, \ V^T x \geq 0_n$$

has no solutions. Using Farkas's lemma this implies that there exists y such that

$$Vy = p, \ y \geq 0_n.$$

This gives the required result.

8.7 FAIR PRICE FOR OPTIONS: FORMULA OF BLACK AND SCHOLES

The result in the previous section implies *the formula of Black and Scholes* determining the fair price of options in the case of one period and two scenarios. The development of this pricing method revolutionized the practice of finance.

A *European call option* on a stock with exercise price w is the right to buy a certain stock at the end of an investment period for the price w. The value of this option at the end of the investment period is $(v-w)_+$, where v denotes the value of the stock after the investment period. Indeed, if $v > w$, then you can buy the option for the price w and sell at at the price v, making a profit $v - w$; if $v \leq w$, then the option will have no value.

Problem 8.7.1 *Formula of Black and Scholes: a fair price for options.* Consider a stock with current price p and with value $v^{(1)}, v^{(2)}$ in two possible scenarios. Assume that the value goes down in the first scenario and up in the second scenario: $v^{(1)} < p < v^{(2)}$. In addition, there is a bond, a riskless asset, with current price 1 and with value 1 in each of the two possible scenarios. We consider the problem of pricing a European call option on a stock with exercise price w. Show that the absence of type A arbitrage determines the

ECONOMIC APPLICATIONS

price p_o of the option. It is given by the following formula of Black and Scholes:

$$p_o = y_1(v^{(1)} - w)_+ + y_2(v^{(2)} - w)_+,$$

where y_1, y_2 are determined by

$$y_1, y_2 \geq 0, \quad y_1 + y_2 = 1,$$

and

$$p = y_1 v^{(1)} + y_2 v^{(2)}.$$

The formula of Black and Scholes can be interpreted as follows:

there exists a unique probability distribution y_1, y_2 for the two possible scenario's for which the price of the stock and the bond are equal to their expected value after the investment period.

Solution. The previous problem, with three assets (stock, bond, and option) and two scenarios, leads to the following system of three linear equations in the unknowns p_o, y_1, y_2:

$$p_o = y_1 v^{(1)} + y_2 v^{(2)},$$

$$1 = y_1 + y_2,$$

$$p_o = y_1(v^{(1)} - w)_+ + y_2(v^{(2)} - w)_+.$$

This is seen to have a unique solution, and this solution has y_1 and y_2 nonnegative.

8.8 ABSENCE OF ARBITRAGE AND EXISTENCE OF A MARTINGALE

There is another type of arbitrage opportunity, and if you forbid this as well, then all coefficients of the vector y are positive. In particular, if the assets contain an asset the price of which is equal to the value in each possible scenario, then we get that all scenarios have positive probability. Let us be precise. We now consider the following type of arbitrage opportunity: the existence of a vector x for which

$$x^T \cdot p = 0, \quad V^T x \geq 0_n, \quad V^T x \neq 0_n.$$

The condition $x^T \cdot p = 0$ means that the initial cost of the investment is zero; the condition $V^T x \geq 0_n$ means that you never have to pay anything, and the additional condition $V^T x \neq 0_n$ gives that in at least one possible scenario you will get something. This is said to be *type B arbitrage*; here you pay nothing and there is a possibility that you get something.

Problem 8.8.1 *Absence of arbitrage and existence of a positive martingale.* Consider again the general one-period investment model. Show that there are no arbitrage opportunities of types A and B precisely if there exists a positive martingale, that is, a probability distribution y on the scenarios for which each scenario has positive probability, $y > 0_n$, and, moreover, the current price of each asset equals the expectation of its value at the end of the investment period with respect to the distribution, $p = Vy$.

This problem can be solved by the same method as the previous two problems. We do not display the solution.

8.9 HOW TO TAKE A PENALTY KICK, AND THE MINIMAX THEOREM

If you have to make certain business decisions very often, a competitor can learn your strategy by observing the decisions you make, and profit from this knowledge at your cost. You can sometimes prevent this by using a nondeterministic strategy. In the problem below we show how to do this in an optimal way for the simplest type of problem of interest.

Problem 8.9.1 *The goalkeeper Hans van Breukelen studied before each game of soccer the statistics of the penalty specialists of the opposing team. Usually a player has a preference either for left or for right if he takes a penalty kick. Thus Hans gained a reputation for having an uncanny intuition for diving in the right direction. But then his method became known somehow. As a consequence this method was not effective any more. One of the last successes of this method was during the final of the European championship EK88 between Russia and Holland, where van Breukelen caused a penalty kick, to be taken by Igor Belanov. How should this penalty kick have been taken if the method had been known already and in which direction should van*

Breukelen have dived in this case?

Solution. Let us formulate a more reasonable question: suppose Igor has to take a series of penalty kicks, with Hans in the goal; what are the best strategies for Hans and for Igor? Well, if they want to optimize their performance, then they need some data on their success rate in each of the four cases. Igor's favorite case is that he aims left and Hans dives in the other direction; then his success rate is $a_{lr} = 0.9$. In the other three cases his success rates are $a_{ll} = 0.6$, $a_{rr} = 0.5$, and $a_{rl} = 0.8$.

Mixed strategies. The common sense solution suggests that Igor should choose two nonnegative numbers x_l and x_r with sum 1 and should aim left (resp. right) in a fraction x_l (resp. x_r) of his kicks. Now let us look from the position of Hans. He wants to keep Igor guessing in which direction he will dive. Therefore, he will also choose two nonnegative numbers y_l and y_r with sum 1, to be used as follows. He will dive left (resp. right)—seen from the direction of Igor—in a fraction y_l (resp. y_r) of the penalty kicks. Then Igor's expected success rate will be $x^T A y$, where $x = (x_l, x_r)^T$, A is the 2×2 matrix

$$\begin{pmatrix} a_{ll} & a_{lr} \\ a_{rl} & a_{rr} \end{pmatrix} = \begin{pmatrix} 0.6 & 0.9 \\ 0.8 & 0.5 \end{pmatrix}$$

and $y = (y_1, y_2)^T$. Let I be the set of nonnegative vectors in $\mathbb{R}^2$ with sum of coordinates equal to 1.

Nash equilibrium. The first problem is to show that there exist strategies $\widehat{x}, \widehat{y} \in I$ such that $\widehat{x}$ is an optimal choice for Igor if Hans chooses $\widehat{y}$ and conversely $\widehat{y}$ is an optimal choice for Hans if Igor chooses $\widehat{x}$ (a *Nash equilibrium*). Then neither of the two is tempted to change strategy given the choice of the other, as the following inequalities hold:

$$x^T A \widehat{y} \leq \widehat{x}^T A \widehat{y} \leq \widehat{x}^T A y \quad \forall x, y \in I.$$

The second problem is to show that the resulting success rate for Igor, $\widehat{x}^T A \widehat{y}$, does not depend on the choices of $\widehat{x}$ and $\widehat{y}$.

Zero sum game. We can view the situation as a special case of a game between two players, where the first player chooses between m moves $i = 1, \ldots, m$ and the second between n moves $j = 1, \ldots, n$, and where the combination of moves (i, j) leads to payment of an amount of money $a_{i,j}$ by the second player to the first player. Here $a_{i,j}$ is allowed to be negative: this means that the first player should pay $-a_{i,j}$ to the second player.

We will prove the following result. Let

$$I_k = \left\{ z \in \mathbb{R}^k \,\Big|\, z \geq 0,\ \sum_{i=1}^k z_i = 1 \right\}$$

for all $k \in \mathbb{N}$. A *convex combination* of vectors $v_1, \ldots, v_k$ from $\mathbb{R}^n$ is a linear combination $\sum_{i=1}^k z_i v_i$ for which the column vector of coefficients $(z_1, \ldots, z_k)^T$ belongs to I_k.

Theorem 8.3 Minimax theorem of von Neumann. *For each $m \times n$ matrix A the following equality holds true:*

$$\max_{x \in I_m} \min_{y \in I_n} x^T A y = \min_{y \in I_n} \max_{x \in I_m} x^T A y$$

and there exists a pair $(\widehat{x}, \widehat{y})$ that is a solution of the minimax and the maximin problem.

Relation between the minimax and Nash equilibrium. This result gives us precisely what we want. Indeed, let us write it out explicitly:

$$\max_{x \in I_m} \min_{y \in I_n} x^T A y = \min_{y \in I_n} \widehat{x}^T A y = \widehat{x}^T A \widehat{y}$$

and

$$\min_{y \in I_n} \max_{x \in I_m} x^T A y = \max_{x \in I_m} x^T A \widehat{y} = \widehat{x}^T A \widehat{y}.$$

This gives that $\widehat{x}$ and $\widehat{y}$ form a *Nash equilibrium*, also called *a saddle point*, in the following sense

$$x^T A \widehat{y} \leq \widehat{x}^T A \widehat{y} \leq \widehat{x}^T A y \quad \forall x \in I_m,\ \forall y \in I_n.$$

This settles our first problem. Now assume that $\bar{x} \in I_m$ and $\bar{y} \in I_n$ form another Nash equilibrium. That is,

$$\max_{x \in I_m} x^T A \bar{y} = \bar{x}^T A \bar{y} = \min_{y \in I_n} \bar{x}^T A y.$$

Then the right- (resp. left-) hand side is less (resp. more) than or equal to

$$\max_{x \in I_m} \min_{y \in I_n} x^T A y \quad (\text{resp. } \min_{y \in I_n} \max_{x \in I_m} x^T A y).$$

Now we use that this maximin and this minimax are both equal to $\widehat{x}^T A \widehat{y}$ by the theorem. It follows that

$$\bar{x}^T A \bar{y} = \widehat{x}^T A \widehat{y}.$$

ECONOMIC APPLICATIONS

This settles our second problem.

Now we give the proof of the minimax theorem.

Proof. We begin by proving that there exists a solution $(\hat{x}, \hat{y})$ of the minimax problem given by the expression

$$\max_{x \in I_m} \min_{y \in I_n} x^T A y.$$

The expression

$$\sup_{x \in I_m} \inf_{y \in I_n} x^T A y$$

equals

$$\sup_{x \in I_m} \min_{k} x^T A e_k,$$

as each $y \in I_n$ is a convex combination of the vectors e_k ($1 \le k \le n$) of the standard basis of $\mathbb{R}^n$, and $x^T A y$ depends linearly on y. This is the optimal value of the optimization problem

$$\xi \to \max, \ \xi \le x^T A e_k, \ 1 \le k \le n, \ x \in I_m.$$

This problem has a solution $\hat{x}$ by the Weierstrass theorem. Let $\hat{y}$ be a solution of the problem

$$\hat{x}^T A y \to \min, \ y \in I_n.$$

This exists by the Weierstrass theorem. Thus $\hat{x}$ and $\hat{y}$ give a solution of our maximin problem. To finish the proof of the theorem, it remains to prove that $\hat{x}$ and $\hat{y}$ also form a solution of the minimax problem given by the expression

$$\min_{y \in I_n} \max_{x \in I_m} x^T A y.$$

The expression

$$\inf_{y \in I_n} \sup_{x \in I_m} x^T A y$$

equals the optimal value of the optimization problem

$$\eta \to \min, \ \eta \ge e_l^T A y, \ 1 \le l \le m, \ y \in I_n.$$

To finish the proof, it is sufficient to verify the KKT conditions for $\eta = \hat{x}^T A \hat{y}$, $y = \hat{y}$, and the following choice of Lagrange multipliers:

$\lambda_0 = 1$ and $(\lambda_1, \ldots, \lambda_m) = \widehat{x}^T$. If we substitute $\lambda = (1, \widehat{x}^T)$ in the Lagrange function, we get

$$\eta + \sum_{l=1}^{m} \widehat{x}_l(e_l^T A y - \eta).$$

If we simplify this expression we get $\widehat{x}^T A y$. Now we are ready to verify the KKT conditions:

1) The minimum condition amounts to the equality

$$\min_{y \in I_n} \widehat{x} A y = \widehat{x} A \widehat{y},$$

which holds by the choice of $\widehat{y}$ above.

2) The nonnegativity condition amounts to $\widehat{x} \geq 0$, which holds by $\widehat{y} \in I$.

3) The complementary slackness condition amounts to

$$\widehat{x}_l(e_l^T A \widehat{y} - \widehat{x}^T A \widehat{y}) = 0$$

for all $l \in \{1, \ldots, m\}$ and this holds as summing all left-hand sides, which we know to be nonpositive, and simplifying the result we get zero. □

The proof of the minimax theorem suggests the following method to calculate optimal strategies $\widehat{x}$ and $\widehat{y}$ for a given matrix A: solve the KKT conditions of the problem

$$\eta \to \min, \ \eta \geq e_l^T A y, \ 1 \leq l \leq m, \ y \in I_n.$$

If $(\bar{\eta}, \bar{y})$ and the Lagrange multiplier $(1, \bar{x}^T)$ form a solution, then $\widehat{x} = \bar{x}$ and $\widehat{y} = \bar{y}$ form optimal strategies.

8.10 THE BEST LUNCH AND THE SECOND WELFARE THEOREM

In the next problem we will present the second welfare theorem from micro-economics. We will offer a short proof of this result, using the KKT theorem. The usual proof uses separation of convex sets and requires some additional arguments. The two welfare theorems give answers to the following two questions. Can one give a theoretical result in support of the "invisible hand," the empirical fact that self-interest sometimes leads to common interest? The *first welfare theorem*, to be given below, is a positive result in this direction: it shows that self-interest leads to an allocation of goods in a society such that it is efficient in some sense. The subject of equity and efficiency leads to the following question: can a price system lead to an allocation that is not only efficient but also fair? For example, if not, then this might suggest that it is necessary to interfere in the labor market to correct for unequal chances due to different backgrounds. The *second welfare theorem* supports the view that price mechanisms can achieve each reasonable allocation of goods without interference, provided one corrects the starting position, for example, by means of education in the case of the labor market.

Problem 8.10.1 *The best lunch (the second welfare theorem). Each one of a group of people has brought some food for lunch and each person intends to eat his own food. Then one of them offers to his neighbor his apple in exchange for her pear. She accepts, as she prefers an apple to a pear. Thus they are both better of by this exchange, because of their difference in taste. Now suppose someone who knows the tastes of the people of the group thinks of a "socially desirable" reallocation of all the food. He wants to realize this using a price system, in the following way. He assigns suitable prices to each type of food, all the food is put on the table, and everyone gets a personal budget to choose his own favorite lunch. This budget need not be the value of the lunch the person in question has brought to the table: the starting position has been corrected. Show that the prices and budgets can indeed be chosen in such a way that the self-interest of all individuals leads precisely to the planned socially desirable allocation, under natural assumptions.*

Solution. The situation can be modeled as follows. We consider n consumers, $1, \ldots, n$, and k goods, $1, \ldots, k$. A nonnegative vector in $\mathbb{R}^k$ is called a *consumption vector*; it represents a bundle of the k goods. The consumers have *initial endowments* $\omega_1, \ldots, \omega_n$, consumption vectors representing the initial lunches. An *allocation* is a sequence of n consumption vectors $x = (x_1, \ldots, x_n)$. An allocation is called *admissible* if

$$\sum_{i=1}^{n} x_i \leq \sum_{i=1}^{n} \omega_i,$$

that is, the allocation can be carried out, given the initial endowments. The consumers have *utility functions* $u_1, \ldots, u_n$, defined on the set of consumption vectors, representing the individual preferences of the consumers as follows: $u_i(x_i) > u_i(x_i')$ means that consumer i prefers consumption vector x_i to x_i' and $u_i(x_i) = u_i(x_i')$ means that he is indifferent. The utility functions are assumed to be strictly quasiconcave and strictly monotonic increasing in each variable. An admissible allocation is called *Pareto efficient* if there is no other admissible allocation that makes everyone better off. This is equivalent to the following more striking property, under the assumptions on the utility functions that have been made: *each allocation that makes one person better off, makes at least one person worse off.* We will not verify the equivalence, as we will only use the first definition. A *price vector* is a nonnegative row vector $p = (p_1, \ldots, p_k)$, representing a choice of prices of the k goods. A budget vector is a nonnegative row vector $b = (b_1, \ldots, b_n)$, representing a choice of budgets for all consumers. For each choice of price vector p, each choice of budget vector b, and each consumer i we consider the problem

$$u_i(x_i) \to \max, \ p \cdot x_i \leq b_i, \ x_i \leq \sum_{j=1}^n \omega_j, \ x_i \in \mathbb{R}_+^k,$$

representing the behavior of consumer i, who chooses the best consumption vector she can get for her budget. This problem has a unique point of maximum $d_i(p, b)$, to be called the *demand* of consumer i for the choice of price vector p and budget b: this follows using the Weierstrass theorem and the assumptions on the utility functions. A nonzero price vector p and a budget vector b with

$$\sum_{i=1}^n b_i = p \cdot \sum_{i=1}^n \omega_i,$$

that is, total budget equals total value of the initial endowments, are said to give a *Walrasian equilibrium* if the sequence of demands

$$d(p, b) = (d_1(p, b), \ldots, d_n(p, b))$$

is an admissible allocation, that is, if the utility maximizing behavior of all individuals does not lead to excess demand.

Theorem 8.4 First welfare theorem. *For each Walrasian equilibrium determined by price p and budget b, the resulting admissible allocation $d(p, b)$ is Pareto efficient.*

Proof. Suppose not, and let x' be an admissible allocation that makes all consumers better off. Then by the optimality property defining

ECONOMIC APPLICATIONS

$d(p, b)$ we have $p \cdot x'_i > b_i$ for all consumers i. Summing gives

$$p \cdot \sum_{i=1}^{n} x'_i > \sum_{i=1}^{n} b_i.$$

The right-hand side of this inequality equals

$$p \cdot \sum_{i=1}^{n} \omega_i$$

by the definition of Walrasian equilibrium. Thus we get a contradiction with the admissibility of the allocation x', that is, with

$$\sum_{j=1}^{n} x'_i \leq \sum_{j=1}^{n} \omega_i.$$

□

Theorem 8.5 Second welfare theorem. *Each Pareto efficient admissible allocation x^* for which all consumers hold a positive amount of each good is a Walrasian equilibrium. To be more precise, there is a nonzero price vector p which gives for the choice of budget vector*

$$b = (p \cdot x^*_1, \ldots, p \cdot x^*_n)$$

a Walrasian equilibrium with

$$d(p, b) = p \cdot x^*.$$

Proof. Let x^* be a Pareto efficient admissible allocation, that is, x^* solves the following minimization problem:

$$f(x) = \max_{i=1,\ldots,n} (u_i(x^*_i) - u_i(x_i)) \to \min, \quad \sum_{j=1}^{n} x_j \leq \sum_{j=1}^{n} x^*_j, \quad x \in \mathbb{R}^n_+.$$

Moreover, assume that all consumers hold a positive amount of each good. Now we apply the KKT theorem, observing that $x = 0$ is a Slater point. It follows that there is a nonnegative row vector $p = (p_1, \ldots, p_n)$, a "Lagrange multiplier," such that the Lagrange function

$$\mathcal{L}(x) = f(x) + p \cdot \sum_{j=1}^{n} (x_j - x^*_j)$$

has a minimum in $x = x^*$. In particular, the vector p cannot be zero: indeed, the function $f(x)$ does not have a minimum, as it is strictly monotonic decreasing in each variable by the assumption that all utility functions are monotonic increasing in all variables.

Now we substitute in the Lagrange function $x_j = x_j^*$ for all j except one, say i. Thus the following function of x_i,

$$-(u_i(x_i) - u_i(x_i^*))_+ + p \cdot (x_i - x_i^*),$$

is minimal in $x_i = x_i^*$. Here we write $t_+ = \max(t, 0)$ for all $t \in \mathbb{R}$. Now we apply again the KKT theorem. It follows that $x_i = x_i^*$ is a point of minimum of the problem

$$-(u_i(x_i) - u_i(x_i^*))^+ \to \min, \ p \cdot x_1 \leq p \cdot x_i^*.$$

This can be reformulated as follows: $x_i = x_i^*$ is a point of maximum of the problem

$$u_i(x_i) \to \max, \ p \cdot x_1 \leq p \cdot x_i^*.$$

Moreover

$$x_i^* \leq \sum_{j=1}^n \omega_j$$

and so the price vector p gives for the choice of budget vector

$$b = (p \cdot x_1^*, \ldots, p \cdot x_n^*)$$

a Walrasian equilibrium with

$$d(p, b) = p \cdot x^*,$$

as required. □

Chapter Nine

Mathematical Applications

Die hohe Bedeutung bestimmter Probleme für den Fortschritt der mathematischen Wissenschaft im algemeinen und die wichtige Rolle, die sie bei der Arbeit des einzelnen Forschers spielen, ist unleugbar. [...] Durch die Lösung von Problemen stält sich die Kraft des Forschers; er findet neue Methoden und Ausblicke, er gewinnt einen weiteren und freieren Horizont.

The deep significance of concrete problems for the advance of mathematical science and the important role that they play in the work of the individual investigator are not to be denied. [...] By solving problems the force of the investigator is steeled; he finds new methods and views, he gains a wider and freeer horizon.

D. Hilbert

In this chapter, we solve problems posed in different times applying the main methods of the general theory, instead of using the special methods of the original solutions.

9.1 FUN AND THE QUEST FOR THE ESSENCE

The next problem is included just for fun. All other problems considered in this chapter will be more: attempts to reach the essence of some phenomenon.

Problem 9.1.1 *Which number is bigger, e^π or π^e?*

Solution. These two numbers are almost equal. It is not hard to guess that e^π is slightly bigger than π^e, by using a calculator. One of the examples of MatLab gives further evidence: it treats this problem by way of a visualization of the graph of the function $x^y - y^x$. However, the real challenge is to give a *proof*. How can optimization methods be of help? Well, taking the natural logarithm of e^π (resp. π^e) and then dividing by the product πe gives $(\ln e)/e$ (resp. $(\ln \pi)/\pi$). Now we use the fact that taking the natural logarithm of two numbers does not change their ordering, as the function ln is monotonically increasing. This shows that to prove that $e^\pi > \pi^e$ it suffices to prove that $(\ln e)/e > (\ln \pi)/\pi$. This formulation suggests considering the function $(\ln x)/x$ (Fig. 9.1):

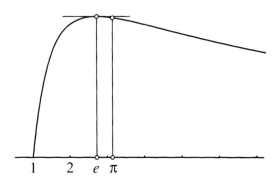

Figure 9.1 The graph of the function $\frac{\ln x}{x}$.

1. $f(x) = (\ln x)/x \to \max$, $x > 0$. Existence of a solution $\widehat{x}$ follows from the Weierstrass theorem, as $f(x) < 0 \;\; \forall x < 1$, $f(1) = 0$, $f(x) > 0 \;\; \forall x > 1$, and $\lim_{x \to +\infty} f(x) = 0$.

2. Fermat: $f'(x) = 0 \Rightarrow (1 - \ln x)/x^2 = 0$.

3. $x = e$, $f(e) = 1/e$.

4. $\widehat{x} = e$, and so $f(e) > f(\pi) \Rightarrow e^\pi > \pi^e$.

9.2 OPTIMIZATION APPROACH TO MATRICES

In this section, we illustrate by three more examples a nonstandard way of proving basic results of matrix theory (see section 9.5 for two other examples). The traditional way to prove results on matrices is by means of the Gaussian elimination algorithm. In view of this, you might find it refreshing to see an alternative way: by optimization methods. Note that most results of matrix theory seem to have nothing to do with optimization.

A symmetric $n \times n$ matrix A is called *positive definite* if $x^T A x > 0$ for all nonzero vectors $x \in \mathbb{R}^n$.

Problem 9.2.1 *Show that if A is a (positive definite) symmetric $n \times n$ matrix, then there exists a (positive) number c such that $x^T A x \geq c|x|^2$ for all $x \in \mathbb{R}^n$.*

Solution. Consider the problem
$$f_0(x) = x^T A x \to \min, \quad f_1(x) = |x| = 1.$$
Existence of a global solution $\widehat{x}$ follows from the Weierstrass theorem, as the solution set of $|x| = 1$ is a sphere, and so it is nonempty and compact (closed and bounded). For each nonzero $x \in \mathbb{R}^n$ the vector $|x|^{-1} x$ is admissible $\Rightarrow |x|^{-2} x^T A x \geq \widehat{x}^T A \widehat{x} = c$.

An $n \times n$ matrix A is said to be *invertible* if there exists for each $y \in \mathbb{R}^n$ a unique vector $x \in \mathbb{R}^n$ with $Ax = y$. Then there is a unique $n \times n$ matrix A^{-1} such that $A^{-1} y = x$ precisely if $Ax = y$. The matrix A^{-1} is called the *inverse* of A. One has $AA^{-1} = A^{-1} A = I_n$.

Problem 9.2.2 *Show that if A is a positive definite symmetric $n \times n$ matrix, then A is invertible.*

Solution. Choose an arbitrary $y \in \mathbb{R}^n$. To prove the existence and uniqueness of a solution of the vector equation $Ax = y$, we consider the following problem:

1. $f_0(x) = x^T A x - 2 x^T y \to \min$. A solution $\widehat{x}$ exists as f is coercive
$$(f_0(x) \stackrel{2.2.2}{\geq} c|x|^2 - 2|x||y| \to +\infty \text{ as } |x| \to +\infty).$$

2. Fermat: $f_0'(x) = 0_n^T \Rightarrow 2x^T A - 2y^T = 0_n^T$.

3. $Ax = y$. It remains to prove uniqueness:
$$Ax_1 = Ax_2 = y \Rightarrow A(x_1 - x_2) = 0_n \Rightarrow (x_1 - x_2)^T A(x_1 - x_2) = 0.$$
This implies $x_1 = x_2$, as A is positive definite.

4. For each $y \in \mathbb{R}^n$ there exists a unique vector $x \in \mathbb{R}^n$ with $Ax = y$.

Problem 9.2.3 *Certificate of redundancy for an equation from a linear system.* A friend of yours wants to know whether, from a given system of homogeneous linear equations, the first one is redundant. This means that it is satisfied by all the solutions of the system consisting of the other equations. However, she does not know how to solve a system of linear equations. She asks you to find out whether the equation is redundant, and then to convince her that your answer is correct. How to do this in an efficient way?

Solution. If it turns out that the equation is not redundant, then give her an example of a vector that is not a solution of the first equation, but which is a solution of all the other equations. If it turns out that the equation is redundant, then write the first equation as a linear combination of the other equations. This linear combination is a *certificate of the redundancy of the equation*, which your friend will certainly accept.

Why does there always exist such a certificate if the equation is redundant? One possible way of showing this would be by an analysis of Gaussian elimination, the usual algorithm to solve a system of linear equations. Here we choose another possibility: we will use the multiplier rule.

Let us write the system as

$$c_i^T x = 0, \ 0 \leq i \leq m,$$

with $c_i \in \mathbb{R}^n$, $0 \leq i \leq m$, given. Assume that the equation $c_0^T x = 0$ is redundant. We want to prove that c_0 is a linear combination of the vectors c_i, $1 \leq i \leq m$. We assume without restricting the generality of the argument that the c_i, $1 \leq i \leq m$, are linearly independent (we can achieve this by going over to a maximal linearly independent subset).

1. We consider the problem

$$f_0(x) = c_0^T x \to \min, \ c_i^T x = 0, \ 1 \leq i \leq m.$$

The value of the problem is 0 by the redundancy assumption. Therefore, the vector $\hat{x} = 0_n$ is a point of global solution.

2. Lagrange function

$$\mathcal{L}(x, \lambda) = \sum_{i=0}^{m} \lambda_i c_i^T \cdot x.$$

Lagrange: $\mathcal{L}_x(0_n, \lambda) = 0_n^T \Rightarrow \sum_{i=0}^{m} \lambda_i c_i^T = 0_n^T$.

3. We put $\lambda_0 = 1$, as we may by the linear independency assumption. Therefore, $c_0 = -\sum_{i=1}^{m} \lambda_i c_i$.

4. The first equation can be written as a linear combination of the other equations.

Remark. The main point of this problem is that it gives an interpretation of the multiplier rule for the case that all functions of the problem are affine. Note that the fact from linear algebra that we have derived above from the multiplier rule was used in the proof of the multiplier rule.

Exercise 9.2.1 *Certificate of nonsolvability for a system of linear equations.* *Consider a system of m linear equations in n variables, written in matrix form as $Ax = b$. Show that if this system has no solutions, then the row vector $(0, \ldots, 0, 1) \in (\mathbb{R}^{n+1})^T$ can be written as a linear combination of the rows of the extended matrix $(A|b)$ (the matrix obtained by adding the column b to A at the end).*

9.3 HOW TO PROVE RESULTS ON LINEAR INEQUALITIES

Plan. There is a well-developed theory of systems of linear inequalities. There are many practical applications of this theory. Here we want to make the point that there is an "algorithm" that produces proofs for the results of this theory. We illustrate this algorithm for Farkas's lemma. This result is useful in situations where it is necessary to show that a system of linear inequalities has no solution. Farkas's lemma helps to reduce this task of showing nonexistence to the easier task of showing the existence of a solution for another system of linear inequalities. Some applications of Farkas's lemma are presented in sections 8.6, 8.7, and 8.8.

The idea of the "algorithm" is to write down the main theorem on convex cones K for a suitable cone. We begin by stating this theorem. A *ray* in $\mathbb{R}^n$ is a set of the form $R_x = \{\alpha x : \alpha \in \mathbb{R}\}$ for some nonzero $x \in \mathbb{R}^n$. A convex set in $\mathbb{R}^n$ that is a union of rays is called a *convex cone*. For each convex cone $K \subseteq \mathbb{R}^n$, the *dual cone* is defined by

$$K^* = \{y \in \mathbb{R}^n : y^T \cdot x \geq 0 \ \forall x \in K\}.$$

Theorem 9.1 *Main theorem on convex cones.* *Let K be a convex cone in $\mathbb{R}^n$. Then $K \neq \mathbb{R}^n$ precisely if $K^* \neq 0$.*

This result is essentially equivalent to the separation theorem for convex sets and to the supporting hyperplane theorem, in the sense that one can derive each of these results easily from each of the other ones.

Theorem 9.2 *Farkas's lemma.* Let A be an $m \times n$-matrix, $c \in \mathbb{R}^n$. The system of inequalities

$$Ax \leq 0_m, \quad c^T \cdot x < 0,$$

and the system of inequalities

$$A^T y + c = 0_n, \quad y \geq 0_m,$$

are strong alternatives. That is, precisely one of the two is solvable.

Proof. We begin by rewriting the second alternative in the form

$$K^* \neq 0$$

for a suitable convex cone K. We choose a sufficiently small $\varepsilon > 0$.

$$\begin{pmatrix} y \\ y_0 \end{pmatrix}^T \cdot \begin{pmatrix} Ax \\ c^T \cdot x \end{pmatrix} = 0 \quad \forall x \in \mathbb{R}^n,$$

$$\begin{pmatrix} y \\ y_0 \end{pmatrix}^T \cdot z \geq 0 \quad \forall z \in \mathbb{R}^{m+1}_+,$$

$$\begin{pmatrix} y \\ y_0 \end{pmatrix}^T \cdot u \geq 0,$$

for all $u \in \mathbb{R}^{m+1}$ that make an angle with

$$\begin{pmatrix} 0_m \\ 1 \end{pmatrix}$$

that is not larger than ε. To see that this is equivalent to the second alternative, note that the last condition is equivalent to the positivity of y_0. Then one can dehomogenize the system of linear inequalities, taking $y_0 = 1$. This gives precisely the system of the second alternative. Now we choose K to be the convex cone spanned by all vectors

MATHEMATICAL APPLICATIONS

$\begin{pmatrix} Ax \\ c^T \cdot x \end{pmatrix}$ with $x \in \mathbb{R}^n$, all vectors from $\mathbb{R}^{m+1}_+$, and all vectors $u \in \mathbb{R}^{m+1}$ that make an angle with

$$\begin{pmatrix} 0_m \\ 1 \end{pmatrix}$$

that is not larger than ε. To prove Farkas's lemma, it remains to write out the condition $K \neq \mathbb{R}^{m+1}$, that is, as

$$v = \begin{pmatrix} 0_m \\ 1 \end{pmatrix}$$

is an interior point of K, the condition that $-v$ does not belong to K. This gives the condition that the system

$$Ax \leq 0_m, \quad c^T \cdot x \leq -1$$

has a solution. This is seen to be equivalent to the first alternative. $\square$

9.4 THE PROBLEM OF APOLLONIUS

9.4.1 The problem of Apollonius: solution

We will solve the problem of Apollonius, which was formulated in section 3.3.8. Let a point P and an ellipse $\mathcal{E}$ in the two-dimensional plane be given. How many normals can one draw from P to $\mathcal{E}$? Equivalently, how many stationary points has the problem to find the point on $\mathcal{E}$ that is closest to P?

Solution

1. The problem to find the point on $\mathcal{E}$ which is closest to P can be modeled as follows:

$$f_0(x) = (x_1 - \xi_1)^2 + (x_2 - \xi_2)^2 \to \min,$$

$$f_1(x) = \left(\frac{x_1}{a_1}\right)^2 + \left(\frac{x_2}{a_2}\right)^2 - 1 = 0.$$

A solution $\widehat{x}$ exists by Weierstrass.

2. Lagrange function

$$\mathcal{L} = \lambda_0((x_1 - \xi_1)^2 + (x_2 - \xi_1)^2) + \lambda_1 \left(\left(\frac{x_1}{a_1}\right)^2 + \left(\frac{x_2}{a_2}\right)^2 - 1\right).$$

Lagrange: $\mathcal{L}_x = 0_2^T \Rightarrow (\lambda_0 = 1, \lambda_1 = \lambda)$,

$$\frac{\partial \mathcal{L}}{\partial x_j} = 2(x_j - \xi_j) + 2\lambda \frac{x_j}{a_j^2} = 0, \ j = 1, 2. \qquad (*)$$

Let us point out the geometric meaning of these equations: the difference vector $\widehat{x} - \xi$ of a stationary point $\widehat{x}$ and the given point ξ is proportional to the gradient of of f_1 at $\widehat{x}$, that is, the line through ξ and $\widehat{x}$ is a normal to the ellipse, in agreement with the discussion above.

3. One has

$$x_j \stackrel{(*)}{=} \frac{\xi_j a_j^2}{a_j^2 + \lambda}, \ j = 1, 2 \Rightarrow \frac{\xi_1^2 a_1^2}{(a_1^2 + \lambda)^2} + \frac{\xi_2^2 a_2^2}{(a_2^2 + \lambda)^2} = 1.$$

The graph of the left-hand side, to be called $\varphi(\lambda)$, is given in Figure 9.2. It shows that this equation has 2 (resp. 3, resp. 4)

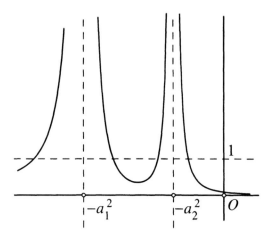

Figure 9.2 Analysis of the problem of Apollonius: the graph of φ.

solutions if the "locally lowest point" R on this graph lies above (resp. on, resp. below) the horizontal line at height 1, that is, if $\varphi(\widehat{\lambda})$ is larger than (resp. equal to, resp. smaller than) 1, where $\widehat{\lambda}$ is the point of local minimum of the function φ. One can find $\widehat{\lambda}$ as the solution of $\varphi'(\lambda) = 0$, using the Fermat theorem. We will now work out the "dividing case" that there are three stationary points: this is the case if $\widehat{\lambda}$ satisfies the equation $\varphi(\lambda) = 1$ as well. Therefore, to find all points ξ for which there are three normals, we have to eliminate λ from the relations

$$\varphi(\lambda) = \frac{\xi_1^2 a_1^2}{(a_1^2 + \lambda)^2} + \frac{\xi_2^2 a_2^2}{(a_2^2 + \lambda)^2} = 1,$$

$$\varphi'(\lambda) = -\frac{\xi_1^2 a_1^2}{(a_1^2 + \lambda)^3} - \frac{\xi_2^2 a_2^2}{(a_2^2 + \lambda)^3} = 0.$$

This elimination is straightforward and is displayed below. To begin with, the second of these relations can be written as an equality of proportions

$$(a_1^2 + \lambda) : (\xi_1 a_1)^{\frac{2}{3}} = (a_2^2 + \lambda) : -(\xi_2 a_2)^{\frac{2}{3}}.$$

Writing A for the constant of proportionality we get

$$a_1^2 + \lambda = A(\xi_1 a_1)^{\frac{2}{3}}, \ a_2^2 + \lambda = -A(\xi_2 a_2)^{\frac{2}{3}}.$$

Subtracting these two relations leads to the following expression for A:

$$A = \frac{a_1^2 - a_2^2}{(\xi_1 a_1)^{\frac{2}{3}} + (\xi_2 a_2)^{\frac{2}{3}}}.$$

Substituting the expressions for $a_1^2 + \lambda$, $a_2^2 + \lambda$ in the equation $\varphi(\lambda) = 1$ gives

$$\frac{\xi_1^2 a_1^2}{A^2 (\xi_1 a_1)^{\frac{4}{3}}} + \frac{\xi_2^2 a_2^2}{A^2 (\xi_2 a_2)^{\frac{4}{3}}} = 1.$$

Simplifying this equation, bringing A to the right-hand side, and substituting the expression for A above gives

$$(\xi_1 a_1)^{\frac{2}{3}} + (\xi_2 a_2)^{\frac{2}{3}} = \frac{(a_1^2 - a_2^2)^2}{((\xi_1 a_1)^{\frac{2}{3}} + (\xi_2 a_2)^{\frac{2}{3}})^2}.$$

Simplifying this equation gives

$$(\xi_1 a_1)^{\frac{2}{3}} + (\xi_2 a_2)^{\frac{2}{3}} = (a_1^2 - a_2^2)^{\frac{2}{3}}.$$

This is the equation of the astroid. These calculations show that $\varphi(\lambda) = 1$ precisely if the point ξ lies on the astroid. In the same way it is shown that $\varphi(\lambda) > 1$ (resp. $\varphi(\lambda) < 1$) precisely if the point lies outside (resp. inside) the astroid. We have been careless at one point: we have assumed in the derivation above that ξ_1 and ξ_2 are nonzero. That is, we have excluded the four vertices of the astroid. Well, it can be seen that for these points ξ there are two normals; we omit this verification.

4. Outside the astroid

$$(\xi_1 a_1)^{\frac{2}{3}} + (\xi_2 a_2)^{\frac{2}{3}} = (a_1^2 - a_2^2)^{\frac{2}{3}}$$

each point has two normals, inside it four, and on the astroid itself three (except at the vertices, where there are two normals).

Remark. We sketch a numerical solution of the problem. One has to solve the equation $\varphi(\lambda) = 1$ numerically. Then one determines for each of these two, three, or four solutions the corresponding points $x(\lambda)$, using the stationarity equations above. These points have to be substituted in the objective function, and finally a comparison of values shows what is the global minimum.

This analysis of the problem of Apollonius leaves one thing to be desired. Which of the λ's corresponding to stationary points ("normals") corresponds to the global minimum ("distance point to ellipse")? We will sketch how the KKT theorem gives an answer to this question in the case that the given point lies *outside the ellipse*. This will illustrate the small advantage of the KKT conditions over the Lagrange equations: the nonnegativity conditions for the Lagrange multipliers.

Problem 9.4.1 *Determine the distance to an ellipse from a point outside the ellipse.*

Solution. We already have modeled this problem and determined the stationary points. This is the solution to the previous problem. It remains to single out the solution(s) among the stationary points. We recall that there are either two, three, or four stationary points. All three possibilities can occur: this follows from the "astroid-rule" and the observation that the astroid is sticking out of the ellipse. Note that we may add the points inside the ellipse to the admissible set. This is the case, as this gives no possibilities to come closer to the given point outside the ellipse. Then we can apply the KKT theorem. We recall that this gives almost the same necessary conditions as the Lagrange multiplier rule. There is a small difference: the KKT conditions give also the condition $\lambda \geq 0$. This small difference is crucial here. This condition excludes all but one of the stationary points, as we will see. We will now give the details.

1. We consider the problem

$$f_0(x) \to \min, \quad f_1(x) \leq 0,$$

with the same f_0 and f_1 as before.

2. As before, but now we add the condition $\lambda \geq 0$.

3. We get as before the equation

$$\frac{\xi_1^2 a_1^2}{(a_1^2 + \lambda)^2} + \frac{\xi_2^2 a_2^2}{(a_2^2 + \lambda)^2} = 1$$

in λ. It remains to note that the left-hand side of the equation has value more than 1 at $\lambda = 0$ (as ξ lies outside the ellipse), and that it is monotonically decreasing for $\lambda \geq 0$ with limit zero. It follows that the equation has a unique positive root.

4. The unique stationary point $(\widehat{x}_1, \widehat{x}_2)$ for which the Lagrange multiplier is positive is the solution of our problem.

9.4.2 The perfect pendulum of Huygens and other variants of Apollonius's rule

Having seen the problem of Apollonius on the number of normals to an ellipse from a point, you might have a taste for considering the same problem for other types of curves. Apollonius himself did this for a parabola and a hyperbola, and much later Christiaan Huygens considered a type of curve called a *cycloid*. We give their motivations below. In all cases, an analysis with the Lagrange multiplier method leads to a rule for the number of normals in terms of the position of the point relative to some intriguing auxiliary curve. We will not display the details. Instead we ask a question. What is behind this phenomenon of auxiliary curves? Is there a more intuitive method to pass from the ellipse directly to the astroid and can this be extended to other curves? The answer is the construction of the *involute*.

Involute of a curve. We define the *involute* of an arbitrary but "well-behaved" curve Γ in the plane. We assume, to begin with, that the curve is smooth in the sense that at each of its points P the tangent is well defined. Then the line through P orthogonal to its tangent is called the *normal* at P. Take an arbitrary point P of Γ and

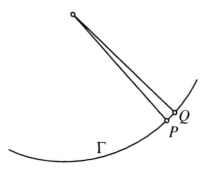

Figure 9.3 The intersection of normals.

fix it; choose a variable point Q on Γ close to P. Consider the point of intersection of the normals to Γ at the points P and Q (Fig. 9.3).

Assume that this point tends to a limit point $S(P)$ if Q tends to P. The curve consisting of the points $S(P)$, where P runs over Γ, is called the *involute* of Γ (Fig. 9.4).

We will use the following properties of the involute: the line through the points P and $S(P)$ is the tangent at $S(P)$ to the involute and the length of the piece of the involute between two points $S(P_1)$ and $S(P_2)$ is equal to the difference of the distance between P_1 and $S(P_1)$ and the distance between P_2 and $S(P_2)$. If you have a parametric description

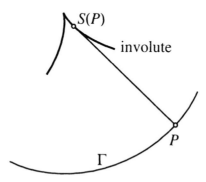

Figure 9.4 The involute.

of a curve, then you can get a parametric description of its involute, as the following exercise shows. We write $D(v, w)$ for the determinant of the 2×2 matrix with rows $v, w \in (\mathbb{R}^2)^T$ and Jv for the vector in the plane $\mathbb{R}^2$ obtained from the vector $v \in \mathbb{R}^2$ by turning it counterclockwise over a right angle, that is, $J(v_1, v_2) = (-v_2, v_1)$.

Exercise 9.4.1 *For a curve in the plane, given in parametric form by $P(b)$, $b \in \mathbb{R}$, its involute has the parametric representation*

$$S(P(b)) = P(b) - D(P''(b), P'(b))^{-1}|P'(b)|^2 JP'(b),$$

provided this expression is defined.

The first thing to do now is to check whether the involute of an ellipse is an astroid.

Exercise 9.4.2 Involute of an ellipse. *Show that the involute of the ellipse*

$$\left(\frac{x_1}{a_1}\right)^2 + \left(\frac{x_2}{a_2}\right)^2 = 1$$

is the astroid

$$(\xi_1 a_1)^{\frac{2}{3}} + (\xi_2 a_2)^{\frac{2}{3}} = (a_1^2 - a_2^2)^{\frac{2}{3}}.$$

To be more precise, show that

$$S(P) = \left(\left(\frac{a_1^2 - a_2^2}{a_1^4}\right) p_1^3, -\left(\frac{a_1^2 - a_2^2}{a_2^4}\right) p_2^3\right)$$

for each point $P = (p_1, p_2)$ of the ellipse.

Hint. The ellipse can be parametrized as follows:

$$P(b) = (a_1 \cos b, a_2 \sin b),\ 0 \leq b < 2\pi.$$

Involutes and the number of normals. We now give the geometric intuition for the role of the involute in the shortest distance from point to curve problem. To be more precise, we will make the following rough version of the rule plausible: the involute is the separating boundary between the set of points in the plane P for which the number of stationary points is at least two and the set for which the number is one. Note that this first set is the set of intersections of two normals, by definition. Therefore, it is plausible that the of this set is the involute: by the definition of this involute of limit-points of this set.

Apollonius's problem for conics. Apollonius solved his problem for all *conics*, as these were the favorite objects of study of the Greek geometers of antiquity. From their point of view, it would be unthinkable to consider a problem for ellipses only and not more generally for conics. A conic or conic section is defined geometrically to be the intersection of a plane in three-dimensional space with the surface of revolution with respect to the vertical axis of a line in the plane which makes an angle of 45° with this axis ("the boundary of a double ice cream cone") (Fig. 9.5). Alternatively, it can be defined geometrically

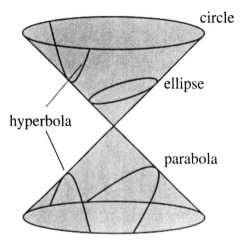

Figure 9.5 Conic sections.

as the set of solutions, in the plane, of an equation $f(x) = 0$, where f is a polynomial of degree two of the variable vector $x = (x_1, x_2)^T \in \mathbb{R}^2$, that is,

$$f(x) = x^T A x + a \cdot x + \alpha,$$

where A is a nonzero symmetric 2×2 matrix, a is a vector in $(\mathbb{R}^2)^T$, and α is a number. The conic sections can be classified into four groups: ellipses (for example, circles), parabolas, hyperbolas, and pairs of lines through one point. A parabola (resp. hyperbola)—in standard form—is the set of solutions of an equation $x_2 = ax_1^2$ with $a \neq 0$ (resp. $(\frac{x_1}{a_1})^2 - (\frac{x_2}{a_2})^2 = 1$ with $a_1, a_2 \neq 0$) for a suitable choice of coordinate axes (Fig. 9.6).

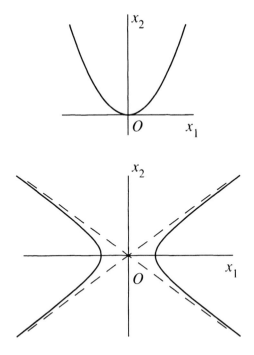

Figure 9.6 A parabola and a hyperbola.

Exercise 9.4.3 Involute of a parabola. *Show that the involute of the parabola* $x_2 = x_1^2$ *is the semicubic parabola—also called parabola of Neil—*

$$9\xi_1^2 = 16\left(\xi_2 - \frac{1}{2}\right)^3.$$

To be more precise, show that

$$S(P) = (-4p_1^3, 3p_2 + \frac{1}{2})^T$$

for each point P of the parabola.

Exercise 9.4.4 Involute of a hyperbola. *Show that the involute of the hyperbola*

$$\left(\frac{x_1}{a_1}\right)^2 - \left(\frac{x_2}{a_2}\right)^2 = 1$$

is the curve

$$(a_1\xi_1)^{\frac{2}{3}} - (a_2\xi_2)^{\frac{2}{3}} = (a_1^2 + a_2^2)^{\frac{2}{3}}.$$

To be more precise, show that

$$S(P) = \left(\frac{a_1^2 + a_2^2}{a_1^4} p_1^3, -\frac{a_1^2 + a_2^2}{a_2^4} p_2^3 \right)$$

for each point P of the hyperbola.

Hint. The hyperbola can be parametrized by means of the *hyperbolic cosine and sine functions*

$$\cosh b = \frac{1}{2}(e^b + e^{-b}) \text{ and } \sinh b = \frac{1}{2}(e^b - e^{-b})$$

in the following way:

$$P(b) = (a_1 \cosh b, a_2 \sinh b), \ b \in \mathbb{R}.$$

The pendulum clock of Huygens. We now turn to the finest application of the involute: to pendulum clocks. Christiaan Huygens (1629–1695) was a leading astronomer, mathematician, and physicist. He built one of the first telescopes and used it to discover the ring—and some of the moons—of Saturn. With his theory that light is a wave he managed to explain the puzzling phenomenon of interference. He is probably most famous for his invention of the pendulum clock. In 1657 he obtained for this clock a patent from the Dutch States. However, Huygens, who has been called the greatest clockmaker of all time, was not satisfied with his invention. Here is why. An important basic fact on which its working was based is the following: if you let a pendulum swing, then the time it takes to swing back to the original position—"a full period"—depends hardly at all on how far you have pulled out the pendulum—"the amplitude." However, it depends a bit on it, and this unfortunately leads to irregularities in the clocks. Therefore, Huygens asked whether there is a perfect pendulum. This should be constructed in such a way that the lower tip of the pendulum would swing on a curve which has the following property. If we let the tip loose at a point of this curve and let it swing following this curve, then it swings back to the original position in a time that does not depend at all on the initial position.

Isochrone. That is, to begin with, Huygens had to solve the problem of finding a curved slide on which a frictionless point mass would always move from any initial point to the lowest point of the curve

MATHEMATICAL APPLICATIONS

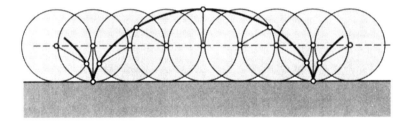

Figure 9.7 The cycloid.

in the same amount of time, regardless how high the initial point is chosen. Such a curve is called an *isochrone* (Latin for "equal time"). Huygens solved this problem: he found that the *cycloid* has this property. This curve can be described as follows: let a wheel roll on the ground and follow the curved path taken by one point of the boundary of the wheel. This is a cycloid (Fig. 9.7). Then turn the curve upside down. This is the isochrone.

Enter the involute. It remained to find a construction that would lead the tip of the pendulum to follow the cycloid without friction. Huygens had the beautiful insight that he could use the involute of the cycloid. He used a flexible pendulum and let it swing between two cheeks that together form the involute of the cycloid (Fig. 9.8). Then

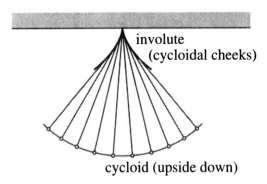

Figure 9.8 The perfect pendulum of Huygens.

the tip of the pendulum followed the cycloid: this is an immediate consequence of the definition of the involute. What is the involute of a cycloid? Huygens found that this is again a cycloid. He published his results in his 1673 treatise *Horologium Oscillatorium* ("the Pendulum Clock")[31] and built many clocks using this device.

The next exercise contains the second of the two mathematical results that Huygens discovered in order to make the perfect pendulum.

Exercise 9.4.5 *Theorem of Huygens on the involute of a cycloid.* Show that the involute of a cycloid is again a cycloid.

Hint. The definition of a cycloid as a superposition of two movements (circular and horizontal linear) shows that the following parametric representation is a cycloid:

$$P(b) = (\cos b - b, -\sin b), \ 0 \leq b < 2\pi.$$

A calculation shows that

$$S(P(b)) = (-\cos b - b, \sin b - 2).$$

This is also a parametric representation of a cycloid.

The Horlogium Oscillatorium is a brilliant book. It is Huygens's masterpiece and is much more than just a book about clocks. It is generally considered to be one of the most important science books ever published, ranking second only to the Principia of Newton. It stands as the foundation of the dynamics of bodies in motion, and contains a theory on the mathematics of curvature.

Apollonius's problem for ellipsoids. Finally, we generalize the problem of Apollonius from ellipses to ellipsoids. An ellipsoid in n-dimensional space $\mathbb{R}^n$ is defined to be the image of the n-dimensional unit sphere under an invertible linear transformation of $\mathbb{R}^n$. Thus ellipsoids are the n-dimensional generalization of ellipses. One can view an ellipsoid as the set of solutions of an equation of type

$$\left(\frac{x_1}{a_1}\right)^2 + \cdots + \left(\frac{x_n}{a_n}\right)^2 = 1,$$

by making a suitable choice of orthogonal axes.

Exercise 9.4.6 *Generalized problem of Apollonius.* Pose the problem to determine in n-dimensional space the distance between a point P and an ellipsoid. What is the geometric meaning of the information given by the Lagrange multiplier method? Characterize the normals to an ellipsoid running through the point P in terms of this optimization problem.

9.5 MINIMIZATION OF A QUADRATIC FUNCTION: SYLVESTER'S CRITERION AND GRAM'S FORMULA

As a corollary of the result of problem 2.4.1 we will deduce Sylvester's criterion for positive definiteness of a symmetric matrix. Let a symmetric $n \times n$ matrix $A = A_n = (a_{ij})_{i,j}^n$ be given. Determinants

$$\det A_k, \ 1 \leq k \leq n, \text{ where } A_k = (a_{ij})_{i,j=1}^k,$$

are called *leading minors* of the matrix A. Denote $Q_k(x) = x^T A_k x$ for all $x \in \mathbb{R}^k$.

Theorem 9.3 Sylvester's criterion. *A symmetric matrix is a positive definite matrix if and only if all its leading minors are positive.*

Proof. We proceed by induction. For $n = 1$, the statement is trivial. Let A_n be a symmetric $n \times n$ matrix for which A_{n-1} is positive definite. To prove the theorem, it suffices to show that A_n is positive definite if and only if $\det A_n > 0$.

We consider the following problem:

$$f_0(y) = Q_n(y, 1) = y^T A_{n-1} y + 2 a_{n-1} \cdot y + a_{nn} \to \min,$$

where $a_{n-1} = (a_{n,1}, \ldots, a_{n,n-1})$. This is a special case of problem 2.4.1. We get that this problem has a unique solution, $\widehat{y} = -A_{n-1}^{-1} a_{n-1}$, and optimal value

$$a_{n,n} - a_{n-1} A_{n-1}^{-1} a_{n-1}^T. \qquad (*)$$

Developing the determinant $\det A_n$ with respect to the last column, we obtain

$$\det A_n = a_{nn} \det A_{n-1} + \sum_{i=1}^{n-1} a_{in} \Delta_i.$$

Developing now all determinants Δ_i with respect to the last row (and using the definition of inverse matrix), we have:

$$\det A_n = \det A_{n-1} (a_{nn} - a_{n-1} A_{n-1}^{-1} a_{n-1}^T). \qquad (**)$$

Therefore, it follows—using $(*)$—that A_n is positive definite if and only if $\det A_n > 0$. $\square$

Now we derive from theorem 9.3 the formula for the distance from a point to a linear subspace in $\mathbb{R}^n$. The special case that the linear subspace is a line was considered in example 1.3.7, in the comments following this example, and the general case was considered, from a different point of view, in problem 2.4.7

Let $\{x^j\}_{j=1}^k$ be a linearly independent system of vectors in $\mathbb{R}^n$. The determinant
$$G(x^1, \ldots, x^k) = ((x^i)^T \cdot x^j)_{i,j=1}^k$$
is called *Gram's determinant*. We denote the distance from a vector x_0 to the space
$$L_k = \operatorname{span}\{x^1, \ldots, x^k\}$$
as
$$d = d(x^0, L_k, \mathbb{R}^n) = \min_{x \in L_k} |x_0 - x|.$$
The problem of the calculation of this distance is equivalent to a minimization problem for a quadratic function of several variables (problem 2.4.1), as
$$|x_0 - x|^2 = x^T x - 2x_0^T x + x_0^T x_0.$$

Theorem 9.4 *Gram's formula.* The following formula holds true:
$$d^2 = \frac{G(x^0, x^1, \ldots, x^k)}{G(x^1, \ldots, x^k)}$$

Note that this extends example 1.3.7.

Proof. Consider the extremal problem
$$f_0(y) = \left| x^0 + \sum_{i=1}^k y_i x^i \right|^2 \to \min, \quad y \in \mathbb{R}^k.$$
It is easy to see that
$$f_0(y) = y^T A_k y + 2a_k \cdot y + b,$$
where
$$A_k = ((x^i)^T \cdot x^j)_{i,j=1}^k, \quad a_k = ((x^0)^T \cdot x^1, \ldots, (x^0)^T \cdot x^k), \quad b = (x^0)^T \cdot x^0.$$
We obtain a problem of the type considered above. It was shown in the proof of theorem 9.3 (see (**)) that
$$d^2 = \min_{y \in \mathbb{R}^k} f_0(y) = d^2 = \frac{G(x^0, x^1, \ldots, x^k)}{G(x^1, \ldots, x^k)}.$$
$\square$

9.6 POLYNOMIALS OF LEAST DEVIATION

Section 2.4.9 contains an introduction to the polynomials of least deviation. In particular, we have given formulas for $T_{np}(\cdot)$ for $p = 1, 2$, and ∞; for $p = 2$ we also have given the proof of the formula. Now we give the proofs of the formulas for $T_{np}(\cdot)$ for $p = \infty$ and $p = 1$. We recall that the polynomial of least deviation $T_{np}(\cdot)$ is the unique solution of the problem

$$f_p(x) = \int_{-1}^{1} \left| t^n + \sum_{k=1}^{n-1} x_k t^{k-1} \right|^p dt \to \min, \quad x = (x_1, \ldots, x_n) \in (\mathbb{R}^n)^T$$

if $1 \leq p < \infty$, and of the problem

$$f_\infty(x) = \max_{t \in [-1,1]} \left| t^n + \sum_{k=1}^{n} x_k t^{k-1} \right| \to \min, \quad x = (x_1, \ldots, x_n) \in (\mathbb{R}^n)^T$$

if $p = \infty$.

9.6.1 Chebyshev polynomials

We consider the case $p = \infty$.

Theorem 9.5 Chebyshev (1854). $T_{n,\infty} = 2^{-(n-1)} \cos n \arccos t$.

Proof.

1. Let us consider a new convex problem without constraints (writing $x_n(t) = t^n$):

$$f_\infty(x) = \max_{t \in [-1,1]} \left| t^n - \sum_{k=1}^{n} x_k t^{k-1} \right|$$

$$= \|x_n(\cdot) - \xi(t, x)\|_{C([-1,1])} \to \min,$$

$$x = (x_1, \ldots, x_n) \in (\mathbb{R}^n)^T.$$

It can be shown that the function $f_\infty(x)$ is coercive; therefore the solution $\hat{x}$ of this problem exists.

2. Fermat for convex problems:

$$0 \in \partial f_\infty(x) = \partial \|x_n(\cdot) - \cdot\|_{C(I)}(\xi(\cdot, x)),$$

where ∂ denotes the subdifferential, and where $I = [-1, 1]$.

3. The polynomial $T_{n\infty}(\cdot) = x_n(\cdot) - \xi(\cdot, x)$ attains its global maxima and minima at $r \leq n+1$ points $-1 \leq \tau_1 < \tau_2 < \ldots < \tau_r \leq 1$. From the F. Riesz theorem on linear functionals on C-spaces, it follows that there exist r nonzero numbers $\{\mu_k\}_{k=1}^r$ such that

$$\sum_{k=1}^{r} \mu_k \xi(\tau_k) = 0 \ \forall \xi(\cdot) \in \mathcal{P}_{n-1} \qquad (i)$$

and $\operatorname{sign}\mu_k = \operatorname{sign} T_{n\infty}(\tau_k)$.

First of all we have to prove that $r = n+1$. In fact, if we assume that $r \leq n$, then substitution of the polynomial

$$\xi_1(t) = \prod_{k=1}^{r-1}(t - \tau_k)$$

into (i) gives $0 = \mu_r \xi_1(\tau_r) \neq 0$ (because μ_r and $\xi_1(\tau_r) \neq 0$). Contradiction.

We have proved that the polynomial $T_{n\infty}(\cdot)$ attains the levels $\pm A$, where

$$A = \|T_{n\infty}(\cdot)\|_{C[-1,1]}$$

at $n+1$ points. This means that $\tau_1 = -1$, $\tau_{n+1} = 1$, and the remaining $n-1$ zeros τ_i belong to the interval $(-1, 1)$.

Denote $y_2(\cdot) = T_{n\infty}(\cdot)$. We see that the two polynomials

$$(1 - t^2)\dot{y}_2^2(t) \text{ and } A^2 - y_2^2(t)$$

(both of degree $2n$) have the same zeros. Hence, they satisfy the equation

$$(1 - t^2)\dot{y}_2^2(t) = n^2(A^2 - y_2^2(t)) \Rightarrow y_2(t) = c_2 \cos n \arccos t,$$

and the n-th degree polynomial c_2 is determined by the condition that the polynomial $y_2^{(n)}(t)$ has leading coefficient one, hence:

4. $T_{n\infty}(t) = 2^{-(n-1)} \cos n \arccos t$.

$\square$

The polynomials

$$T_n(t) = \cos n \arccos t$$

are called *Chebyshev polynomials*.

9.6.2 Chebyshev polynomials of the second type

We consider the case $p = 1$.

Theorem 9.6
$$T_{n1}(t) = \dot{T}_{n+1\infty}(t)/(n+1) = \frac{\sin((n+1)\arccos t)}{2^n(n+1)\sqrt{1-t^2}}.$$

Proof.

1. Consider the problem
$$f_1(x) = \int_{-1}^{1} \left| t^n + \sum_{k=1}^{n} x_k t^{k-1} \right| dt \to \min,$$
$$x = (x_1, \ldots, x_n) \in (\mathbb{R}^n)^T.$$

2. Fermat:
$$f_1'(\hat{x}) = 0 \Rightarrow \int_{-1}^{1} \operatorname{sgn} T_{n1}(t) t^k dt = 0, \ 0 \le k \le n-1. \qquad (ii)$$

3. Equations (ii) are necessary and sufficient conditions for the polynomial $T_{n1}(\cdot)$ to be a polynomial of the least deviation in L_1 norm. Let us check that the function
$$\operatorname{sgn}\dot{T}_{n+1\infty}(\cdot)$$
satisfies (ii). Changing variables $t = \cos\tau$ and using the equality
$$\dot{T}_{n+1\infty}(t) = \frac{(n+1)\sin(n+1)\arccos t}{\sqrt{1-t^2}},$$
we have
$$\int_{-1}^{1} \operatorname{sgn}\dot{T}_{n+1\infty}(t) t^{k-1} dt = \int_{0}^{\pi} \operatorname{sgn}\sin(n+1)\tau \cos^{k-1}\tau \sin\tau d\tau$$
$$= \frac{1}{2} \int_{-\pi}^{\pi} \operatorname{sgn}\sin(n+1)\tau \cos^{k-1}\tau \sin\tau d\tau,$$
and it is possible to show that this integral equals zero.

4. Hence
$$T_{n1}(t) = \frac{\sin(n+1)\arccos t}{2^n(n+1)\sqrt{1-t^2}}.$$

This result was published by Korkin and Zolotarev (1886), but in an implicit form it is due to Chebyshev.

9.7 BERNSTEIN INEQUALITY

Theorem 9.7 *For an arbitrary trigonometric polynomial $x(\cdot)$ of degree n the following inequality holds true:*
$$\|\dot{x}(\cdot)\|_{C([-\pi,\pi])} \leq n\|x(\cdot)\|_{C([-\pi,\pi])}.$$

Proof. We begin by solving the following problem ($\mathcal{J}_n$ is the set of trigonometric polynomials of degree $\leq n$):
$$\dot{x}(0) \to \min, \|x(\cdot)\|_{C([-\pi,\pi])} \leq 1, \ x(\cdot) \in \mathcal{J}_n. \quad (P_{9.4})$$

We rewrite $(P_{9.4})$ as a convex problem. We have
$$x(t) = x_0 + \sum_{k=1}^{n}(x_k \cos kt + x_{n=k} \sin kt).$$

1. Then problem $(P_{9.4})$ takes the following form:
$$f_0(x) = \sum_{k=1}^{n} kx_k \to \min, \ f_1(x) = \max_{t\in[-1,1]} f(t,x) \leq 1, \quad (P'_{9.4})$$

where
$$f(t,x) = \left|x_0 + \sum_{k=1}^{n}(x_k \cos kt + x_{n+k} \sin kt)\right|.$$

The set $\{x \mid f_1(x) \leq 1\}$ is closed and bounded, and so, by the Weierstrass theorem, a solution of problem $(P'_{9.4})$ exists. We denote it as $\widehat{x} = (\widehat{x}_0, \widehat{x}_1, \ldots, \widehat{x}_{2n})$, and we denote the polynomial
$$\widehat{x}_0 + \sum_{k=1}^{n}(\widehat{x}_k \cos kt + \widehat{x}_{n+k} \sin kt)$$
as $\widehat{x}(t)$.

2. KKT: in problem $(P'_{9.4})$ the Slater condition is satisfied, and so, in correspondence with the KKT theorem, there exists a number $\lambda > 0$ for which the function
$$x \to \mathcal{L}(x,1,\lambda) = f_0(x) + \lambda f_1(x)$$
attains its minimum at the point $\widehat{x}$ in the problem
$$\mathcal{L}(x,\lambda,1) \to \min$$

without constraints. Then, by the Fermat theorem for convex problems,

$$0 \in \partial \mathcal{L}(\widehat{x}, \lambda, 1).$$

3. Let $-\pi \leq \tau_1 < \tau_2 < \cdots < \tau_r < \pi$ be points where

$$|\widehat{x}(\tau)| = \|\widehat{x}(\cdot)\|_{C([-\pi,\pi])} = 1.$$

Of such points there cannot be more than $2n$ (as they are roots of the derivative of $\widehat{x}(\cdot)$). In this way, the function attains its minimum at the point $\widehat{x}$. Applying the theorem of F. Riesz, we obtain that there exist $r \leq 2n$ nonzero numbers μ_j,

$$\sum_{j=1}^{r} |\mu_j| = 1,$$

such that

$$\dot{x}(0) + \sum_{j=1}^{r} \mu_j x(\tau_j) = 0 \ \forall x(\cdot) \in \mathcal{J}_n. \qquad (iii)$$

We have proved that $r \leq 2n$. If we now assume that $r < 2n$, then we add to $\{\tau_i\}_{i=1}^{r}$ the point zero (zero was not among $\{\tau_i\}_{i=1}^{r}$; indeed, otherwise the value of the problem $(P'_{9.4})$ would be zero, which would be absurd) and moreover, a series of points to make the total number $2n$, and we consider the polynomial $\bar{x}(\cdot)$ with zeros at the points $\{\tau_j\}_{j=1}^{2n}$. Then $\dot{\bar{x}}(0) \neq 0$, indeed, otherwise the polynomial $\bar{x}(\cdot)$ would have, by Rolle's theorem, more than $2n$ zeros.

However, then, substituting $\bar{x}(\cdot)$ in the equality (iii) we come to a contradiction:

$$0 \neq \dot{\bar{x}}(0) = \sum_{j=1}^{2n} \mu_j \bar{x}(\tau_j) = 0.$$

This gives $r = 2n$, and so the polynomials $\dot{\widehat{x}}^2(\cdot)$ of degree $2n$ and $1 - \widehat{x}^2(\cdot)$ have identical zeros (double at the points $\{\tau_i\}_{i=1}^{n}$). Comparing leading coefficients, we come to the following equation (for $y(t) = \widehat{x}(t)$):

$$y'^2 = n^2(1 - y^2).$$

Integrating it, we obtain $\widehat{x}(t) = \sin nt$. In this way
$$|\dot{x}(0)| \leq n\|x(\cdot)\|_{C([-\pi,\pi])} \ \forall x(\cdot) \in \mathcal{J}_n.$$
We denote $T_\tau x(t) = x(t+\tau)$. Then
$$|\dot{x}(\tau)| = |\frac{d}{dt}T_\tau x(0)| \leq n\|(T_\tau)x\| = n\|x(\cdot)\|_{C([-\pi,\pi])}.$$
4. $\Rightarrow \|\dot{x}(\cdot)\|_{C([-\pi,\pi])} \leq n\|x(\cdot)\|_{C([-\pi,\pi])}.$

$\square$

Remark. We have proved the main identity
$$\dot{x}(0) = \sum_{i=-n}^{n-1} \mu_i x(\tau_i) \ \forall x(\cdot) \in \mathcal{T}_n$$
(τ_i are zeros of the polynomial obtained by expressing $\sin nx$ in $\sin x = t$). Substituting into this identity the polynomial $l_j(\cdot)$ such that $l_j(\tau_i) = \delta_{ij}$, we obtain the explicit expressions of μ_i and as a result we are led to the interpolation formula of M. Riesz:
$$\dot{x}(t) = \frac{4}{n}\sum_{k=-n}^{n-1} \frac{(-1)^{k-1}}{\sin \tau_k/2} x(\tau_k) \ \forall x(\cdot) \in \mathcal{T}_n, \ \tau_k = \frac{\pi k}{n} + \frac{\pi k}{2n}.$$

Chapter Ten

Mixed Smooth-Convex Problems

- What are the necessary conditions for problems that are partly smooth and partly convex?

But definite concrete problems were first conquered in their undivided complexity, singlehanded by brute force, so to speak. Only afterwards the axiomaticians came along and stated: Instead of breaking in the door with all your might and bruising your hands, you should have constructed such and such a key of skills, and by it you would have been able to open the door quite smoothly. But they can construct the key only because they are able, after the breaking in was successful, to study the lock from within and without. Before you generalize, formalize, and axiomatize, there must be mathematical substance.

H. Weyl

10.1 INTRODUCTION

How do the main theorems of the differential and convex calculus—the tangent space theorem and the supporting hyperplane theorem—cooperate to establish necessary and sufficient conditions for mixed smooth-convex problems? These are problems that are partly smooth and partly convex. The simplest example is the smooth programming problem ($P_{4.1}$), where the convexity is the mere presence of inequality constraints and the smoothness is given by differentiability assumptions. For this type of problem we stated—but did not prove—the John theorem 4.13. Another example is optimal control problems—the latter type will be discussed briefly in chapter twelve. In this relatively long chapter, we offer the insights we have obtained in our search for the essence of the cooperation of the tangent space theorem and the supporting hyperplane theorem. In particular, we will give—and prove—necessary conditions for three types of mixed smooth-convex problems of increasing generality. For the most general type, problems from smooth-convex perturbations, these conditions—to be called the principle of Fermat-Lagrange—unify all necessary conditions that are used to solve concrete optimization problems. In particular, this type of problem contains semidefinite optimization problems. The principle of Fermat-Lagrange can be extended to the infinite-dimensional case, and then the unification includes, for example, Pontryagin's maximum principle (cf. [12]).

We believe that a simplification of the presentation in this chapter of the cooperation of the smooth and convex calculus will be possible. This is in contrast with the treatment of the material in the first five chapters, where we reached the essence of the matter.

The following three types of problems will be considered in this section.

- **Smooth problems with inequality constraints.** These are problems $(P_{4.1})$ for which all functions involved are smooth enough but these functions are not necessarily convex; here the convexity of the problem is the mere presence of inequality constraints. For this type of problem, the necessary condition is the John theorem 4.13.

- **Conic smooth-convex problems.** The type of problems given above is too narrow: for example, it does not include the class of semidefinite programming problems. Therefore, we consider a wider class, conic smooth-convex problems. An advantage of this type of problem is that it allows us to model a complicated nonlinear problem as much as possible as a linear problem and then to analyze it with powerful linear methods. To be more specific, all the convexity of the problem will be modeled by a sort of inequality constraint

$$x \succeq 0_n,$$

where $\succeq$ is a *cone ordering*, to be defined below. In particular, all convex optimization problems can be modeled—and often handled—in the same way as LP problems. The most excellent example is the class of semidefinite programming problems.

- **A smooth-convex perturbation.** The possibility of giving necessary conditions of the "multiplier type" depends on the existence of meaningful perturbations of the problem. The shadow price interpretation of the multiplier rule, given in section 3.5.3, makes this clear. Therefore, it is natural to attempt to give necessary conditions for an arbitrary perturbation of an arbitrary smooth-convex problem. This is a problem that is embedded in a family of problems $(P_y)_{y \in \mathbb{R}^m}$, where the "variable of optimization" is of the type (x, u), with $x \in \mathbb{R}^n$ and $u \in \mathcal{U}$, where $\mathcal{U}$ is some set. We call such a family a "perturbation," as it formalizes the intuitive notion "perturbation of an optimization problem." The success of this attempt is as follows. We will formulate a simple "candidate necessary condition"—to be called the condition of Fermat-Lagrange—without making any assumptions on the family. These conditions are, to begin with, of value for a heuristic analysis of a concrete problem: you will "always" find that the solution of the problem is contained among the solutions of this condition. This condition can be *proved* to be necessary for optimality, under suitable assumptions of smoothness with respect to x and of convexity with respect to u. It unifies all first order conditions that are used to solve concrete problems. We expect that perturbations and the condition of Fermat-Lagrange mark a natural boundary of how far multiplier-type necessary conditions for smooth-convex problems can be pushed.

10.2 CONSTRAINTS GIVEN BY INCLUSION IN A CONE

Plan. The aim of this section is to explain what is a cone ordering $\succeq$ on $\mathbb{R}^n$; our interest in this concept is that it allows us to introduce a new type of constraint, $x \succeq 0_n$. This constraint will be used in the definition of the second of our three types of smooth-convex problems: for this problem, all its convexity will be modeled as a constraint $x \succeq 0_n$.

Conditions on cones. The most convenient way to define cone orderings is by means of *closed, pointed, solid, convex cones*. A *cone* in $\mathbb{R}^n$ is a nonempty set in $\mathbb{R}^n$ that is the union of rays. A *ray* is defined as the half-line consisting of all nonnegative scalar multiples of one nonzero vector. A convex cone is a cone that is a convex set. Alternatively, a convex cone $C \subseteq \mathbb{R}^n$ can be characterized as a nonempty subset of $\mathbb{R}^n$ that has the following two properties: it is closed under taking sums and under taking nonnegative scalar multiples, that is,

$$x, y \in C \Rightarrow x + y \in C$$

and

$$\rho \geq 0, \ x \in C \Rightarrow \rho x \in C.$$

A convex cone is called *pointed* if it does not contain any straight line through the origin, that is, if it does not contain simultaneously v and $-v$ for any nonzero vector $v \in \mathbb{R}^n$ (geometrically: "the endpoint 0_n of the convex cone looks like the point of a pencil"). A convex cone is called *solid* if it contains some open ball ("it has positive n-dimensional volume" or, equivalently, "it is not contained in any hyperplane").

Cone orderings. For each closed, pointed, solid, convex cone C in $\mathbb{R}^n$ we define the *cone ordering* $\succeq_C$ by writing $x \succeq_C y$ for all $x, y \in \mathbb{R}^n$ for which $x - y \in C$. We write $x \succ_C y$ if $x - y$ lies in the interior of C.

An alternative way to define cone orderings is by the following list of axioms:

- $x \succeq y$, $u \succeq v \Rightarrow x + u \succeq y + v$,

- $x \succeq y$, $\alpha \geq 0 \Rightarrow \alpha x \succeq \alpha y$,

- $x \succeq y$, $y \succeq x \Rightarrow x = y$,

- for each x there are $y, z \succeq 0_n$ for which $x = y - z$,

- if $\lim_{k \to \infty} x_k = x$, $\lim_{k \to \infty} y_k = y$, $x_k \succeq y_k$ for all k, then $x \succeq y$.

Then the set of $x \in \mathbb{R}^n$ with $x \succeq 0_n$ form a closed, pointed, solid, convex cone. We illustrate cone orderings by some examples. The first example shows that the ordering of numbers on the real line is a special case of a cone ordering.

Example 10.2.1 *We choose C to be the set of nonnegative numbers in $\mathbb{R}$. This is a closed, pointed, solid, convex cone; it gives the usual ordering of the real numbers.*

The next example shows that cone orderings are in general only partial orderings.

Example 10.2.2 *We choose C to be the first orthant in $\mathbb{R}^n$, that is,*

$$\mathbb{R}^n_+ = \{x \in \mathbb{R}^n : x_i \geq 0,\ 1 \leq i \leq n\}.$$

This is a closed, pointed, solid, convex cone; it gives the coordinatewise ordering,

$$x \geq y \text{ if and only if } x_i \geq y_i \text{ for all } 1 \leq i \leq n.$$

This ordering is indeed only partial: for example, if $n = 2$, then $(2, 3)^T \not\succeq (4, 1)^T$ and $(4, 1)^T \not\succeq (2, 3)^T$.

Cone orderings can be used to model nonlinear constraints "as if they were linear constraints," as the following example shows.

Example 10.2.3 *We choose C to be the second order cone in $\mathbb{R}^n$, also called the Lorenz cone or ice cream cone, that is,*

$$C = \{x \in \mathbb{R}^n \mid x_n \geq (x_1^2 + \cdots + x_{n-1}^2)^{\frac{1}{2}}\}.$$

This is a closed, pointed, solid, convex cone and so it gives a cone ordering, called the ice cream ordering. Thus, the nonlinear inequality constraint $x_n \geq (x_1^2 + \cdots + x_{n-1}^2)^{\frac{1}{2}}$ is modeled as follows: $x \succeq_C 0_n$.

The ice cream cone for $n = 4$ can be used to throw light on the relativity of time, as we will see in section 10.5.1.

The following example shows that cone orderings arise naturally from the preferences of a consumer.

Example 10.2.4 *Consider the preferences of a consumer who is offered the choice between pairs of consumption vectors in $\mathbb{R}^n$. We write that $v \succeq w$ if he prefers v to w or if $v = w$. We assume that the following properties hold:*

- *if $t \succeq u$ and $v \succeq w$, then $t + v \succeq u + w$,*
- *if $v \succeq 0_n$ and $\rho \geq 0$, then $\rho v \succeq 0_n$.*

Then the set of vectors v with $v \succeq 0_n$ is a pointed convex cone. If this cone is moreover closed and solid, which would not be unreasonable, then we have a cone ordering associated to the preference of the consumer.

Our final example is the most interesting one.

Example 10.2.5 *We choose C to be the set of positive semidefinite $n \times n$ matrices. This is a closed, pointed, solid, convex cone in the vector space of symmetric $n \times n$ matrices. It gives the semidefinite ordering on the vector space of symmetric $n \times n$ matrices,*

$$A \succeq B \text{ if and only if } x^T A x \geq x^T B x \text{ for all } x \in \mathbb{R}^n.$$

Remark. Note that the positive semidefinite $n \times n$ matrices can be characterized inside the symmetric $n \times n$ matrices as the solution set of a finite system of inequality constraints (see the minor criteria in appendix A; these criteria are related to the Sylvester criterion from theorem 9.3). However, this characterization is not convenient. To begin with, this system is large—it consists of 2^n inequalities—and complicated. Moreover, it is worth noting that the functions defining it are not all convex, but that the set of all positive semidefinite matrices is convex.

Example 10.2.6 *Give the minor criteria for positive semidefiniteness of symmetric 2×2 matrices A and show that not all functions defining this system are convex.*

Solution. A symmetric 2×2 matrix A is positive semidefinite if and only if

$$a_{11} \geq 0, \ a_{22} \geq 0, \ a_{11}a_{22} - a_{12}^2 \geq 0.$$

The function $a_{11}a_{22} - a_{12}^2$ is not convex, as the quadratic function of one variable $a_{12} \to -a_{12}^2$ is not convex.

Conclusion to section 10.2. We have established the concept of ordering vectors by means of a convex cone.

10.3 MAIN RESULT: NECESSARY CONDITIONS FOR MIXED SMOOTH-CONVEX PROBLEMS

Plan. We will formulate three variants of the main result, corresponding to the three mixed smooth-convex problem types introduced in section 10.1.

10.3.1 Smooth problems with inequality constraints

Plan. We will give the first order necessary conditions for the first type of mixed smooth-convex problem: problems with inequality constraints $(P_{4.1})$, under assumptions of smoothness.

We recall the definition of problem $(P_{4.1})$ and John's theorem 4.13, also called the KKT theorem for smooth problems. Let $\widehat{x}$ be a point of $\mathbb{R}^n$ and f_i, $0 \leq i \leq m$, continuous functions of n variables, defined on an open ball $U_n(\widehat{x}, \varepsilon)$ with center $\widehat{x}$, such that $f_i(\widehat{x}) \leq 0$, $1 \leq i \leq m$.
Definition 10.1 *The problem*

$$f_0(x) \to \min, \quad f_i(x) \leq 0, \ 1 \leq i \leq m \qquad (P_{4.1})$$

is called a finite-dimensional problem with inequality constraints, or a nonlinear programming problem.

We recall that the Lagrange function is defined to be

$$\mathcal{L}(x,\lambda) = \sum_{i=0}^{m} \lambda_i f_i(x).$$

Theorem 10.2 *John theorem—necessary conditions for smooth nonlinear problems.* *Consider a problem of type* $(P_{4.1})$. *Assume that the problem is smooth at* $\hat{x}$ *in the following "strong" sense:* f_0 *is differentiable at* $\hat{x}$ *and the functions* f_i, $1 \leq i \leq m$, *are continuously differentiable at* $\hat{x}$. *Then, if* $\hat{x}$ *is a point of local minimum, there exists a nonzero selection of Lagrange multipliers* λ *such that the KKT conditions hold at* $\hat{x}$:

$$\mathcal{L}_x(\hat{x},\lambda) = 0_n^T,\ \lambda_i \geq 0,\ 0 \leq i \leq m,\ \lambda_i f_i(\hat{x}) = 0,\ 1 \leq i \leq m.$$

Conclusion. We have given the first order necessary conditions for the first type of mixed smooth-convex problems: problems with inequality constraints under smoothness conditions.

10.3.2 Conic smooth-convex problems

Plan. We will give the first order necessary conditions for the second type of mixed smooth-convex problems: conic smooth-convex problems.

Conic problems. A set in $\mathbb{R}^n$ is called *open* if it contains with each of its points also an open ball with center this point. Let U be an open set in $\mathbb{R}^n$, let f_0 be a linear function on U, say $f_0(x) = c^T \cdot x$ for all $x \in U$ for a fixed vector $c \in \mathbb{R}^n$, let $F = (f_1, \ldots, f_m)^T$ be a vector function on U taking values in $\mathbb{R}^m$, and let $\succeq_C$ be a conic ordering on $\mathbb{R}^n$. Often, we will omit C from the notation $\succeq_C$.

Definition 10.3 *The problem*

$$f_0(x) \to \min,\ F(x) = 0_m,\ x \succeq 0_n \qquad (P_{10.1})$$

is called a problem with one vector equality constraint and one vector inequality constraint, or also, a conic (smooth-convex) problem.

A good way to view this problem type is as a straightforward extension of the well-known linear programming problem in standard form. The following example clarifies this point of view.

Example 10.3.1 *Show that a linear programming problem in standard form*

$$c^T \cdot x \to \min,\ Ax = b,\ x \geq 0_n,$$

is a conic problem.

Solution. Choose $f_0(x) = c^T \cdot x$, $F(x) = Ax - b$, $C = \mathbb{R}_+^n$.

In fact, LP problems can be characterized as conic problems ($P_{10.1}$) in which f_0 and F are affine ("linear + constant") and the cone is the first orthant. If we replace here the first orthant by the ice cream cone (resp. semidefinite cone), then the problem is called a quadratic (resp. semidefinite) programming problem. Explicitly, a semidefinite programming problem is defined to be a problem of the following form (define $\langle A, B \rangle = \sum_{ij} a_{ij} b_{ij}$):

$$\langle C, X \rangle \to \min, \ \langle A_k, X \rangle = b_k, \ 1 \le k \le m, \ X \succeq 0_{n,n}.$$

Here C and A_k, $1 \le k \le m$ are given symmetric $n \times n$ matrices, b_k, $1 \le k \le m$, are given numbers, X is a variable $n \times n$ matrix and $\succeq$ is the semidefinite ordering. We recall from remark 10.6 that it is not convenient to model semidefinite problems as problems with inequality constraints ($P_{4.1}$). For completeness' sake, we mention the easy fact that, more generally, each convex optimization problem can be reformulated as a conic problem with f_0 and F affine. This matter is discussed in detail in section 10.5.4.

Example 10.3.2 *Show that the problem with inequality constraints ($P_{4.1}$) is a special case of problem ($P_{10.1}$).*

Solution. By introducing dummy variables s_i, $1 \le i \le m$, we can split each inequality constraint $f_i(x) \le 0$ into an equality constraint $f_i(x) + s_i = 0$ and a nonnegativity constraint $s_i \ge 0$; going over to vector notation leads to a problem of type ($P_{10.1}$) (to be more precise, we should in addition replace each variable x_i by the difference of two nonnegative variables, $x_i = y_i - z_i$).

A vector $x \in U$ for which $F(x) = 0_m$ is called admissible. The vector $\hat{x}$ is called *a point of local minimum of the problem* ($P_{10.1}$) if it is a point of global minimum for the problem that you get by adding to the problem the constraint that x is contained in a sufficiently small open ball with center $\hat{x}$.

A *selection of Lagrange multipliers* is a row vector $\lambda = (\lambda_0, \ldots, \lambda_m)$. The *Lagrange function* is defined to be

$$\mathcal{L} = \mathcal{L}(x, \lambda) = \lambda_0 f_0(x) + (\lambda_1, \ldots, \lambda_m) \cdot F(x).$$

MIXED SMOOTH-CONVEX PROBLEMS

Theorem 10.4 *First order necessary conditions for conic smooth-convex problems.* Consider a problem of type $(P_{10.1})$ and an admissible point $\widehat{x}$. Assume that f_0 is differentiable at $\widehat{x}$ and that F is continuously differentiable at $\widehat{x}$. Then, if $\widehat{x}$ is a point of local minimum of $(P_{10.1})$, there exists a nonzero selection of Lagrange multipliers λ with $\lambda_0 \geq 0$ for which the following minimum principle holds:

$$\min_{h \succeq -\widehat{x}} \mathcal{L}_x(\widehat{x}, \lambda) \cdot h = 0.$$

This theorem unifies the necessary conditions from the first four chapters. To illustrate that it also gives something new, we apply it to semidefinite programming problems. We will use the easy fact that a symmetric $n \times n$ matrix A is positive semidefinite precisely if the inner product

$$\langle A, B \rangle = \sum_{i,j} a_{ij} b_{ij}$$

is nonnegative for all positive semidefinite $n \times n$ matrices B. That is, the cone of positive semidefinite matrices is *self-dual* (equals its conjugate cone).

Example 10.3.3 *Write down the necessary conditions for a semidefinite programming problem.*

Solution. If $\widehat{X}$ is a point of minimum of the semidefinite programming problem, then there exists a nonzero selection of Lagrange multipliers $\lambda = (\lambda_0, \ldots, \lambda_m) \in (\mathbb{R}^m)^T$ with $\lambda_0 \geq 0$ such that

$$\lambda_0 C + \sum_{k=1}^{m} \lambda_k A_k$$

is positive semidefinite and

$$\left\langle \lambda_0 C + \sum_{k=1}^{m} \lambda_k A_k, \widehat{X} \right\rangle = 0.$$

Here we use that for convex problems, the first order conditions are sufficient for optimality, and we use the self-duality of the cone of positive semidefinite matrices.

Conclusion. We have given the first order necessary conditions for the second type of mixed smooth-convex problems: conic smooth-convex problems.

10.3.3 Problems with smooth-convex perturbations

Plan. We will give the first order necessary conditions for the third type of mixed smooth-convex problems: problems with smooth-convex perturbations.

We will extend the principle of Lagrange to the case of a perturbation, a family of problems indexed by certain parameters, such that giving the parameters the value zero, we get the problem of current interest. The value of this extension is to give additional insight. Let an element $F_y(x, u)$ in $\mathbb{R} \cup \{+\infty\}$, the reals extended with the symbol $+\infty$, be defined for each $y \in \mathbb{R}^m$, $x \in \mathbb{R}^n$, and $u \in \mathcal{U}$, for an arbitrary set $\mathcal{U}$, not necessarily contained in a vector space. We call this a *perturbation*. The following example illustrates this notion. It also illustrates the formal device of hiding the constraints of a minimization problem by use of the $+\infty$ symbol. The idea of this is to give the objective function the value $+\infty$ outside the admissible set; this does not change the problem essentially.

Example 10.3.4 *Perturbation of an LP problem. Consider the LP problem*

$$c^T \cdot x \to \min, \ Ax \leq b, \quad (LP)$$

where $c \in \mathbb{R}^n$, A is an $m \times n$ matrix, $b \in \mathbb{R}^m$, and the variable vector x runs over $\mathbb{R}^n$. We embed it in a family of problems $(P_y)_{y \in \mathbb{R}^m}$, replacing the inequality constraint $Ax \leq b$ by $Ax \leq b + y$ for some vector y, giving

$$c^T \cdot x \to \min, \ Ax \leq b + y. \quad (LP_y)$$

Give a set $\mathcal{U}$ and a perturbation

$$(x, u, y) \to F_y(x, u)$$

for which the problem $F_y(x, u) \to \min$, $x \in \mathbb{R}^n$, $u \in \mathcal{U}$ is equivalent to (LP_y) for all $y \in Y$.

Solution. Define $\mathcal{U} = \mathbb{R}^m_+$ and

$$F_y(x, u) = c^T \cdot x \text{ for } Ax + u = b + y, \ u \geq 0_n \text{ (and } +\infty \text{ otherwise).}$$

In the problem

$$F_y(x, u) \to \min, \ x \in \mathbb{R}^n, \ u \in \mathcal{U},$$

we can omit from consideration pairs (x,u) for which $F_y(x,u) = +\infty$ without changing the problem. The resulting problem is precisely (P_y), as required.

More generally, to each perturbation $(x,u,y) \to F_y(x,u)$ we associate a problem
$$F_{0_m}(x,u) \to \min; \quad x \in \mathbb{R}^n, \; u \in \mathcal{U}, \qquad (P_{10.2})$$
and we embed it in a family of problems $(P_y)_{y \in \mathbb{R}^m}$, defined by
$$F_y(x,u) \to \min, \quad x \in \mathbb{R}^n, \; u \in \mathcal{U}. \qquad (P_y)$$
for each $y \in \mathbb{R}^m$. Let $\widehat{x} \in \mathbb{R}^n$, $\widehat{u} \in \mathcal{U}$ be such that $F_{0_m}(\widehat{x},\widehat{u}) \ne +\infty$. A *Lagrange multiplier* is defined to be a vector
$$\lambda = (\lambda_0, \tilde\lambda) \in (\mathbb{R}^{m+1})^T$$
with $\lambda_0 \in \mathbb{R}$ and $\tilde\lambda \in (\mathbb{R}^m)^T$.

The *Fermat-Lagrange condition* is said to hold if there exists a nonzero Lagrange multiplier λ such that
- $\lambda(y', r') \le 0$ for each tangent vector (x', y', r') to the graph of the vector function $(x,y) \to F_y(x,\widehat{u})$ at the point "above" $(\widehat{x}, 0_m)$.
- $\lambda(y'', r'') \ge 0$ for each point (x'', y'', r'') of the epigraph of the function $(u,y) \to F_y(\widehat{x}, u) - F_{0_m}(\widehat{x}, \widehat{u})$.

The geometrical sense of this condition is that the sets S (resp. C) of all these points (y', r') (resp. (y'', r'')) can be separated by a hyperplane through the origin. Below, we will make assumptions on the perturbation which imply that S is a linear subspace of $\mathbb{R}^m \times \mathbb{R}$; then $\lambda(y', r') = 0$ for all elements $(y', r') \in S$ of this set.

Geometrical illustration Fermat-Lagrange. In the special case of a perturbation $(x,y) \to f(x,y)$, where f is a differentiable function of two variables, the principle of Fermat-Lagrange condition can be illustrated in three-dimensional space (Fig. 10.1).

Definition 10.5 *A pair $(\widehat{x}, \widehat{u})$ is called a strong minimum of the problem $(P_{10.2})$ if it is a global minimum when we add the constraint $x \in U_n(\widehat{x}, \varepsilon)$ for a sufficiently small $\varepsilon > 0$.*

We will use the following notation. Γ_F is the graph of the vector function
$$(x,u,y) \to F_y(x,u),$$

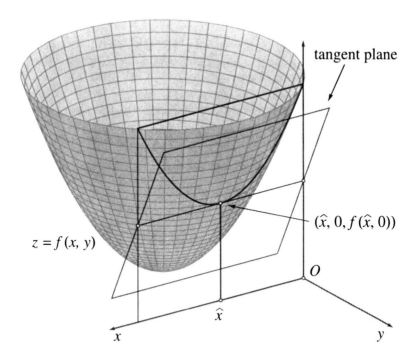

Figure 10.1 Illustration of the Fermat-Lagrange principle.

that is, it is the set of (x, u, y, r) with $x \in \mathbb{R}^n$, $u \in \mathcal{U}$, $y \in \mathbb{R}^m$, $r \in \mathbb{R}$, for which $r = F_y(x, u)$. Note that r is not allowed to be equal to $+\infty$. For a vector function g on a set A taking values in $\mathbb{R}^M$, we write zero(g) for the set of elements in A which have as g-value the zero vector in $\mathbb{R}^M$.

Theorem 10.6 *Fermat-Lagrange—necessary conditions for smooth-convex perturbations (geometrical formulation).* Consider a perturbation

$$(x, u, y) \to F_y(x, u),$$

that is, a function on $\mathbb{R}^n \times \mathcal{U} \times \mathbb{R}^m$, taking values in $\mathbb{R} \cup \{+\infty\}$. Let the following assumptions on the perturbation hold true.

- For each $x \in \mathbb{R}^n$, the set of pairs $(y, r) \in \mathbb{R}^m \times \mathbb{R}$ for which $r \geq F_y(x, u)$ for some $u \in \mathcal{U}$, is convex. Moreover, there is a vector function $g : \mathbb{R}^n \times \mathcal{U} \times \mathbb{R}^m \times \mathbb{R} \to \mathbb{R}^s$ for which

MIXED SMOOTH-CONVEX PROBLEMS

- $\Gamma_F = \text{zero}(g)$.

- *For each $u \in \mathcal{U}$ the vector function $(x, y, r) \to g(x, u, y, r)$ is continuously differentiable.*

- *If $g(\widehat{x}, \bar{u}, \bar{y}, \bar{r}) = 0_s$, then the rows of the derivative of the vector function*

$$(y, r) \to g(\widehat{x}, \bar{u}, y, r)$$

at $(\bar{y}, \bar{r})$ are linearly independent.

Then, if $(\widehat{x}, \widehat{u})$ is a strong minimum of the problem $(P_{10.2})$, there exists a nonzero Lagrange multiplier λ such that the conditions of Fermat-Lagrange hold.

Note that the assumptions of the theorem imply that S is a linear subspace of $\mathbb{R}^m \times \mathbb{R}$, using the tangent space theorem.

Then the Lagrange multiplier λ can be interpreted as the slope of the tangent plane, provided $\lambda_0 = 1$. The principle of Fermat-Lagrange implies all known necessary conditions that can be used to solve concrete finite-dimensional problems. The following example illustrates this for the Fermat theorem.

Example 10.3.5 *Show that the principle of Fermat-Lagrange implies the Fermat theorem for one variable.*

Solution. Assume that $\widehat{x}$ is a point of local minimum for a function f which is differentiable at a point $\widehat{x}$. Then the tangent space to the graph of f at $(\widehat{x}, f(\widehat{x}))$ consists of the vector $(1, f'(\widehat{x}))$ and all its scalar multiples (Fig. 10.2).

The principle of Fermat-Lagrange gives the existence of a nonzero number λ with $\lambda f'(\widehat{x}) = 0$. This gives $f'(\widehat{x}) = 0$, as required.

We give the result above in an analytical form as well. We will write $\widehat{g}_x$ for the derivative of $x \to g(x, \widehat{u}, 0_m, \widehat{r})$ at $x = \widehat{x}$, where $\widehat{r}$ denotes $F_{0_m}(\widehat{x}, \widehat{u})$; in the same way, we will use the notation $\widehat{g}_{y,r}$ and $\widehat{g}_r$.

Theorem 10.7 Principle of Fermat-Lagrange (analytical formulation). *Consider the same situation and assumptions as in theorem 10.6. Then, if $(\widehat{x}, \widehat{u})$ is a strong minimum of the problem $(P_{10.2})$, there exists a nonzero $\zeta \in (\mathbb{R}^s)^T$ such that*

$$\zeta(\widehat{g}_x)^T = 0_m^T \text{ and } \min_{c \in C}(\zeta(\widehat{g}_{y,r})^T)(c) = 0.$$

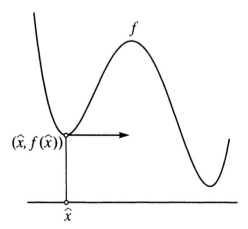

Figure 10.2 One-variable Fermat theorem as a special case of the principle of Fermat-Lagrange.

Conclusion. We have given the first order necessary conditions for the third type of mixed smooth-convex problems: problems with smooth-convex perturbations.

10.4 PROOF OF THE NECESSARY CONDITIONS

Plan. The proof of the first order necessary conditions of mixed smooth-convex problems requires a combination of the tangent space theorem and the supporting hyperplane theorem. We will sketch the proof for the third, most general type of smooth-convex problem. The proofs for the first two types—that is, the proofs of the John theorem and of the necessary conditions for conic smooth-convex problems—are given in appendix F, "Crash Course on Optimization Theory: Analytical Style."

Now we sketch the proof of the necessary conditions for perturbations (theorem 10.6). The details of the general proof are given in [12].

MIXED SMOOTH-CONVEX PROBLEMS

Proof. We assume $\hat{r} = 0$, as we may, without restricting the generality of the argument. We choose arbitrary points $(\tilde{y}, \tilde{r}) \in S$ and $(\bar{y}, \bar{r}) \in C$ with the same first coordinate, $\tilde{y} = \bar{y}$. To prove the theorem, it suffices to prove that $\bar{r} \geq \tilde{r}$: then the separation theorem of convex sets shows that S and C can be separated. We make the following choices:

- $\bar{x} \in \mathbb{R}^n$ for which $g_{x,y,r}(\hat{x}, \hat{u}, 0_m, 0)(\bar{x}, \bar{y}, \bar{r}) = 0_s$.

- $\bar{u} \in U$ for which $g(\hat{x}, \bar{u}, \bar{y}, \bar{r}) = 0_s$, as we may, without restricting the generality of the argument (by making $\bar{r}$ smaller if necessary).

- $(\tilde{y}, \tilde{r})$ with $g_{x,y,r}(\hat{x}, \bar{u}, \bar{y}, \bar{r})(-\bar{x}, \tilde{y}, \tilde{r}) = 0_s$.

Then one has the following system of equalities:

$$g(\hat{x} - t\bar{x}, \hat{u}, -t\bar{y}, -t\bar{r}) = o(t),$$

$$g(\hat{x} - t\bar{x}, \bar{u}, \bar{y} + t\tilde{y}, \bar{r} + t\tilde{r}) = o(t),$$

$$(1-t)(-t\bar{y}) + t(\bar{y} + t\tilde{y}) = o(t).$$

The tangent space theorem can be applied to this system, as can readily be checked. This gives the following system of equalities:

$$g(\hat{x} - t\bar{x} - r(t), \hat{u}, -t\bar{y} - \rho_1(t), -t\bar{r} - \sigma_1(t)) = 0_s,$$

$$g(\hat{x} - t\bar{x} - r(t), \bar{u}, \bar{y} + t\tilde{y} + \rho_2(t), \bar{r} + t\tilde{r} + \sigma_2(t)) = 0_s,$$

$$(1 - t - \tau(t))(-t\bar{y} - \rho_1(t)) + (t + \tau(t))(\bar{y} + t\tilde{y} + \rho_2(t)) = 0,$$

where $r(\cdot), \rho_i(\cdot), \sigma_i(\cdot), i = 1, 2, \tau(\cdot)$ are all $o(t)$.

This system implies that for $t > 0$ sufficiently small there exists $u_t \in U$ such that

$$(1 - t - \tau(t))(-t\bar{r} - \sigma_1(t)) + (t + \tau(t))(\bar{r} + t\tilde{r} + \sigma_2(t)) = 0_s,$$

using the convexity assumption (i) of the theorem for $x = \hat{x} - t\bar{x} + r(t)$. This implies that

$$(1-t-\tau(t))(-t\bar{r}-\sigma_1(t)) + (t+\tau(t))(\bar{r}+t\tilde{r}+\sigma_2(t)) \geq F_{0_m}(\hat{x}-t\bar{x}+r(t), u_t),$$

for $t > 0$ sufficiently small, using that $(\hat{x}, \hat{u})$ is a strong minimum of $(P_{10.2})$. On expanding, this gives

$$(\bar{r} - \tilde{r})t + o(t) \geq 0,$$

and this implies $\bar{\bar{r}} - \bar{r} \geq 0$ as required. □

Conclusion to section 10.4. The tangent space theorem and the supporting hyperplane theorem work together to prove the first order necessary conditions for mixed smooth-convex optimization problems.

10.5 DISCUSSION AND COMMENTS

10.5.1 Conic orderings and the relativity of time

Plan. As an illustration of the theoretical concept conic ordering, we give an application of it. We show that it helps to illuminate a fundamental question about the nature of time.

The ice cream ordering, with $n = 4$, plays a role in physics. We explain this, as it is an interesting example of a convex cone that arises naturally and that clarifies an interesting problem. However, this example has nothing to do with optimization. At first sight, you might think that of two *events* which are not simultaneous, one of the two must be earlier than the other. However, if you take a fundamental point of view, this is not true and in fact it is not even possible to define the concept "simultaneous events" in a satisfactory way. It can happen that for one observer an event A precedes an event B, but for another one B precedes A or they are simultaneous. This strange phenomenon can occur if the observers move relative to each other with high speed; for example, one stands on earth and the other one is in a supersonic airplane. Their different observations are not due to imprecise measurements, but they have a fundamental reason. It might seem that there is a simple way out of this paradox: you should only consider observers who are standing still. However, this does not solve the paradox, as such observers do not exist: the earth is also moving, for example, and so observers on earth are not standing still. Therefore one is forced to accept that time and in particular the concepts "simultaneous" and "earlier" can only be considered relative to an observer. However, many pairs of events—in our everyday experience all pairs—do occur in the same order for all observers. Relativity theory gives a simple criterion for when, from a pair of events, one precedes the other for all observers. The description and meaning of this criterion, to be given below, do not require familiarity with relativity theory. Note first that each event A takes place at a certain "moment" and at a certain "place." The main point will be that moments of different events can sometimes be compared, independently of the observer ("one precedes the other (for all observers)"), but sometimes not.

Example 10.5.1 *An event A is said to precede an event B if a light signal sent at the moment A occurs reaches the place of B before event B occurs. Note that this definition does not involve observers. This concept of "precede" turns out to have the required property: precisely in this case, event A precedes event B for all observers. Moreover, it has the following interpretation: A precedes B precisely if A could be the cause of B. This is so, as the fastest way to send a message is by means of a light signal.*

Model this concept of one event preceding another one by means of a suitable conic ordering, by choosing one observer.

Solution. Choose coordinates of place x_1, x_2, x_3 and time t, that is, "choose one observer." For convenience we normalize the units of length and time in such a way that the speed of light is 1. We denote the coordinates of place (resp. moment) of an event A by x_A (resp. t_A), so $x_A \in \mathbb{R}^3$ and $t_A \in \mathbb{R}$. Consider the Lorenz cone C in $\mathbb{R}^4$. It is readily seen that an event A precedes an event B precisely if

$$(x_B, t_B) \succ_C (x_A, t_A).$$

Indeed, writing out this condition explicitly gives

$$t_B - t_A > |x_B - x_A|.$$

This means that for our observer event B happens after event A and, moreover, for him the time between the two events $t_B - t_A$ is greater than the time it takes to go from the place of A to the place of B with speed 1.

Conclusion. The puzzling fact that simultaneity does not exist on a fundamental level can be clarified completely using conic orderings.

10.5.2 Second order conditions

Plan. We will supplement the first order necessary conditions with second order conditions, just as was done in chapter five.

We recall from chapter five that the main interest of second order conditions is that they give insight. For mixed smooth-convex problems this is in fact the only interest: in general it is not so easy to verify the second order conditions for a given concrete problem. Here we will give the second order conditions for conic problems.

Theorem 10.8 *Second order conditions for conic problems.*
Consider a problem of type $(P_{10.1})$ *and an admissible point* $\widehat{x}$. *Assume that* f_0 *is twice differentiable at* $\widehat{x}$, *that* F *is twice continuously differentiable on* $U_n(\widehat{x}, \varepsilon)$ *and that the* $m \times n$ *matrix* $F'(\widehat{x})$ *has rank* m.

1. **Necessary conditions.** *Suppose that* $\widehat{x}$ *is a point of local minimum of* $(P_{10.1})$. *Then there exists* $\lambda = (1, \lambda_1, \ldots, \lambda_m)$ *such that*

$$\min_{x \succeq -\widehat{x}} \mathcal{L}_x(\widehat{x}, \lambda) x = 0,$$

and

$$\mathcal{L}_{xx}(\widehat{x}, \lambda)[\xi, \xi] \geq 0$$

for all $\xi \in \ker F'(\widehat{x})$ *that can be written as* $x - \rho \widehat{x}$ *with* $x \in \operatorname{int} C$ *and* $\rho > 0$.

2. **Sufficient conditions.** *If there exists* λ *with* $\lambda_0 = 1$ *such that*

$$\min_{x \succeq -\widehat{x}} \mathcal{L}_x(\widehat{x}, \lambda) = 0,$$

and

$$\mathcal{L}_{xx}(\widehat{x}, \lambda)[\xi, \xi] > 0,$$

for all nonzero $\xi \in \ker F'(\widehat{x})$ *that can be written as* $x - \rho \widehat{x}$ *with* $x \in \operatorname{int} C$ *and* $\rho > 0$, *then* $\widehat{x}$ *is a point of local minimum of* $(P_{10.1})$.

The proof is given in appendix F, "Crash Course on Optimization Theory: Analytical Style."

Remark. One can eliminate the Lagrange multipliers from the second order conditions if $\operatorname{rank} F'(\widehat{x}) = m$. Indeed,

$$f_0'(\widehat{x}) + (\lambda_1, \ldots, \lambda_m) F'(\widehat{x}) = 0_m$$

implies the formula

$$(\lambda_1, \ldots, \lambda_m) = -f_0'(\widehat{x}) F'(\widehat{x}) (F'(\widehat{x}) F'(\widehat{x})^T)^{-1},$$

and this can be used to eliminate the Lagrange multipliers from

$$\mathcal{L}_{xx}(x, \lambda)[\xi, \xi],$$

giving the quadratic form

$$Q(x) = (f_0''(\widehat{x}) - f_0'(\widehat{x}) F'(\widehat{x})^T (F'(\widehat{x}) F'(\widehat{x})^T)^{-1} F''(\widehat{x}))[\xi, \xi].$$

This result implies all usual second order conditions for all types of n-dimensional optimization problems as special cases.

Conclusion. The theory of second order necessary and sufficient conditions extends to mixed smooth-convex problems.

10.5.3 Second order conditions for standard problems

Plan. We display the second order necessary conditions for all types of standard problems. We give the version without Lagrange multipliers. These conditions are special cases of theorem 10.8.

One variable without constraints. The quadratic form is

$$Q(x) = f_0''(\widehat{x})x^2$$

and the second order condition for the conic problem is $f_0''(\widehat{x})x^2 \geq 0$ for all $x \in \mathbb{R}$. This gives the condition $f''(\widehat{x}) \geq 0$.

More than one variable without constraints. The quadratic form is

$$Q(x) = x^T f_0''(\widehat{x}) x$$

and the second order condition for the conic problem is $x^T f_0''(\widehat{x})x \geq 0$ for all $x \in \mathbb{R}^n$. That is, we get the condition that the Hessian matrix $f''(\widehat{x})$ is positive semidefinite.

Equality constraints. The quadratic form is

$$Q(x) = x^T f_0''(\widehat{x})x - f_0'(\widehat{x})F'(\widehat{x})^T (F'(\widehat{x})F'(\widehat{x})^T)^{-1} F''(\widehat{x})[x,x] \quad \forall x \in \mathbb{R}^n,$$

and the second order condition for the conic problem is $Q(x) \geq 0$ for all $x \in \mathbb{R}^n$. That is, we get the condition that the underlying symmetric $n \times n$ matrix of the quadratic form defined by

$$Q(x) = -f_0'(\widehat{x})F'(\widehat{x})^T (F'(\widehat{x})F'(\widehat{x})^T)^{-1} F''(\widehat{x})[x,x] + x^T f_0''(\widehat{x})x$$

is positive semidefinite.

Smooth inequality constraints. Here, the second order condition for the conic problem gives that the underlying symmetric $n \times n$ matrix of the quadratic form $Q(x)$ defined by the formula

$$x^T f_0''(\widehat{x})x - (f_0'(\widehat{x}) \cdot f_i'(\widehat{x})^T)_{1 \leq i \leq m}^T (F'(\widehat{x})F'(\widehat{x})^T + I_m)^{-1} (x^T f_i''(\widehat{x})x)_{1 \leq i \leq m}$$

is positive semidefinite.

Convex inequality constraints. Here the second order condition for the conic problem turns out to be a trivial condition which always holds. This is in agreement with the fact that for linear conic problems the first order necessary conditions are also sufficient.

Conclusion. The second order conditions of mixed smooth-convex problems project down to the usual second order necessary conditions for all types of standard problems.

10.5.4 Linear conic problems

Plan. We consider a special type of conic problems, linear ones, and we show that this class coincides with the class of convex problems. If the cone is the nonnegative orthant, then we get precisely the class of LP problems.

A conic problem $(P_{10.1})$ with f_0 and F affine (= linear + constant) is called a *linear conic problem*. For example, linear programming problems are linear conic problems with underlying cone the nonnegative orthant, as we have seen in example 10.3.1 and in the text following it. However, many *nonlinear* optimization problems can be modeled as linear conic problems as well. The following key example illustrates this: semidefinite programming problems. These problems have come to the center of attention during the last decade.

Example 10.5.2 *Show that a semidefinite programming problem*

$$\langle C, X \rangle = \sum_{i,j} c_{ij} x_{ij} \to \min, \ \langle A_k, X \rangle = b_k, \ 1 \le k \le m, \ X \succeq 0_{n,n}$$

is a linear conic problem. Here C and A_k, (resp. X), $1 \le k \le m$, are given (resp. variable) symmetric $n \times n$ matrices, b_k, $1 \le k \le m$, are given numbers and $\succeq$ is the semidefinite ordering.

Solution. Choose $f_0(X) = \langle C, X \rangle$, $F(X) = (\langle A_k, X \rangle - b_k)_k \in \mathbb{R}^m$, where X runs over the vector space of symmetric $n \times n$ matrices, and take, as the cone, the positive semidefinite $n \times n$ matrices.

The following more general fact holds true. We call a convex problem *closed* if its admissible set is closed.
- The class of linear conic problems coincides with the class of closed convex problems.

We illustrate this with an example.

Example 10.5.3 *Model as a linear conic problem, the problem to find the point in a given closed bounded set A with nonempty interior in the plane $\mathbb{R}^2$ with the smallest second coordinate.*

Solution. We view the plane as the floor and lift up the set A by one unit. This gives the set A_1 of points $(a_1, a_2, 1)$ with $(a_1, a_2) \in A$. Thus, this set is a copy of A. It lies precisely above A, in the plane that is parallel to the floor and lies 1 unit above the floor. Now we take all rays through the points of this lifted-up copy of A, starting from the origin in space. Their union is a closed, pointed, solid convex cone C in three-dimensional space (Fig. 10.3). Thus we get that the

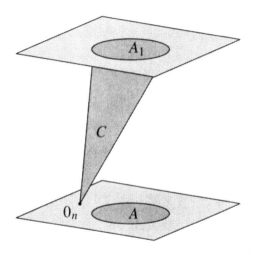

Figure 10.3 Associating a convex cone C to a convex set A.

following conic problem is equivalent to the given problem,

$$x_2 \to \min, \ x_3 = 1, \ x \succeq_C 0_3.$$

Sketch of the conic modeling of arbitrary closed convex problems. The idea of this example can be extended, to begin with, to the case that A is not necessarily bounded: this requires replacement of the constructed cone by the smallest closed convex cone containing it (note that this can only add points on the floor). More generally, the idea of this example extends to the general case. To see this, note that each convex problem can be written in the form $x_n \to \min, \ x \in A$, for some convex set A in $\mathbb{R}^n$. Indeed, if f is a convex function on the convex set $B \in \mathbb{R}^k$, then the minimization of it, can be written as

$$x_{k+1} \to \min, x \in A,$$

where A is the set in $\mathbb{R}^{k+1}$ determined by

$$x_{k+1} \geq f(x_1, \ldots, x_k), \; (x_1, \ldots, x_k) \in B.$$

We give one more example of the conic modeling of closed convex problems. This time, we will do this using formulas, and not, as above, using a picture. Of course, the construction is the same: it is just a matter of language.

Example 10.5.4 *Model as a linear conic problem the problem of finding the point closest to $(2,3)$ with first coordinate not larger than 2 in absolute value.*

Solution. We start with the modeling given in the traditional way,

$$(x_1 - 2)^2 + (x_2 - 3)^2 \to \min, \; x_1^2 \leq 4,$$

and observe that the functions $(x_1 - 2)^2 + (x_2 - 3)^2$ and $x_2^2 - 4$ are strictly convex. We replace the objective function by x_3 and add the inequality constraint

$$x_3 \geq (x_1 - 2)^2 + (x_2 - 3)^2,$$

as we may. This leads to the equivalent problem

$$x_3 \to \min, \; x_3 \geq (x_1 - 2)^2 + (x_2 - 3)^2, \; x_1^2 \leq 4.$$

Then we *homogenize* the system of inequality constraints

$$x_3 \geq (x_1 - 2)^2 + (x_2 - 3)^2, \; x_1^2 \leq 4.$$

That is, we divide each variable by a new one, x_4, which is allowed to run over the positive numbers. This gives the following system of inequalities:

$$\frac{x_3}{x_4} \geq \left(\frac{x_1}{x_4} - 2\right)^2 + \left(\frac{x_2}{x_4} - 3\right)^2, \; \left(\frac{x_1}{x_4}\right)^2 \leq 4, \; x_4 > 0.$$

Multiplying by x_4^2 to remove the denominators, we get the homogenized system of inequalities

$$x_3 x_4 \geq (x_1 - 2x_4)^2 + (x_2 - 3x_4)^2, \; x_1^2 \leq 4x_4^2, \; x_4 > 0.$$

MIXED SMOOTH-CONVEX PROBLEMS

Finally, we replace in this system $x_4 > 0$ by $x_4 \geq 0$. The solution set of the resulting system of inequalities,

$$x_3 x_4 \geq (x_1 - 2x_4)^2 + (x_2 - 3x_4)^2, \ x_1^2 \leq 4x_4^2, \ x_4 \geq 0,$$

can be checked to be a closed, pointed, solid convex cone C in $\mathbb{R}^4$. We can dehomogenize by putting x_4 equal to 1. Therefore, we get the following four-dimensional modeling of our problem as a linear conic problem,

$$f_0(x) = x_3 \to \min, \ F(x) = x_4 - 1 = 0, \ x \succeq_C 0_4.$$

Here $x = (x_1, \ldots, x_4)^T$.

Conclusion. The notion linear conic problem clarifies the relation between two well-known types of optimization problems, linear programming problems and convex problems.

- The class of closed convex problems is the same as the class of linear conic problems.

- Moreover, a linear programming problem is just a linear conic problem where the underlying cone is the nonnegative orthant.

10.5.5 Variants of the conic model

We note that we have made a choice of format for defining conic problems. The reason for this choice is that we want to model all smoothness of a problem by one constraint, $F(x) = 0_m$, and all convexity by another constraint, $x \succeq 0_n$. Another attractive choice of format which gives the same concept would be the following problem type, which has only one constraint

$$f_0(x) \to \min, \ F(x) \succeq 0_m.$$

Furthermore, sometimes the choice of format

$$f_0(x) \to \min, \ F(x) \succeq 0_m, \ x \geq 0_n$$

is the most convenient one. For example, all linear programming problems can be written as

$$c^T \cdot x \to \min, \ Ax \leq b$$

or as

$$c^T \cdot x \to \min, \ Ax \leq b, \ x \geq 0_n,$$

and these alternative formats for LP problems are used very often as well.

Chapter Eleven

Dynamic Programming in Discrete Time

> Today is the first day of the rest of your life.
>
> *On the wall in many homes.*

- How to solve a discrete time dynamic programming problem? We consider problems of the following type:

$$F(\bar{x}, \bar{u}) = \sum_{k=0}^{n} f_k(x_k, u_k) \to \min,$$

$$\bar{x} = (x_k)_{0 \le k \le n+1}, \quad \bar{u} = (u_k)_{0 \le k \le n},$$

$$x_k \in X, \ k = 0, \ldots, n+1, \ u_k \in U, \ k = 0, \ldots n,$$

$$x_{k+1} = g_k(x_k, u_k), \ k = 0, \ldots, n, \ x_0 = a, \ x_{n+1} \in \mathcal{T}.$$

11.0 SUMMARY

We consider multidecision problems, where a number of related decisions have to be taken sequentially. The problem to take the last decision in an optimal way is usually easy. Moreover, if we know how to solve the multidecision problem with r decisions still to be taken, then this can be used to give an easy solution of the multidecision problem with $r+1$ decisions still to be taken. The reason for this is a simple observation, the *optimality principle*: for an optimal chain of decisions, each end chain is also optimal. Here are three examples.

1. **Shortest route problem.** How to find the shortest route? Dynamic programming reduces the work as it does not require that you search through all possible routes.

2. **Dynamic investment problem.** In making investment decisions you often have to look ahead in order to make an optimal trade-off between high return and low risk.

3. Monopolist competing with himself over time. Even a monopolist has to deal with competition. He competes against himself. For example, if he sells an improved version of his product, he is competing with the old version and maybe also with the improved version the customers expect in the future. You get a similar dynamic problem if a new product of a monopolist has been on the market for some time and he wants to lower the price in a few steps.

DYNAMIC PROGRAMMING IN DISCRETE TIME

11.1 INTRODUCTION

Tomorrow's breakfast. The previous chapter marks the end of our book. It is dinner, the last meal of the day. However, tomorrow there will be another day. In this chapter and the next one, we offer already a taste of tomorrow's breakfast. In this chapter we will consider discrete time *dynamic programming*.

Domino idea. The main idea behind dynamic programming is, roughly speaking, as follows. View a given problem as one of a family of problems. Often, one of these problems is easy, and moreover, for two "neighboring" problems, it is easy to solve one if the other one is already solved. Then you get a "domino effect." You begin by solving the easy problem ("the first domino falls"). Then you solve a neighboring problem ("the second domino falls"). You go on in this way till you reach and solve the given problem ("the last domino falls").

Optimality principle. To be more precise, let a multidecision problem be given, where the decisions have to be taken sequentially. Such a problem can be viewed as one of a family of problems, by viewing each of the decisions as the first one of a new multi-decision problem. The epigraph of this chapter—*"Today is the first day of the rest of your life"*—is in this spirit. The problem of taking the last decision in an optimal way is usually easy. Moreover, if we know how to solve the multidecision problem with r decisions still to be taken, then this can be used to give an easy solution of the multidecision problem with $r+1$ decisions still to be taken. The reason for this is the following simple but powerful observation.

The optimality principle: for an optimal sequence of decisions, each end sequence is also optimal for the corresponding problem.

This method of solution is called *dynamic programming*. It was developed by Richard Bellman in the 1950s. Its continuous time version can be traced back to work of Hamilton and Jacobi, and even to Huygens (on the wave character of light).

Royal road: you can read, without any preparation, all the concrete applications in section 11.3 and look at the exercises in section 11.4. In particular, you can skip the next section, as almost all problems are solved by applying the optimality principle directly and we use the name of the formal version of the optimality principle, the Hamilton-Jacobi-Bellman equation.

11.2 MAIN RESULT: HAMILTON-JACOBI-BELLMAN EQUATION

Plan. We present the formalism of dynamic programming problems and the main result, the Hamilton-Jacobi-Bellman equation, also called the Bellman equation. This formal version of the optimality principle gives some insight, but it is not necessary to solve the problems using this relatively heavy formalism. It is usually more convenient to solve concrete problems by applying the optimality principle directly.

Let U and X be sets, T a subset of X, $a \in X$, $(x_k, u_k) \to f_k(x_k, u_k)$ (resp. $(x_k, u_k) \to g_k(x_k, u_k)$) functions (resp. mappings taking values in X), where the variables run as follows: $x_k \in X$, $u_k \in U$, $k = 0, \ldots, n$.

Definition 11.1 *The problem following problem is called a deterministic, discrete time, dynamic programming problem:*

$$F(\bar{x}, \bar{u}) = \sum_{k=0}^{n} f_k(x_k, u_k) \to \min,$$

$$\bar{x} = (x_k)_{0 \leq k \leq n+1}, \bar{u} = (u_k)_{0 \leq k \leq n},$$

$$x_k \in X, \ k = 0, \ldots, n+1, \ u_k \in U, \ k = 0, \ldots n,$$

$$x_{k+1} = g_k(x_k, u_k), \ k = 0, \ldots, n,$$

$$x_0 = a, \ x_{n+1} \in T \quad (P_{11.1})$$

Consider the following family of problems $(P_{r,x})$—r decisions to go with initial state a—with $x \in X$ and $r \in \{0, 1, \ldots, n\}$,

$$F_{r,x}(\bar{x}, \bar{u}) = \sum_{k=n-r}^{n} f_k(x_k, u_k) \to \min,$$

$$\bar{x} = (x_k)_{n-r \leq k \leq n+1}, \bar{u} = (u_k)_{n-r \leq k \leq n},$$

$$x_k \in X, \ k = n-r, \ldots, n+1, \ u_k \in U, \ k = n-r, \ldots n,$$

$$x_{k+1} = g_k(x_k, u_k), \ k = n-r, \ldots, n,$$

DYNAMIC PROGRAMMING IN DISCRETE TIME

$$x_{n-r} = x, \ x_{n+1} \in \mathcal{T}. \qquad (P_{r,x})$$

We write $J_r(x)$ for the value of the problem $(P_{r,x})$, where $x \in X$ and $r = 0, \ldots, n+1$. We note that $(P_{n,a})$ is the given problem $(P_{11.1})$. The following fact is just a formal description of the optimality principle.

Theorem 11.2 *Hamilton-Jacobi-Bellman equation—the necessary and sufficient conditions for a deterministic, discrete time, dynamic programming problem. The function J can be characterized recursively by the following properties:*

$$J_{r+1}(x) = \inf_{u \in U} (f_{n-r}(x,u) + J_r(g_{n-r}(x,u)))$$

for $r = 0, \ldots, n, \ x \in X$, and

$$J_0(x) = 0 \quad \text{for } x \in \mathcal{T}, \ J_0(x) = \infty \text{ otherwise}.$$

Remark. In all concrete problems we will find a solution, say $\widehat{u}(r,x)$, of the auxiliary problem

$$h_{r,x}(u) = f_{n-r}(x,u) + J_r(g_{n-r}(x,u)) \to \min, \ u \in U$$

for all $r = 0, 1, \ldots, n$ and $x \in X$. Then the mapping $(r,x) \to \widehat{u}(r,x)$ is called a solution in *feedback form* (also called *closed loop form*) for the family of problems $(P_{r,x})_{r,x}$. It is readily seen that a solution in feedback form can be rewritten as solutions in ordinary form (also called *open loop form*) for the problems of this family. We will do this in all concrete examples, but we will not display the formal transition from open loop to closed loop.

The majority of the applications of dynamic programming problems $(P_{11.1})$ has suggested the following standard terminology: the variable k (resp. x, resp. u) is called *time* (resp. *state*, resp. *control*), the constraints are said to form a *dynamic system*. The "true" variables of optimization are the controls ("decisions"); note that a choice of controls u_k, $0 \le k \le n$, determines corresponding states x_k, $0 \le k \le n+1$, by means of the dynamic system. We emphasize that there are various applications where the variable k does not model time.

Conclusion. We have presented the standard concepts, notations and the main result of dynamic programming, the HJB equation.

11.3 APPLICATIONS TO CONCRETE PROBLEMS

In concrete problems, it is usually convenient to write down the HJB equation immediately, using the principle of optimality, without a formal specification of time, state, control, and dynamic system. We will write down all solutions in the style of the four-step method.

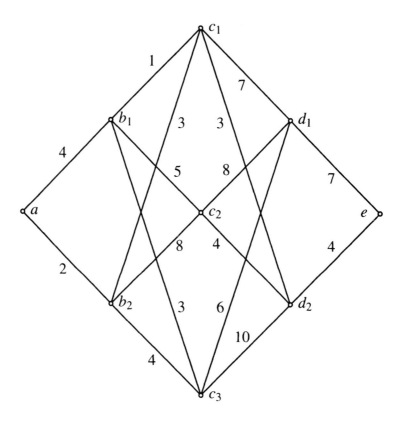

Figure 11.1 Maximum cargo problem.

11.3.1 Maximum cargo problem

We begin with the basic example, the maximal cargo problem. Here, it is of interest to note how dynamic programming manages to avoid searching through the finite but often large collection of all routes.

Problem 11.3.1 *A ship makes a journey from a to e. On its route it makes stops in three ports (one of the two b's, one of the three c's,*

DYNAMIC PROGRAMMING IN DISCRETE TIME

and one of the two d's) to deliver and pick up cargo. These cargos are indicated in Figure 11.1 in units of hundred tons.

The aim is to transport as much as possible during the entire journey.

What we could do is to consider all possible routes, calculate the total transport of each route, and compare the outcomes. The one with the highest outcome is the solution of the problem. However, even in this simple network, the number of the possible routes is relatively high (12), and this number grows very fast if the complexity of the network increases. Therefore, this method is not attractive.

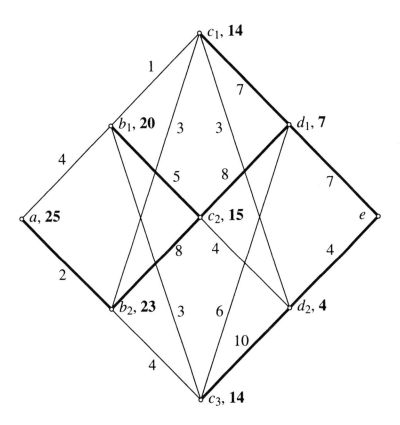

Figure 11.2 Solution by dynamic programming.

Solution

1. Instead, we consider not only the problem of how to go in an optimal way from a to e, but also how, for each of the ports, to go in an optimal way from that port to e.

2. HJB: we have to solve recursively all the problems of finding an optimal route from any point of the network to e.

3. In Figure 11.2, we display the solutions of all these problems. For example, at the point b_2, the number 23 is written, and this indicates that the total cargo for the optimal route from b_2 to e is 23.

   This number has been calculated by taking the maximum of 3+14=17, 8+15=23 and 4+14=18. Moreover, if we start our trip in port b_2, then the optimal first decision is to go to port c_2, as indicated.

4. The optimal route is $a \to b_2 \to c_2 \to d_1 \to e$, and then the cargo that can be transported during the journey is 25.

Remark. If we had been greedy for immediate profit, we would have been misled at the very start, going from a to b_1, as $4 > 2$. This greedy strategy would have led to a total profit of 24. The difference between the optimal solution and the result of the greedy strategy is usually much more substantial.

11.3.2 Wine at the wedding of the crown prince

A particularly common type of dynamic programming problem is the *distribution of effort problem*. For this type of problem, there is just one kind of *resource*, and it has to be allocated to a number of different *activities*. The objective is to determine how to distribute the effort ("the resource") among the activities in the most effective way. The next problem illustrates this type of dynamic programming problem.

Problem 11.3.2 *The celebrations of the wedding of the crown prince will last for many days. The main attraction is the collection of 100 barrels filled with exquisite wine. Each day, some barrels will be fetched from the cellars, at least one. At the end of the day, the entire contents of these barrels will be consumed by the insatiable guests. How many days should the celebrations last, and how should the 100 barrels be distributed over the days to maximize the total pleasure, which is taken to be the product of all the daily wine consumption, measured in barrels? In case two distributions give the same pleasure, then the crown prince prefers the solution with the maximal number of days.*

Solution

1. "Replace 100 by n" and write $J(n)$ for the optimal pleasure if n barrels are available for the celebrations.

2. HJB:
$$J(n) = \max_{k \in \{1,\ldots,n\}} kJ(n-k) \quad \text{and} \quad J(0) = 1.$$

3. Make the following three observations.

   (a) It cannot be optimal to have $r > 3$ barrels on one day: it is better to split r into 2 and $r - 2$ as $2(r-2) \geq r$ for $r > 3$.

   (b) It cannot be optimal to have 1 barrel on one day, unless the total number of barrels is 1.

   (c) It cannot be optimal to have more than two days 2 barrels: three days 2 barrels gives less pleasure than two days 3 barrels, as $2^3 < 3^2$.

   These three properties characterize the optimal celebrations, as each number $n > 1$ can be written in a unique way as $3a + 2b$, where a is a nonnegative integer and b is $0, 1$ or 2. Indeed, if n is a multiple of 3, then $b = 0$, if $n + 1$ is a multiple of 3, then $a = 1$ and if $n - 1$ is a multiple of 3, then $b = 2$. In particular, for $n = 100$ we get $a = 32$ and $b = 2$.

4. The crown prince will prefer the following way to organize the wedding over all other ones: the celebrations last 34 days, there are 3 barrels on 32 of these days and 2 barrels on the remaining two days.

11.3.3 Towers of Hanoi

We will consider a problem where the recursion of the Bellman equation brings order in a seemingly complicated situation.

Problem 11.3.3 *In a temple in Hanoi, monks are displacing day and night 64 golden disks with a hole in the middle—all of different sizes—on three ivory sticks. They observe the following rules:*

- *it is only allowed to move one disk at a time from one stick to another one,*

- *it is forbidden to put a disk on top of a smaller disk.*

At the beginning of time, all 64 disks formed a tower on one of the sticks. The aim of the monks is to displace this tower to the last stick. According to legend, at the moment of completion of this task, the world will perish. How long will this take, if the monks make a move each second and if they follow the strategy with the least number of moves?

Solution

1. Replace 64 by n and write $J(n)$ for the minimal number of moves in which a tower of n disks can be moved from one stick to another.

2. HJB:
$$J(n) = 2J(n-1) + 1 \quad \text{and} \quad J(1) = 1.$$

   This can be derived from the following observation: during the move that the largest disk is moved, the other $n-1$ disks form a tower on the remaining stick.

3. $J(n) = 2^n - 1$.

4. It will take $2^{64} - 1$ moves, that is, more than 500 billion years, before the monks will finish their task.

11.3.4 Tower of Pisa

The leaning tower of Pisa is known for a defect. One can experiment with how far a tower can lean, using playing cards, beer coasters or cd-boxes at the edge of a table. Many different strategies can be attempted. To begin with, there is the greedy approach: you can let the first beer coaster stick out as far as possible; but this greedy beginning turns out to be a bad one as it makes further attempts to get away from the table impossible. From this attempt you learn that you should "hold back" at the beginning. A next try could be to make a straight tower: then the problem is what is the best slope. Here it turns out that there is no universal best slope: the best slope depends on the number of beer coasters. Another tempting possibility is to build a tower that is supported by several auxiliary towers that give some counterweight. The best achievement we know of is by some students in econometrics. One evening, after a lecture on dynamic programming, they built a tower with free postcards from the student restaurant with top one sticking out the length of two and a half postcards. The question is now, how much better can you do? What is the best building strategy if you want a tower that leans as much as possible but does not fall down?

This problem is a metaphor for a multidecision problem. Each beer coaster that you put down represents a decision. When taking a decision you have to look ahead. Indeed, the end result depends on the whole sequence of decisions. Moreover, when taking a decision you might limit the choice of decisions in the future.

Problem 11.3.4 (i) *How can you make a tower from a collection of 52 playing cards (or beer coasters or cd-boxes) such that the top of the tower sticks out of the table in a horizontal direction.* (ii) *How much does the top stick out?*

Solution

1. Replace 52 by n and write $J(n)$ for how far the top of an optimal n-tower sticks out.

2. HJB: $J(n) = J(n-1) + x(n)$, where $x(n)$ is characterized by the equation

$$1 \cdot \left(x(n) - \frac{1}{2}\right) + (n-1) \cdot x(n) = 0.$$

Here $x(n)$ is how far the bottom card sticks out over the edge of the table. This follows from the following characterization of optimal n-towers: for $1 \leq k \leq n$, the center of gravity of the tower formed by the top k cards is situated exactly above the border of the $(k+1)$-st card from the top. For $k = n-1$ this gives the equality above, $1 \cdot (x(n) - \frac{1}{2}) + (n-1) \cdot x(n) = 0$. To see this, one has to use the formula for the center of gravity given in section 1.4.2, but now one should work in the horizontal direction and not in the vertical direction.

3. $x(n) = \frac{1}{2n}$ and so $J(n) = \sum_{k=1}^{n} \frac{1}{2k}$.

4. (i) The optimal n tower can be built as follows. Make a vertical tower of all the cards to begin with. Then shift the top card as much as possible, after this, shift the top two cards as much as possible *without altering their relative position*, and so on. The

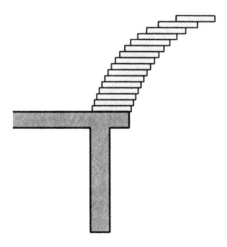

Figure 11.3 Optimal tower of Pisa.

resulting tower goes up initially almost vertically, and near the top it veers away from the table (Fig. 11.3).

(ii) The top of the tower sticks out a distance $\sum_{k=1}^{52} 1/2k$, where we take the length of a card as a unit of length. If we use the inequalities

$$1 + \int_1^n \frac{1}{x}dx \geq \sum_{k=1}^{n}(1/k) \geq \int_1^{n+1} \frac{1}{x}dx$$

—which follows by comparing integrals with lower and upper Riemann sums—we get the inequalities

$$1 + \ln n \geq \sum_{k=1}^{n}(1/k) \geq \ln(n+1). \qquad (*)$$

This gives the estimate

$$\sum_{k=1}^{52} 1/2k \approx \frac{1}{2}\ln 53 \approx 2.$$

By the way, if one has enough cards, one can get arbitrarily far away from the table in the horizontal direction, as

$$\sum_{k=1}^{\infty} \frac{1}{2k} = \infty$$

by (∗). In theory, at least.

Conclusion. The tower of Pisa problem illustrates all aspects of the dynamic programming method. To begin with, we replace the task of solving the given optimization problem ("52 cards") by that of solving a family of related problems that include the given problem: make an optimal tower of Pisa of n cards. Moreover, our method transforms the problem of taking a sequence of *linked* decisions into the problem of taking a sequence of *unlinked* decisions: in this case by building the tower from top to bottom instead of from bottom to top. Finally, the idea of solving the problem backward is seen in action here.

11.3.5 A dynamic investment problem

It is possible to incorporate *risk* into the method of dynamic programming. We will illustrate this by examples in problems 11.5, 11.6, and 11.7, but we will not display *stochastic* dynamic programming problems and the corresponding HJB equation formally.

Problem 11.3.5 *A person has 5,000 euros that she wants to invest. She will have at the beginning of each of the next three years an opportunity to invest that amount for one year in either of two investments (A or B). Both investments have uncertain returns. For investment A, she will either double her money—with high probability 0.7—or lose it entirely, with small probability 0.3. For investment B, she will either double her money—with very low probability 0.1—or get her money back, with very high probability 0.9. She is allowed to make (at most) one investment each year, and she can invest only 5,000 each time—any additional money accumulated is left idle.*

(i) Which investment policy maximizes the expected amount of money she will have after three years?

(ii) Now supposes that her aim is to buy something after three years that costs 10,000 euro. Which investment policy maximizes the probability that she will have at least 10,000 euro after three years?

We will make use of the elements of conditional probability in the solution of this problem.

Solution. (*i*)

1. Replace 3 by n with $1 \leq n \leq 3$ and write $J_n(K)$ for the expected amount of money after three years, given in multiples of 5,000

euros, when the investment policy is chosen that maximizes this expectation.

2. We do not display the HJB equations.

3. Note that $J_n(0) = 0$ for all n.

- HJB for 1 year to go gives that $J_1(5)$ is the maximum of
  - $0.3 \cdot 0 + 0.7 \cdot 10 = 7$ (A),
  - $0.9 \cdot 5 + 0.1 \cdot 10 = 5.5$ (B).

  Therefore, $J_1(5) = 7$ and the optimal investment choice is A. This gives $J_1(k) = k + 2$ for all $k \geq 5$ with optimal investment choice A.

- HJB for 2 years to go gives that $J_2(5)$ is the maximum of
  - $0.3 \cdot J_1(0) + 0.7 J_1(10) = 8.4$ (A),
  - $0.9 J_1(5) + 0.1 J_1(10) = 7.5$ (B).

  Therefore, $J_2(5) = 8.4$ and the optimal investment choice is A. It gives as well that $J_2(10)$ is the maximum of
  - $0.3 J_1(5) + 0.7 J_1(15) = 14$ (A),
  - $0.9 J_1(10) + 0.1 J_1(15) = 12.5$ (B).

  Therefore, $J_2(10) = 14$ and the optimal investment choice is A.

- HJB for 3 years to go gives that $J_3(5)$ is the maximum of
  - $0.3 J_2(0) + 0.7 J_2(10) = 9.8$ (A),
  - $0.9 J_2(5) + 0.1 J_2(10) = 8.96$ (B).

  Therefore, $J_3(5) = 9.8$ and the optimal investment choice is A.

4. The optimal investment policy is to invest 5,000 euro each year in investment A, provided you are not broke. This strategy gives the following expected amount of money after three years: 9,800 euros.

(ii)

1. Replace 3 by n with $1 \leq n \leq 3$ and write $p_n(K)$ for the probability that you will have at least 10,000 euros after three years for initial capital K, if you follow the optimal strategy.

2. We do not display the HJB equations.

3. Note that $p_n(0) = 0$ and $p_n(10) = 1$ for all n.

- HJB for 1 year to go gives that $p_1(5)$ is the maximum of 0.7 (A) and 0.1 (B); therefore, $p_1(5) = 0.7$ and the optimal investment choice is A.
- HJB for 2 years to go gives that $p_2(5)$ is the maximum of
  - $0.3p_1(0) + 0.7p_1(10) = 0.7$ (A),
  - $0.9p_1(5) + 0.1p_1(10) = 0.73$ (B).

  Therefore, $p_2(5) = 0.73$ and the optimal investment choice is B.
- HJB for 3 years to go gives that $p_3(5)$ is the maximum of
  - $0.3p_2(0) + 0.7p_2(10) = 0.7$ (A),
  - $0.9p_2(5) + 0.1p_2(10) = 0.757$ (B).

  Therefore, $p_3(5) = 0.757$ and the optimal investment choice is B.

4. The optimal investment policy is as follows: the person should choose B the first two years; if she has precisely 5,000 euros after this, then she should choose A the last year. Otherwise she will have at least 10,000 euros, and then she should choose B the last year as well. This strategy gives the following probability that she will have at least 10,000 euros after 3 years: 0.757.

Conclusion. We have seen that "best investment strategy" can be interpreted in more than one way. It depends on the aim of the investor. Different aims can lead to different optimal strategies.

11.3.6 A monopolist who competes against himself over time

Consider a monopolist of a durable good. He will be confronted with competition, although he is the only person in the world who sells this good. Here is why. Suppose he brings a new exclusive car on the market and decides to sell it for 500,000 dollars. One year later, most of the people who wanted to pay this price have bought the car. Therefore, he lowers the price to 400,000 dollars. As a result, he again sells many cars. One year later he again lowers the price, to 300,000 dollars. The problem with this story is that it is too simple. In reality, many of the people who are prepared to pay 500,000 dollars will not buy the car, because they expect that the price will go down. As a consequence, the monopolist has to compete against himself over time. The following problem illustrates the principle of this competition in a variant of this problem.

Problem 11.3.6 *A seller gives a buyer three times in succession the opportunity to buy a good. The buyer has decided once and for all on a maximum price that he is prepared to pay. If the seller proposes a price that is not higher than this, then the buyer will accept. Otherwise, there is no deal. The seller knows this but he does not know the maximum price the buyer is prepared to pay. His information about this price can be formulated as follows: it is uniformly distributed on the interval $[\frac{1}{2}, 1]$ and it never changes, also not if the buyer buys the good. The seller values the good at 0. What is is the optimal selling strategy if the seller wants to maximize the expected profit from sales?*

The essential point is that each time an offer is accepted or refused, this gives the seller information on the buyer's price, and later the seller should try to use this information to his advantage.

Solution

1. Again, instead of solving just one problem, we solve a family of problems $(P_{n,a,b})_{n,a,b}$, where $n = 1, 2$, or 3, and where a and b are two real numbers with $0 \leq a \leq b \leq 1$. The problem $P_{n,a,b}$ is defined to be the problem to find the selling strategy that maximizes the expected profit for the seller if he gives the buyer the opportunity to buy n times, and if the initial information of the seller on the buyer's price is that it is uniformly distributed on the interval $[a, b]$, called *the uncertainty interval*. The given problem is $P_{3, \frac{1}{2}, 1}$. Note that if we start with uncertainty interval $[\frac{1}{2}, 1]$, then each subsequent uncertainty interval $[a, b]$ is a subinterval of $[\frac{1}{2}, 1]$; therefore, $1 \leq b/a \leq 2$. We write $J_n(a, b)$ for value of the optimization problem $P_{n,a,b}$; that is, it is the maximal expected profit for the seller for the case of n opportunities and initial interval of uncertainty $[a, b]$.

2. We do not display the HJB equations.

3. - HJB for one offer gives that $J_1(a, b)$, which is defined above, is the value of the problem

$$\frac{b-p}{b-a} p \to \max, \ a \leq p \leq b.$$

Indeed, if the seller asks price p, then the probability that this is not higher than the buyer's price is $\frac{b-p}{b-a}$ and so the expected profit is $\frac{b-p}{b-a}p$. Using properties of parabolas, it follows that the solution is

$$\widehat{p}_1(a,b) = a \text{ as } 1 \leq \frac{b}{a} \leq 2;$$

therefore, $J_1(a,b) = a$. Indeed, then a is the point of the admissible set $[a,b]$ that lies closest to the unconstrained minimum $\frac{b}{2}$, which lies outside the admissible set.

- HJB for two offers gives that $J_2(a,b)$ is the value of the problem

$$\frac{p-a}{b-a}(0 + J_1(a,p)) + \frac{b-p}{b-a}(p + J_1(p,b)) \to \max, \ a \leq p \leq b.$$

This problem can be solved as well using properties of parabolas and the formula for J_1 above. We display the solution:

  - if $\frac{b}{a} \geq \frac{3}{2}$, then
    * $p_2(a,b) = \frac{1}{4}a + \frac{1}{2}b$,
    * $J_2(a,b) = \frac{1}{b-a}\left(-\frac{7}{8}a^2 + \frac{1}{2}ab + \frac{1}{2}b^2\right)$;
  - if $1 \leq \frac{b}{a} \leq \frac{3}{2}$, then
    * $p_2(a,b) = a$,
    * $J_2(a,b) = 2a$.

- HJB for three offers gives that $J_3(\frac{1}{2}, 1)$ is the value of the problem

$$\frac{p - \frac{1}{2}}{1 - \frac{1}{2}}\left(0 + J_2\left(\frac{1}{2}, p\right)\right) + \frac{1-p}{1 - \frac{1}{2}}(p + J_2(p, 1)) \to \max,$$

$$a \leq p \leq b.$$

This leads to the solution $p_3(\frac{1}{2}, 1) = \frac{2}{3}$. This involves distinction of the cases $1 \leq 2p \leq \frac{3}{2}$ (which turns out to be "the winning case") and $2p \geq \frac{3}{2}$.

4. In all, the optimal strategy for the seller is as follows: ask the price $\frac{2}{3}$ the first time; if this offer is accepted (resp. refused),

then the next two times ask $\frac{2}{3}$ (resp. $\frac{1}{2}$). The expected value of this strategy is

$$\frac{\frac{2}{3}-\frac{1}{2}}{1-\frac{1}{2}}\cdot 2\cdot\frac{1}{2}+\frac{1-\frac{2}{3}}{1-\frac{1}{2}}\cdot 3\cdot\frac{2}{3}=\frac{5}{3}.$$

11.3.7 Spider and insect

A spider spots an insect, and wants to catch it as quickly as possible. The insect is not aware of the spider, and continues to wander around aimlessly till it meets its fate. We want to gain some insight into the minimal expected time needed to capture the insect, and into the optimal strategy for the spider ("sit" or "move").

Problem 11.3.7 *Spider catching insect. Let us consider the simplest nontrivial model of this problem. The insect and the spider move along a straight line, and their initial distance is n cm, where n is an integer ($n = 0, 1, 2, 3, \ldots$). Each second, the insect moves either 1 cm to the left—with probability p—or 1 cm to the right—with probability p—or stays put—with probability $1 - 2p$; here $0 \leq p \leq \frac{1}{2}$. Simultaneously, each second the spider observes the position of the insect, and then it decides whether to move 1 cm to the left, 1 cm to the right, or stay put. The very moment the spider and the insect are at the same position, the spider catches the insect. The spider wants to choose a strategy that minimizes the expected time of the hunt (in seconds). What is this optimal strategy, and what is the expected time $J(n)$ for this strategy?*

Solution. The first question can be answered almost completely without a precise analysis of the problem: if the spider sees that the distance to the insect is at least 2 cm, it is clearly optimal to make a move in the direction of its prey. This leaves only one strategic problem: what to do if this distance is 1 cm—wait for his prey to move towards him, or take the initiative and move to the position where the insect is now? Intuitively, it is clear that the spider should move toward the insect if p is small ("lazy insect") and that it should stay put and wait for its prey to come to him if p is large ("active insect"). To determine the precise value p_0 such that the spider should move if $p < p_0$ and stay put if $p > p_0$, we turn to the Bellman equation.

Now we specify the meaning of the concepts time, state, control, and value problem in the present situation.

DYNAMIC PROGRAMMING IN DISCRETE TIME

time: we consider all moments $0, 1, 2, 3, \ldots$ (in minutes after the beginning of the hunt).
state: the distance $x \geq 0$ between spider and insect.
control: the decision of the spider whether to move toward the insect or to stay put.
value problem: expected time of the optimal hunt $J(i)$.

We write down the Bellman equations, taking into account the following insights into the optimal strategy for the spider: it is never optimal to move away from the insect, and if $i > 1$, it is not optimal to stay put.

Bellman equations

$$J(x) = 1 + pJ(x) + (1-2p)J(x-1) + pJ(x-2) \quad \forall x \geq 2,$$
$$J(1) = 1 + \min\left[2pJ(1), (1-2p)J(1) + pJ(2)\right],$$
$$J(0) = 0.$$

Note that the first equation can be simplified:

$$J(x) = \frac{1}{1-p} + \frac{1-2p}{1-p}J(x-1) + \frac{p}{1-p}J(x-2) \quad \forall x \geq 2.$$

Let us formulate these three equations in words.

1. The case $x \geq 2$. The expected time $J(x)$ of the optimal hunt if the distance between spider and insect is $x \geq 2$ is equal to the sum of 1—the first second—plus the expected remaining time. During the first second the spider will move toward the insect. After this first second, three states are possible: either it is x with probability p ("insect has moved away from spider"), or it is $x-1$ with probability $1-2p$ ("insect has stayed put"), or it is $x-2$ with probability p ("insect has moved toward spider"). Therefore, this expected remaining time equals

$$pJ(x) + (1-2p)J(x-1) + pJ(x-2).$$

2. The case $x = 1$. The expected time $J(x)$ of the optimal hunt if the distance between spider and insect is 1 is equal to the sum of 1—the first second—plus the expected remaining time. During the first second, the spider will either move to the present position of the insect or stay put. In the first case, either the insect will stay put and then the spider will eat it—this will happen

with probability $1 - 2p$—or the insect will make a move to the left or to the right, and then the distance between spider and insect is again 1—this will happen with probability $2p$. This leads to the following formula for the expected remaining time: $2pJ(1)$. In the second case, either the insect will move to the position of the spider and then the spider will eat it—this will happen with probability p—or the insect will stay put and then the distance between spider and insect is again 1—this will happen with probability $1 - 2p$—or the insect will move away from the spider and then the distance between spider and insect will increase to 2—this will happen with probability p. This leads to the following formula for the remaining time:

$$(1 - 2p)J(1) + pJ(2).$$

What will the spider do? Well this depends on p. It will move to the present position of the insect if this gives the smallest expected remaining optimal hunt, that is, if

$$2pJ(1) < (1 - 2p)J(1) + pJ(2).$$

However it will stay put if that gives the smallest expected remaining time for the optimal hunt, that is, if

$$(1 - 2p)J(1) + pJ(2) < 2pJ(1).$$

Finally, if both of these decisions give the same expected remaining time for the optimal hunt, that is,

$$2pJ(1) = (1 - 2p)J(1) + pJ(2),$$

then both decisions are optimal.

3. The case $x = 0$. If the spider and the insect are at the same position, then the spider will eat the insect immediately, so $J(0) = 0$.

Application of the Bellman equations. We distinguish two cases.

Case 1 ("lazy insect"): p is so small that it is optimal for the spider to move toward the insect also if the distance is 1. That is,

$$2pJ(1) \leq pJ(2) + (1 - 2p)J(1).$$

Therefore the Bellman equation for $x = 1$ is

$$J(1) = 1 + 2pJ(1).$$

As $J(0) = 0$, this gives
$$J(1) = \frac{1}{1-2p}, \quad J(2) = \frac{2}{1-p},$$
and $p \leq \frac{1}{3}$, as can be checked.

Case 2 ("active insect"): p is so big that it is optimal for the spider to stay put if the distance is 1. That is,
$$2pJ(1) \geq pJ(2) + (1-2p)J(1).$$
Then the following formulas hold:
$$J(1) = 1 + pJ(2) + (1-2p)J(1),$$
$$J(2) = 1 + pJ(2) + (1-2p)J(1) + pJ(0).$$
As $J(0) = 0$, this gives
$$J(1) = \frac{1}{p}, \quad J(2) = \frac{1}{p},$$
and $p \geq \frac{1}{3}$.

Conclusion. The optimal strategy for the spider is to move toward the insect if the distance is > 1 or if $p \leq \frac{1}{3}$. If the distance is 1 and $p \geq \frac{1}{3}$ it is optimal to stay put. The expected time of the optimal hunt if the distance between spider and insect is x, that is, $J(x)$, can be calculated as follows.

Case 1: If $p \leq \frac{1}{3}$, then $J(0) = 0$, $J(1) = \frac{1}{1-2p}$ and
$$J(x) = \frac{1}{1-p} + \frac{1-2p}{1-p}J(x-1) + \frac{p}{1-p}J(x-2)$$
for all $x \geq 2$.

Case 2: If $p \geq \frac{1}{3}$, then $J(0) = 0$, $J(1) = \frac{1}{p}$ and
$$J(x) = \frac{1}{1-p} + \frac{1-2p}{1-p}J(x-1) + \frac{p}{1-p}J(x-2)$$
for all $x \geq 2$.

11.3.8 A production and inventory control problem

The problem of planning production and inventory control can be quite complicated even if you know the demand for the product some time in advance. Each month the capacities of production and storage might be different. The same might be true for the costs of production and storage. Dynamic programming can help to minimize total costs of production and inventory control.

Problem 11.3.8 *Suppose the demand for a certain product for the coming three months is known. Each month, the level of production can be chosen, taking into account the maximal capacity for production that month. There is the possibility of holding the product in storage for some time, taking into account each month the maximal capacity for storage. All demand should be satisfied the same month. The aim is to choose the monthly levels of production in such a way that the total variable costs of production and holding over the whole planning period of three months is minimized. The data of this problem are given in the following table. The holding costs in month n are dependent on the stored amount at the end of month n.*

Data		Capacity		Cost per unit	
Month	Demand	Production	Storage	Production	Holding
n	D_n	P_n	W_n	C_n	H_n
Jan ($n=1$)	2	3	2	175	30
Feb ($n=2$)	3	2	3	150	30
Mar ($n=3$)	3	3	2	200	40
Apr ($n=4$)					

The beginning inventory for January is one unit.

Solution
Description as a dynamic programming problem

We have a *dynamic system*:

- The *state* of this system at the beginning of the n-th month ($n = 1, 2, 3, 4$) is denoted by x_n; this is the initial inventory for the n-th month.

- The *control* of this system at the n-th month ($n = 1, 2, 3$) is denoted by d_n; this is the *decision* which is taken in the n-th month: the level of production in this month.

- The *state equation* of this system at the n-th month ($n = 1, 2, 3$)

DYNAMIC PROGRAMMING IN DISCRETE TIME

describes the *dynamics* of the system: how the state changes by the decision taken in this month:

$$x_{n+1} = x_n + d_n - D_n.$$

In words, the inventory increases because the production is added to it, but it also decreases as the demand is taken away from it.

- The *restrictions on the control* at the n-th month ($n = 1, 2, 3$) are

$$0 \leq d_n \leq P_n$$

In words, the restriction on the production capacity has to be satisfied.

- The *restrictions on the state* at the $(n+1)$-th month ($n = 1, 2, 3$) are

$$0 \leq x_{n+1} \leq W_n.$$

In words, the demand has to be satisfied (that is, $0 \leq x_{n+1}$) and the restriction on the storage capacity has to be satisfied (that is, $x_{n+1} \leq W_n$)

- The initial state is given, $x_1 = 1$.

We have just described a dynamic process consisting of three stages:

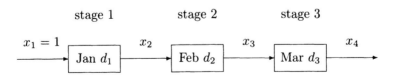

$$x_1 = 1, \quad x_{n+1} = x_n + d_n - D_n \ (n = 1, 2, 3),$$

$$0 \leq d_n \leq P_n \ (n = 1, 2, 3), \quad 0 \leq x_{n+1} \leq W_n \ (n = 1, 2, 3).$$

We have the following grip on this dynamic process. We can *control* it by choosing d_1, d_2, and d_3, ("we can set the levels of production"), provided we make our choice in such a way that the last two restrictions are satisfied.

Now we model the objective function of this problem. This is the total cost over the three months. The cost during the first month is the cost of production and holding in month n ($n = 1, 2, 3$). This is

$$C_n d_n + H_n x_{n+1}.$$

Now we give a formal description of this *dynamic programming problem*.

$$\begin{array}{rl} \text{minimize} & \sum_{n=1}^{3}(C_n d_n + H_n x_{n+1}) \\ \text{s.t.} & x_1 = 1, \\ & x_4 = 0, \\ & x_{n+1} = x_n + d_n - D_n \ (n = 1, 2, 3), \\ & 0 \leq d_n \leq P_n \ (n = 1, 2, 3), \\ & 0 \leq x_{n+1} \leq W_n \ (n = 1, 2, 3). \end{array}$$

Remark. We have added the restriction $x_4 = 0$ as it is not optimal to leave inventory at the end.

We are going to solve this problem in the usual way. Instead of solving only this problem we consider *a collection of* problems containing this problem: for each of the months January, February, and March, and for each possible initial inventory for that month, we pose the following dynamic programming problem:

How to choose the controls in the *remaining time*, such that the total costs of production and holding during the *remaining time* are minimal? We solve these problems *backward*.

Beginning of March. We start with the problem (Q_{x_3}), where $0 \leq x_3 \leq 3$. In words, it is at the beginning of March and the initial inventory is x_3.

$$\begin{array}{rl} \text{minimize} & g_3(d_3) = 200 d_3 + 40 x_4 \\ \text{s.t.} & x_4 = 0, \\ & x_4 = x_3 + d_3 - 3, \\ & 0 \leq d_3 \leq 3, \\ & 0 \leq x_3 \leq 3. \end{array}$$

Solution of (Q_{x_3}): the two equalities in the problem give $d_3 = 3 - x_3$. That is, the solution of (Q_{x_3}) is $d_3 = 3 - x_3$, and its optimal value is $200 \times (3 - x_3) = 600 - 200 x_3$.

DYNAMIC PROGRAMMING IN DISCRETE TIME

Beginning of February. Now we consider the problem (Q_{x_2}), where $0 \leq x_2 \leq 2$. In words, it is the beginning of February and the initial inventory is x_2.

$$\begin{aligned}
\text{minimize} \quad & g_2(d_2) = 150 d_2 + 30 x_3 + (600 - 200 x_3) = 150 d_2 - 170 x_3 + 600 \\
\text{s.t.} \quad & x_3 = x_2 + d_2 - 3, \\
& 0 \leq d_2 \leq 2, \\
& 0 \leq x_2 \leq 2, \\
& 0 \leq x_3 \leq 3.
\end{aligned}$$

Solution of (Q_{x_2}): we begin by eliminating x_3. This gives

$$\begin{aligned}
\text{minimize} \quad & 150 d_2 - 170(x_2 + d_2 - 3) + 600 = 1110 - 20 d_2 - 170 x_2 \\
\text{s.t.} \quad & 0 \leq d_2 \leq 2, \\
& 0 \leq x_2 \leq 2, \\
& 0 \leq x_2 + d_2 - 3 \leq 3.
\end{aligned}$$

This can also be written as

$$\begin{aligned}
\text{minimize} \quad & 1110 - 20 d_2 - 170 x_2 \\
\text{s.t.} \quad & 0 \leq d_2 \leq 2, \\
& 0 \leq x_2 \leq 2, \\
& 3 - x_2 \leq d_2 \leq 6 - x_2.
\end{aligned}$$

This problem is feasible precisely if

$$\max(0, 3 - x_2) \leq \min(2, 6 - x_2),$$

that is, if $1 \leq x_2 \leq 6$. Then the problem (Q_{x_2}) takes the following simpler form.

$$\begin{aligned}
\text{minimize} \quad & 1110 - 20 d_2 - 170 x_2 \\
\text{s.t.} \quad & 0 \leq d_2 \leq 2.
\end{aligned}$$

Conclusion: the problem (Q_{x_2}) is feasible if $1 \leq x_2 \leq 6$; then it has solution $d_2 = 2$ and its optimal value is $1070 - 170 x_2$.

Beginning of January. Finally we consider the original problem. This reduces by working-backward to the following problem:

$$\begin{aligned}
\text{minimize} \quad & 175 d_1 + 30 x_2 + (1070 - 170 x_2) \\
\text{s.t.} \quad & x_2 = 1 + d_1 - 2 = d_1 - 1, \\
& 0 \leq d_1 \leq 3, \\
& 1 \leq x_2 \leq 2.
\end{aligned}$$

Again we give the solution of this problem. We begin by eliminating the variable x_2.

$$\begin{aligned}\text{minimize} \quad & 175d_1 - 140(d_1-1) + 1070 = 35d_1 + 1210 \\ \text{s.t.} \quad & 0 \leq d_1 \leq 3, \\ & 1 \leq d_1 - 1 \leq 2.\end{aligned}$$

This can also be written as

$$\begin{aligned}\text{minimize} \quad & 35d_1 + 1210 \\ \text{s.t.} \quad & 0 \leq d_1 \leq 3, \\ & 2 \leq d_1 \leq 3.\end{aligned}$$

That is, the problem can be rewritten in the following simpler form:

$$\begin{aligned}\text{minimize} \quad & 35d_1 + 1210 \\ \text{s.t.} \quad & 2 \leq d_1 \leq 3.\end{aligned}$$

Conclusion: the problem has solution $d_1 = 2$ and optimal value $35 \times 2 + 1210 = 1280$.

Now we can give the optimal production levels d_1^*, d_2^*, d_3^*.

$d_1^* = 2$; then $x_2^* = d_1^* - 1 = 1$.
So $d_2^* = 2$, $x_3^* = x_2^* + d_2^* - 3 = 0$ and $d_3^* = 3 - x_3^* = 3$.

The following table gives all details of the optimal solution.

Month	Beginning inventory	Production	Production cost	Ending inventory	Holding cost	Total monthly cost
n	x_n^*	d_n^*	$c_n d_n^*$	x_{n+1}^*	$W_n x_{n+1}^*$	$c_n d_n^* + W_n x_{n+1}^*$
Jan	1	2	$350	1	$30	$380
Feb	1	2	$300	0	$0	$300
Mar	0	3	$600	0	$0	$600
			$1250		$30	$1280

11.3.9 A knapsack problem

The following problem is an illustration of a celebrated type of problems, knapsack problems. One is given a collection of objects of different sizes and values. One has a knapsack of fixed size and one has to fill the knapsack with objects in such a way that the total value is maximal. This problem is in general very hard to solve. Dynamic programming can solve relatively small instances of knapsack problems. This is illustrated in the following problem.

DYNAMIC PROGRAMMING IN DISCRETE TIME

Problem 11.3.9 *A trainee on a ship has the following four types of jobs to do during a ten hour shift:*

1. *making coffee (4×)*

2. *checking the position (look at the plan, route and compass)(3×)*

3. *to check—with the engine room—the stability of the ship (2×)*

4. *to help the captain with such things as the transport documents and the bills of lading, and with communication with the agent in port (2×)*

However, there is not enough time to do each type of job the required number of times in this ten hour shift, as the following table shows:

Job category	Estimated completion time per job (in hours)
1	1
2	3
3	4
4	7

Therefore, the trainee must make a choice which jobs he will do and how often he will do these. He wants to make his choice in such a way that his work will be appreciated as much as possible. To do this, he uses the following table:

Job category	Value rating per job
1	2
2	8
3	11
4	20

Solution. We only sketch the first two steps; the others are similar.

Step 1. We consider the following 11 problems: the trainee has $0, 1, 2, \ldots$, or 10 hours to work and he only is allowed to do job 1 ("making coffee" up to 4 times). How often should he do job 1 and to which "score" does this lead? The solutions of these simple problems are displayed in the following table.

hours to work	number of times d_1^* job 1 is done	"score"
0	0	0
1	1	2
2	2	4
3	3	6
4	4	8
5	4	8
6	4	8
7	4	8
8	4	8
9	4	8
10	4	8

Step 2. Now we consider the following 11 problems: the trainee has $0, 1, \ldots$, or 10 hours to work and he is only allowed to do job 1 ("making coffee" up to 4 times) and job 2 ("checking the position" up to 3 times). How often should he do each of these jobs to maximize his "score"? Let us look at an example: the trainee has to work for 6 hours. Now he has three possibilities.

1. He does not do job 2. Then he has 6 hours for job 1. In the table above, one can see that he can score 8 points.

2. He does do job 2 once; this gives him 8 points. In the remaining $6 - 3 = 3$ hours he can score 6 points with job 1 as one can see in the table above. This gives a total of $8 + 6 = 14$ points.

3. He does do job 2 twice. This gives him 16 points; no time remains.

Comparing these three possibilities we see that the last one is the best one: $16 > 8$ and $16 > 14$.

In this way one can solve each of the following 11 problems: the trainee has $0, 1, \ldots$, or 10 hours to work and he is only allowed to do job 1 ("making coffee" up to 4 times) and job 2 ("checking the position" up to 3 times).

Now we display the solutions of these 11 problems in a table. The leading column represents the 11 problems, working either $0, 1, \ldots, 10$ hours. The top row starts with 0, 1, 2, 3: this represents the number of times the trainee can choose to do job 2; it continues with d_2^*, s, u, where d_2^* stands for the optimal number of times to do job 2, s stands for the optimal total score, and u stands for the number of hours that the trainee does not do any task in the optimal situation.

DYNAMIC PROGRAMMING IN DISCRETE TIME   469

For example, the results of the solution of the problem for 6 hours, which was derived above, is given in the underlined row. This starts with 6, denoting that we consider the problem for 6 hours. The next three numbers, 8, 14, 16, stand for the total score of the trainee if the number of times he chooses to do task 2 is 0, 1, or 2. The number 2 in the d_2^*-column is the optimal number of times the trainee should do task 2. The number 16 in the s-column is the optimal score, and the number 0 in the u-column indicates that in the optimal situation he is busy all the time.

These calculations are given in the underlined row in the following table. In particular, the last item 0 means that the trainee is 0 hours idle. The columns with headings 0 to 3 give the maximum scores he can get if the number of times he does job 2 is 0, 1, 2 or 3. The sixth column gives the optimal number of times he does job 2 has to be done, and the seventh column gives the optimal score belonging to this number of times.

	0	1	2	3	d_2^*	s	u
0	0	-	-	-	0	0	0
1	2	-	-	-	0	2	1
2	4	-	-	-	0	4	2
3	6	8	-	-	1	8	0
4	8	10	-	-	1	10	1
5	8	12	-	-	1	12	2
<u>6</u>	<u>8</u>	<u>14</u>	<u>16</u>	-	<u>2</u>	<u>16</u>	<u>0</u>
7	8	16	18	-	2	18	1
8	8	16	20	-	2	20	2
9	8	16	22	24	3	24	0
10	8	16	24	26	3	26	1

Continuing in this way, one has to do two more steps. We restrict ourselves here to the display of the two tables that are the result of these two steps.

	0	1	2	d_3^*		
0	*0*	-	-	0	0	0
1	*2*	-	-	0	2	1
2	*4*	-	-	0	4	2
3	*8*	-	-	0	8	3
4	10	*11*	-	1	11	0
5	12	*13*	-	1	13	1
6	*16*	15	-	0	16	6
7	18	*19*	-	1	19	3
8	20	21	*22*	2	22	0
9	*24*	23	*24*	0;2	24	9;1
10	26	*27*	26	1	27	6

	0	1	d_4^*		
10	27	*28*	1	28	3

Now, one can read off immediately from these four tables the solution of the original optimization problem, that is, the optimal decisions d_1^*, d_2^*, d_3^*, and d_4^*.

Decision	Return
$d_1^* = 0$	0
$d_2^* = 1$	8
$d_3^* = 0$	0
$d_4^* = 1$	20
Total	28

Indeed, the fourth table above shows that $d_4^* = 1$. That is, the trainee should do task 4 once. This leaves him with $10 - 7 \times d_4^* = 3$ hours for the other three tasks. The third table above shows that in this situation one has $d_3^* = 0$. That is, the trainee should not do task 3 at all. This leaves him still with $3 - 4 \times d_3^* = 3$ hours for the first two tasks. The second table above shows that in this situation one has $d_2^* = 1$. That is, the trainee should do task 2 once. This leaves him with $3 - 3 \times d_2^* = 0$ hours for the first task. Now it is clear that $d_1^* = 0$, that is, the trainee should not do task 1 at all. For completeness' sake we mention that this last obvious conclusion also follows alternatively from the first table above.

We have solved this problem of the trainee by using the relatively sophisticated dynamic programming method. Was this really necessary? What would a trainee who just uses his common sense have done? He would probably compare a quotient for each of the four jobs: that of the value rating for that job and the estimated completion time for that job (in hours). This quotient gives the value rating per hour for that job. The higher this quotient, the more attractive is the job for the trainee. Thus he would get the following table.

Job category	Value rating per hour
1	$2/1 = 2$
2	$8/3 = 2.67$
3	$11/4 = 2.75$
4	$20/7 = 2.86$

The trainee will do the most attractive job, that is job 4, as often as he can. He can do it only once: he has 10 hours and job 4 takes 7 hours. In the remaining 3 hours he does not have enough time to do the second most attractive job, that is job 3, as this takes 4 hours. However, in these remaining 3 hours he can just do the third most attractive job once.

Conclusion. Using his common sense the trainee arrives easily at the optimal solution, which we determined above with considerable effort using the dynamic programming method. Does this discredit the dynamic programming method? Not at all! The toy problem we just considered was very simple. Real problems from practice are much more complicated; for these the common sense approach will lead to a result that is usually far from optimal.

11.4 EXERCISES

Exercise 11.4.1 *Interpret the graph in Figure 11.1 as a shortest route problem, and solve it by dynamic programming.*

Exercise 11.4.2 *Someone wants to sell a house and plans to auction it in four weeks time, unless she can sell it privately before then. She assumes that the house will realize 500,000 euros at auction. In each of the intervening weeks there is probability 0.5 that a prospective buyer will appear. Each buyer makes a single offer, the value of which is uniformly distributed over the range 500,000 to 600,000 euros, and which the seller either accepts or rejects. Determine rules for accepting or rejecting offers so as to maximize the seller's expected return.*

Exercise 11.4.3 *A manufacturing company has received an order to* supply one item of a particular type. However, the customer has specified such stringent quality requirements that the manufacturer may have to produce more than one item to obtain an item that is acceptable. To produce a number of extra items in a production run is common practice when producing for a custom order, and it seems advisable in this case.

The manufacturer estimates that each item of this type that is produced will be acceptable with probability $\frac{1}{2}$ and defective (without possibility for rework) with probability $\frac{1}{2}$.

Marginal production costs for this product are estimated to be 100 dollars per item (even if defective), and excess items are worthless. In addition, a setup cost of 300 dollars must be incurred whenever the production process is set up for this product, and a completely new setup at this same cost is required for each subsequent production run, if a lengthy inspection procedure reveals that a completed lot has not yielded an acceptable item. The manufacturer has time to make no more than three production runs. If an acceptable item has not been obtained by the end of the third production run, the cost to the manufacturer in lost sales income and penalty costs will be 1,600 dollars.

The objective is to determine the policy regarding the lot size (one plus reject allowance) for the required production run(s) that minimize expected total cost for the manufacturer.

Exercise 11.4.4 *Reconsider the manufacturer's problem from the previous exercise. Suppose that the situation has changed somewhat. After a more careful analysis, you now estimate that each item produced will be acceptable with probability $\frac{2}{3}$, rather than $\frac{1}{2}$. Furthermore, there now is only enough time available to make two production runs. Find the new optimal policy for this problem.*

Dynamic programming can be used to solve some LP problems and even nonlinear programming problems. To do this you have to view the choice of the variables as decisions.

Exercise 11.4.5 *Solve the following LP problem by dynamic programming:*

$$3x_1 + 5x_2 \to \max, \quad x_1 \leq 4, \quad 2x_2 \leq 12, \quad 3x_1 + 2x_2 \leq 18, \quad x_1, x_2 \geq 0.$$

Exercise 11.4.6 *Solve the following nonlinear programming problem by dynamic programming:*

$$18x_1 - x_1^2 + 20x_2 + 10x_3 \to \max, \quad 2x_1 + 4x_2 + 3x_3 \leq 11, \quad x_1, x_2, x_3 \geq 0,$$

where we allow the variables to take only integers as values.

DYNAMIC PROGRAMMING IN DISCRETE TIME 473

The optimal strategy for two-person games, consisting of a sequence of moves, can often be found using dynamic programming.

Exercise 11.4.7 *Game with matches.* *Each of two players can take in turn some matches from a bunch of matches, at least one and at most half of the remaining matches. The player who is confronted with the last match wins (alternatively, you can choose the rule that this player loses). Find the optimal strategy for playing this game.*

Exercise 11.4.8 * *Nim.* *Some rows of coins are lying on the table. Each of two players in turn selects a row and then takes one coin or two adjacent coins from this row (note that this will often increase the number of rows). The player who takes the last coin wins. Find the optimal strategy for playing this game.*

Exercise 11.4.9 * *Counterfeit coins.* *A number of the same coins are given. One of them is heavier than the other ones. A balance beam is given. Find a weighing scheme to identify the counterfeit coin in as few weighings as possible. Note that there is more than one way to interpret this problem: for example, you can minimize the number of weighings in the worst case, or you can minimize the expected number of weighings.*

Chapter Twelve

Dynamic Optimization in Continuous Time

If one considers motion with the same initial and terminal points then, the shortest distance between them being a straight line, one might think that the motion along it needs least time. It turns out that this is not so.

G. Galilei

- How to find the extrema of an infinite-dimensional problem with constraints, that is, how to solve the problem

$$J(x(\cdot), u(\cdot)) = \int_{t_0}^{t_1} f(t, x(t), u(t))dt \to \text{extr}, \ \dot{x} = \varphi(t, x, u),$$

$$u(t) \in U \ \forall t, \ x(t_i) = x_i, \ i = 0, 1.$$

12.1 INTRODUCTION

Plan. We will give a brief historical introduction to the calculus of variations, optimal control, and dynamic programming.

Euler. Galileo was the first natural philosopher of the Renaissance period. He had infinite-dimensional problems in mind. Our epigraph shows that he was thinking about the problem of the *brachistochrone*, the slide of quickest descent without friction. A modern application of the solution is to the design of slides in aqua-parks. Galileo did not solve this problem himself. Johann Bernoulli (1667–1748) settled the problem in 1696, maybe under the influence of Galileo. He showed that the optimal curve is a *cycloid* (Fig. 12.1); we have already come across this curve, in chapter nine, in the analysis of the perfect pendulum of Huygens. This marks the date of birth of a new branch of optimization, which came to be called the *calculus of variations*. The way in which this result was brought to the attention of the scientific world was as follows. Bernoulli published a note in *Acta Eruditorum*,

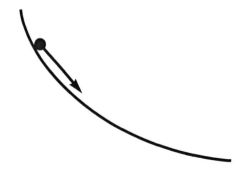

Figure 12.1 Brachistochrone.

the first scientific journal, which began publication in 1682. The title was intriguing, "A New Problem that Mathematicians are Invited to Solve." A quarter of a century later, Bernoulli posed to his student Leonard Euler (1707–1783) the task of finding general methods for solving problems that are similar to that of the brachistochrone. Euler was equal to this task and began to develop these methods.

Lagrange. In 1759 Euler received a letter from an unknown author. This letter contained a great leap forward for a theory that until then Euler had been developing on his own. The author of this letter was Lagrange, then twenty-three years old. Lagrange proposed for the proof of necessary conditions a new method, the method of *variations*. Moreover, the letter contained the germ of the multiplier rule for problems of the variational calculus with constraints. Euler was elated about the letter of the young author. He wrote him as an answer that he himself had also succeeded in making progress in the new calculus (which he later called the *calculus of variations*, terminology which is preserved to our time), but he refrained from publication of his results until the young scholar had obtained from his method all that he could. An unprecedented case of generosity.

The development of the calculus of variations lasted until the 1930s.
Pontryagin. We make a jump in time, two centuries forward. In the mid-1950s, Lev Semenovich Pontryagin (1908–1988) started a seminar on the mathematical theory of control. At this seminar, specialists were invited to present concrete problems that occurred

in practice (Feldbaum, Lerner, Rozonoer, and others). As a result, Pontryagin and his students defined a class of problems, the problems of *optimal control*, containing all variational problems, and also the majority of the special cases considered by the seminar. Necessary conditions were obtained for this problem, and these conditions came to be called the *maximum principle of Pontryagin*.

Bellman. Around the same time, Richard Bellman (1920–) became interested in the theory of multidecision processes. His invention in 1953 of *dynamic programming* set the stage for the application of functional equation techniques in a wide spectrum of fields. The central idea of dynamic programming is *the principle of optimality*. With hindsight some ideas of dynamic programming can be traced back to the works of Hamilton and Jacobi and even Huygens.

Royal road. If you want a shortcut to the applications, then it suffices to read the following problems, their necessary conditions (just the equations), and the examples illustrating these conditions.

- The simplest problem of the calculus of variations ($P_{12.1}$), the Euler equation and the transversality conditions for free final state and for free final time are given in theorem 12.2, and illustrated in examples 12.2.1 and 12.2.2.

- The Lagrange problem of the calculus of variations ($P_{12.2}$), the Euler-Lagrange equation and the transversality conditions for free final state and for free final time are given in theorem 12.4, and illustrated in example 12.2.3.

- The problem of optimal control ($P_{12.3}$) and the Pontryagin maximum principle are given in theorem 12.6, and illustrated in examples 12.2.4, 12.2.5 and 12.2.6.

- The continuous time dynamic programming problem and the Hamilton-Jacobi-Bellman equation are not given explicitly, but the idea is given in the form of a story about a boat, in section 12.2.4.

Conclusion. The calculus of variations, optimal control and dynamic programming have been developed from around 1700 till around 1950 by many researchers, including Euler, Lagrange, Pontryagin, and Bellman.

12.2 MAIN RESULTS: NECESSARY CONDITIONS OF EULER, LAGRANGE, PONTRYAGIN, AND BELLMAN

12.2.1 Calculus of variations: Euler equation and transversality conditions

Let $L: \mathbb{R}^3 \to \mathbb{R}$ $(L = L(t, x, \dot{x}))$ be a continuous function of three variables.

The problem

$$J(x(\cdot)) = \int_{t_0}^{t_1} L(t, x(t), \dot{x}(t)) dt \to \min, \quad x(t_i) = x_i, \ i = 0, 1 \quad (P_{12.1})$$

is called *the simplest problem of the calculus of variations*.

The problem $(P_{12.1})$ will be considered in the space $C^1([t_0, t_1])$, equipped with the C^1-norm $\|\cdot\|_{C^1([t_0,t_1])}$, defined by

$$\|x(\cdot)\|_{C^1([t_0,t_1])} = \max_{t \in [t_0,t_1]} \max(|x(t)|, |\dot{x}(t)|).$$

Definition 12.1 *A local minimum $\widehat{x}(\cdot)$ of the problem $(P_{12.1})$ in the space $C^1([t_0,t_1])$ is called a weak minimum of $(P_{12.1})$. This means that there exists $\varepsilon > 0$ such that $J(x(\cdot)) \geq J(\widehat{x}(\cdot))$ for all $x(\cdot) \in C^1([t_0,t_1])$ with $x(t_i) = x_i$, $i = 0, 1$ and*

$$\|x(\cdot) - \widehat{x}(\cdot)\|_{C^1([t_0,t_1])} \leq \varepsilon.$$

Theorem 12.2 *Euler equation and transversality condition.*
Let a function $\widehat{x}(\cdot)$ be admissible in the problem $(P_{12.1})$

(this means that $\widehat{x}(\cdot) \in C^1([t_0, t_1])$ and $\widehat{x}(t_i) = x_i$, $i = 0, 1$),

and L, L_x, $L_{\dot{x}}$ are continuous in a neighborhood of the curve (in three dimensional space $\mathbb{R}^3$)

$$t \mapsto (t, \widehat{x}(t), \dot{\widehat{x}}(t)), \ t \in [t_0, t_1].$$

If the function $\widehat{x}(\cdot)$ is a weak minimum in the problem $(P_{12.1})$, then the function $\widehat{L}_{\dot{x}}(\cdot)$ is continuously differentiable and the Euler equation

$$-\frac{d}{dt}\widehat{L}_{\dot{x}}(t) + \widehat{L}_x(t) = 0 \qquad (1)$$

is satisfied.

If we consider the problem with final state x_1 free, then the following transversality condition holds true:

$$\widehat{L}_{\dot{x}}(t_1) = 0;$$

DYNAMIC OPTIMIZATION IN CONTINUOUS TIME

if the final time t_1 is free, then the following transversality condition holds true:

$$(\widehat{L}_{\dot{x}}\dot{\widehat{x}} - \widehat{L})(t_1) = 0.$$

(We have used the abbreviations

$$\widehat{L}_{\dot{x}}(t) = L_{\dot{x}}(t, \widehat{x}(t), \dot{\widehat{x}}(t)), \widehat{L}_x(t) = L_x(t, \widehat{x}(t), \dot{\widehat{x}}(t)).$$

Similar abbreviations will appear further on.)

Direct verification of optimality. It is not so easy to extend the Weierstrass theorem to problems of the calculus of variations and of optimal control. We recommend the following elementary method to complement the analysis, which is not so well known, but which is often surprisingly effective: *verify directly that the candidate solution is a global optimum*. In the examples and applications in this chapter we will assume tacitly that all candidate solutions are global optima. We will display the direct verification in the first example only.

The following examples illustrate the theorem.

Example 12.2.1 *Solve the simplest problem of the calculus of variations in the following numerical case:*

$$t_0 = 0, \ t_1 = 2, \ x_0 = 0, \ x_1 = 8, \ L(t, x, \dot{x}) = 12tx + \dot{x}^2.$$

Solution. We apply the four-step method:

1. $J(x(\cdot)) = \int_0^2 (12tx + \dot{x}^2) dt \to \min, \ x(0) = 0, \ x(2) = 8.$

2. Euler: $-\frac{d}{dt}\widehat{L}_{\dot{x}}(t) + \widehat{L}_x(t) = 0 \Rightarrow -2\ddot{x} + 12t = 0.$

3. $x = t^3 + at + b$. The terminal conditions $x(0) = 0, \ x(2) = 8$ lead to $a = 0, \ b = 0$. That is, $\widehat{x}(t) = t^3$. Direct verification: for each admissible $x(\cdot)$, write $h(\cdot) = x(\cdot) - \widehat{x}(\cdot)$; then

$$J(x(\cdot)) - J(\widehat{x}(\cdot)) = \int_0^2 [(12t(\widehat{x}+h) + (\dot{\widehat{x}}+\dot{h})^2) - (12t\widehat{x}+\dot{\widehat{x}}^2)]dt,$$

which—after expanding and canceling terms—can be simplified to

$$\int_0^2 (12th + 2\dot{\widehat{x}}\dot{h} + \dot{h}^2)dt.$$

Partial integration gives

$$(\dot{\widehat{x}}h)\mid_0^2 + \int_0^2 (12th - 2\ddot{\widehat{x}}h + \dot{h}^2)dt = \int_0^2 \dot{h}^2 dt$$

(use that $h(0) = h(2) = 0$, as $h(\cdot)$ is the difference of two admissible functions, and that $12t - 2\ddot{\hat{x}} = 0$ as $\hat{x}(t) = t^3$). Finally, note that this outcome is nonnegative, as squares are always nonnegative and the integral of a nonnegative-valued function is nonnegative.

4. The problem has a unique global minimum $\hat{x}(t) = t^3$.

Now we consider the same problem, but with final state free.

Example 12.2.2 *Solve the simplest problem of the calculus of variations in the following numerical case:*

$$t_0 = 0, \ t_1 = 2, \ x_0 = 0, \ x_1 = \text{free}, \ L(t, x, \dot{x}) = 12tx + \dot{x}^2.$$

Solution. We apply the four-step method:

1. $J(x(\cdot)) = \int_0^2 (12tx + \dot{x}^2) dt \to \min, \ x(0) = 0, \ x(2) = \text{free}$.

2. Euler: $-\frac{d}{dt}\hat{L}_{\dot{x}}(t) + \hat{L}_x(t) = 0 \Rightarrow -2\ddot{x} + 12t = 0$,

    transversality $\hat{L}_{\dot{x}}(2) = 0 \Rightarrow 2\dot{x}(2) = 0$.

3. Euler $\Rightarrow x = t^3 + at + b$ and transversality $\Rightarrow a = -12$. The initial condition $x(0) = 0$ gives $b = 0$. The analysis can be completed by direct verification.

4. The problem has a unique global minimum $\hat{x}(t) = t^3 - 12t$.

12.2.2 Calculus of variations: Euler-Lagrange equation

Let

$$f : [t_0, t_1] \times \mathbb{R}^n \times \mathbb{R}^r \to \mathbb{R} \ (f = f(t, x, u))$$

be a continuous function, and

$$\varphi : [t_0, t_1] \times \mathbb{R}^n \times \mathbb{R}^r \to \mathbb{R}^n \ (\varphi = \varphi(t, x, u))$$

a continuous mapping.

The problem

$$J((x(\cdot), u(\cdot))) = \int_{t_0}^{t_1} f(t, x(t), u(t)) dt \to \min, \ \dot{x} = \varphi(t, x, u),$$

$$x(t_i) = x_i, \ i = 0, 1 \qquad (P_{12.2})$$

is called *the Lagrange problem of the calculus of variations* (in Pontryagin's form). We consider this problem in the space

$$X_1 = C^1([t_0, t_1], \mathbb{R}^n) \times C([t_0, t_1], \mathbb{R}^r),$$

equipped with the norm $\|\cdot\|_{X_1}$ defined by

$$\|(x(\cdot), u(\cdot))\|_{X_1} = \max(\|x(\cdot)\|_{C^1([t_0,t_1],\mathbb{R}^n)}, \|u(\cdot)\|_{C([t_0,t_1],\mathbb{R}^r)}).$$

A pair of functions $(x(\cdot), u(\cdot))$ that satisfies the constraints of a problem $(P_{12.2})$ is called an admissible one for this problem.

Definition 12.3 *A local minimum of the problem* $(P_{12.2})$ *in the space* X_1 *is called a weak minimum of the Lagrange problem* $(P_{12.2})$.

The function

$$\mathcal{L}((x(\cdot), u(\cdot)), \lambda) = \int_{t_0}^{t_1} L(t, x(t), \dot{x}(t), u(t)) dt,$$

where

$$L(t, x, \dot{x}, u) = \lambda_0 f(t, x, u) + p(t) \cdot (\dot{x} - \varphi(t, x, u)),$$

is called *the Lagrange function* for the problem $(P_{12.2})$, and $L(t, x, \dot{x}, u)$ is called *the Lagrangian*.

A *selection of Lagrange multipliers* consists of a pair $(p(\cdot), \lambda_0)$, the first coordinate of which is a vector function $p : [t_0, t_1] \to (\mathbb{R}^n)'$ and the second is a nonnegative number λ_0.

Theorem 12.4 Euler-Lagrange equations for problem $(P_{12.2})$
Let the pair $(\widehat{x}(\cdot), \widehat{u}(\cdot))$ be admissible in $(P_{12.2})$ and let f, f_x, f_u, φ, φ_x, φ_u be continuous in a neighborhood of the curve

$$t \mapsto (t, \widehat{x}(t), \widehat{u}(t)), \ t \in [t_0, t_1]$$

in $\mathbb{R} \times \mathbb{R}^n \times \mathbb{R}^r$. If the pair $(\widehat{x}(\cdot), \widehat{u}(\cdot))$ is a weak minimum in the problem $(P_{12.2})$, then there exist a number λ_0 and a vector function

$$p(\cdot) \in C^1([t_0, t_1], (\mathbb{R}^n)'),$$

not both zero, such that the Euler equations over x:

$$-\frac{d}{dt}\widehat{L}_{\dot{x}}(t) + \widehat{L}_x(t) = 0 \Leftrightarrow \dot{p} = -p\widehat{\varphi}_x(t) + \lambda_0 \widehat{f}_x(t) \quad (12.2_a)$$

and over u:

$$\widehat{L}_u(t) = 0 \Leftrightarrow p(t)\widehat{\varphi}_u(t) = \lambda_0 \widehat{f}_u(t) \quad (12.2_b)$$

are satisfied.

If we consider the problem with final state x_1 free, then the following transversality condition holds,

$$\widehat{L}_{\dot{x}}(t_1) = 0;$$

if the final time t_1 is free, then the following transversality condition holds:

$$(\widehat{L}_{\dot{x}} \cdot \dot{\widehat{x}} - \widehat{L})(t_1) = 0.$$

The formulae (12.2$_a$) and (12.2$_b$) are called *the Euler-Lagrange equations* for the Lagrange problem ($P_{12.2}$).

Most types of problems of the calculus of variations can be reformulated as a Lagrange problem. This is illustrated by the following example.

Example 12.2.3 *Solve the following problem:*

$$\int_0^1 \dot{x}^2 dt \to \min, \quad \int_0^1 x dt = 0, \quad x(0) = 0, \quad x(1) = 1.$$

Solution. The constraint $\int_0^1 x dt = 0$ is equivalent to the existence of a function $y(\cdot)$ for which $\dot{y} = x$, $y(0) = 0$, $y(1) = 0$. Therefore, writing $x_1 = x$, $x_2 = y$, and $u = \dot{x}_1$ gives that the given problem is equivalent to the Lagrange problem (where we use again the symbol x, but now to denote $x = (x_1, x_2)^T$)

$$J(x(\cdot), u(\cdot)) = \int_0^1 u^2 dt \min, \quad \dot{x} = (u, x_1)^T,$$

$$x(0) = (0, 0)^T, \quad x(1) = (1, 0)^T.$$

This problem can be solved with the Euler-Lagrange equations and direct verification. This leads to the following solution for the given problem $\widehat{x} = 3t^2 - 2t$.

12.2.3 Optimal control: Pontryagin's Maximum Principle

Plan. We consider the same situation as for the Lagrange problem ($P_{12.2}$), but now we add the constraint $u(t) \in U$ $\forall t$ and we allow $u(\cdot)$ to have a finite number of points of discontinuity.

Let U be a subset of $\mathbb{R}^r$,
$$f : [t_0, t_1] \times \mathbb{R}^n \times U \to \mathbb{R} \ (f = f(t, x, u))$$
be a continuous function, and
$$\varphi : [t_0, t_1] \times \mathbb{R}^n \times U \to \mathbb{R}^n \ (\varphi = \varphi(t, x, u))$$
a continuous mapping.

The problem
$$J((x(\cdot), u(\cdot))) = \int_{t_0}^{t_1} f(t, x(t), u(t)) dt \to \min, \ \dot{x} = \varphi(t, x, u),$$

$$(u(t) \in U \ \forall t), \ x(t_i) = x_i, \ i = 0, 1 \quad (P_{12.3})$$

is called *the problem of optimal control*.

We consider problem $(P_{12.3})$ in the space
$$X_2 = PC^1([t_0, t_1], \mathbb{R}^n) \times PC([t_0, t_1], \mathbb{R}^r),$$
equipped with the norm
$$\|\cdot\|_{C([t_0,t_1], \mathbb{R}^n)}$$
on $PC^1([t_0, t_1], \mathbb{R}^n)$ (that is, $x(\cdot)$ is a piecewise continuously differentiable vector function and $u(\cdot)$ is a piecewise continuous vector function).

A pair of functions $(x(\cdot), u(\cdot))$ that satisfies the constraints of problem $(P_{12.3})$ is called an admissible one for this problem.

Definition 12.5 *We say that an admissible pair $(\widehat{x}(\cdot), \widehat{u}(\cdot))$ in $(P_{12.3})$ is a strong minimum of the problem of optimal control $(P_{12.3})$ if there exists $\varepsilon > 0$ such that*
$$J(x(\cdot), u(\cdot)) \geq J(\widehat{x}(\cdot), \widehat{u}(\cdot))$$
for all admissible pairs $(x(\cdot), u(\cdot))$ in $(P_{12.3})$ with
$$\|(x(\cdot) - \widehat{x}(\cdot)\|_{C([t_0,t_1], \mathbb{R}^n)} \leq \varepsilon.$$

The function
$$L(t, x, \dot{x}, u) = \lambda_0 f(t, x, u) + p(t) \cdot (\dot{x} - \varphi(t, x, u))$$
is called *the Lagrangian* and the function
$$\mathcal{L}((x(\cdot), u(\cdot))) = \int_{t_0}^{t_1} L(t, x(t), \dot{x}(t), u(t)) dt,$$

is called *the Lagrange function* for the problem $(P_{12.3})$.

A *selection of Lagrange multipliers* consists here of a pair $(p(\cdot), \lambda_0)$, the first coordinate of which is a vector function $p : [t_0, t_1] \to (\mathbb{R}^n)'$ and the second is a nonnegative number λ_0.

Theorem 12.6 *Pontryagin maximum principle for problem* $(P_{12.3})$. *Let the pair $(\widehat{x}(\cdot), \widehat{u}(\cdot))$ be admissible in $(P_{12.3})$ and let f, f_x, φ, φ_x be continuous in a neighborhood of the curve*

$$t \mapsto (t, \widehat{x}(t), \widehat{u}(t)), \ t \in [t_0, t_1]$$

in $\mathbb{R} \times \mathbb{R}^n \times U$. If the pair $(\widehat{x}(\cdot), \widehat{u}(\cdot))$ is a strong minimum in the problem $(P_{12.3})$, then there exist a nonnegative number λ_0 and a vector function

$$p(\cdot) \in PC^1([t_0, t_1], (\mathbb{R}^n)'),$$

not both zero, such that the Euler equations over x,

$$-\frac{d}{dt}\widehat{L}_{\dot{x}}(t) + \widehat{L}_x(t) = 0 \Leftrightarrow \dot{p} = -p\widehat{\varphi}_x(t) + \lambda_0 \widehat{f}_x(t), \qquad (12.3_a)$$

and the minimum condition over u,

$$\min_u L(t, \widehat{x}(t), \dot{\widehat{x}}(t), u) = \widehat{L}(t), \qquad (12.3_b)$$

for each point t of continuity of $u(\cdot)$, are satisfied.

If we consider the problem with final state x_1 free, then the following transversality condition holds true:

$$\widehat{L}_{\dot{x}}(t_1) = 0;$$

if the final time t_1 is free, then the following transversality condition holds true:

$$(\widehat{L}_{\dot{x}} \cdot \dot{\widehat{x}} - \widehat{L})(t_1) = 0.$$

Now we will give the usual formulation of the necessary conditions for Lagrange problems and optimal control problems, in terms of Hamiltonians.

Consider the setup of problem $(P_{12.2})$ or problem $(P_{12.3})$. The *Hamiltonian* is defined to be

$$H(t, x, u) = p\varphi(t, x, u) - \lambda_0 f(t, x, u).$$

Necessary conditions in Hamiltonian form.

- The necessary conditions of problem $(P_{12.2})$ together with its constraints can be written as follows: there exist a nonnegative number λ_0 and a function $p(\cdot) \in C^1[t_0, t_1]$, not both zero, such that

$$\dot{\widehat{x}} = \widehat{H}_p,$$

$$\dot{\widehat{p}} = -\widehat{H}_x,$$

$$\widehat{H}_u = 0,$$

$$\widehat{x}(t_i) = x_i, \ i = 0, 1.$$

- The necessary conditions of problem $(P_{12.3})$ together with its constraints can be written as follows: there exist a nonnegative number λ_0 and a function $p(\cdot) \in PC^1[t_0, t_1]$, not both zero, such that,

$$\dot{\widehat{x}} = \widehat{H}_p,$$

$$\dot{\widehat{p}} = -\widehat{H}_x,$$

$$\widehat{H}(t) = \max_{u \in U} H(t, \widehat{x}, u, p, \lambda),$$

$$\widehat{x}(t_i) = x_i, \ i = 0, 1.$$

If we consider either of these problems with free final state x_1, then the following transversality condition holds true: $\widehat{p}(t_1) = 0$; if the final time t_1 is free, then the following transversality condition holds true: $\widehat{H}(t_1) = 0$.

Transition from Lagrangian to Hamiltonian. Let us make some remarks about this transition from L to H. The new formulation is simpler in the following two respects: it does not involve $\dot{x}$ and the formulation of the transversality conditions is simpler. Moreover, we note that the transformation from $\dot{x} \mapsto L(t, x, \dot{x}, u)$ to $u \mapsto H(t, x, u)$ is natural in the following sense: it is essentially the Legendre transform, the differentiable analog of the Fenchel transform considered in appendix B. Indeed, the problem

$$g(\xi) = \xi \cdot \dot{x} - L(t, x, \dot{x}) \rightarrow \text{extr}$$

has stationarity equation

$$\xi - L_{\dot{x}}(t, x, \dot{x}) = 0$$

and so it has stationary value

$$L_{\dot{x}}(t, x, \dot{x}) \cdot \dot{x} - L(t, x, \dot{x}) = p \cdot \varphi(t, x, u) - \lambda_0 f(t, x, u) = H(t, x, u).$$

The following three examples illustrate the use of Pontryagin's maximum principle (PMP). Note in particular the ease with which the precise description of discontinuous solutions is found.

Hamiltonian linear in the control. In the first two examples f and φ are linear in the variable u, and this variable is restricted to some closed interval. As a consequence, the Hamiltonian is linear in u, and so PMP leads to the maximization of a linear function of one variable on a closed interval. The result of this maximization is, of course, that the right-hand (resp. left-hand) side of the interval is the point of maximum if the slope of the linear function is positive (resp. negative)

Example 12.2.4 *Solve the optimal control problem in the following numerical case:*

$$t_0 = 0,\ t_1 = 2,\ x_0 = 4,\ x_1 = \text{free},\ f(t, x, u) = 3u - 2x,\ \varphi(t, x, u) = x + u,$$

$$U = [0, 2].$$

Solution

1. $J(x(\cdot), u(\cdot)) = \int_0^2 (3u - 2x) dt \to \min,\ \dot{x} = x + u,\ x(0) = 4,$

$$u(t) \in [0, 2].$$

2. Hamiltonian $H = p\varphi - \lambda_0 f = p(x + u) - \lambda_0(3u - 2x).$

   Pontryagin:

   (a) $\dot{\widehat{x}} = \widehat{x} + \widehat{u},$
   (b) $\dot{\widehat{p}} = -(p + 2\lambda_0),$
   (c) $u \mapsto \widehat{p}(\widehat{x} + u) - \widehat{\lambda}_0(3u - 2\widehat{x})$ assumes its maximum value at $u = \widehat{u}$ for all $t \in [0, 2],$
   (d) $\widehat{x}(0) = 4,\ p(2) = 0.$

3.
- $\lambda_0 \ne 0$: indeed, if $\lambda_0 = 0$, then $\dot{p} = -p$, $p(2) = 0 \Rightarrow p \equiv 0$, so we are led to a contradiction, as the Lagrange multipliers λ_0 and $p(\cdot)$ cannot be both zero.

  We put $\lambda_0 = 1$.

- Simplification of the maximum condition. The maximum condition gives

$$\hat{u} = \text{bang}[0, 2; p - 3].$$

  This notation means that $\hat{u}(t) = 0$ for all t with $p(t) - 3 < 0$ and $\hat{u}(t) = 2$ for all t with $p(t) - 3 > 0$.

  To see this, one has to apply the following well-known fact to the minimum principle: an affine function $x \to cx + d$ on a closed interval $[a, b]$ assumes its maximum value at a if $c < 0$ and at b if $c > 0$. This is possible in this example as the Hamiltonian depends affine on u—with coefficient of u equal to $p - 3$—and, moreover, U is the closed interval $[0, 2]$.

PMP in shortened form (and omitting the hats):

(a) $\dot{x} = x + u$,

(b) $\dot{p} = -p - 2$,

(c) $u = \text{bang}[0, 2; p - 3]$,

(d) $x(0) = 4, p(2) = 0$.

- Determination of p. $\dot{p} = -p - 2$, $p(2) = 0 \Rightarrow p(t) = 2e^{2-t} - 2$.

- Determination of u. The expression for u above leads to: $u(t) = 2$ for $t < \tau$ and $= 0$ for $t > \tau$, where τ is the solution of $p(t) - 3 = 0$, which is seen to be $2 - \ln(\frac{5}{2}) \approx 1.096$.

- Determination of x. On $[0, \tau]$ we get

$$\dot{x} = x + 2, x(0) = 4 \Rightarrow x(t) = 6e^t - 2.$$

  In particular, at the endpoint τ of the interval $[0, \tau]$, we have $x(\tau) = (\frac{12}{5})e^2 - 2$. On $[\tau, 2]$ we get

$$\dot{x} = x, \ x(\tau) = (\frac{12}{5})e^2 - 2 \Rightarrow x(t) = (6 - 5e^{-2})e^t.$$

4. The problem has the following unique solution:
$$\widehat{u}(t) = 2, \ \widehat{x}(t) = 6e^t - 2 \ \forall t < \tau$$
and
$$\widehat{u}(t) = 0, \ \widehat{x}(t) = (6 - 5e^{-2})e^t \ \forall t > \tau,$$
where $\tau = (12/5)e^2 - 2 \approx 1.096$.

Example 12.2.5 *Solve the problem*
$$J(x(\cdot), u(\cdot), T) = T \rightarrow \min, \ \dot{x} = x + u, \ x(0) = 5, \ x(T) = 11,$$
$$u(t) \in [-1, 1].$$

Solution

1. We can write this problem as an optimal control problem ($P_{12.3}$) by rewriting the objective function as $J(x(\cdot), u(\cdot), T) = \int_0^T 1 dt$.

2. Hamiltonian $H = p\varphi - \lambda_0 f = p(x+u) - \lambda_0$.
   Pontryagin:
   $$\dot{\widehat{x}} = \widehat{x} + \widehat{u},$$
   $$\dot{p} = -p,$$
   $$u \mapsto p(\widehat{x} + u) - \lambda_0$$
   assumes its maximum value at $u = \widehat{u}(t)$ for all $t \in [0, T]$,
   $$x(0) = 5, \ x(T) = 11, \ p(T)(\widehat{x}(T) + \widehat{u}(T)) - \lambda_0 = 0.$$

3. We put $\lambda_0 = 1$, as we may, we omit the verification, and we simplify the maximum condition to $\widehat{u} = \text{bang}[-1, 1; p]$.
   PMP in shortened form (and omitting the hats):
   (a) $\dot{x} = x + u$,
   (b) $\dot{p} = -p$,
   (c) $u = \text{bang}[-1, 1; p]$,
   (d) $x(0) = 5, \ x(T) = 11, \ p(T)(x(T) + u(T)) = 1.$

- Determination of p. $\dot{p} = -p \Rightarrow p(t) = Ae^t$. Moreover,
  $$p(T)(x(T)+u(T)) = 1, \ x(T) = 11, \ u(T) = \pm 1 \Rightarrow p(T) > 0,$$
  and so $A > 0$.
- Determination of u. The expression for u above leads to $u \equiv 1$.
- Determination of x. We have $\dot{x} = x + 1$, $x(0) = 5 \Rightarrow x = 6e^t - 1$.
- Determination of T. We have $x(T) = 11 \Rightarrow e^T = 2$. This gives $T = \ln 2$.

4. The problem has the following unique solution:
$$\widehat{x}(t) = 6e^t - 11, \ \widehat{u}(t) = 1, \ T = \ln 2.$$

Hamiltonian quadratic in the control. In the next example one of the functions f and φ is quadratic in u, the other one is at most quadratic, and again u is restricted to a closed interval. As a consequence, the Hamiltonian is quadratic in u. Moreover, the coefficient of u^2 will happen to be negative for all choices of t and x in this example. This leads to problems of maximizing a mountain parabola on a closed interval. Here, three cases have to be distinguished, according to whether the top lies in the interval, left of it, or right of it. In the first case the top is the solution; otherwise the endpoint of the interval closest to the top is the solution.

Example 12.2.6 *Solve the following problem:*
$$J(x(\cdot)) = \int_0^4 (\dot{x}^2 + x)dt \to \min, \ |\dot{x}| \leq 1, \ x(0) = 0.$$

Solution. We solve the problem as an optimal control problem.

1. $J(x(\cdot), u(\cdot)) = \int_0^4 (u^2 + x)dt \to \min, \ \dot{x} = u, \ x(0) = 0,$
$$u \in [-1, 1].$$

2. Hamiltonian: $H = p\varphi - \lambda_0 f = pu - \lambda_0(u^2 + x).$

   Pontryagin:
   $$\dot{\widehat{x}} = \widehat{u},$$
   $$\dot{p} = \lambda_0,$$

$$u \mapsto pu - \lambda_0(u^2 + x)$$

assumes its maximal value at $u = \widehat{u}(t)$ for all $t \in [0, 4]$,

$$\widehat{x}(0) = 0, \ p(4) = 0.$$

3. To begin with, we put $\lambda_0 = 1$, as we may, we omit the verification, and we simplify the maximum condition to

$$\widehat{u} = \text{sat}\left[\frac{1}{2}p; -1, 1\right].$$

This notation means $\widehat{u}(t) = \frac{1}{2}p(t)$ provided $\frac{1}{2}p(t) \in [-1, 1]$; otherwise $\widehat{u}(t)$ equals the endpoint of the interval $[-1, 1]$ that lies closest to $\frac{1}{2}p$ ("sat" is short for saturation).

PMP in shortened form (and omitting the hats):

(a) $\dot{x} = u$,
(b) $\dot{p} = 1$,
(c) $u = \text{sat}[\frac{1}{2}p; -1, 1]$,
(d) $x(0) = 0, \ p(4) = 0$.

- Determination of p. $\dot{p} = 1$, $p(4) = 0 \Rightarrow p = t - 4$.
- Determination of u. The graph of $\frac{1}{2}p = \frac{1}{2}t - 2$ shows that $u(t) = -1$ for $t < 2$, and $u(t) = \frac{1}{2}t - 2$ for $t > 2$.
- Determination of x. On $[0, 1]$: $\dot{x} = -1$, $x(0) = 0 \Rightarrow x(t) = -t$. On $[2, 4]$: $\dot{x} = \frac{1}{2}t - 2$, $x(2) = 2 \Rightarrow x(t) = \frac{1}{4}t^2 - 2t + 1$.

4. The problem has the following unique solution

$\widehat{x}(t) = -t$ for $t \in [0, 2]$ and $\widehat{x}(t) = \frac{1}{4}t^2 - 2t + 1$ for $t \in [2, 4]$,

$\widehat{u}(t) = -1$ for $t \in [0, 2]$ and $u(t) = \frac{1}{2}t - 2$ for $t \in [2, 4]$.

12.2.4 Dynamic programming: Hamilton-Jacobi-Bellman equation

Plan. The method of dynamic programming—the HJB-equation—in continuous time is very simple, and the proof is also very simple, and intuitive; we will present it below, using an allegory. However, the *formalism* of the continuous time dynamic programming problem and its HJB equation might look at first glance relatively heavy. Therefore, we will not display the formal description of the HJB equation in continuous time. Dynamic programming has many applications of interest, and a particular useful feature of it that it can be readily extended to *stochastic* problems.

Boat and dynamic programming. A wealthy man takes a day rest from his business and takes his bookkeeper with him on a sailing boat trip to an island. The bookkeeper knows that on the island landing rights have to be paid for, so he plans how to sail to the place on the island where the landing rights are lowest. Having made this plan he takes the required amount of money out of his wallet, in order not to lose any moment at the landing, when the payment is due. He starts telling his boss how to sail. The boss ignores his instructions and sails in another direction. As he has taken a day off in order to relax, he just wants to sail a bit on the sea before heading for the island. To his dismay, the bookkeeper realizes that, because of the strong wind, certain rocks in the sea, and the tide, they cannot reach the cheapest landing place any more. He does not flinch, however, and quickly finds the cheapest landing place that can be reached from their new position. Then he takes some more money out of his wallet, in order to have the right amount of money ready for the new landing place. He points out the unnecessary additional costs to his boss and asks him not to ignore his sailing instructions. However, the boss is unmoved, and the bookkeeper is forced to keep on making new plans and taking additional money out of his wallet. He gets increasingly excited about this waste of money, and finally his boss takes pity on him. From this moment on, the boss follows the instructions of the bookkeeper and as a result the bookkeeper does not have to take additional money out of his wallet. At the landing place, the bookkeeper pays immediately; this is precisely the amount of money he has in his hand.

Translation of the boat story into the Bellman equation. This story gives the idea of dynamic programming; it is easy to turn this story into a rigorous derivation of the Hamilton-Jacobi-Bellman equation, which is essentially a partial differential equation. To see how to do this, note that the money the bookkeeper has in his hand at any time t_0 and at any position x_0 on the sea is the value of an optimization problem. We refer to the problem of minimizing the landing costs if you start your boat trip at time t_0 at position x_0. Let us denote it by $V(t_0, x_0)$. Let us denote the position of the boat at time t by $x(t)$ for all t. Then the function $t \mapsto V(t, x(t))$ is monotonic nondecreasing: this represents the simple insight that the bookkeeper will never put money back into his wallet. Moreover, if the function

$t \mapsto V(t, x(t))$ is constant for some time, then this is the same as saying that the boss follows the instructions of his bookkeeper during this time. It remains to use the fact that the function $t \mapsto V(t, x(t))$ is constant (resp. monotonic nondecreasing) for some time if its derivative

$$\frac{dV}{dt}(t, x(t)) = \frac{\partial V}{\partial t}(t, x(t)) + \frac{\partial V}{\partial x}(t, x(t))\dot{x}(t)$$

is zero (resp. nonnegative) during this time. This fact is the essence of the Bellman equation, also called Hamilton-Jacobi-Bellman equation (HJB).

Conclusion. We have explained the essence of dynamic programming in continuous time in terms of a simple story about a boat.

12.3 APPLICATIONS TO CONCRETE PROBLEMS

12.3.1 Illustration Euler equation: growth theory and Ramsey's model

How much should a nation save? Two possible answers are: *nothing* "Après nous le déluge," Madame de Pompadour, mistress of Louis XV) and *everything* ("Yes, they live on rations, they deny themselves everything. [...] But with this gold new factories will be built [...] a guarantee for future plentifulness." from the novel *Children of the Arbat* of Anatoly Rybakov, illustrating an aspect of the economic policy of Stalin). A third answer is given by a model by Frank P. Ramsey: choose the golden mean; save something, but consume (enjoy) something now as well. Ramsey's paper on optimal social saving behavior is among the very first applications of the calculus of variations to economics. This paper has exerted an enormous if delayed influence on the current literature on optimal economic growth. A simple version of this model is the following optimization model:

$$I(C(\cdot), k(\cdot)) = \int_0^\infty \mathcal{U}(C)e^{-\theta t}dt \to \max, \ \dot{k} = F(k) - C.$$

Here
  $C = C(t) =$ the rate of consumption at time t,
  $\mathcal{U}(C) =$ the utility of consumption C,
  $\theta =$ the discount rate,
  $k = k(t) =$ the capital stock at time t,
  $F(k) =$ the rate of production when capital stock is k.

It is usual to assume $\mathcal{U}(C) = (1-\rho)^{-1}C^{1-\rho}$ for some $\rho \in (0,1)$ and $F(k) = Ak^{1/2}$ for some positive constant A. Then the solution of the problem cannot be given explicitly; however, a qualitative analysis shows that it is optimal to let consumption grow asymptotically to some finite level. Now let us consider a modern variant of this model from [66] and [65]. The intuition behind the model above allows one to model the production function as $F(k) = Ak$ for some positive constant A instead of $F(k) = Ak^{1/2}$. Now we apply Euler's result to this problem. To this end we eliminate C, rewriting the equation $\dot{k} = F(k) - C$ as $C = F(k) - \dot{k}$; the result is the problem

$$J(k(\cdot)) = \int_0^\infty -(1-\rho)^{-1}(Ak - \dot{k})^{1-\rho} e^{-\theta t} dt \to \min.$$

Let $\widehat{k}(\cdot)$ be a solution of this problem and write $\widehat{C}(\cdot)$ for the corresponding consumption function. The Euler equation gives

$$-A\widehat{C}^{-\rho}e^{-\theta t} - \frac{d}{dt}(\widehat{C}^{-\rho}e^{-\theta t}) = 0.$$

This implies $\widehat{C}^{-\rho}e^{-\theta t} = re^{-At}$ for some constant r. Therefore,

$$\widehat{C} = C_0 e^{\frac{A-\theta}{\rho}t}.$$

Therefore, this modern version has a more upbeat conclusion: there is an explicit formula for the solution of the problem and moreover, consumption can continue to grow forever to unlimited levels.

12.3.2 Illustration of the Euler equation: the road from Moscow to St. Petersburg

Once the czar ordered a road to be built from Moscow to St. Petersburg. The optimal choice of the precise trajectory was a very difficult problem due to obstacles such as villages and swamps. Finally, in desperation, the engineers in charge decided to consult the czar himself. The czar flew into a rage that he was bothered about such a trifle. He asked for a map of the region, took a ruler, and drew a straight line from Moscow to St. Petersburg. The authors of this book verified this story and one day drove to St. Petersburg and back. Indeed, you drive in a straight line, "as the crow flies", as the saying goes. That is, almost: somewhere there is a small deviation from the straight line. The full story explains this as well. The reason is the czar's finger. He pressed the ruler to the map with one hand when he drew the line, but one of his fingers was in the way, and so the line is not completely straight at one point. The engineers were too intimidated to discuss the matter, and included "the finger of the czar" in the trajectory. The following problem illustrates the sense of the solution of the czar: it is the ancient wisdom that a straight line is the shortest connection between two points.

Problem 12.3.1 *Find the shortest connection between two points.*

Solution. We will model the problem as a simple problem of the calculus of variations and then solve it, using the Euler equation. We will put the following intuitive insight about the optimal solution into this model: it is the graph of a continuously differentiable function. Moreover, we will use the following formula for the length $J(x(\cdot))$ of the graph of a function $x(\cdot)$ between the points (t_0, x_0) and (t_1, x_1):

$$\int_{t_0}^{t_1} (1 + \dot{x}^2)^{\frac{1}{2}} dt.$$

This can be derived by approximating the function by a continuous piecewise linear function, using the theorem of Pythagoras, and then letting the approximation converge.

We apply the four-step method.

1. $J(x(\cdot)) = \int_{t_0}^{t_1} (1 + \dot{x}^2)^{\frac{1}{2}} dt$, $x(t_0) = x_0$, $x(t_1) = x_1$.

2. Euler: $-\frac{d}{dt}\widehat{L}_{\dot{x}}(t) + \widehat{L}_x(t) = 0 \Rightarrow -\frac{d}{dt}(\frac{1}{2}(1 + \dot{x}^2)^{-\frac{1}{2}} 2\dot{x}) + 0 = 0$.

3. $\dot{x}(1 + \dot{x}^2)^{-\frac{1}{2}}$ is constant $\Rightarrow \dot{x}$ is constant. It follows that the graph of $x(\cdot)$ is a straight line. A direct verification shows that this is indeed optimal.

4. The straight line between two given points is the unique shortest connection between them.

12.3.3 Pontryagin's maximum principle: Newton's aerodynamic problem

In his *Philosophiae Naturalis Principia Mathematica* "Mathematical Principles of Natural Philosophy" ([59]), arguably the greatest scientific work of all ages, Newton considers the following problem: *"figures may be compared together as to their resistance; and those may be found which are most apt to continue their motions in resisting mediums."*

Newton proposed a solution for this problem (cf. Fig. 12.2). However, this solution was not understood until relatively recently (see [74] for one example—the discussion in [76]—and [2] for another one– *"the solution of this problem has an internal singularity, of which Newton was aware, but of which his publishers in the 20th century were not apparently aware and smoothed out a figure"*). It has generally been considered an example of a mistake by a genius. One of the formalizations is the following:

$$J(u(\cdot)) = \int_0^T \frac{tdt}{1+u^2} \to \min, \ u(\cdot) \in PC([0,T]),$$

$$\int_0^T udt = \xi, \ u \geq 0. \tag{P}$$

The relation of problem (P) with Newton's problem can be described as follows. Let $\hat{u}(\cdot)$ be a solution of (P). Take the primitive $\hat{x}(\cdot)$ of $\hat{u}(\cdot)$ that has $\hat{x}(0) = 0$. Its graph is a curve in the t, x-plane. Now we take the surface of revolution of this curve around the x-axis. This is precisely the shape of the front of the optimal figure in Newton's problem. The details of this relation are given in [74]. The constraint $u(\cdot) \geq 0$ (the monotonicity of $x(\cdot)$) was not made explicit by Newton. We stress once more that precisely this type of constraint can be dealt with very well by Pontryagin's maximum principle, but not by the calculus of variations. For this problem the conditions of PMP have the following form. There are constants $\lambda_0 \geq 0$ and λ, not both zero, such that $\hat{u}(\cdot)$ is a solution of the following auxiliary problem

$$I(u(\cdot)) = \int_0^T \frac{tdt}{1+u^2} + \lambda \left[\left(\int_0^T udt\right) - \xi\right] \to \min,$$

$$u(\cdot) \in PC[0,T], \ u \geq 0. \tag{Q}$$

It is intuitively clear that a piecewise continuous function $\hat{u}(\cdot)$ is a solution of (Q) precisely if for all points t of continuity of $\hat{u}(\cdot)$ the nonnegative value of u that minimizes the integrand

$$g_t(u) = \lambda_0 \frac{t}{1+u^2} + \lambda u$$

is $u = \hat{u}(t)$. In fact, it is not difficult to give a rigorous proof of this claim. Thus the problem has been reduced to the minimization of differentiable functions $g_t(u)$ of one nonnegative variable u. Clearly, for each t the function $g_t(u)$ is minimal either at $u = 0$ or at a solution of the stationarity equation $\frac{d}{du}g_t(u) = 0$. Now a straightforward calculation leads to an explicit determination of the—unique—solution of the problem (Q). One can verify directly that this is also a solution of (P). The resulting optimal shape is given in Figure 12.2 (in cross-section). We observe in particular that it has kinks.

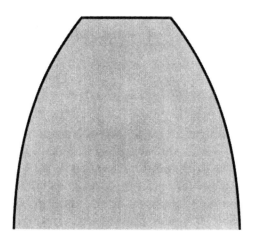

Figure 12.2 Optimal shape of spacecraft.

This is precisely the solution that was proposed by Newton. Also in another respect Newton was ahead of his time here: his solution has been used to design the optimal shape of spacecraft.

Screwdriver. Very recently this story has taken an unexpected turn. It is not yet at an end: a shape has been discovered that is better than the one proposed by Newton. How can this be possible? For hundreds of years all investigators analyzing this problem have made the plausible assumption that the optimal shape should be *rotation symmetric*. However, the new and better shape that has been discovered looks like a 'screwdriver' and so is not rotation symmetric. It is not yet clear whether this shape is optimal or whether there are even better shapes.

12.3.4 Time (in)consistency.

Introduction. In 2004 the Nobel prize in economics was awarded to Finn Kydland and Edward Prescott. One of their ideas can be explained in terms of the difference between solutions between dynamic optimization problems by the following two different methods: Pontryagin's maximum principle and the Bellman equation. The former gives an *open loop solution*: this specifies all decisions to be taken during the planning period (in technical terms: a control function $u(t)$ of time t). The latter gives a *closed loop solution* (or *feedback solution*): this solution gives a contingency plan, that is, a strategy how to make decisions during the planning period depending on how the situation will be in the future (in technical terms: a control function $u(t,x)$ of time t and the current state x). At first sight it might seem that a closed loop solution is always preferable because of its greater flexibility. However, this is not always the case, as the model from [39] below demonstrates. If a decision-maker can make the commitment to carry out a Pontryagin solution, she is in a strong position. A decision-maker who uses a Bellman solution is vulnerable to manipulations. Models of Kydland and Prescott that captured this insight played a role in creating independent central banks. This is believed to have been a crucial factor in a long period with low inflation.

The model. At $t = 0$ an investor decides whether or not to invest in a country. At $t = 1$ the government decides on the tax rate on capital. We assume the payoff structure depicted in Figure 12.3.

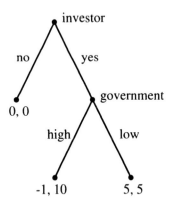

Figure 12.3 Payoff structure.

If the investor does not invest in the country, both the investor and the government get payoffs normalized to zero. If the investor does invest in the country and the government implements a low tax rate, the investor and the government both get a payoff equal to 5. If, however, the investor invests and the government at $t = 1$ chooses a high tax rate, the investor loses and gets -1 while the government gains a big tax receipt and gets 10.

If the government can commit at $t = 0$ to the action it takes at $t = 1$, it would commit to a low tax rate. Commitment can be achieved by making a law stating that the tax rate is low (that is, changing the law leads to a cost bigger than 5, making the switch to a high tax rate unprofitable later on). Or the government may try to get a reputation for low tax rates. In case of a government trying to fight inflation, the commitment comes from creating a central bank that is independent.

This commitment solution corresponds to the open loop solution or the Pontryagin solution to a dynamic optimization problem. In the problem above the government would commit to choosing the low tax rate at $t = 1$.

If the government cannot commit, it will choose its optimal plan at each moment it has to make a decision. That implies that once the investor has invested, the government chooses the high tax rate since a payoff of 10 exceeds a payoff of 5. The investor (using backward induction) sees this happening and chooses not to invest because a payoff of 0 exceeds a payoff of -1. The case of no commitment corresponds to the Bellman solution or closed loop solution: at each moment in time you choose your best action.

Clearly, the commitment solution is never worse than the no-commitment solution (because you can always commit to the Bellman solution) and in the case above it is clearly better. That is, commitment has value. An example often used here is Ulysses tied to the mast to listen to the Sirens. In that way, he committed not to go to them while he could still listen.

Also note that in a deterministic world commitment is always better than no commitment. In a world with uncertainty, there is trade-off. With commitment one cannot respond to unexpected events. In the example above, you may have committed to a low tax rate, but the country may end up in fiscal crisis and you cannot respond by increasing the tax rate.

12.4 DISCUSSION AND COMMENTS

Proofs of theorems 12.2, 12.4, and 12.6 are given in appendix G, "Conditions of Extremum from Fermat to Pontryagin." These proofs are simple but they are given from an advanced point of view.

12.4.1 Euler equation as a special case of the Fermat theorem

The following facts might come as a surprise: the Euler equation and the transversality condition are just decoded versions of the condition of the Fermat theorem ("derivative equal to zero").

Explanation. We give here an intuitive explanation of this point of view for the case of the Euler equation. We want to decode the condition $J'(\widehat{x}(\cdot)) = 0$. To begin with, how can the derivative $J'(\widehat{x}(\cdot))$ be defined in this context? According to the approximation approach to the derivative, given in definitions 1.9 and 2.2, we have to approximate the difference

$$J(\widehat{x}(\cdot) + h(\cdot)) - J(\widehat{x}(\cdot))$$

by an expression that is linear in $h(\cdot)$. Here $h(\cdot)$ is a function with $h(0) = h(1) = 0$—then $\widehat{x}(\cdot)$ and $\widehat{x}(\cdot) + h(\cdot)$ both satisfy the endpoint conditions. Well, substituting the linear approximation

$$L(t, \widehat{x} + h, \dot{\widehat{x}} + \dot{h}) \approx L(t, \widehat{x}, \dot{\widehat{x}}) + L_x(t, \widehat{x}, \dot{\widehat{x}})h + L_{\dot{x}}(t, \widehat{x}, \dot{\widehat{x}})\dot{h}$$

under the integral sign, we get the following approximation for the difference $J(\widehat{x}(\cdot) + h(\cdot)) - J(\widehat{x}(\cdot))$:

$$\int_{t_0}^{t_1} (L_x(t, \widehat{x}, \dot{\widehat{x}})h + L_{\dot{x}}(t, \widehat{x}, \dot{\widehat{x}})\dot{h})dt.$$

Then, integrating by parts, and using $h(t_0) = h(t_1) = 0$, we get the following expression:

$$\int_{t_0}^{t_1} (L_x(t, \widehat{x}(t), \dot{\widehat{x}}(t)) - \frac{d}{dt}L_{\dot{x}}(t, \widehat{x}(t), \dot{\widehat{x}}(t)))h(t)dt.$$

This expression depends linearly on $h(\cdot)$, as required. This makes it plausible that this is the derivative $J'(\widehat{x}(\cdot))$. Now we are ready to decode the condition $J'(\widehat{x}(\cdot)) = 0$. That is, this condition means that

$$\int_{t_0}^{t_1} \left(L_x(t, \widehat{x}(t), \dot{\widehat{x}}(t)) - \frac{d}{dt}L_{\dot{x}}(t, \widehat{x}(t), \dot{\widehat{x}}(t)) \right) h(t)dt = 0$$

for all C^1-functions $h(\cdot)$ on $[t_0, t_1]$ for which $h(t_0) = h(t_1) = 0$. For each $\tau \in (t_0, t_1)$ we can choose a C^1-function on $[t_0, t_1]$ that is equal to 1 on a sufficiently small interval containing τ, and that decreases sufficiently rapidly to zero outside this interval. The choice of such functions $h(\cdot)$ leads to the conclusion that the function

$$L_x(t, \widehat{x}(t), \dot{\widehat{x}}(t)) - \frac{d}{dt}L_{\dot{x}}(t, \widehat{x}(t), \dot{\widehat{x}}(t))$$

has value zero at $t = \tau$. As $\tau \in (t_0, t_1)$ is arbitrary, it follows that the Euler equation holds on (t_0, t_1). By continuity, it holds at the endpoints t_0 and t_1 as well. One can go on in this way to derive the transversality condition $L_{\dot{x}}(t_1) = 0$ for the problem where $x(t_1)$ is free.

12.4.2 Derivation of Pontryagin's maximum principle from the discrete version

Plan. PMP is not very intuitive at first sight. Here we make it plausible in the following way. The discrete time version of the PMP is just a special case of the discrete time HJB theorem. By a transition from discrete time to continuous time we get the formulation of the continuous time PMP.

Let $\hat{x} = (\hat{x}_0, \ldots, \hat{x}_{n+1})$ and $\hat{u} = (\hat{u}_0, \ldots, \hat{u}_n)$ be two row vectors $(x_k, u_k) \to f_k(x_k, u_k)$ and $(x_k, u_k) \to \varphi(x_k, u_k)$, with $k = 0, \ldots, n$, continuously differentiable functions of two variables, and U a set of real numbers.

Definition 12.7 *The problem*

$$F(x, u) = \sum_{k=0}^{n} f_k(x_k, u_k) \to \min,$$

$$x_{k+1} - x_k = \varphi_k(x_k, u_k), \ u_k \in U, \ k = 0, \ldots, n,$$

$$x_0 = a, \ x_{n+1} = b \quad (P_{12.4})$$

is called a discrete optimal control problem.

One says that $(\hat{x}, \hat{u})$ is a *point of strong minimum* of $(P_{12.4})$ if there exists a sufficiently small positive number $\varepsilon > 0$ such that $(\hat{x}, \hat{u})$ is a point of global minimum on the set of those admissible points (x, u) of the problem for which $|x - \hat{x}| < \varepsilon$.

The functions

$$H_k(x_k, u_k, p_k, \lambda_0) = -\lambda_0 f_k(x_k, u_k) + p_k \varphi_k(x_k, u_k), \ k = 0, \ldots, n,$$

are called *the Hamiltonian functions* of $(P_{12.1})$ and the variables λ_0 and p_k, $0 \le k \le n$, are called *Lagrange multipliers* of $(P_{12.4})$. We will call the row vector

$$p = (\lambda_0, p_0, \ldots, p_n)$$

a *selection of Lagrange multipliers*. We will write

$$\frac{\partial \widehat{H}_k}{\partial \lambda_k} \quad (\text{resp.} \ \frac{\partial \widehat{H}_k}{\partial x_k})$$

for the value at $(\widehat{x}_k, \widehat{u}_k, \widehat{p}_k, \widehat{\lambda}_0)$ of the partial derivative of H_k with respect to λ_k (resp. x_k) for $0 \leq k \leq n$. We will write $\widehat{H}_k$ for the number $H_k(\widehat{x}_k, \widehat{u}_k, \widehat{p}_k, \widehat{\lambda}_0)$ for $0 \leq k \leq n$.

Theorem 12.8 Pontryagin's maximum principle: discrete time. Consider a problem of type $(P_{12.4})$. If $(\widehat{x}, \widehat{u})$ is a point of strong minimum of $(P_{12.4})$, then there is a nonzero selection of Lagrange multipliers $(\widehat{\lambda}_0, \widehat{p}_0, \ldots, \widehat{p}_n)$ such that the following conditions hold true for $0 \leq k \leq n$:

$$\widehat{x}_{k+1} - \widehat{x}_k = \frac{\partial \widehat{H}_k}{\partial \lambda_k},$$

$$\widehat{\lambda}_{k-1} - \widehat{\lambda}_k = \frac{\partial \widehat{H}_k}{\partial x_k},$$

$$\widehat{H}_k = \max_{u_k \in U} H_k(\widehat{x}_k, u_k, \widehat{p}_k).$$

This result is a formal consequence of the Bellman equation in discrete time. Transition from discrete time to continuous time is readily seen to give PMP in continuous time.

We emphasize that we have not proved anything. We have only made PMP plausible by means of the transition from discrete to continuous time.

Conclusion. For discrete time optimal control problems, one can apply the discrete time HJB theorem. If we go over from discrete to continuous time, then we get a condition that happens to be the necessary condition for the continuous time dynamic optimization problem. This heuristic approach makes it plausible that the resulting condition is indeed necessary.

Appendix A

On Linear Algebra: Vector and Matrix Calculus

- Throughout the book we profit from the convenience of vector and matrix notation and methods. This allows us to carry out the analysis of optimization problems involving two or more variables in a transparent way, without messy computations.

A.1 INTRODUCTION

The main advantage of linear algebra is conceptual. A collection of objects is viewed as one object, as a *vector* or as a *matrix*. This makes it often possible to write down data and formulas and to manipulate these in such a way that the structure is presented in a transparent way.

The main objects of study of linear algebra are a system of linear equations and its set of solutions. There are three main questions. What is the efficient way to solve such a system? Which insight can be gained into the relation between systems and their solution sets? When is the solution unique, and can you give a formula for it in this case? These questions have completely satisfactory answers, as we will see. Linear algebra is one of the success stories of mathematics. If you can model a problem as a linear problem, it is almost certain that it can be solved in a straightforward way.

A.2 ZERO-SWEEPING OR GAUSSIAN ELIMINATION, AND A FORMULA FOR THE DIMENSION OF THE SOLUTION SET

We will answer the first two questions above. Whenever you want to solve a concrete problem of linear algebra such as solving a numerical system of linear equations, then the best way to do this is by a method called zero-sweeping or Gaussian elimination. In fact, this holds more generally for any concrete numerical problem of linear algebra, such as for example the computation of determinants and inverses of matrices. The basic idea, known from high-school, is successive elimination of variables. We will now describe the basic idea of this algorithm and give the promised insight.

Example A.2.1 *Solve the following system of linear equations:*

$$2x_1 + 3x_2 = -6,$$

$$5x_1 - 2x_2 = 23.$$

Solution. Choose one of the two equations, say, the first one, and one of the two variables, say, x_1. Use the first equation to express x_1 in the remaining variable $x_1 = -3 - \frac{3}{2}x_2$. Substitute this expression in the other equation. This gives

$$5(-3 - \frac{3}{2}x_2) - 2x_2 = 23,$$

which can be simplified to $\frac{-19}{2}x_2 = 38$ and this gives $x_2 = -4$, and so $x_1 = -3 - \frac{3}{2}(-4) = 3$.

In general, if you have a system of m linear equations in n variables, then one step of the elimination process reduces both the number of equations and the number of variables by one. This gives a system of $m-1$ equations in $n-1$ variables. Continuing in this way, you will end up with one equation in $n-m+1$ variables, provided $m \leq n$ and provided you do not run into complications. You can choose $n-m$ of the variables freely and the remaining one can be expressed in these free variables. Then you can go on to achieve by substitution that all variables are expressed in the free variables. This finishes the description of the algorithm in case there are no complications. As a byproduct we get the following insight:

the number of free variables in the general solution equals $n - m$, the number of variables minus the number of equations.

However, in general two types of complications can arise: it could happen that during this process you come across an equation $0 = 0$ (such an equation should be omitted and "should not be counted") or an equation $0 = a$ for some nonzero number a (in this case you should stop: then the system has no solutions). The following examples illustrate these complications.

Example A.2.2 *Solve the following system of linear equations:*

$$2x_1 + 3x_2 = -6,$$

$$6x_1 + 9x_2 = -18.$$

Solution. Three times the first equation is the second equation. Therefore, you could say that the true number of equations is only one. This is spotted by the algorithm: if you apply the algorithm, then using the first equation to express x_1 in x_2 and substituting this expression in the second equation, you get the equation $0 = 0$.

Example A.2.3 *Solve the following system of linear equations:*

$$2x_1 + 3x_2 = -6,$$

$$6x_1 + 9x_2 = -20.$$

Solution. Three times the first equation is $6x_1 + 9x_2 = -18$, and this contradicts the second equation. Therefore, there are no solutions. This is spotted by the algorithm: if you apply the algorithm, then using the first equation to express x_1 in x_2 and substituting this expression into the second equation, you get the equation $0 = -2$.

Remark. It is possible to analyze these three problems from a geometrical point of view as well. In all cases we have two straight lines in the plane. Then there are three possibilities: the two lines intersect in one point (see example A.1, where this point is $\begin{pmatrix} 3 \\ -4 \end{pmatrix}$, or the two lines coincide (see example A.2), or they are parallel (see example A.3).

Taking these complications into account leads to the following general rule:

the number of free variables in the general solution equals the number of variables minus the "true" number of equations, provided the system has any solutions at all.

This is essentially all that can be said about the two questions above. What remains to be done is to give professional notation to do the bookkeeping of the algorithm in a convenient way, and to introduce professional concepts to formulate the rule above. This notation makes use of matrices.

The following example illustrates the professional notation to do the bookkeeping.

Example A.2.4 *Solve the system of linear equations,*

$$2x_1 + 3x_2 = -6,$$

$$5x_1 - 2x_2 = 23.$$

Solution. We write the system in matrix notation, using a 2×3 matrix,

$$\begin{pmatrix} 2 & 3 & -6 \\ 5 & -2 & 23 \end{pmatrix}.$$

This is called the—extended—matrix of the system. We carry out the steps of the algorithm again, now using this notation. We divide the first row by 2:

$$\begin{pmatrix} 1 & 3/2 & -3 \\ 5 & -2 & 23 \end{pmatrix}.$$

Then we subtract from the second row 5 times the first row:

$$\begin{pmatrix} 1 & 3/2 & -3 \\ 0 & -19/2 & 38 \end{pmatrix}.$$

After this we divide the second row by $-19/2$:

$$\begin{pmatrix} 1 & 3/2 & -3 \\ 0 & 1 & -4 \end{pmatrix}.$$

Finally, we subtract from the first row $3/2$ times the second row:

$$\begin{pmatrix} 1 & 0 & 3 \\ 0 & 1 & -4 \end{pmatrix}.$$

It remains to go back to the usual notation: $x_1 = 3$ and $x_2 = -4$. Written in vector notation, $x = \begin{pmatrix} x_1 \\ x_2 \end{pmatrix} = \begin{pmatrix} 3 \\ -4 \end{pmatrix}$.

Two matrices of the same size can be added coordinatewise, and one can multiply a matrix by a scalar. Vectors can be seen as special matrices: row vectors are $1 \times s$ matrices and column vectors are $r \times 1$ vectors.

Finally, we will introduce the promised professional concepts needed to formulate the rule above. If a system of m linear equations in n variables is given, and $a, v_1, \ldots, v_k$ are vectors in $\mathbb{R}^n$ such that each solution of the system can be written in a unique way as a linear combination of these vectors with coefficient of a equal to 1, that is, as $a + \mu_1 v_1 + \cdots + \mu_k v_k$ for a unique selection of numbers $\mu_1, \ldots, \mu_k$, then we call $k \geq 0$ the *dimension of the solution set of the system*. This is the professional concept for the intuitive concept "number of free variables in the general solution." It can be shown that for every solvable system such vectors $a, v_1, \ldots, v_k$ can be chosen and that for each choice we have that k, the number of v's, is the same.

For each matrix A, we can choose a collection of rows such that each row of the matrix can be expressed as a linear combination of the rows of this collection in a unique way. The number of rows in this collection is called *the rank of the matrix A*. It can be shown that this number does not depend on the choice of the collection. The rank of the coefficient matrix of a given system is the professional concept for the intuitive concept "true number of equations." Thus the rule above can be given its precise formulation:

the dimension of the solution set equals the number of variables minus the rank of the extended matrix, provided the system has any solutions at all.

A.3 CRAMER'S RULE

In this section we answer the third question. We begin with an experiment and solve a system of two linear equations in two variables in the general case,

$$a_{11}x_1 + a_{12}x_2 = b_1,$$

$$a_{21}x_1 + a_{21}x_2 = b_2.$$

A calculation gives the following result:

$$x_1 = \frac{b_1 a_{22} - b_2 a_{12}}{a_{11} a_{22} - a_{12} a_{21}}, \quad x_2 = \frac{a_{11} b_2 - a_{21} b_1}{a_{11} a_{22} - a_{12} a_{21}},$$

provided $a_{11}a_{22} - a_{12}a_{21} \neq 0$.

This is called *Cramer's rule* for the given system. This extends in principle to systems of n linear equations in n variables, but one gets unwieldy formulas. In order to control these somehow, we observe that in the case $n = 2$, the numerators and denominators all have the same structure. If we call the expression

$$a_{11}a_{22} - a_{12}a_{21}$$

in the denominators the *determinant* of the matrix $\begin{pmatrix} a_{11} & a_{12} \\ a_{21} & a_{22} \end{pmatrix}$ and denote it as $\det \begin{pmatrix} a_{11} & a_{12} \\ a_{21} & a_{22} \end{pmatrix}$, then Cramer's rule for the case $n = 2$ can be written as follows:

$$x_1 = \frac{\det \begin{pmatrix} b_1 & a_{12} \\ b_2 & a_{22} \end{pmatrix}}{\det \begin{pmatrix} a_{11} & a_{12} \\ a_{21} & a_{22} \end{pmatrix}}, \quad x_2 = \frac{\det \begin{pmatrix} a_{11} & b_1 \\ a_{21} & b_2 \end{pmatrix}}{\det \begin{pmatrix} a_{11} & a_{12} \\ a_{21} & a_{22} \end{pmatrix}},$$

provided $\det \begin{pmatrix} a_{11} & a_{12} \\ a_{21} & a_{22} \end{pmatrix} \neq 0$.

Now, the structure of the formula for the solution is transparent. This way of presenting the result, including the definition of the concept of determinant, can be extended to the case of general n. This extension is called Cramer's rule for a system of n linear equations in n variables.

Remark. It is possible to give a geometrical interpretation of the determinant in terms of the transformation $y = Ax$. We consider the case $n = 2$. If the determinant is zero, then the transformation is degenerate: the plane is mapped onto a line or even onto the origin. Otherwise, it is convenient to interpret the absolute value and the sign of the determinant separately. The transformation multiplies areas by the factor $|\det A|$ and it preserves (resp. reverses) the orientation if $\det A$ is positive (resp. negative). Preserving (resp. reversing) the orientation means that a clockwise turn is transformed into a clockwise (resp. anticlockwise) turn. For $n = 3$ one can give a similar interpretation, using volume instead of area. These interpretations can be extended to general n, using n-dimensional volume.

Remark. The determinant function can be characterized by the following properties:

1. $\det(AB) = \det A \det B$ for all $n \times n$ matrices A and B.

2. $\det D = \prod_{i=1}^{n} d_i$ if D is a diagonal $n \times n$ matrix with diagonal elements $d_1, \ldots, d_n$.

A.4 SOLUTION USING THE INVERSE MATRIX

We give an alternative way of presenting Cramer's rule. This requires some concepts.

Matrix multiplication. The definition of matrix multiplication is as follows: if B is an $m \times n$ matrix and A an $n \times p$ matrix, then the product matrix BA is the $m \times p$ matrix that has on the intersection of the i-th row and the j-th column the number

$$b_{i1}a_{1j} + \cdots + b_{in}a_{nj}.$$

In the special case of the multiplication of a row vector by a column vector, we always write a dot:

$$a \cdot b = a_1 b_1 + \cdots + a_n b_n$$

if a (resp. b) is an n-dimensional row (resp. column) vector. This is usually called the—standard—*inner product* of the column vectors a^T and x (a formal definition of column vectors is given in section A.7).

In all other cases of matrix multiplication, we will always omit the dot. Matrix multiplication has, for example, the property

$$(CB)A = C(BA),$$

as you would expect from a multiplication. However AB is usually not equal to BA. We explain the idea behind this multiplication by an example.

Example A.4.1 *Super-factory. A factory makes three products, using two ingredients, and the output depends on the input in a linear way, that is,*

$$y_1 = a_{11}x_1 + a_{12}x_2,$$

$$y_2 = a_{21}x_1 + a_{22}x_2,$$

$$y_3 = a_{31}x_1 + a_{32}x_2.$$

These three products are intermediate products: they are used by a second factory to make two end products. Here the dependence of the output on the input is also linear, that is,

$$z_1 = b_{11}y_1 + b_{12}y_2 + b_{13}y_3,$$

$$z_2 = b_{21}y_1 + b_{22}y_2 + b_{23}y_3.$$

Now we view these two factories together as one super-factory with inputs x_1, x_2 and outputs z_1, z_2. We are asked to give the dependence of output on input for the super-factory.

Solution. One should take the expressions for y_1, y_2, y_3 above and substitute these into the expressions for z_1, z_2. Expansion of the brackets and collecting terms gives the required expressions of z_1, z_2 in x_1, x_2. The calculations and the resulting formulas are relatively complicated. We do not display them. Instead, we will use matrix multiplication. This allows us, to begin with, to write the given relations in matrix form $y = Ax$ and $z = By$, where

$$y = \begin{pmatrix} y_1 \\ y_2 \\ y_3 \end{pmatrix}, \quad A = \begin{pmatrix} a_{11} & a_{12} \\ a_{21} & a_{22} \\ a_{31} & a_{32} \end{pmatrix}, \quad x = \begin{pmatrix} x_1 \\ x_2 \end{pmatrix},$$

$$z = \begin{pmatrix} z_1 \\ z_2 \end{pmatrix}, \quad B = \begin{pmatrix} b_{11} & b_{12} & b_{13} \\ b_{21} & b_{22} & b_{23} \end{pmatrix}.$$

Moreover, it allows us to do the calculations without any effort:

$$z = By = B(Ax) = (BA)x.$$

That is, the answer to the question is

the dependence for the super-factory is given by the *matrix product* of B and A.

Inverse matrix. The $n \times n$ matrix that has 1 everywhere on the main diagonal and 0 everywhere else is called the identity matrix, and it is denoted by I_n. It has the property that $I_n A = A$ and $BI_n = B$ for all $n \times n$ matrices A and B. We define the inverse A^{-1} of an $n \times n$ matrix A by the following property, which characterizes it, $AA^{-1} = I_n$, or, equivalently, $A^{-1}A = I_n$. Here a warning is in place: not every nonzero $n \times n$ matrix has an inverse. The following criterion holds true: an $n \times n$ matrix has an inverse precisely if its determinant is not zero, or, equivalently, if its rank is n.

Solution using the inverse matrix. Now we are ready to give the promised alternative presentation of Cramer's rule:

the system $Ax = b$ has a unique solution precisely if $\det A \neq 0$; then this solution is $x = A^{-1}b$.

A.5 SYMMETRIC MATRICES

An $n \times n$ matrix A is called *symmetric* if $a_{ij} = a_{ji}$ for all i, j. These matrices are of special interest for the following two related reasons.

Each quadratic function of n variables $x_1, \ldots, x_n$—this is the simplest type of function apart from linear functions—can be written in matrix form as

$$x^T A x + 2 a^T \cdot x + \alpha,$$

with A a symmetric $n \times n$ matrix, a an n-dimensional column vector, and α a number. If we want to find the extrema of such a function, we are led by the Fermat theorem to the system $Ax = -a$. The other reason is that the Hessian matrix or second derivative

$$\left(\frac{\partial^2 f}{\partial x_i \partial x_j}\right)_{ij}$$

is symmetric for all functions f of $x_1, \ldots, x_n$ for which all second order partial derivatives exist and are continuous. For our purposes of optimization, a central property of a symmetric matrix is its *definiteness*. This plays a role in the second order conditions and also in the subject of algorithms. A symmetric $n \times n$ matrix is called *positive definite (resp. semidefinite)* if the number $x^T A x$ is positive (resp. nonnegative) for each nonzero vector $x \in \mathbb{R}^n$. How to determine the definiteness of a given symmetric matrix? It turns out that each matrix can be written as $M^T D M$ for some invertible matrix M and some diagonal matrix $D = (d_{ij})$—that is, $d_{ij} = 0$ for $i \neq j$. In fact, we can achieve moreover that the matrix M is an *orthogonal matrix* (that is, $MM^T = I_n$, or equivalently, the columns of M are orthogonal to each other and have length 1). Then A is positive (resp. nonnegative) definite if all diagonal elements of D are positive (resp. nonnegative), as the vector equation

$$x^T A x = (M x)^T D (M x)$$

shows. These observations can be turned into an efficient algorithm, to be called "symmetric zero-sweeping." The idea of it is to let each operation on the rows be followed by the same operation on the columns in order to keep the matrix symmetric, continuing till you obtain a diagonal matrix D.

Minor criteria. Moreover, the following *minor criteria* for the definiteness of a symmetric $n \times n$ matrix A are sometimes useful. The matrix A is positive semidefinite if and only if all its minors are nonnegative. A *minor* of A is defined to be the determinant of the matrix that you get from A by omitting k rows and the corresponding columns with $0 \leq k \leq n - 1$. The matrix A is positive definite if and only if all its leading minors are positive. A *leading minor* of A is defined to be the determinant of the matrix that you get from A by omitting the last k rows and columns with $0 \le k \le n-1$. A proof of this latter criterion, which is also called the Sylvester criterion, is given in section 9.5.

A.6 MATRICES OF MAXIMAL RANK

Consider an $m \times n$ matrix A of rank m. Then the $m \times m$ matrix AA^T is invertible. Moreover, we will need that there exists a constant $C > 0$ for which

$$|Ax| \ge C|x| \quad \forall x \in (\ker A)^{\perp}.$$

Here $\ker A$, the kernel of A, is defined to be the solution set of the linear system of equations $Ax = 0_m$. Moreover, $(\ker A)^{\perp}$, the orthogonal complement of $\ker A$, is defined to consist of all vectors in $\mathbb{R}^n$ that are orthogonal to all vectors of $\ker A$. Two vectors, $x, y \in \mathbb{R}^n$ are orthogonal if $x^T \cdot y = x_1 y_1 + \cdots + x_n y_n = 0$; then one writes $x \perp y$.

A.7 VECTOR NOTATION

The space $\mathbb{R}^n$ of n-variables is the space of n-dimensional column vectors

$$x = \begin{pmatrix} x_1 \\ \vdots \\ x_n \end{pmatrix},$$

where each coordinate x_i is a real number. Such vectors can be added coordinatewise, and they can be multiplied coordinatewise by a number ("scalar multiplication"):

$$\text{if } x = \begin{pmatrix} x_1 \\ \vdots \\ x_n \end{pmatrix}, \ y = \begin{pmatrix} y_1 \\ \vdots \\ y_n \end{pmatrix}, \ \alpha \in \mathbb{R},$$

$$\text{then } x + y = \begin{pmatrix} x_1 + y_1 \\ \vdots \\ x_n + y_n \end{pmatrix}, \ \alpha x = \begin{pmatrix} \alpha x_1 \\ \vdots \\ \alpha x_n \end{pmatrix}.$$

Along with column vectors, we will also consider row vectors, partly for typographical reasons. For each column vector

$$x = \begin{pmatrix} x_1 \\ \vdots \\ x_n \end{pmatrix},$$

its *transpose* is defined to be the row vector $x^T = (x_1, \ldots, x_n)$. We write $(\mathbb{R}^n)^T$ or $(\mathbb{R}^n)'$ for the space of n-dimensional row vectors. A vector $x \in \mathbb{R}^n$ may be considered as a matrix with one column and n rows, a vector $y \in (\mathbb{R}^n)^T$ as a matrix with one row and n columns.

The transpose can be extended to matrices: the transpose A^T of an $m \times n$ matrix $A = (a_{ij})$ is defined to be the $n \times m$ matrix that has at the (ji) position the number a_{ij}. The transpose operation turns rows into columns, and conversely. Then

$$(AB)^T = B^T A^T$$

if A is an $m \times n$ matrix and B an $n \times p$ matrix. Moreover,

$$A^{TT} = A$$

for all matrices A. The definition of symmetric $n \times n$ matrices A can be written as $A^T = A$.

Geometrical interpretation of vectors. For small n, one has a concrete geometrical interpretation. For $n = 1$ one gets $\mathbb{R}^1 = \mathbb{R}$, the real line, consisting of all real numbers, with the usual addition and multiplication. For $n = 2$ (resp. $n = 3$) the column vector x represents the point in the two-dimensional plane (resp. three-dimensional space) with coordinates x_1 and x_2 (resp. x_1, x_2, and x_3); alternatively, it can be seen as the "vector-arrow" with head at the point x and tail at the origin. Here, addition of vectors is given by the parallelogram law or, equivalently, by placing the two vector-arrows head-to-tail. Multiplication of a vector by a number $\alpha > 1$ ($\alpha < 1$) makes the vector-arrow stretch (shrink) but leaves its direction the same; multiplication by -1 changes the direction into the opposite one.

A.8 COORDINATE FREE APPROACH TO VECTORS AND MATRICES

Often we come across objects that are essentially vector spaces $\mathbb{R}^n$. Here are some examples.

1. The collection of all $k \times l$ matrices is essentially $\mathbb{R}^{kl}$ (by stacking matrices as vectors).

2. The collection of symmetric $k \times k$ matrices is essentially $\mathbb{R}^{1/2(k^2+k)}$.

3. The set of solutions of $Ax = 0_k$, where A is a $k \times l$ matrix, is essentially $\mathbb{R}^{l-r}$, where r is the rank of A.

4. The set of polynomials $a_0 + a_1 x + \cdots + a_k x^k$ in one variable x of degree at most k, is essentially $\mathbb{R}^{k+1}$.

It is convenient to view such objects as vector spaces, and not just as "essentially vector spaces." To do this, one needs a coordinate-free approach.

The most elegant way to do this is by drawing up a suitable list of properties of the addition of vectors in $\mathbb{R}^n$ and the multiplication of vectors in $\mathbb{R}^n$ by scalars. Each "collection of objects" with an addition and a scalar multiplication for which these properties hold is then by definition a vector space.

For all our purposes, the following more concrete approach is adequate. Call the set of all functions on a given set I a vector space, with the usual addition and scalar multiplication of functions:

$$(f + g)(x) = f(x) + g(x), \quad (\rho f)(x) = \rho f(x),$$

for all scalars ρ and all vectors x. Moreover, call every subset W of this vector space that has the following three properties also a vector space:

1. W contains the zero-function.

2. W is closed under addition, $v, w \in W \Rightarrow v + w \in W$.

3. W is closed under scalar interpretation, $\rho \in \mathbb{R}$, $w \in W \Rightarrow \rho w \in W$.

Example A.8.1 *Verify that $\mathbb{R}^n$ and all the examples given above are vector spaces according to this concrete definition.*

Solution. We can identify $\mathbb{R}^n$ with the set of all functions on $I = \{1, 2, \ldots, n\}$ by identifying $x \in \mathbb{R}^n$ with the function $i \to x_i$. Now we consider the examples above.

1. The collection of all $k \times l$ matrices can be identified with the set of all functions on the set

$$I = \{(i, j) : 1 \leq i \leq k, \ 1 \leq j \leq l\}$$

if one associates to a matrix $A = (a_{ij})$ the function $I \to \mathbb{R}$ defined by $(i, j) \mapsto a_{ij}$.

2. The collection of symmetric $k \times k$ matrices is a subset of the collection of all $k \times k$ matrices, so it suffices to verify the three properties above: the zero matrix is symmetric, the sum of two symmetric matrices is symmetric, and each scalar multiple of a symmetric matrix is again symmetric.

3. The set of solutions of $Ax = 0_k$, where A is a $k \times l$ matrix, is a subset of $\mathbb{R}^l$, so it suffices to verify the three properties above: 0_l is a solution, the sum of two solutions is a solution,

$$Ax = Ax' = 0_k \Rightarrow A(x + x') = Ax + Ax' = 0_k + 0_k = 0_k,$$

and each scalar multiple of a solution is a solution,

$$Ax = 0_k \Rightarrow A(\rho x) = \rho Ax = \rho 0_k = 0_k.$$

4. The set of polynomials $a_0 + a_1 x + \cdots + a_k x^k$ in one variable of degree at most k is a subset of the set of all functions of one variable ("$I = \mathbb{R}$"), so it suffices to verify the three properties above: the zero function is the zero polynomial, the sum of two polynomials of degree at most k is a polynomial of degree at most k,

$$(a_0 + a_1 x + \cdots + a_k x^k) + (b_0 + b_1 x + \cdots + b_k x^k)$$

$$= (a_0 + b_0) + (a_1 + b_1)x + \cdots + (a_k + b_k)x^k,$$

and each scalar multiple of a polynomial of degree at most k is a polynomial of degree at most k,

$$\rho(a_0 + a_1 x + \cdots + a_k x^k) = \rho a_0 + \rho a_1 x + \cdots + \rho a_k x^k.$$

We note that there are vector spaces that are essentially different from any $\mathbb{R}^n$: for example, the space $C(\mathbb{R}^k)$ of continuous functions of k variables. These are called infinite-dimensional vector spaces.

We offer a list of definitions of some vector concepts in the coordinate-free ("high-level") language (we have written in parentheses the corresponding concept in the "low-level" language of $\mathbb{R}^n$ and systems of linear equations).

1. **Linear operator** ("matrix"). A mapping T from a vector space V to a vector space W is called a *linear operator* if it preserves addition and scalar multiplication, $T(v + v') = T(v) + T(v')$ for all $v, v' \in V$ and $T(\rho v) = \rho T(v)$ for all $\rho \in \mathbb{R}$ and $v \in V$.

2. **Linear independence.** A subset S of a vector space is *linearly independent* if no vector of S can be written as a finite linear combination of other elements of S.

3. **Rank** ("rank of a matrix"). Let S be a subset of a vector space. The *rank* of S is the number of elements in a maximal linearly independent subset of S. The rank can be infinite.

4. **Dimension** ("number of degrees of freedom in general solution system of linear equations"). The *dimension* of a vector space is its rank.

5. **Linear subspace** ("lines and planes in space that contain the origin"). A subset of a vector space is called a *linear subspace* if it is itself a vector space with respect to the addition and scalar multiplication of the given vector space.

6. **Kernel** ("solution set of a homogeneous system of linear equations $Ax = 0_k$"). The kernel of a linear operator $T : V \to W$ is the set $\{v \in V : Tv = 0_W\}$. This is clearly a linear subspace of V. It is denoted by $\ker T$.

7. **Linear span** The linear span of a subset S of a vector space V is the smallest linear subspace of V that contains S. This consists of all finite linear combinations of elements of S.

We come across these concepts when we establish the methods to solve concrete optimization problems. For example, when considering problems with equality constraints

$$f_0(x) \to \min, \ F(x) = 0_m, \qquad (P_{3.1})$$

with solution $\widehat{x}$, we are led to define the derivative $F'(\widehat{x})$, which is a linear operator from $\mathbb{R}^n$ to $\mathbb{R}^m$. Indeed, in the regular case that the rank of $F'(\widehat{x})$ is m, the elements of the kernel of the derivative $F'(\widehat{x})$ are precisely the direction vectors of admissible variations of $\widehat{x}$, by virtue of the tangent space theorem. This is the key step in the proof of the Lagrange multiplier rule, the central result of the method to solve problems of type $(P_{3.1})$.

Working with vector spaces rather than just with $\mathbb{R}^n$ gives a greater flexibility, which can be very useful. For example, it gives the possibility to carry out induction arguments with respect to the dimension. The following advanced exercise illustrates this convenience.

Exercise A.8.1 * *Orthogonal diagonalization of symmetric matrices.* Derive theorem 3.12 from the result of problem 3.7.

Appendix B

On Real Analysis

- The—optimal—value of an optimization problem is a central concept. The introduction of this concept forces us to go into the foundations of the real numbers and to consider the completeness property.

- In the proof of the envelope theorem, we use the inverse function theorem.

- One of the two types of good behavior of optimization problems is convexity, as we see in chapter four. We give some results on convexity, for example, we describe methods to verify whether a function is convex.

B.1 COMPLETENESS OF THE REAL NUMBERS

To begin with, we have to address the following question. What is a real number? We will give an intuitive answer to this question.

Real numbers. Let us start with the rational numbers, that is, the numbers of the form r/s, where r and s are integers and $s \neq 0$. If we write such numbers in decimal notation, then we get an infinite *periodic* decimal expansion. For example, $31/6 = 5.1666\ldots$ and $-2/37 = -0.054054054\ldots$. Conversely, each infinite periodic decimal expansion arises by the choice of a suitable rational number. If the repetitive part is 0, then the zeros are usually omitted and so we get a finite decimal expansion. For example, one writes $43/25 = 1.92$. We note that if the repetitive part is 0, then there is an alternative decimal expansion, with repetitive part 9 and preceding digit lowered by one. For example $1.92 = 1.91999\ldots$. We define a real number by its decimal expansion. There are many infinite decimal expansions that are not periodic. One of the simplest examples is $0.101001000100001\ldots$. These nonperiodic decimal expansions represent real numbers that are not rational numbers. For example, the real numbers $\sqrt{2}, \pi$, and e are not rational numbers. There is another, geometrical way to view the real numbers $\mathbb{R}$: as the points on a line, called the *real line*.

Conclusion: Instead of giving precise definitions, we have given two intuitive models for the real numbers: decimal expansions and points on a line.

Infimum and supremum. For our purposes, the following phenomenon deserves some attention. "Unfortunately," not every subset of $\mathbb{R}$ has a minimum and a maximum. For example, the set of positive numbers has no minimum. This fact has led to the introduction of the concept of infimum (resp. supremum), to be thought of as a 'surrogate' minimum (resp. maximum), which has the advantage that it always exists. We call a number $r \in \mathbb{R}$ a lower bound of a subset S if $r \leq s$ for all $s \in S$. If a nonempty subset S of $\mathbb{R}$ has a lower bound, then it has a largest lower bound. This is readily seen by considering the decimal expansions of the real numbers that belong to S. This largest lower bound is called the infimum of S and is denoted by $\inf S$. If S is empty, then we define $\inf S = +\infty$ and if S has no lower bound, then we define $\inf S = -\infty$. Thus we have defined for an arbitrary subset S of $\mathbb{R}$ its infimum $\inf S \in [-\infty, \infty]$. If a subset $S \subseteq \mathbb{R}$ has a minimum, then this is the infimum of S. That is, the concept of infimum is an extension of the concept of minimum.

In the same way as the infimum one defines the supremum of a subset $S \subseteq \mathbb{R}$, the "surrogate" maximum, to be the smallest upper bound of S. Clearly $\sup S = -\inf(-S)$, where $-S$ denotes the set of numbers $-s$, $s \in S$.

Value of a problem. Thus prepared, we are ready to define, for each minimization (resp. maximization) problem (P) its *value* $v(P)$ often written as $S_{\min}$ (resp. $S_{\max}$)—also called its *optimal value*. It is the infimum (resp. supremum) of all values that the objective function f assumes on the admissible set. This is a central concept for optimization problems. We observe that these definitions allow the following reformulation of the definition of the concept of global solution of an optimization problem. An admissible element $\widehat{x}$ is a global solution precisely if $v(P) \in \mathbb{R}$ and $f(\widehat{x}) = v(P)$. Three things can prevent a minimization problem from having a solution:

1. $v(P) = \infty$, that is, there are no admissible elements,

2. $v(P) = -\infty$, that is, the objective function is not bounded below,

3. $v(P) \in \mathbb{R}$, but this finite value is not assumed.

Example B.1.1   *The problem $f(x) = x^{-1} \to \min$, $x > 0$, has value 0, but this value is not assumed.*

On the completeness of the real numbers. The most intricate property of the real numbers is the *completeness property*. This property can be expressed in many equivalent ways. Let us mention some of these with some comments, omitting the straightforward—but technical—verification of their equivalence, and without proving them.

The first expression of the completeness of $\mathbb{R}$ makes it possible to define a basic notion of optimization, the value of a problem, as we have seen above.

(i) **Existence of a supremum and infimum.** Each bounded set of real numbers has a supremum and an infimum.

Convergence of a sequence is a subtle property; however, it is straightforward if the sequence is monotonic, as the following fact shows.

(ii) **Monotonic convergence.** Each monotonic nondecreasing or nonincreasing sequence of real numbers on a finite interval $[a, b]$ converges, that is, it tends to a limit.

This formulation of the completeness property is perhaps the most intuitive one: it looks like a self-evident property if we view the real numbers as decimal fractions or as points on the real line.

The following fact illustrates the powerful idea of "reducing convergence to monotonic convergence."

(iii) **Monotonic subsequence.** Each infinite sequence of real numbers $(r_k)_k$ on a finite interval $[a, b]$ has a monotonic—and so convergent—subsequence $(r_{k_i})_i$ with $k_1 < k_2 < \cdots < k_i < \cdots$.

(iv) **Convergent subsequence.** Each infinite sequence of vectors from a bounded subset of $\mathbb{R}^n$ has a convergent subsequence. A subset S of $\mathbb{R}^n$ is called *bounded* if there exists a number $N > 0$ for which all coordinates of all points of S are—in absolute value—at most N.

The usual way to prove convergence of a sequence is to prove explicitly that the terms tend to a certain number, its limit. Sometimes we will want to prove convergence of a sequence without possessing an explicit candidate for the limit and then we will use the following criterion.

(v) **Cauchy sequence.** The following internal criterion for convergence holds true for a sequence of real numbers $(x_k)_k$. For each

$\varepsilon > 0$, there exists a number $N > 0$ such that $|x_{N+k} - x_N| < \varepsilon$ for all $k \geq 0$.

Most optimization problems that have to be solved for pragmatic reasons cannot be solved analytically but only numerically. The following fact guarantees that one of the most basic optimization algorithms converges.

(vi) **"Bisection."** Let an interval be given, say $[0,1]$. Divide it into two halves $[0, \frac{1}{2}]$ and $[\frac{1}{2}, 1]$. Choose one of them. Divide this again in two halves and choose one of them. Continue in this way without end. The infinite collection of chosen intervals which arises in this way has a unique point in common.

Finally, we give the expression for completeness that is the most useful one for our purposes.

(vii) **Lemma on centered systems of closed subsets.** For a collection of bounded, closed subsets of $\mathbb{R}$, the following criterion holds for the existence of a common point of all sets of the collection. Each finite subcollection has a common point: that is, the collection satisfies the finite intersection property.

This criterion can be extended to dimension n, to subsets of $\mathbb{R}^n$. For most applications the following special case of the lemma on centered systems of closed sets suffices: each infinite descending chain

$$C_1 \supseteq C_2 \supseteq \cdots C_k \supseteq \cdots$$

of nonempty closed sets in $\mathbb{R}^n$ has a common point. This special case is intuitively clear.

The solution of the following problem about n-dimensional space is for $n = 2$ an immediate consequence of the completeness property of the real numbers. For $n = 3$ it is a formidable brainteaser for every professional mathematician, which, however, could be solved "without using any methods" by a beginning high school student, at least in principle. We could not resist the temptation to give it here. The case $n > 3$ follows immediately from the case $n = 3$.

Exercise B.1.1 *** *Can n-dimensional Euclidean space be viewed as the disjoint union of circles? Hint: show that the answer is yes for $n > 2$, by giving an explicit construction for $n = 3$ (and so for $n > 3$), and no for $n = 2$, by using the completeness property; for $n = 1$ the answer is no for a trivial reason.*

B.2 CALCULUS OF DIFFERENTIATION

B.2.1 The one-variable case

The calculus of differentiation consists of lists of derivatives of standard functions and rules "how to obtain new derivatives from old ones." Here is a list:

$$(x^n)' = nx^{n-1}, \ (\sin x)' = \cos x, \ (\cos x)' = -\sin x,$$

$$(e^x)' = e^x, \ (\ln x)' = x^{-1}.$$

The main rule is the chain rule,

$$\frac{dz}{dx} = \frac{dz}{dy}\frac{dy}{dx},$$

if z is a function of y and y a function of x. Another useful rule is the product rule,

$$(fg)' = f'g + fg'.$$

In addition, one uses the following obvious rules: $(f+g)' = f'+g'$ and $(cf)' = cf'$. Furthermore, quotients f/g can be differentiated either by writing them as a product fg^{-1} and applying the product rule, or by applying immediately the quotient rule,

$$(f/g)' = f'g - fg'/g^2.$$

Example B.2.1 *Solve the problem* $x^x \to \min$.

Solution. A computation gives

$$(x^x)' = (e^{(x\ln x)})' = e^{(x\ln x)}(x\ln x)'$$

$$= e^{(x\ln x)}(1\ln x + xx^{-1}) = e^{(x\ln x)}(\ln x + 1).$$

This is zero for $x = e^{-1}$, negative for $x < e^{-1}$, and positive for $x > e^{-1}$. It follows that $x = e^{-1}$ is the unique point of global minimum.

B.2.2 Partial derivatives

We give the definition of the partial derivative and of its use in the calculation of derivatives of functions of more than one variable.

The i-th partial derivative of f at $\widehat{x}$, to be denoted as $\frac{\partial f}{\partial x_i}(\widehat{x})$ or sometimes as $f_{x_i}(\widehat{x})$, is defined to be the derivative of the function of one variable x_i,

$$f(\widehat{x}_1, \ldots, \widehat{x}_{i-1}, x_i, \widehat{x}_{i+1}, \ldots, \widehat{x}_n),$$

at $x_i = \widehat{x}_i$, for $1 \leq i \leq n$. Here we have written the arguments of the function f in a row for convenience of typography, though we should have written them in a column, strictly speaking. If these partial derivatives exist for all $\widehat{x} \in \mathbb{R}^n$, then the resulting functions $\frac{\partial f}{\partial x_i}$ of n variables are called the *partial derivatives* of f. The symbol ∂ is used exclusively if one deals with a function of *more than one* variable that is differentiated with respect to only one of these variables, leaving the values of all the other variables fixed.

Partial derivatives play a crucial role in the solution of optimization problems of functions f of n variables, because of the following connection with the derivative of such functions. In chapter two the properties "differentiability of f at a point $\widehat{x}$" and "the derivative $f'(\widehat{x}) \in (\mathbb{R}^n)'$ of f at $\widehat{x}$" are defined for functions f of n variables. If f is differentiable at $\widehat{x}$, then we write $f \in D^1(\widehat{x})$. Then

$$f'(\widehat{x}) = \left(\frac{\partial f}{\partial x_1}(\widehat{x}), \ldots, \frac{\partial f}{\partial x_n}(\widehat{x}) \right). \qquad (*)$$

Moreover, it turns out that the converse implication holds true under a continuity assumption.

Definition B.1 *A function g of n variables is called continuous at a point $\bar{x}$ if*

$$\lim_{|h| \to 0} g(\bar{x} + h) = g(\bar{x}).$$

The geometric sense of continuity at $\bar{x}$ is that g makes no "jump" at $\bar{x}$. We formulate the promised converse implication.

If all partial derivatives of f exist at all points of an open ball with center $\widehat{x}$, and, moreover, are continuous at $\widehat{x}$, then f is differentiable at $\widehat{x}$ and formula $(*)$ holds true.

This fact is often used in applications.

The differential calculus. In general, it is not advisable to calculate the derivative directly from the definition given in chapter two. Now we sketch two methods of calculating derivatives that are more convenient.

Calculating derivatives by means of partial derivatives. For functions of one variable the derivative is calculated by using lists of basic examples such as $\sin' x = \cos x$ and rules such as the product rule and the chain rule, as we have seen. For functions of n variables the calculation of the derivative can be reduced to that of the derivative of a function of *one* variable by virtue of formula (∗). We illustrate this method by an example.

Example B.2.2 *Calculate the derivative of the modulus function $f(x) = |x| = \sqrt{x_1^2 + \cdots + x_n^2}$ at a nonzero point. Show that*

$$\frac{d}{dx}|x| = |x|^{-1} x^T.$$

Solution. Calculation of the partial derivatives of f gives

$$\frac{\partial f}{\partial x_i}(x) = x_i / \sqrt{x_1^2 + \cdots + x_n^2}$$

for all $i \in \{1, \ldots, n\}$. These functions are seen to be continuous outside the origin 0_2 (at 0_2 we run into trouble, "as division by zero is not allowed"). It follows that f is differentiable at all nonzero $x \in \mathbb{R}^n$, and

$$f'(x) = (x_1 / \sqrt{x_1^2 + \cdots + x_n^2}, \ldots, x_n / \sqrt{x_1^2 + \cdots + x_n^2}) = |x|^{-1} x^T.$$

Derivative for vector functions. The definition of the derivative can be extended in a straightforward way to *vector functions* (these are functions taking vectors as values rather than numbers). This extension is sometimes useful in solving problems; moreover, the formulation of this extension in terms of matrix notation is another convincing example of the convenience of this notation. We write $f \in D^1(\widehat{x}, \mathbb{R}^m)$ if, for some $\varepsilon > 0$, $f : U_n(\widehat{x}, \varepsilon) \to \mathbb{R}^m$ is a vector function on an open ball with center $\widehat{x} \in \mathbb{R}^n$ taking values in $\mathbb{R}^m$ that is differentiable at $\widehat{x}$. Then the derivative, or *Jacobi matrix*, $f'(\widehat{x})$ is the $m \times n$ matrix that is characterized by

$$f(\widehat{x} + h) = f(\widehat{x}) + f'(\widehat{x})h + o(h)$$

for all $h \in \mathbb{R}^n$. This concept allows us to extend the chain rule.

Lemma B.2 *Chain rule for vector functions (or superposition principle). Let $\widehat{x} \in \mathbb{R}^n$, $f \in D^1(\widehat{x}, \mathbb{R}^m)$, and $g \in D^1(f(\widehat{x}), \mathbb{R}^p)$ be given. Then $g \circ f \in D^1(\widehat{x}, \mathbb{R}^p)$ and*

$$(g \circ f)'(\widehat{x}) = g'(f(\widehat{x})) f'(\widehat{x}),$$

the matrix product of $g'(f(\widehat{x}))$ and $f'(\widehat{x})$.

B.2.3 Inverse function theorem

The main theorem of the differential calculus can be given in various forms, each of which can easily be derived from each of the other ones. The one that is the most intuitive and the most useful for optimization is the tangent space theorem. It is formulated and proved in chapter three. The most popular is the implicit function theorem, but in the present book we make no use of it. A third is the inverse function theorem; we use it in the proof of the envelope theorem 1.23. We need more general neighborhoods than just open balls,

$$U_n(u, \rho) = \{x \in \mathbb{R}^n : |x - u| < \rho\},$$

with $u \in \mathbb{R}^n$ and $\rho > 0$. To this end we make some definitions.

1. **Open set.** A subset U of $\mathbb{R}^n$ is called an *open set* if for each $u \in U$ there exists $\rho > 0$ such that the open ball with center u and radius ρ is contained in U, that is, $U_n(u, \rho) \subseteq U$. The set of all open sets in $\mathbb{R}^n$ is denoted by $\mathcal{O}(\mathbb{R}^n)$.

2. **Neighborhood.** A *neighborhood* U of a point $\widehat{x} \in \mathbb{R}^n$ is an open set in $\mathbb{R}^n$ that contains $\widehat{x}$. The set of all neighborhoods of $\widehat{x} \in \mathbb{R}^n$ is denoted by $\mathcal{O}(\widehat{x}, \mathbb{R}^n)$.

3. **Continuous differentiability.** A vector function

$$F = (f_1, \ldots, f_m) : U \to V$$

from an open set U in $\mathbb{R}^n$ to an open set V in $\mathbb{R}^m$ is called *continuously differentiable* if all partial derivatives $\frac{\partial f_i}{\partial x_j}$ exist and are continuous on U.

4. **Inverse function.** A vector function $F : U \to V$ from a set $U \in \mathbb{R}^n$ to a set $V \in \mathbb{R}^n$ is said to be *invertible* if there exists for each $y \in V$ a unique $x \in U$ for which

$$F(x) = y.$$

Then *the inverse function* $F^{-1} : V \to U$ of F is defined by $F^{-1}(y) = x$ precisely if $F(x) = y$.

Theorem B.3 *Inverse function theorem.* Let $U, V \in \mathcal{O}(0_n, \mathbb{R}^n)$ and $F = (f_1, \ldots, f_n)^T \in C^1(U, V)$ be given such that the $n \times n$ matrix

$$F'(0_n) = (\frac{\partial f_i}{\partial x_j}(0_n))_{i,j=1}^n$$

has maximal rank n. *Then we can achieve by a suitable shrinking of* U *and* V *to smaller neighborhoods of* 0_n *that* $F : U \to V$ *is invertible and that* $F^{-1} \in C^1(V, U)$ *(Fig. B.1).*

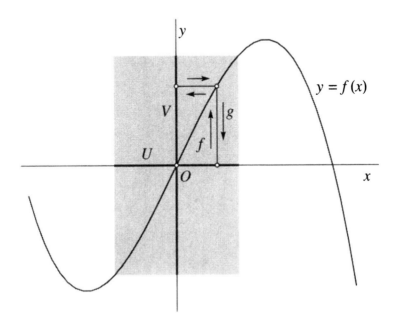

Figure B.1 The sets U and V and the function $g = f^{-1} : V \to U$.

This is a convenient place to give the definitions of some related concepts.

1. **Interior.** The *interior* of a set A in $\mathbb{R}^n$ is the largest open set contained in A. It is denoted by int$A$. A point in int$A$ is called an *interior point* of A.

2. **Closed set.** A set G in $\mathbb{R}^n$ is a *closed* set if it contains with each convergent sequence also its limit.

3. **Closure.** The *closure* of a set A in $\mathbb{R}^n$ is the smallest closed set containing A. It is denoted by cl$A$.

4. **Boundary.** The *boundary* of a set A in $\mathbb{R}^n$ is the set of all points b for which each open ball with center b contains a point of A and a point that is not in A. It is denoted by ∂A. A point in ∂A is called a *boundary point*.

Finally, we record the relation between the concepts open and closed. These concepts are complementary: a subset $A \subseteq \mathbb{R}^n$ is open precisely if the complement $\mathbb{R} \setminus A$ is closed.

B.3 CONVEXITY

We give the basic definitions and properties of convexity.

A set C in $\mathbb{R}^n$ is called a *convex set* if it contains for each two of its points all intermediate points:

$$a, b \in C,\ 0 \leq \rho \leq 1 \Rightarrow (1-\rho)a + \rho b \in C.$$

A function $f : A \to \mathbb{R}$ defined on a set $A \subseteq \mathbb{R}^n$ is called a *convex function* if its *epigraph*

$$\{(a, \rho) : a \in A,\ \rho \geq f(a)\}$$

is a convex set; this implies that A is a convex set. This definition is equivalent to the usual one.

Theorem B.4 *A function $f : A \to \mathbb{R}$ defined on a set $A \subseteq \mathbb{R}$ is convex precisely if the set A is convex, and, moreover, the following inequality holds true:*

$$f((1-\rho)a + \rho b) \leq (1-\rho)f(a) + \rho f(b)$$

for all $a, b \in A$ and $0 \leq \rho \leq 1$

The continuity of convex functions is intuitively obvious. However, at boundary points discontinuity is possible.

Theorem B.5 *A convex function $f : A \to \mathbb{R}$ is continuous at all interior points of its domain A.*

The qualification "interior" cannot be omitted here, as the following example illustrates.

Example B.3.1 *Take an arbitrary function $g : S_{n-1} \to \mathbb{R}_+$ on the unit sphere*

$$S_{n-1} = \{x \in \mathbb{R}^n : |x| = 1\}.$$

Extend it to a function f on the unit ball

$$B_n = \{x \in \mathbb{R}^n : |x| \leq 1\}$$

by defining $f(x) = 0$ if $|x| < 1$ and $f(x) = g(x)$ if $|x| = 1$. Then f is a convex function; this function is continuous precisely if $g(x) = 0 \ \forall x$.

A convex function need not be differentiable, as the following example illustrates.

Example B.3.2 Each norm on $\mathbb{R}^n$ is a convex function. Norms are always nondifferentiable at the origin. In particular, the absolute value function is convex and nondifferentiable at zero.

Let an open subset $U \subseteq \mathbb{R}^n$, a function $f : U \to \mathbb{R}$, and a point $\widehat{x} \in U$ be given.

Definition B.6 We say that $f : U \to \mathbb{R}$ is twice continuously differentiable if all second order partial derivatives of f exist and are continuous on U. Then the $n \times n$ matrix

$$f''(x) = \left(\frac{\partial^2 f}{\partial x_i \partial x_j}(x) \right)_{i,j=1}^n$$

is called the second derivative—or the Hessian (matrix)—of the function $f : U \to \mathbb{R}$ at x for all $x \in \mathbb{R}^n$. We write in this case $f \in C^2(U)$.

A function f defined on a convex set A is called a *strictly convex function* if

$$f((1-\rho)a + \rho b) < (1-\rho)f(a) + \rho f(b)$$

for all $a, b \in A$ and $0 < \rho < 1$.

Criterion. We are ready to give the following criterion.

Theorem B.7 Let an open convex set $U \subseteq \mathbb{R}^n$ and a twice continuously differentiable function $f : U \to \mathbb{R}$ be given.
 (i) The matrix $f''(x)$ is symmetric for all $x \in U$.
 (ii) The function $f : U \to \mathbb{R}$ is convex if and only if the symmetric matrix $f''(x)$ is nonnegative definite for all $x \in U$.
 (iii) If the symmetric matrix $f''(x)$ is positive definite for all $x \in U$, then the function $f : U \to \mathbb{R}$ is strictly convex.

Note that in (ii) the opposite implication does not hold. For example, the function $f(x) = x^4$ is strictly convex, but $f''(0) = 0$.

Example B.3.3 *The function $f(x) = x^2$ is strictly convex. This follows immediately from the theorem B.7 as $f''(x) = 2 > 0 \; \forall x$.*

We display the definiteness conditions of theorem B.7 for a symmetric 2×2 matrix A.

1. A is positive semidefinite precisely if
$$a_{11} \geq 0, \; a_{22} \geq 0, \; a_{11}a_{22} - a_{12}^2 \geq 0.$$

2. A is positive definite precisely if
$$a_{11} > 0, \; a_{11}a_{22} - a_{12}^2 > 0.$$

The central results on convex sets (and so on convex functions) are the two separation theorems.

Theorem B.8 *On separation of convex sets*

1. **Geometrical version.** *Two disjoint open convex sets A and B in $\mathbb{R}^n$ can be separated by a hyperplane. That is, A lies on one side of this hyperplane and B lies on the other side of it (here it is allowed that points of A and B lie on the hyperplane).*

2. **Analytical version.** *For any two disjoint open convex sets A and B in $\mathbb{R}^n$, there are a nonzero linear function L on $\mathbb{R}^n$ and a number ρ such that $L(a) \leq \rho \leq L(b)$ for all $a \in A$ and $b \in B$ (then the hyperplane with equation $L(x) = \rho$ separates A and B).*

Theorem B.9 *On strict separation of convex sets*

1. **Geometrical version.** *Two disjoint closed convex sets A and B in $\mathbb{R}^n$, at least one of which is bounded, can be separated strictly by a hyperplane. That is, there exist two different parallel hyperplanes that separate A and B.*

2. **Analytical version.** *For any two disjoint closed bounded convex sets A and B in $\mathbb{R}^n$, at least one of which is bounded, there are a nonzero linear function L on $\mathbb{R}^n$ and a number ρ such that $L(a) < \rho < L(b)$ for all $a \in A$ and $b \in B$ (then the hyperplane with equation $L(x) = \rho$ separates A and B strictly).*

The following variant is used in the proofs of optimization theory.

Theorem B.10 *Supporting hyperplane theorem. A convex set A in $\mathbb{R}^n$ can be supported at each boundary point $a \in A$ by a hyperplane. That is, there exists a hyperplane that contains a and has A on one of its two sides.*

Proving boundedness of a convex function is slightly easier than proving boundedness in general, by virtue of the following criterion. The same is true for the coercivity property.

Theorem B.11 1. *A convex set A is bounded precisely if it does not contain any half-line.*

2. *A convex function f is coercive precisely if it does not have any half-line of descent (that is, if its restriction to each half-line is coercive).*

A *convex problem* is an optimization problem of the following form:

$$f(x) \to \min, \ x \in A,$$

where A is a convex set in $\mathbb{R}^n$ and f is a convex function on A. Often A is given by systems of inequalities $f_i(x) \leq 0$, $1 \leq i \leq m$, where the f_i are convex functions.

The following theorem lists the three main advantages of convex optimization problems.

Theorem B.12 *Consider a convex problem $f(x) \to \min, \ x \in A.$*

1. *There is no difference between local and global minima.*

2. *Assume $f \in D^1(\widehat{x})$. Then $f'(\widehat{x}) = 0_n^T$ precisely if $\widehat{x}$ is a point of minimum.*

3. *If the function f is strictly convex, then there is at most one point of minimum.*

Now we formulate the analog for convex problems of the Fermat theorem for differentiable functions, which is given in chapter two. To this end, we define the *subdifferential* $\partial f(\widehat{x})$ of a convex function $f: A \to \mathbb{R}$ on a convex set $A \subseteq \mathbb{R}^n$ at a point $\widehat{x} \in A$ to be the set of $a \in (\mathbb{R}^n)'$ for which

$$f(x) - f(\widehat{x}) \geq a \cdot (x - \widehat{x}).$$

Such elements a are called *subgradients* of f at $\widehat{x}$. Now we can state the promised analog of the Fermat theorem for convex functions: if $\widehat{x}$ is a minimum of f, then $0_n^T \in \partial f(\widehat{x})$. There is a calculus to compute subdifferentials. We do not display these rules as we have not made use of them in solving concrete problems. We only mention that all rules of this calculus follow from the supporting hyperplane theorem.

For a convex function g, the function $-g$ is called *concave*.

Example B.3.4 *The pleasure of consuming a quantity x of some product can be modeled by an increasing, concave function $U(x)$, called the utility function. Each consumer has his own taste, and so his own utility function. The assumption that U is increasing represents the usual situation that more consumption gives more—or at least not less—pleasure. That U is also assumed to be concave represents the satiation of consumers: additional units of consumption give less and less additional pleasure. Popular utility functions include the functions $U_1(x) = \ln x$ and $U_2(x) = x^a$, with $0 < a < 1$. From the shapes of the graphs of these functions, one can see that they are increasing and concave for $x > 0$, as required.*

In addition, we give another—not so well-known—criterion for the convexity of functions, along with the criterion in terms of the Hessian matrix. To this end, we associate to each function $f : A \to \mathbb{R}$ on a nonempty subset A of $\mathbb{R}^n$ a function $f^* : A' \to \mathbb{R}$, called the *conjugate function of f* or the *(Young-)Fenchel transform* of f, on a subset A' of $\mathbb{R}^n$ as follows:

$$f^*(y) = \sup_{x \in A}(x^T \cdot y - f(x))$$

and A' is the set of all $y \in \mathbb{R}^n$ for which this supremum is finite, that is, not equal to $+\infty$ or $-\infty$. The function $f^* : A' \to \mathbb{R}$ is lower semicontinuous and convex, as its epigraph is by definition the intersection of closed half-spaces.

Theorem B.13 *A continuous function f on a nonempty subset of $\mathbb{R}^n$ is convex if and only if it is equal to the conjugate of a function.*

Let us give an example of a function for which it would not be so simple to verify its convexity by the criterion involving second order derivatives. However, an ingenious use of the criterion of theorem B.13 gives a quick verification.

Example B.3.5 *Show that the function $f(x) = \ln(e^{x_1} + \cdots + e^{x_n})$ is convex.*

Solution. To prove the convexity of f it suffices to show that f is the conjugate of the function g defined on the set of all positive vectors $y \in \mathbb{R}^n$ with $\sum_i y_i = 1$ by $g(y) = \sum_i y_i \ln y_i$. Note that g is strictly convex: this follows from the fact that the second derivative of the function of one variable $\eta \ln \eta$ equals η^{-1} and so is positive for $\eta > 0$. Therefore, the problem

$$h(y) = x^T y - g(y) \to \max, \quad \sum_i y_i = 1, \; y > 0 \qquad (Q_x)$$

is a convex problem. If we write the stationarity conditions of this problem, solve them, and substitute the solutions in the objective function, then the outcome turns out to be $f(x)$. On the other hand, it is by definition $g^*(x)$. This proves the identity $f = g^*$, as promised. Therefore, by theorem B.13, the function f is convex.

How the formula for g was found. This method of proving convexity depends on the function g, which seems to come out of the blue. However, it arises in a natural way by the following reckless method of calculating f^*. In order to compute $f^*(y)$, we have to consider the problem $h(x) = x^T \cdot y - f(x) \to \max$. The stationarity conditions are

$$y_i - (e^{x_1} + \cdots + e^{x_n})^{-1} e^{x_i} = 0$$

for $1 \leq i \leq n$. These imply the following properties of y:

$$\sum_{i=1}^n y_i = 1 \text{ and } y_i > 0$$

for $1 \leq i \leq n$. For each such vector y, the following choice of x is a solution of the stationarity conditions: $x_i = \ln y_i$ for $1 \leq i \leq n$. This is a point of global maximum of the problem under consideration, as f is convex and so $x^T \cdot y - f(x)$ is a concave function of x. The number $f^*(y)$ is defined to be the optimal value of the problem under consideration; therefore, it equals the result of substituting this point in the objective function h, which is seen to be

$$\sum_{i=1}^n y_i \ln y_i.$$

Vicious circle. We seem to have calculated f^*, and to have shown that $f^*(y)$ is defined precisely if $\sum_{i=1}^n y_i = 1$ and $y_i > 0$ for $1 \leq i \leq n$, and then $f^*(y) = \sum_{i=1}^n y_i \ln y_i$. However, this reckless calculation is a vicious circle. Our main aim is to prove that f is convex, but in we have used that f is convex in our calculation!

There is a way out: our reckless calculation leads to an explicit function
$$g(y) = \sum_{i=1}^n y_i \ln y_i,$$
defined if $\sum_{i=1}^n y_i = 1$ and $y_i > 0$ for $1 \leq i \leq n$. Its status is as follows: if f is convex, then $g = f^*$ and so $g^* = f$. This suggests a way out of the vicious circle: check that $g^* = f$; this implies that f is convex. This is precisely what we have done above.

Now we turn to the problem of verifying the convexity of *sets*. Each level set of a convex function is a convex set. If we combine this with the fact that the intersection of convex sets is again convex, we get the following observation.

Example B.3.6 Let $f_1, \ldots, f_m$ be convex functions on a convex subset A of $\mathbb{R}^n$. Then the solution set of the following system of inequality constraints is a convex set
$$f_i(x) \leq 0, \ 1 \leq i \leq m.$$

This is a rich source of convex sets, as convex functions are easy to get by, using theorem B.7 above.

Exercise B.3.1 *Verify that the following functions are convex and compute for each of them its conjugate function:*

1. $ax + b$,

2. $x^2 + ax + b$,

3. $|x|$,

4. e^x,

5. $-\ln x$, $(x > 0)$,

6. $-x^a$ $(x > 0)$ *with* $0 \leq a \leq 1$,

7. x^a $(x > 0)$ *with* $a \notin (0, 1)$,

8. $\sqrt{1+x^2}$,

9. $x \ln x$ $(x > 0)$.

Exercise B.3.2 *Let a and b be positive numbers with $a + b \leq 1$. Verify that the function $f(x, y) = x^a y^b$, $x \geq 0$, $y \geq 0$, is concave.*

Exercise B.3.3 *Show that a quadratic function f on $\mathbb{R}^n$, given by $f(x) = x^T A x + b \cdot x + c$, is convex if A is positive semidefinite.*

B.4 DIFFERENTIATION AND INTEGRATION

Always when we calculate an integral, we use the fundamental fact that differentiation and integration are inverse operations, well known from high school. Let us state this result precisely.

Theorem B.14 *Newton-Leibniz formula or fundamental theorem of analysis.* *For each differentiable function $f : [a, b] \to \mathbb{R}$ the following formula holds true:*

$$f(b) - f(a) = \int_a^b f'(t)dt.$$

Appendix C

The Weierstrass Theorem on Existence of Global Solutions

- The most convenient way to complete the analysis of any concrete finite-dimensional optimization problem is by establishing the existence of global solutions using the Weierstrass theorem. This requires a verification of the assumptions of this theorem: the objective function has to be continuous and the admissible set has to be closed, bounded, and nonempty. Moreover, sometimes one has to check that the objective function is coercive, for example, if the admissible set is unbounded. We will present the calculus for carrying out these verifications. Furthermore, we provide insight into why the Weierstrass theorem holds.

C.1 ON THE USE OF THE WEIERSTRASS THEOREM

Why (not to) read this appendix. In this appendix, we present the Weierstrass theorem: a continuous function on a nonempty, closed, bounded subset of $\mathbb{R}^n$ has global minima and maxima. As a consequence of this, a continuous function on the whole $\mathbb{R}^n$ that is coercive for minimization (respectively maximization) has global minima (respectively maxima) ("coercive" means that $f(x) \to +\infty$ (respectively $f(x) \to -\infty$) if $|x| \to +\infty$). Experience has shown that this method to establish existence of solutions almost always works. If you are already used to the concepts "continuous function" and "closed and bounded subset," then you can skip this appendix.

Motivation for the Weierstrass theorem. If you have decided to go on reading, you might want a more fundamental question to be answered first: why do you need to establish the existence of a solution of a concrete problem at all? At first sight, the existence might seem evident in all concrete cases, whether you are trying, for example, to find minimal costs, maximal profits, optimal trade-offs, or quickest paths of light. Now we consider a problem where the existence question is not so obvious at first sight (it is a physical interpretation of problem 2.5).

Example C.1.1 *Three holes in a table. Drill three holes in a table, take three equal weights, attach them to three strings, pull the strings up through the holes in the table, and tie the three loose ends in a knot above the table. Then, let the system go and see what happens. By a law of physics, the position of the knot will keep changing as long as this leads to a lowering of the center of gravity of the system. It seems evident from a physical point of view that the knot will tend to some specific position. However, can we exclude with absolute certainty that the knot keeps making movements forever without tending to some definite point?*

We will give the answer to this question later in this appendix. What will be the reward of your efforts if you invest some time in learning to use the Weierstrass theorem? It will be that you have absolute certainty that what you have found using the Fermat theorem, differential calculus, and comparison of candidate solutions is indeed a solution of the given optimization problem. This might not look very exciting. The reward for learning to work with the Fermat method and with the differential calculus is much more spectacular. They allow you to proceed from a state of complete ignorance about a given optimization problem to a state where you have a specific "candidate solution" for it. In practice, this candidate solution is usually the true solution.

Therefore, a practitioner might well decide to stop her analysis at this point. However, in this subsection we show that the price for obtaining absolute certainty is not very high. If you are prepared to accept the truth of the Weierstrass theorem, to be given below, and to familiarize yourself with some calculus rules for *continuity*, then you will have the pleasure of achieving a completely reliable analysis of your optimization problem.

Finally, we want to emphasize two additional points in favor of the Weierstrass theorem.

1) A pragmatic or physical optimization problem is a model of reality, but not reality itself. If you cannot prove the existence of a solution, there is usually something wrong with your model. This use of the Weierstrass theorem as a rough check on the model is handy in practice.

2) The following two alternative ways to complete the analysis of optimization problems are popular, but less convenient: the method of the sign of the derivative ("high school method") and the method

of second order sufficient conditions.

The Weierstrass theorem. The formulation of the main result requires the definitions of the concepts "bounded set," "closed subset," and "continuous function." Their formulation is a simple matter. However, you need to get used to these concepts to be able to work with them.

Definition C.1 *A subset A of $\mathbb{R}^n$ is called bounded if it is included in the open ball $U_n(0_n, R)$ for large enough $R > 0$ (or, equivalently, if there is a constant $M > 0$ such that all coordinates of all points of A are in absolute value smaller than M).*

Definition C.2 *A subset A of $\mathbb{R}^n$ is called closed if for each point $p \in \mathbb{R}^n$ outside A there exists $\varepsilon > 0$ such that the ball $U_n(p, \varepsilon)$ contains no points of A.*

The geometric sense of a closed set is that it contains all points of its boundary. The following two equivalent definitions are useful.

1) A subset A of $\mathbb{R}^n$ is closed precisely if its complement $\mathbb{R}^n \setminus A$ is *open*, that is, if each point in the complement is the center of an open ball that is entirely contained in the complement.

2) A subset A of $\mathbb{R}^n$ is closed precisely if, for each sequence of elements of A that converges to an element of $\mathbb{R}^n$, this element belongs to A as well.

Consider a function $f : A \to \mathbb{R}$ on a subset $A \subseteq \mathbb{R}^n$. The *graph of f* is defined to be the set of points $(x_1, \ldots, x_n, y)$ in $(n+1)$-dimensional space, where $x = (x_1, \ldots, x_n)$ runs over A and $y = f(x)$. For $n = 1$, this is a curved line in a two-dimensional plane, for $n = 2$, this is a curved surface in three-dimensional space.

Definition C.3 *A function $f : A \to \mathbb{R}$ on a subset A of $\mathbb{R}^n$ is called continuous if it is continuous at each point $\bar{x} \in A$, that is, if $\lim_{|h| \to 0} f(\bar{x} + h) = f(\bar{x})$ for all $\bar{x} \in A$.*

The geometric sense of continuity is that the graph of the function f makes no jumps.

Now we are ready to present the key result of the method.

Theorem C.4 *The Weierstrass theorem—existence of global solutions.* Let A be a nonempty, closed, and bounded subset of $\mathbb{R}^n$ and let $f : A \to \mathbb{R}$ be a continuous function on A. Then the problem $f(x) \to \min$, $x \in A$ has a point of global minimum.

This result is also called the *extreme value theorem*. How to apply this result to a concrete problem? To do this, you have to check carefully whether all assumptions are satisfied.

Calculus for continuity. Let us begin with the calculus to check the continuity of functions. The idea of this calculus is to observe that all basic functions occurring in the formula defining the function are known to be continuous. Let us be more precise. You start with a list of basic functions of one variable x which are known to be continuous: for example,

$$c \ (c \in \mathbb{R}), \ x, \ |x|, \ e^x, \ \sin x, \ \cos x,$$

$$\ln x \ (x > 0), \text{ and } x^r \ (x > 0) \text{ for all } r \in \mathbb{R}.$$

Then you use the following rules. For given continuous functions $f, g : [a, b] \to \mathbb{R}$, the sum $f + g$ and the product fg are also continuous on $[a, b]$; moreover, the quotient $f(x)/g(x)$ is continuous on $[a, b]$, provided $g(x) \neq 0$ for all $x \in [a, b]$. Furthermore, if g (resp. f) is a continuous function on $[a, b]$ (resp. on $[c, d]$), then the *composition* $f \circ g$, that is, $x \to f(g(x))$, the result of substituting g in f, is continuous on $[a, b]$, provided $g(x) \in [c, d]$ for all $x \in [a, b]$. Let us illustrate this method of verifying continuity. Most of the functions in the following example are from concrete problems from chapter one.

Example C.1.2 *Show that the following functions of one variable are continuous:*

$$x^3, \ |\sin x|, \ \frac{\sin x}{|\sin x|} \ (x \neq k\pi), \ x + 5x^{-1} \ (x \neq 0),$$

$$x(x+6), \ |x| + |x-1| + |x-3| + |x-10|, \ \sqrt{x^2 + 1} - \frac{1}{2}x, \ \sqrt{-(\sin x)^2}.$$

Solution. One has $x^3 = x \cdot x \cdot x$, so this function is made from the standard continuous function x by repeated multiplication. One can view $|\sin x|$ as the composition $f(g(x))$, where $f(x) = |x|$ and $g(x) = \sin x$. Then it follows that the quotient $\sin x/|\sin x|$ is continuous except at the solutions of $|\sin x| = 0$, that is, of $\sin x = 0$; these solutions are the points $k\pi$ with k an integer, $k \in \mathbb{Z}$. In the same way, we can also check the continuity of the other functions. They can all be made by starting from basic continuous functions and then repeatedly taking sums, products, quotients, and compositions. The last example is included as a warning: the calculus rules show that it is continuous, but it is an unusual function. It is defined only at all multiples of π, as one cannot take the square root of a negative number within the real numbers, and it takes the value 0 at these points.

In addition, we illustrate the method for some of the functions of several variables that are considered in chapter two.

Example C.1.3 *Show that the following functions are continuous:*

$$\frac{x^2}{2} + \frac{y^2}{4} + \frac{(1-x-y)^2}{2}, \; x, y \in \mathbb{R}, \; |x-a| + |x-b| + |x-c|, \; x \in \mathbb{R}^2,$$

$$(a, b, c \in \mathbb{R}^2 \; given)$$

$$\sum_{i=1}^{5}(y_i - ax_i - b)^2, \; a, b \in \mathbb{R} \; (x_i, y_i \in \mathbb{R} \; given),$$

$$x^T A x + 2b \cdot x + c, \; x \in \mathbb{R}^n$$

$(A$ a given $n \times n$ matrix, $b \in (\mathbb{R}^n)^T$ given, $c \in \mathbb{R}$ given),

$$|Ax|/|x|, \; x \in \mathbb{R}^n \setminus \{0_n\} \; (A \; a \; given \; m \times n \; matrix).$$

Solution. Let us begin with the fourth example. It is built up from the standard continuous functions x_i, $1 \le i \le m$, and the constant functions by repeated addition and multiplication. The first and third example can be seen to be special cases of it. For the second example, one has to use that the Euclidean norm is a continuous function. Let us give the details for the fifth example. The function $x \to |Ax|$ is the composition of the linear—and so continuous—mapping $x \to Ax$ and the Euclidean norm; therefore, it is continuous. It follows that $|Ax|/|x|$ is the quotient of two continuous functions; therefore, it is continuous for all x for which the denominator $|x|$ is not zero, that is, for $x \ne 0_n$.

Showing that a subset of $\mathbb{R}^n$ is closed. Here the following facts can be used. Let f_i, $1 \le i \le m$, be continuous functions on $\mathbb{R}^n$. Then the set of solutions of the system of *nonstrict* inequalities

$$f_i(x) \le 0, \; 1 \le i \le m,$$

is a closed subset of $\mathbb{R}^n$. As a special case we get that the set of solutions of the system of equations

$$f_i(x) = 0, \ 1 \leq i \leq m,$$

is closed. Indeed, observe that an equation $g(x) = 0$ is equivalent to the following system of two inequalities: $g(x) \leq 0$ and $-g(x) \leq 0$.

Showing that a subset of $\mathbb{R}^n$ is bounded. This is the trickiest point. At first sight, there seems to be a very simple method to check that a subset is bounded: "coordinatewise verification." The following example shows that this "method" is not at all reliable.

Example C.1.4 *Let A be the subset of $\mathbb{R}^2$ of all solutions x of the inequality $|x_1 - x_2| \leq 1$. Why do the following two facts not imply that A is bounded?*

1. *For each number x_1 the set of numbers $\{x_2 : (x_1, x_2)^T \in A\}$ is bounded, as this is the interval $[x_1 - 1, x_1 + 1]$.*

2. *Similarly, for each number x_2, the set of all numbers x_1 for which $\{x_2 : (x_1, x_2)^T \in A\}$ is bounded, as this is the interval $[x_2 - 1, x_2 + 1]$.*

Solution. The set A is not bounded: for example, it contains the entire line $x_1 = x_2$.

As a consequence, a correct verification of boundedness can sometimes require some technical details.

We will illustrate the main strategy of the correct method to check boundedness for the case that the subset A is a level set of a function f of n variables. Then it is recommended to find a function g of one variable t for which $\lim_{t \to +\infty} g(t) = +\infty$, and for which one has the lower bound $|f(x)| \geq g(|x|)$ for all $x \in \mathbb{R}^n$. This implies that all level sets of f are bounded.

Example C.1.5 *Show that the set of solutions of the inequality $|x_1|^p + \cdots + |x_n|^p \leq 1$ is bounded for each positive number p.*

Solution. We have the lower bound

$$|x_1|^p + \cdots + |x_n|^p \geq \max_{i=1}^{n} |x_i|^p = (\max_{i=1}^{n} |x_i|^2)^{p/2}$$

$$\geq \left(n^{-1} \sum_{i=1}^{n} |x_i|^2 \right)^{p/2} = n^{-p/2} |x|^p$$

for all $x \in \mathbb{R}^n$. Moreover, $\lim_{t \to +\infty} t^p = +\infty$. It follows that the set of solutions of the inequality $|x_1|^p + \cdots + |x_n|^p \leq 1$ is bounded for each positive number p.

An alternative convenient method of verifying that a subset $A \subseteq \mathbb{R}^n$ is bounded is by showing that there are constants $c \leq d$ such that $c \leq x_i \leq d$ for all $x \in A$ and all $1 \leq i \leq n$. We now illustrate this method with the same example: the inequality $|x_1|^p + \cdots + |x_n|^p \leq 1$ implies $|x_i| \leq 1$ for all $1 \leq i \leq n$.

Existence for the three holes in the table problem. Finally, we answer the question about the three holes in the table problem. The three holes in the table mark a triangle: this is the admissible set of the optimization problem, and it is clear that it is a nonempty, closed, and bounded subset of the plane. The position of the center of gravity of the system consisting of the three weights and strings depends continuously on the position of the knot. That is, the assumptions of the Weierstrass theorem hold true. Therefore, the conclusion holds: the problem has a global solution. For details about this solution see the analysis of problem 2.5.

Variants of the method. However, there is a problem with this method. In many applications, the admissible set A is not bounded (resp. closed). At first sight, the method for proving existence cannot be used in such cases, but a closer look reveals that there is a way out. Fortunately, it turns out that one can usually reduce "somehow" to the case that A is bounded (resp. closed). For example, if A is not bounded, then one can often do this in the following way.

Definition C.5 *A function $f : \mathbb{R}^n \to \mathbb{R}$ on the whole $\mathbb{R}^n$ is called coercive for minimization (resp. for maximization) if*

$$\lim_{|x| \to +\infty} f(x) = +\infty \quad (\text{resp.} \quad \lim_{|x| \to +\infty} f(x) = -\infty).$$

We will write just *coercive* if it is clear from the context whether we mean "for minimization" or "for maximization."

Corollary C.6 *Let $f : \mathbb{R}^n \to \mathbb{R}$ be a continuous function. Then the problem $f(x) \to$ extr has a point of global minimum (resp. maximum) if f is coercive for minimization (resp. maximization).*

Proof. If we add the constraint $|x| \leq M$ for large enough M if f is coercive for minimization (resp. maximization) and apply the Weierstrass theorem, we get that the given problem has a point of global minimum (resp. maximum). □

How to show coerciveness of a function f on $\mathbb{R}^n$? We have to beware of the same trap as in the case of verifying that subsets of $\mathbb{R}^n$ are bounded.

Example C.1.6 *Consider the function of two variables defined by $f(x) = |x_1 - x_2|$ for all $x \in \mathbb{R}^2$. Then, if we vary in $f(x)$ one of the coordinates, for any given value of the other one, we get a coercive function. However, f itself is not coercive. Show this.*

Solution. For example, f is identically zero on the entire line $x_1 = x_2$.

To show that a function $f : \mathbb{R}^n \to \mathbb{R}$ is coercive, it is recommended to find a function g of one variable for which $\lim_{t \to +\infty} g(t) = +\infty$ and $|f(x)| \geq g(|x|)$ for all $x \in \mathbb{R}^n$. Then it follows that f is coercive. Some flexibility in applying this method can lead to simplifications, as the following example illustrates.

Example C.1.7 *Show the existence of a solution for the following problem: find a point in the plane $\mathbb{R}^2$ for which the sum of the squares of the distances to three given points in the plane, a, b, and c, is minimal (example 2.11).*

Solution. This problem can be modeled as follows:

$$f(x) = |x - a|^2 + |x - b|^2 + |x - c|^2 \to \min, \; x \in \mathbb{R}^2.$$

Existence of a solution follows as f is continuous and coercive (note that $f(x) \approx 3|x|^2$ for sufficiently large $|x|$).

C.2 DERIVATION OF THE WEIERSTRASS THEOREM

Finally, we provide some insight into why the Weierstrass theorem holds. We will not prove it, but have the more modest aim to "derive" it. What is the matter? A proof would get us deeper into the intricate foundations of the real numbers than is necessary for our purpose. In appendix B we give an intuitive account of the real numbers and of their most intricate property, the *completeness property*; in particular, we list several equivalent ways to express this property. Here we give a quick derivation of the Weierstrass theorem from one of the formulations of the completeness property, the *lemma on centered systems* of closed subsets.

Derivation 1 of the Weierstrass theorem. Let $f : A \to \mathbb{R}$ be a continuous on a nonempty, closed, and bounded subset A of $\mathbb{R}^n$ and consider the problem
$$f(x) \to \min, \ x \in A.$$
Consider the collection of all nonempty level sets
$$\mathcal{L}_f(r) = \{x \in A \mid f(x) \leq r\}$$
of $f : A \to \mathbb{R}$. This collection clearly satisfies the finite intersection property. Moreover, these level sets are closed. That is, every finite subcollection of it has a common point. Therefore, the lemma on centered systems of closed sets gives that there exists a common point $\widehat{x}$ of all the sets of the collection. Then $\widehat{x}$ must be a point of global minimum. Indeed, for any $x \in A$, the level set $\mathcal{L}_f(f(x))$ is nonempty, as it contains x. Therefore, it must contain $\widehat{x}$ by the choice of $\widehat{x}$. This means that $f(\widehat{x}) \leq f(x)$. As x is an arbitrary point of A, this proves that $\widehat{x}$ is a point of global minimum, as required.

It is of interest to rephrase this slick derivation in terms of *descent*. Then we need another way to express the completeness property of the real numbers: the convergent subsequence property.

Derivation 2 of the Weierstrass theorem. Let $f : A \to \mathbb{R}$ be a continuous function on a nonempty, closed, and bounded subset A of $\mathbb{R}^n$, and consider the problem
$$f(x) \to \min, \ x \in A.$$
Choose an infinite sequence $x_1, x_2, \ldots$ in A such that f descends on it to the value of the problem, that is, to $\inf\{f(x) : x \in A\}$. This is possible by the definition of the concept of infimum. We assume that this infinite sequence is convergent, say to $\widehat{x} \in \mathbb{R}^n$, as we may, without restricting the generality of the argument: by going over to a convergent subsequence. We have $\widehat{x} \in A$, as A is closed. Moreover, the continuity of $f : A \to \mathbb{R}$ gives that $f(x_k)$ tends to $f(\widehat{x})$ for $k \to +\infty$. In all, we have that $\widehat{x} \in A$ and that
$$f(\widehat{x}) = \inf\{f(x) : x \in A\}.$$
This means that $\widehat{x}$ is a point of global minimum, as required.

Appendix D

Crash Course on Problem Solving

All concrete optimization problems that can be solved analytically (that is, by a formula) at all can be solved in one and the same brief and transparent way, involving four steps:

1. Model, and establish existence global extrema.

2. Write down first order conditions.

3. Solve first order conditions and compare all candidate extrema (including boundary points and points of nondifferentiability).

4. Conclusion.

D.1 ONE VARIABLE WITHOUT CONSTRAINTS

- **Fermat theorem.** Let $f : (\hat{x} - \varepsilon, \hat{x} + \varepsilon) \to \mathbb{R}$ be a function of one variable that is differentiable at $\hat{x}$. If $\hat{x}$ is a local solution of
$$f(x) \to \text{extr}, \ x \in (\hat{x}, \hat{x} + \varepsilon),$$
then it is a solution of the stationarity equation $f'(x) = 0$.

- **Weierstrass theorem.** A continuous function $f : [a, b] \to \mathbb{R}$ on a closed interval attains its global extremal values.

- **Weierstrass theorem for coercive functions.** A continuous function $f : \mathbb{R} \to \mathbb{R}$ on the whole real line that is coercive (for minimization) (that is, "$f(-\infty) = f(+\infty) = +\infty$") attains its global minimal value.

Example D.1.1 $f(x) = x^4 + x^3 - \frac{1}{2}x^2 \to \min, \ x \in \mathbb{R}.$

Solution

1. f coercive ($f(x) \approx x^4$ for $|x|$ sufficiently large) $\Rightarrow$ existence of global minimum $\hat{x}$.

2. Fermat: $f'(x) = 0 \Rightarrow 4x^3 + 3x^2 - x = 0$.

3. $x(4x-1)(x+1) = 0 \Rightarrow x \in \{0, \frac{1}{4}, -1\}$.
   Compare: $f(0) = 0$, $f(\frac{1}{4}) = -\frac{3}{256}$, $f(-1) = -\frac{1}{2}$.

4. $\widehat{x} = -1$.

Exercise D.1.1 $x^3 - 12x^2 + 36x + 8 \to \min$, $x > 1$.

Exercise D.1.2 $-2x^2 + 4x + 8 \to \max$, $x \in \mathbb{R}$.

Exercise D.1.3 $x + 1/x \to \min$, $x > 0$.

Exercise D.1.4 $x^x \to \min$, $x > 0$.

Exercise D.1.5 $x^3 + 4x^2 + x - 5 \to \min$, $x \in \mathbb{R}$.

D.2 SEVERAL VARIABLES WITHOUT CONSTRAINTS

- **Fermat theorem.** Let $f : U_n(\widehat{x}, \varepsilon) \to \mathbb{R}$ be a function of n variables that is differentiable at $\widehat{x}$. If $\widehat{x}$ is a local solution of the problem
$$f(x) \to \text{extr}, \ x \in U_n(\widehat{x}, \varepsilon),$$
then it is a solution of the stationarity vector equation
$$f'(x) = 0_n^T (\Leftrightarrow \frac{\partial f(x)}{\partial x_i} = 0, \ 1 \le i \le n).$$

- **Weierstrass theorem.** A continuous function $f : A \to \mathbb{R}$ on a nonempty compact (= closed and bounded) set $A \subseteq \mathbb{R}^n$ attains its global extremal values ("closed" means that all points on the boundary of A belong to A).

- **Weierstrass theorem for coercive functions.** A continuous function $f : \mathbb{R}^n \to \mathbb{R}$ on the whole $\mathbb{R}^n$ that is coercive (for minimization) (that is, $f(x) \to +\infty$ for $|x| \to +\infty$) attains its global minimal value.

Remark. A sufficient condition for coercivity of $f : \mathbb{R}^n \to \mathbb{R}$ is the existence of a function $g : [0, \infty) \to \mathbb{R}$ for which $f(x) \ge g(|x|)$ if $|x|$ is sufficiently large and $g(r) \to +\infty$ for $r \to +\infty$. Finding such a function g is a matter of trial and error.

Example D.2.1 $f(x) = e^{2x_1} - 2x_1 + 2x_2^2 + 3 \to \min$, $x \in \mathbb{R}^2$.

Solution

1. f coercive (one can take $g(r) = |r|$) $\Rightarrow$ existence global minimum $\widehat{x}$.

2. Fermat: $f'(x) = (0,0) \Rightarrow$
$$\frac{\partial f}{\partial x_1} = 2e^{2x_1} - 2 = 0, \quad \frac{\partial f}{\partial x_2} = 4x_2 = 0.$$

3. $x_1 = x_2 = 0$.

4. $\widehat{x} = 0_2$.

Exercise D.2.1 $x_1^2 + x_1 x_2 + 2x_2^2 + 3 \to \min$, $x \in \mathbb{R}^2$.

Exercise D.2.2 $-x_1^2 + x_1 x_2 - x_2^2 + 2x_1 + x_2 \to \max$, $x \in \mathbb{R}^2$.

Exercise D.2.3 $(x_1^2 - x_2)^2 + (x_2 - 1)^4 \to \min$, $x \in \mathbb{R}^2$.

Exercise D.2.4 $x_1^2 + x_2^2 + x_3^2 - x_1 x_2 + x_1 - 2x_3 \to \min$, $x \in \mathbb{R}^3$.

Exercise D.2.5 * $3x_1 x_2 - x_1^2 x_2 - x_1 x_2^2 \to \min$, $x \in \mathbb{R}^2$.

D.3 SEVERAL VARIABLES UNDER EQUALITY CONSTRAINTS

- **Lagrange multiplier rule.** Let $f_i : U_n(\widehat{x}, \varepsilon) \to \mathbb{R}$, $0 \leq i \leq m$ be functions of n variables that are continuously differentiable at $\widehat{x}$. If $\widehat{x}$ is a local solution of the problem

$$f_0(x) \to \text{extr}, \ x \in U_n(\widehat{x}, \varepsilon), \ f_i(x) = 0, \ 1 \leq i \leq m,$$

then it is a solution of the system of equations

$$\sum_{i=0}^{m} \lambda_i \frac{\partial f_i}{\partial x_j}(x) = 0, \ 1 \leq j \leq n,$$

for a suitable choice of $\lambda_0, \ldots, \lambda_m$, not all zero.

Remark. Note that we get a system of equations where the number of unknowns is one more than the number of equations. However, we can get rid of one unknown: we always can assume that λ_0 is 0 or 1, without restricting generality, as the stationarity equations are homogeneous in the λ's. We will always put $\lambda_0 = 1$ and not display the routine verification that $\lambda_0 = 0$ is impossible. The "bad case" $\lambda_0 = 0$ can occur, but only in artificially concocted examples.

Example D.3.1 $f_0(x) = (x_1 - 1)^2 + (x_2 - 2)^2 \to \min$, $f_1(x) = 2x_1 + 3x_2 - 1 = 0$, $x \in \mathbb{R}^2$.

Solution

1. f_0 is coercive (one can take $g(r) = r^2$) $\Rightarrow$ existence global minimum $\widehat{x}$.

2. Lagrange function
$$\mathcal{L} = \lambda_0((x_1 - 1)^2 + (x_2 - 2)^2) + \lambda_1(2x_1 + 3x_2 - 1).$$
Lagrange: $\mathcal{L}_x = 0 \Rightarrow$
$$\frac{\partial \mathcal{L}}{\partial x_1} = 2\lambda_0(x_1 - 1) + 2\lambda_1 = 0, \quad \frac{\partial \mathcal{L}}{\partial x_2} = 2\lambda_0(x_2 - 2) + 3\lambda_1 = 0.$$
We put $\lambda_0 = 1$, as we may.

3. Eliminate λ_1: $6(x_1 - 1) = 4(x_2 - 2)$. Use equality constraint: $x = (\frac{-1}{13}, \frac{5}{13})^T$.

4. $\widehat{x} = (\frac{-1}{13}, \frac{5}{13})^T$.

Exercise D.3.1 $e^{x_1 x_2} \to \max$, $x_1 + x_2 - 1 = 0$, $x \in \mathbb{R}^2$.

Exercise D.3.2 $x_1 x_2 x_3 \to \min$, $x_1 + x_2 + x_3 = 0$, $x_1^2 + x_2^2 + x_3^2 = 1$, $x \in \mathbb{R}^3$.

Exercise D.3.3 $x_1 - 3x_2 - x_1 x_2 \to \min$, $x_1 + x_2 - 6 = 0$, $x \in \mathbb{R}^2$.

Exercise D.3.4 $x_1^2 + x_2^2 \to \min$, $x_1 + 4x_2 - 2 = 0$, $x \in \mathbb{R}^2$.

Exercise D.3.5 $e^{x_1 x_2} \to \min$, $x_1 + x_2 = 1$, $x \in \mathbb{R}^2$.

D.4 INEQUALITY CONSTRAINTS AND CONVEXITY

- **Karush-Kuhn-Tucker theorem.** Let $f_i : U_n(\widehat{x}, \varepsilon) \to \mathbb{R}$, $0 \leq i \leq m$ be functions that are continuously differentiable at $\widehat{x}$. If $\widehat{x}$ is a local solution of the problem
$$f_0(x) \to \min, \quad f_i(x) \leq 0, \quad 1 \leq i \leq m,$$

then it is a solution of the system of equations

$$\sum_{i=0}^{m} \lambda_i \frac{\partial f_i}{\partial x_j}(x) = 0, \ 1 \le j \le n$$

for a suitable choice of *nonnegative* numbers $\lambda_0, \ldots, \lambda_m$, not all zero, and such that $\lambda_i = 0$ for all $i \in \{1, \ldots, m\}$ for which the constraint $f_i(\hat{x}) < 0$ is not binding.

Moreover, assume that the functions f_i, $0 \le i \le m$, are in fact defined on some convex set $A \supseteq U_n(\hat{x}, \varepsilon)$, and assume that these functions $f_i : A \to \mathbb{R}$, $0 \le i \le m$, are convex. Then the conditions above are sufficient for global optimality on A provided $\lambda_0 \ne 0$. Here, all partial derivatives of all functions should exist and be continuous, and all functions should be convex (or at least quasiconvex). If, moreover, f_0 is strictly convex, then there can be at most one solution.

Remark. If you drop either the assumptions of differentiability or those of convexity, then weaker versions of this result hold true. In the first case, you have to replace the stationarity condition by the condition that $\hat{x}$ is a point of minimum of the Lagrange function

$$\mathcal{L} = \sum_{i=0}^{m} \lambda_i f_i(x).$$

In the second case, you lose the sufficiency property. One can deal with equality constraints by replacing them by two inequality constraint $(a = b \Leftrightarrow a \le b, \ b \le a)$.

Example D.4.1 $\quad x_1^2 + x_2^2 \to \min, \ x_1 x_2 \ge 25, \ x_1 \ge 0, \ x_2 \ge 0$.

Solution

1. $f_0(x) = x_1^2 + x_2^2 \to \min$, $f_1(x) = 25 - x_1 x_2 \le 0$, $f_2(x) = -x_1 \le 0$ (note that $x_1 x_2 \ge 25$, $x_1 \ge 0 \Rightarrow x_2 \ge 0$),

   f_0 coercive (one can take $g(r) = r^2$) $\Rightarrow$ existence of a global minimum $\hat{x}$.

2. Lagrange function

$$\mathcal{L} = \lambda_0 (x_1^2 + x_2^2) + \lambda_1 (25 - x_1 x_2) + \lambda_2 (-x_1).$$

KKT:

- $\mathcal{L}_x = 0_2^T \Rightarrow$

$$\frac{\partial \mathcal{L}}{\partial x_1} = \lambda_0(2x_1) + \lambda_1(-x_2) + \lambda_2(-1) = 0,$$

$$\frac{\partial \mathcal{L}}{\partial x_2} = \lambda_0(2x_1) + \lambda_1(-x_1) = 0,$$

- $\lambda_1, \lambda_2 \geq 0$,
- $\lambda_1(25 - x_1 x_2) = 0$, $\lambda_2(-x_1) = 0$.

We take $\lambda_0 = 1$, as we may.

3. We distinguish four cases, depending on whether the constraints are binding or not. We start with the most promising case: $x_1 x_2 = 25$ and $x_1 > 0$. $\Rightarrow \lambda_2 = 0$. Eliminate λ_1: $x_1^2 = x_2^2$. We get: $x_1 = x_2 = 5$. This satisfies all KKT conditions, so we do not have to consider the other cases.

4. $\hat{x} = (5, 5)^T$.

Exercise D.4.1 $\sum_{i=1}^n x_i^2 \to \min$, $\sum_{i=1}^n x_i^4 \leq 1$, $x \in \mathbb{R}^n$.

Exercise D.4.2 $\sum_{i=1}^n x_i^2 \to \max$, $\sum_{i=1}^n x_i^4 \leq 1$, $x \in \mathbb{R}^n$.

Exercise D.4.3 $2x_1^2 + 2x_1 + 4x_2 - 3x_3 \to \min$,
$8x_1 - 3x_2 + 3x_3 \leq 40$, $2x_1 - x_2 + x_3 - 3 = 0$, $x_2 \geq 0$, $x \in \mathbb{R}^3$.

Exercise D.4.4 $x_1 + x_2 \to \min$, $x_1^2 + x_2 \geq 9$, $x_1 x_2 \leq 8$, $x_1, x_2 \geq 0$, $x \in \mathbb{R}^2$.

Exercise D.4.5 $2x_1^2 + 2x_1 + 4x_4 - 3x_3 \to \max$, $8x_1 - 3x_2 \leq 40$, $2x_1 - x_2 + x_3 = 3$, $x_2 \geq 0$, $x \in \mathbb{R}^3$.

Appendix E

Crash Course on Optimization Theory: Geometrical Style

Plan. This crash course summarizes in words the theoretical results used to solve analytically finite-dimensional optimization problems, as well as the geometric ideas of the proofs of the necessary conditions for finite-dimensional problems.

E.1 THE MAIN POINTS

- Two types of problems: unconstrained and constrained.

- Two principles: Fermat and Lagrange.

    - Fermat: $f'(\hat{x}) = 0$ or $0 \in \partial f(\hat{x})$.
    - Lagrange: construct the Lagrange function and then use Fermat (or the necessary condition for the problem of minimization of the Lagrange function "as if the variables are independent."

- The Lagrange principle is universal; it can be applied to all problems from Fermat to Pontryagin.

- The proof of the Lagrange principle is based on two sources: in the smooth case on Newton's method, in the convex case on the separation method. In the mixed smooth-convex case there is a mixing of smooth and convex methods (for example, smooth problems with inequalities or problems of optimal control are smooth-convex).

- Existence of solutions is based on compactness (Weierstrass theorem). Weierstrass gives a key for the proof of the Lagrange principle in the finite-dimensional case.

- Algorithms are based on Newton's method, methods of reasonable descent, barrier methods, cutting methods.

E.2 UNCONSTRAINED PROBLEMS

One variable. The method for maximizing/minimizing a function of one variable f, using the derivative, is well known. Each local solution is a solution of the stationarity equation $f'(x) = 0$ (Fermat theorem), but not conversely: inflection points satisfy the equation as well. Then one determines the course of the sign of f' and compares f-values of local solutions. Here it can be used that a continuous function which has different signs at two points should be zero at some intermediate point (intermediate value theorem). This usually leads to the solution of the problem. In addition one can sometimes use that a continuous function on an interval $[a, b]$ assumes its maximal and minimal value (Weierstrass theorem). This fact, as well as the intermediate value theorem, are geometrically obvious and we will accept their truth without proof. Otherwise we would have to delve into the foundations of the real numbers.

Two or more variables. Most of the method above extends to the case of two or more variables $f(x_1, \ldots, x_n)$. Each local solution is a solution of the stationarity equations

$$\frac{\partial f}{\partial x_i}(x_1, \ldots, x_n) = 0, \ 1 \le i \le n.$$

Moreover, the Weierstrass theorem can be extended: you have to replace "interval $[a, b]$" by "a nonempty compact" in $\mathbb{R}^n$—a compact is a closed and bounded set. "Closed" means that all points on the boundary of A belong to A.

E.3 CONVEX PROBLEMS

One variable. A continuous function of one variable f is *convex* if for each two numbers x_1 and x_2,

$$f(\frac{1}{2}x_1 + \frac{1}{2}x_2) \le \frac{1}{2}f(x_1) + \frac{1}{2}f(x_2).$$

If this inequality holds strictly for all x, then the function is called *strictly convex*. A convenient criterion for convexity is that $f''(x) \ge 0$ for all x—provided that the second derivative f'' exists, of course. Moreover the following condition is sufficient for strict convexity:

$f''(x) > 0$ for all x. Problems of minimizing convex functions have two properties.

- A local minimum is always a global minimum.

- The stationarity condition $f'(x) = 0$ is also sufficient for minimality.

The significance of strict convex functions is that they have at most one minimum.

Two or more variables. The same as the case of one variable. Only the criterion for convexity is more involved. However, we will not need it, so we will not consider it.

E.4 EQUALITY CONSTRAINTS

Each local solution of the problem to minimize or maximize $f_0(x_1, \ldots, x_n)$ under the equality constraints $f_i(x_1, \ldots, x_n) = 0$, $1 \leq i \leq m$, is a solution of the system of equations

$$\sum_{i=0}^{m} \lambda_i \frac{\partial f_i}{\partial x_j}(x) = 0, \ 1 \leq j \leq n$$

for a suitable choice of $\lambda_0, \ldots, \lambda_m$, not all zero (Lagrange multiplier rule). Here all partial derivatives of all functions should exist and be continuous. Note that we get a system of equations where the number of unknowns is one more than the number of equations. However, we can get rid of one unknown: we always can assume that λ_0 is 0 or 1, without restricting generality, as the stationarity equations are homogeneous in the λ's. We will always put $\lambda_0 = 1$ and not display the routine verification that $\lambda_0 = 0$ is impossible. The case $\lambda_0 = 0$ can occur, but only in artificially concocted examples.

We sketch the idea of the proof of the Lagrange multiplier rule.

Special case. We consider the problem of minimizing $f(x_1, x_2)$ under the equality constraint $g(x_1, x_2) = 0$. Let $(\hat{x}_1, \hat{x}_2)$ be a local minimum. By translation we can achieve that it is the origin $(0,0)$. The crux of the matter is to show that the solution set of the equation $g(x_1, x_2) = 0$ looks near the point $(0,0)$ like a curved line that has the line through $(0,0)$ orthogonal to the gradient vector

$$\nabla g(0,0) = (\frac{\partial g}{\partial x_1}(0,0), \frac{\partial g}{\partial x_2}(0,0))$$

as tangent line, provided this gradient is nonzero ("tangent space theorem"). We assume that the gradient of g is continuous at the origin, and therefore almost constant near the origin. Then, using $g(0,0) = 0$ and the intermediate value theorem, we get that the function g has on each line which passes close to the origin with direction $\nabla g(0,0)$, a point where it is zero, *very close* to the line l, as required. You can prove a corresponding fact on the equation $f(x_1, x_2) - f(0,0) = 0$ and the point $(0,0)$. Then it is readily seen that local minimality of $(0,0)$ implies that the curves $g(x_1, x_2) = 0$ and $f(x_1, x_2) = f(0,0)$ are tangent at $(0,0)$. Therefore, $\nabla f(0,0)$ is a scalar multiple of $\nabla g(0,0)$. This establishes the Lagrange equations.

General case. The idea of the proof is the same as in the special case, but instead of applying the intermediate value theorem, one should apply the Weierstrass theorem. We sketch this for the case of one equality constraint. We apply the Weierstrass theorem to the problem of minimizing $|g(x)|$ on lines passing close to the origin and having direction $\nabla g(0, \ldots, 0)$. To be more precise, here one allows x to run only in a small ball with center the origin. The stationarity equations have no solution, so the solution of the problem must be at a point of nondifferentiability of $|g(x)|$. That is, the solution of this problem must satisfy the equation $g(x) = 0$.

E.5 INEQUALITY CONSTRAINTS

Karush-Kuhn-Tucker conditions. Each solution of the problem to minimize or maximize $f_0(x_1, \ldots, x_n)$ under the inequality constraints $f_i(x_1, \ldots, x_n) \leq 0$, $1 \leq i \leq m$ is a solution of the system of equations

$$\sum_{i=0}^{m} \lambda_i \frac{\partial f_i}{\partial x_j}(x) = 0, \ 1 \leq j \leq n$$

for a suitable choice of nonnegative numbers $\lambda_0, \ldots, \lambda_m$, not all zero and such that $\lambda_i = 0$ for all $i \in \{1, \ldots, m\}$ for which the constraint is not binding, $f_i(\widehat{x}) < 0$ (Karush-Kuhn-Tucker theorem). Moreover, this condition is sufficient for optimality provided $\lambda_0 \neq 0$. Here all partial derivatives of all functions should exist and be continuous and all functions should be continuous. If you drop either the assumption of differentiability or those of convexity, then weaker versions of this result hold true. In the first case you have to replace the stationarity condition by the condition that $\widehat{x}$ is a point of minimum of the Lagrange function $\sum_{i=0}^{m} \lambda_i f_i(x)$. In the second case you lose the sufficiency property.

Idea of the proof of the KKT-conditions. The basis of the proof is the following separation theorem: a closed convex set C can be separated from any point p outside this set by a hyperplane. This can be proved by choosing a point in C which lies closest to p—this exists by the Weierstrass theorem—and verifying that the hyperplane orthogonal to the line through p and c which contains the midpoint $\frac{1}{2}p + \frac{1}{2}c$, is a separating hyperplane. A corollary of this theorem is that two convex sets which have no common interior point can be separated by a hyperplane. We are ready to prove the KKT theorem. Separate the following convex sets by a hyperplane:

$$\{(x, y, z) \mid f_i(x) \leq y_i,\ 1 \leq i \leq m,\ f_0(x) \leq z\}$$

and

$$\{(x, 0_m, z) \mid f_i(x) \leq 0,\ 1 \leq i \leq m,\ z \leq f_0(\widehat{x})\},$$

using the separation theorem for convex sets. The equation of the separating hyperplane is seen to be of the following form:

$$\lambda_0(z - f_0(\widehat{x})) + \sum_{i=1}^{m} \lambda_i(x_i - f_i(\widehat{x})) = 0.$$

Writing out explicitly what this means gives the KKT conditions.

E.6 TRANSITION TO INFINITELY MANY VARIABLES

The two results that form the backbone of the finite-dimensional theory, the tangent space theorem and the supporting hyperplane theorem—or, equivalently, the separation theorems—can be extended to the infinite-dimensional setting. However, different proofs have to be given. The reason is that the proofs above use the Weierstrass theorem; this involves the concept of compactness, and this does not agree with closed and bounded in the infinite-dimensional case. Therefore the Weierstrass theorem cannot be used to prove the required extensions.

We will sketch the idea of alternative proofs, which can be extended.

Alternative proof of the tangent space theorem. You carry out the proof as above, but at the moment that you apply the intermediate value theorem (or the Weierstrass theorem) to find a point where g is zero, you do the following instead. You apply the modified Newton method with slope determined by $\nabla g(0, 0)$. This is an algorithm to find the solution of a system of k equations in k unknowns.

For $k = 1$ the idea of the algorithm is given in Figure 6.5; the main point of the algorithm is that all slanted lines have the same slope, and that this slope is roughly equal to the derivative in the zero point. The known convergence properties of the algorithm give the required conclusion.

Alternative proof of the supporting hyperplane theorem. It is most convenient to give this proof for the following separation result, which is equivalent to the separating hyperplane theorem. A convex cone—that is, a convex set which consists of rays with endpoint the origin—can be separated from the origin by a hyperplane. This can be proved by induction with respect to the dimension.

We begin with the case of dimension two: let a convex cone in the plane having an interior point be given. Choose an interior point of the convex cone. Take the line through this point and the origin. Then let the line rotate around the origin. Stop at the first position of the line which contains no interior point of the convex cone. This is a separating line.

We proceed to consider the case of dimension three: let a convex cone in three-dimensional space having an interior point be given. Choose an interior point of the convex cone. Take a plane H through this point and the origin. Consider the intersection of the convex cone with this plane. Separate it by a line l in the plane from the origin. Now start rotating the plane H around the line l. Stop at the first position of the plane which contains no interior point of the given convex cone. This is a separating hyperplane.

Continuing in this way, one establishes the separation theorem in dimension n for all n.

Motivation for the modified Newton method. The idea of proving the tangent space theorem (or the related theorems on inverse and implicit functions and mappings) by means of the modified Newton method is very natural. For example, consider the proof that a continuously differentiable function of one variable f with $f(0) = 0$ and $f'(0) \neq 0$ has an inverse function in the neighborhood of $\hat{x} = 0$. Here the problem is to show that for each $y \approx 0$ the equation $f(x) = y$ has a unique solution with $x \approx 0$. To show this it might seem a good idea to apply the Newton method to this equation with starting point $\hat{x} = 0$ and to show that this method converges to the unique solution. However, the following modification of the idea is needed to make it work. One should not adjust the direction of the slanted lines: these should all be taken parallel to the first one. This modification is more stable than the Newton method.

Contraction principle compared to the modified Newton method. There are—as far as we know—only two essentially different proofs of the inverse function theorem and all related results: a proof using the Weierstrass theorem and a proof using the modified Newton method. The usual proof of the inverse function theorem given in textbooks is based on the *contraction principle*. We will explain that it is essentially the same proof as the one using the modified Newton method. A mapping F from a ball in $\mathbb{R}^n$ to itself is called a *contraction* if it reduces all distances at least by a factor $\theta < 1$. That is, there is a number $\theta \in (0,1)$ with

$$|F(x_1) - F(x_2)| \leq \theta |x_1 - x_2|$$

for all x_1, x_2. The contraction principle states that a contraction has a unique *fixed point*—that is, a solution of $F(x) = x$. Moreover, each sequence defined by repeated application of F converges to this unique fixed point. The contraction principle is illustrated by the following example.

Example E.6.1 *Take a—possibly folded—map of Rotterdam, and throw it on the pavement somewhere in Rotterdam. Then there exists one and only one point on the plan that lies precisely on top of the point of Rotterdam that it represents.*

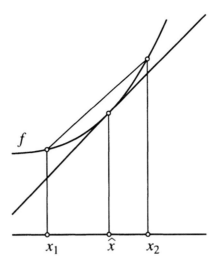

Figure E.1 Illustrating strict differentiability.

The following fact can readily be checked: the recursion defining the modified Newton method is a contraction for $y \approx 0$. This makes clear that the proof of the inverse function theorem by means of the modified Newton method is essentially a proof by means of the contraction principle.

Finally, we note that the fact above can be formalized as follows. Call a differentiable function f of one variable *strictly differentiable* at $\widehat{x}$ if for each $\varepsilon > 0$ there exists $\delta > 0$ such that

$$|f(x_1) - f(x_2) - f'(\widehat{x})(x_1 - x_2)| \leq \varepsilon |x_1 - x_2|$$

if $|x_i - \widehat{x}| \leq \delta$, $i = 1, 2$ (Fig. E.1).

This can be written as a contraction property for the modified Newton method:

$$|(x_1 - f'(\widehat{x})^{-1} f(x_1)) - (x_2 - f'(\widehat{x})^{-1} f(x_2))| \leq \varepsilon f'(\widehat{x})^{-1} |x_1 - x_2|.$$

Thus the fact above can be formulated as follows: continuous differentiability at a point implies strict differentiability at this point.

Appendix F

Crash Course on Optimization Theory: Analytical Style

> All styles are fine, except the boring one
>
> *French saying*

- What are the formulas of optimization and how to prove them?

Plan. This crash course is meant for readers who have read this book, but advanced readers can begin with this crash course. We give the precise statements of the necessary and sufficient conditions that are used for solving analytically finite-dimensional optimization problems. We do this under the natural assumptions (for example, of strict differentiability rather than continuous differentiability) and in full generality (for example, we consider problems with three types of constraints simultaneously: equality, inequality, and inclusion constraints). Then we give full analytical proofs of all results, without pictures, but including all details. Thus the style is completely different from the style in the main text, where we supported the intuition for each of the theoretical results by simple geometric figures. In the proofs below, we use the tangent space theorem and the supporting hyperplane theorem. We give the statement of these two main theorems of smooth and convex calculus, and briefly recall the two possible ways of proving each of these.

F.1 PROBLEM TYPES

We will use the following notation.

- $X = \mathbb{R}^n = \left\{ x : x = \begin{pmatrix} x_1 \\ \vdots \\ x_n \end{pmatrix}, \ x_i \in \mathbb{R}, \ 1 \leq i \leq n \right\}$.

- $x^T = (x_1, \ldots, x_n)$ for $x \in X$.

- $X' = (\mathbb{R}^n)' = \{x' : x' = (x'_1, \ldots, x'_n),\ x'_i \in \mathbb{R},\ 1 \leq i \leq n\}$ (an alternative notation for X' is X^T).

- $x' \cdot x = x'_1 x_1 + \cdots + x'_n x_n$ for $x' \in X'$ and $x \in X$.

- $|x| = \sqrt{x^T \cdot x}$ for $x \in X$.

- $\mathcal{O}(X)$ is the collection of open sets of X.

- $\mathcal{O}(\widehat{x}, X)$ is the collection of neighborhoods of $\widehat{x}$—that is, open sets in X containing $\widehat{x}$—for a given $\widehat{x} \in X$.

- For a given set C in X and a given function $f : C \to \mathbb{R}$, we denote the corresponding extremal problem as

$$f(x) \to \text{extr},\ x \in C. \qquad (P)$$

- Local extrema are denoted and defined as follows:
$\widehat{x} \in \text{locmin(max)}(P)$ if

$$\exists U \in \mathcal{O}(\widehat{x}, X) : f(x) \geq f(\widehat{x})\ (\leq f(\widehat{x}))\ \forall x \in C \cap U.$$

- $\text{Cone}(X)$ is the collection of all convex cones in X.

We consider the following problem types (with definitions of Lagrange functions and selections of Lagrange multipliers—if applicable). Let $X = \mathbb{R}^n$ and $U \in \mathcal{O}(X)$.

1. **Problem without constraints** of $f_0 : U \to \mathbb{R}$:

$$f_0(x) \to \text{extr},\ x \in U. \qquad (P_1)$$

2. **Problem with equality constraints** of $f_0 : U \to \mathbb{R}$ and $F = (f_1, \ldots, f_m)^T : U \to \mathbb{R}^m$:

$$f_0(x) \to \text{extr},\ x \in U,\ F(x) = 0_m. \qquad (P_2)$$

A selection of Lagrange multipliers is an element $\lambda = (\lambda_0, y)$, where $\lambda_0 \in \mathbb{R}$ and $y \in (\mathbb{R}^m)'$, not both zero. The Lagrange function is

$$\mathcal{L}(x, \lambda) = \lambda_0 f_0(x) + y \cdot F(x).$$

3. **Problem with equality and inequality constraints** of $f_i : U \to \mathbb{R}$, $0 \leq i \leq k$, $F : U \to \mathbb{R}^{m-k}$, and $A \subseteq X$:

$$f_0(x) \to \min, \ x \in U, \ f_i(x) \leq 0, \ 1 \leq i \leq k, \ F(x) = 0_{m-k}, \ x \in A. \quad (P_3)$$

A selection of Lagrange multipliers is an element $\lambda = (\lambda_0, \mu, y)$, where $\lambda_0 \in \mathbb{R}_+$, $\mu \in (\mathbb{R}^k)'$ and $y \in (\mathbb{R}^{m-k})'$, not all three zero. The Lagrange function is

$$\mathcal{L}(x, \lambda) = \lambda_0 f_0(x) + \sum_{i=1}^{k} \mu_i f_i(x) + y \cdot F(x).$$

4. **Conic smooth-convex problem** of $f_0 : U \to \mathbb{R}$, $F : U \to \mathbb{R}^m$, and $C \in \text{Cone}(X)$:

$$f_0(x) \to \min, \ x \in U, \ F(x) = 0_m, \ x \in C. \quad (P_4)$$

A selection of Lagrange multipliers is an element $\lambda = (\lambda_0, y)$, where $\lambda_0 \in \mathbb{R}_+$ and $y \in (\mathbb{R}^m)'$, not both zero. The Lagrange function is

$$\mathcal{L}(x, \lambda) = \lambda_0 f_0(x) + y \cdot F(x).$$

Remark. Each type is essentially a special case of the next one. In chapter ten we consider an even more general type, "the problem of a smooth-convex perturbation."

F.2 DEFINITIONS OF DIFFERENTIABILITY

Let $X = \mathbb{R}^n$, $\widehat{x} \in X$, $U \in \mathcal{O}(\widehat{x}, X)$, and $Y = \mathbb{R}^m$.

Definition F.1 *Differentiability of functions.* *A function $f : U \to \mathbb{R}$ is called differentiable at $\widehat{x}$ if there exists $x' \in X'$ such that*

$$f(\widehat{x} + x) = f(\widehat{x}) + x' \cdot x + o(|x|).$$

Then the vector x' is uniquely determined, it is called the derivative of f at $\widehat{x}$, it is denoted by $f'(\widehat{x})$, and we write $f \in \mathcal{D}^1(\widehat{x})$.

Definition F.2 *Differentiability of vector functions.* *A vector function $F = (f_1, \ldots, f_m)^T : U \to Y$ is called differentiable at $\widehat{x}$ if $f_i \in \mathcal{D}^1(\widehat{x})$, $1 \leq i \leq m$. Then the $m \times n$ matrix $\begin{pmatrix} f_1'(\widehat{x}) \\ \vdots \\ f_m'(\widehat{x}) \end{pmatrix}$ is* called the derivative of F at $\hat{x}$, it is denoted by $F'(\hat{x})$, and we write $F \in \mathcal{D}^1(\hat{x}, Y)$.

Equivalently, we call a vector function $F : U \to Y$ differentiable at $\hat{x}$ if there exists an $m \times n$ matrix C such that

$$F(\hat{x} + x) = F(\hat{x}) + Cx + o(|x|).$$

Then $C = F'(\hat{x})$.

Definition F.3 Twice differentiability of functions. *A function $f : U \to \mathbb{R}$ is called twice differentiable at $\hat{x}$ if $f \in \mathcal{D}^1(\hat{x})$ and if, moreover, there exists a symmetric $n \times n$ matrix A such that*

$$f(\hat{x} + x) = f(\hat{x}) + f'(\hat{x}) \cdot x + \frac{1}{2} x^T A x + o(|x|^2).$$

Then the matrix A is uniquely determined, it is called the second derivative of f at $\hat{x}$, we denote it by $f''(\hat{x})$, and we write $f \in \mathcal{D}^2(\hat{x})$. It is convenient to write

$$f''(\hat{x})[h, k] = h^T A k$$

for all $h, k \in X$, that is, to view $f''(\hat{x})$ as a bilinear function.

Definition F.4 Twice differentiability of vector functions. *A vector function $F = (f_1, \ldots, f_m)^T : U \to Y$ is called twice differentiable at $\hat{x}$ if $F \in \mathcal{D}^1(\hat{x}, Y)$ and if, moreover, $f_i \in \mathcal{D}^2(\hat{x})$, $1 \leq i \leq m$. Then the second derivative is defined to be the bilinear vector-valued function*

$$\begin{pmatrix} f_1''(\hat{x}) \\ \vdots \\ f_m''(\hat{x}) \end{pmatrix},$$

it is denoted by $F''(\hat{x})$, and we write $F \in \mathcal{D}^2(\hat{x}, Y)$. That is,

$$F''(\hat{x})[h, k] = (f_1''(\hat{x})[h, k], \ldots, f_m''(\hat{x})[h, k])^T,$$

for all $h, k \in X$.

Remark. This definition of the second derivative is the most useful one, though it would be more natural to define the second derivative as the derivative of the derivative. The two definitions agree if the second derivative according to one of the two definitions is continuous.

We now define two additional classes of differentiable vector functions. The first one is very convenient for the presentation of the proofs.

Definition F.5 *Strict differentiability of vector functions.* *A vector function $F : U \to Y$ is called strictly differentiable at the point $\widehat{x}$ if $F \in \mathcal{D}^1(\widehat{x}, Y)$ and if, moreover,*

$$\forall \varepsilon > 0 \; \exists \delta > 0 \; \text{s.t.} \; |x_i - \widehat{x}| < \delta, \; i = 1, 2 \Rightarrow$$

$$|F(x_2) - F(x_1) - F'(\widehat{x})(x_2 - x_1)| \leq \varepsilon |x_2 - x_1|.$$

Then we write $F \in \mathcal{SD}^1(\widehat{x}, Y)$.

Strict differentiability at a point $\widehat{x}$ implies continuity at a neighborhood of $\widehat{x}$ and differentiability at $\widehat{x}$. This concept is illustrated in Figure ??.

The second additional class is convenient as it contains all functions that occur in concrete problems—and no functions with pathological behavior—and moreover, membership in this class can be checked easily.

Definition F.6 *Continuous differentiability of vector functions.* *A vector function $F : U \to Y$ is called continuously differentiable at the point $\widehat{x}$ if $F \in \mathcal{D}^1(x, Y)$ for all $x \in U$ and if, moreover, $F'(x)$ depends continuously on x at $x = \widehat{x}$. Then we write $F \in C^1(\widehat{x}, Y)$.*

The good properties of these two classes can be combined by virtue of the following fact: $C^1(\widehat{x}, Y) \subseteq \mathcal{SD}^1(\widehat{x}, Y)$. Thus, we can present the proofs for strictly differentiable vector functions and then apply these results to continuously differentiable vector functions.

F.3 MAIN THEOREMS OF DIFFERENTIAL AND CONVEX CALCULUS

Roughly speaking, the sense of the main theorem of the differential (resp. convex) calculus is that it gives the possibility to approximate differentiable (resp. convex) nonlinear objects by linear ones.

Again, we let $X = \mathbb{R}^n$, $\widehat{x} \in X$, $U \in \mathcal{O}(\widehat{x}, X)$, and $Y = \mathbb{R}^m$. We will use the following notation.

- $\ker A = \{x \in X : Ax = 0_m\}$ for an $m \times n$ matrix A ("the kernel of A").

- $x \perp y$ for $x, y \in X$ if $x^T \cdot y = 0$ ("x and y are orthogonal").

- $L^\perp = \{x \in X : x \perp y \ \forall y \in L\}$ for a linear subspace L of X ("$L^\perp$ is called the orthogonal complement of L").

- The boundary ∂C of a set $C \subseteq X$ is the set of points b of X for which each ball with center b contains a point of C as well as a point that is not in C.

First, we give the main theorem of the differential calculus.

Theorem F.7 *Tangent space theorem.* Let a vector-valued function $F : U \to Y$ have the following properties: $F \in SD^1(\widehat{x}, Y)$, $\mathrm{rank}\, F'(\widehat{x}) = m$ and $F(\widehat{x}) = 0_m$. Then there is a vector function

$$r : \ker F'(\widehat{x}) \to (\ker F'(\widehat{x}))^\perp$$

with $|r(\xi)| = o(|\xi|)$ for which the set of solutions of the vector equation $F(x) = 0_m$ coincides in a neighborhood of $\widehat{x}$ with the set of points $\widehat{x} + \xi + r(\xi)$ where ξ runs over $\ker F'(\widehat{x})$.

There are several equivalent variants of it; the best known is the implicit function theorem. There are two, *related* ways to prove the tangent space theorem. It can be derived from the Weierstrass theorem. Alternatively, it can be established by using an algorithm, the modified Newton method—or, equivalently, by the contraction principle. This second proof is the universal one: it can be extended to the infinite-dimensional case (which is needed for optimization problems of the calculus of variations and of optimal control). Details of these two proofs and their relation are given in chapter three.

Second, we give the main result of convex calculus.

Theorem F.8 *Supporting hyperplane theorem.* Let C be a convex set in X and let $b \in \partial C$. Then there exists a nonzero $\lambda \in X'$ such that $\lambda \cdot (x - b) \geq 0$ for all $x \in C$.

There are several equivalent variants of this result as well; the best known are the separation theorems for convex sets. There are two, *different* ways to prove the supporting hyperplane theorem. It can be derived from the Weierstrass theorem. Alternatively, it can be established by induction. Indeed, the core of the matter is the case where the dimension is two: dimension one is trivial, and the step from dimension n to dimension $n+1$ follows from the two-dimensional case. This second proof is the universal one: it can be extended to the infinite-dimensional case (which is needed for optimization problems of optimal control). Details of these two proofs are given in chapter four .

F.4 CONDITIONS THAT ARE NECESSARY AND/OR SUFFICIENT

We will give the formulation for minimization problems only.

Theorem F.9 *Conditions that are necessary and/or sufficient*

1. **Problems without constraint.** *Consider a problem of type* (P_1) *with a point* $\widehat{x} \in U$. *Assume* $f_0 \in \mathcal{D}^1(\widehat{x})$.

    (a) *First order necessary conditions ("Fermat theorem"). If* $\widehat{x} \in \mathrm{locmin}(P_1)$, *then* $f_0'(\widehat{x}) = 0_n^T$ *("stationarity").*

    Assume for the next two statements that $f_0 \in \mathcal{D}^2(\widehat{x})$.

    (b) *Second order necessary conditions. If* $\widehat{x} \in \mathrm{locmin}(P_1)$, *then*
    $$f_0''(\widehat{x})[\xi, \xi] \geq 0 \ \forall \xi \in X$$
    (" $f_0''(\widehat{x})$ positive semidefinite").

    (c) *Sufficient conditions. If* $f_0'(\widehat{x}) = 0_n^T$ *and*
    $$f_0''(\widehat{x})[\xi, \xi] > 0 \forall \xi \in X \setminus \{0_n\}$$
    (" $f_0''(\widehat{x})$ positive definite"), then $\widehat{x} \in \mathrm{strictlocmin}(P_1)$.

2. **Problems with equality constraints.** *Consider a problem of type* (P_2) *with an admissible point* $\widehat{x}$. *Assume* $f_0 \in \mathcal{D}^1(\widehat{x})$ *and* $F \in \mathcal{SD}^1(\widehat{x}, \mathbb{R}^m)$.

    (a) *First order necessary conditions ("Lagrange multiplier rule"). If* $\widehat{x} \in \mathrm{locmin}(P_2)$, *then there exists a selection of Lagrange multipliers* λ *for which*
    $$\mathcal{L}_x(\widehat{x}, \lambda) = 0_n^T$$

("Lagrange equations").
For the next two statements make the following additional assumptions: $f_0 \in \mathcal{D}^2(\widehat{x})$ and $F \in \mathcal{D}^2(\widehat{x}, \mathbb{R}^m)$.

(b) **First and second order necessary conditions.** Assume that $\text{rank} F'(\widehat{x}) = m$. If $\widehat{x} \in \text{locmin}(P_2)$, then there exists a selection of Lagrange multipliers $\lambda = (1, y)$ for which not only the first order conditions, but also the second order conditions

$$\mathcal{L}_{xx}(\widehat{x}, \lambda)[\xi, \xi] \geq 0 \text{ for all } \xi \in \ker F'(\widehat{x})$$

hold *("$\mathcal{L}_{xx}(\widehat{x}, \lambda)$ positive semidefinite on $\ker F'(\widehat{x})$")*.

(c) **Sufficient conditions.** If there exists a selection of Lagrange multipliers $\lambda = (1, y)$ for which

$$\mathcal{L}_x(\widehat{x}, \lambda) = 0_n^T \text{ and } \mathcal{L}_{xx}(\widehat{x}, \lambda)[\xi, \xi] > 0$$

for all nonzero $\xi \in \ker F'(\widehat{x})$ *("$\mathcal{L}_{xx}(\widehat{x}, \lambda)$ positive semidefinite on $\ker F'(\widehat{x})$")*,

then $\widehat{x} \in \text{strictlocmin}(P_2)$.

3. **Problems with equality and inequality constraints.** Consider a problem of type (P_3) with an admissible point $\widehat{x}$.

   (a) **Convex case**

   Make the following assumptions: the set A is convex, the functions f_i, $0 \leq i \leq k$, are convex, and the vector function F is affine *("linear plus constant")*.

   i. **Necessary conditions** *("Karush-Kuhn-Tucker theorem")*. If $\widehat{x} \in \text{absmin}(P_3)$, then there exists a selection of Lagrange multipliers λ for which the following conditions *("KKT conditions")* hold:

   A. $\min_{x \in A} \mathcal{L}(x, \lambda) = \mathcal{L}(\widehat{x}, \lambda)$ *("minimum condition" or "minimum principle")*,

   B. $\mu_i \geq 0$, $1 \leq i \leq k$ *("nonnegativity conditions")*,

   C. $\mu_i f_i(\widehat{x}) = 0$, $1 \leq i \leq k$ *("complementarity")*.

   ii. **Sufficient conditions.** If there exists a selection of Lagrange multipliers $\lambda = (1, \mu, y)$ for which the KKT conditions hold, then $\widehat{x} \in \text{absmin}(P_3)$.

(b) **Differentiable case**
Make the following assumptions: $A = X$, $f_0 \in \mathcal{D}^1(\widehat{x})$, $f_i \in SD^1(\widehat{x})$, $1 \leq i \leq k$, $F \in SD^1(\widehat{x}, \mathbb{R}^{m-k})$.

i. *First order necessary conditions ("John theorem")*. If $\widehat{x} \in \text{locmin}(P_3)$, then there exists a selection of Lagrange multipliers λ such that
   A. $\mathcal{L}_x(\widehat{x}, \lambda) = 0_n^T$ ("stationarity condition"),
   B. $\mu_i \geq 0$, $1 \leq i \leq k$ ("nonnegativity conditions"),
   C. $\mu_i f_i(\widehat{x}) = 0$, $1 \leq i \leq k$ ("complementarity").

For the next two statements, make the following additional assumptions: $f_i \in \mathcal{D}^2(\widehat{x})$, $1 \leq i \leq k$, and $F \in \mathcal{D}^2(\widehat{x}, \mathbb{R}^{m-k})$.

ii. *First and second order necessary conditions.* Let the regularity assumptions
$$\text{rank} F'(\widehat{x}) = m - k$$
and
$$\exists \xi \in \mathbb{R}^n : f_i'(\widehat{x}) \cdot \xi < 0 \ \forall i \in I, \ F'(\widehat{x})\xi = 0_{m-k}$$
be satisfied. Here, I is the set of numbers $i \in \{1, \ldots, k\}$ for which $f_i(\widehat{x}) = 0$. If $\widehat{x} \in \text{locmin}(P_3)$, then there exists a selection of Lagrange multipliers $\lambda = (1, \mu, y)$ for which not only the first order conditions, but also the second order conditions
$$\mathcal{L}_{xx}(\widehat{x}, \lambda)[\xi, \xi] \geq 0$$
holds true for all $\xi \in \ker F'(\widehat{x})$ for which $f_i'(\widehat{x}) \cdot \xi \leq 0$, $1 \leq i \leq k$.

iii. *Sufficient conditions.* If there exists a selection of Lagrange multipliers $\lambda = (1, \mu, y)$ for which not only the first order conditions, but also the second order conditions
$$\mathcal{L}_{xx}(\widehat{x}, \lambda)[\xi, \xi] > 0$$
hold true for all nonzero $\xi \in \ker F'(\widehat{x})$ which satisfy the inequality $f_i'(\widehat{x}) \cdot \xi \leq 0$, $1 \leq i \leq k$, then it follows that $\widehat{x} \in \text{strictlocmin}(P_3)$.

4. **Conic smooth-convex problems.** *Consider a problem of type (P_4) with an admissible point $\widehat{x}$.*

   Make the following assumptions: $f_0 \in D^1(\widehat{x})$, $F \in \mathcal{SD}^1(\widehat{x}, \mathbb{R}^m)$, and the convex cone C is solid—that is, contains interior points—and closed.

   (a) *First order necessary conditions. If $\widehat{x} \in \mathrm{locmin}(P_4)$, then there exists a selection of Lagrange multipliers λ for which*
   $$\min\nolimits_{x \succeq -\widehat{x}} \mathcal{L}_x(\widehat{x}, \lambda) x = 0.$$

   For the next two statements, make the following additional assumption: $F \in \mathcal{D}^2(\widehat{x}, \mathbb{R}^m)$.

   (b) *First and second order necessary conditions. Let the regularity assumptions*
   $$\mathrm{rank} F'(\widehat{x}) = m - k$$
   and
   $$\exists \xi \in \ker F'(\widehat{x}) : \widehat{x} + \xi \in \mathrm{int} C$$
   be satisfied. If $\widehat{x} \in \mathrm{locmin}(P_4)$, then there exists a selection of Lagrange multipliers λ for which not only the first order conditions, but also the second order conditions
   $$\mathcal{L}_{xx}(\widehat{x}, \lambda)[\xi, \xi] \geq 0$$
   hold for all nonzero $\xi \in \ker F'(\widehat{x})$ for which $\widehat{x} + \xi \in \mathrm{int} C$.

   (c) *Sufficient conditions. If there exists a selection of Lagrange multipliers λ for which not only the first order conditions, but also the second order conditions*
   $$\mathcal{L}_{xx}(\widehat{x}, \lambda)[\xi, \xi] > 0$$
   hold for all nonzero $\xi \in \ker F'(\widehat{x})$ for which $\widehat{x} + \xi \in \mathrm{int} C$.

Remark. We note that for problem type (P_2) one can eliminate the Lagrange multiplier from the first and second order conditions, if the regularity assumption holds. We now show how this can be done. Then the Lagrange equations (with $\lambda_0 = 1$)
$$f_0'(\widehat{x}) + y F'(\widehat{x}) = 0_n^T$$

CRASH COURSE ON OPTIMIZATION THEORY: ANALYTICAL STYLE 571

give, after multiplication on the right by the transpose $F'(\widehat{x})^T$, the equation

$$f'_0(\widehat{x})F'(\widehat{x})^T + yF'(\widehat{x})F'(\widehat{x})^T = 0_m^T.$$

Finally, using that $F'(\widehat{x})F'(\widehat{x})^T$ is invertible (as $\mathrm{rank} F'(\widehat{x}) = m$), we get the following expression for y:

$$y = -f'_0(\widehat{x})F'(\widehat{x})^T(F'(\widehat{x})F'(\widehat{x})^T)^{-1}.$$

This expression can be substituted in the first and second order conditions. A similar elimination is possible for problem (P_3) in the differentiable case.

Remark. An admissible point $\bar{x} \in X$ of a convex problem of type (P_3) is called a Slater point if the inequality constraints hold strictly: $f_i(\bar{x}) < 0$, $1 \leq i \leq k$. The existence of a Slater point implies $\lambda_0 \neq 0$. We now verify this, arguing by contradiction. Assume $\lambda_0 = 0$; then

$$0 = \mathcal{L}(\widehat{x}, \lambda) \leq \mathcal{L}(\bar{x}, \lambda) = \sum_{i=1}^n \mu_i f_i(\bar{x}) < 0,$$

which is the required contradiction. For problem type (P_4), the regularity assumption implies $\lambda_0 \neq 0$ as well. This will be clear from the proof.

F.5 PROOFS

We assume $\widehat{x} = 0_n$ (except for problem type (P_4)), as we may without restricting the generality of the proof.

1. **Problems without constraints**

   (a) Take an arbitrary $\xi \in X$. Consider the variation $(t\xi)_{t \in \mathbb{R}}$ and observe that $f_0(t\xi) \geq f_0(0_n)$ if $|t|$ is sufficiently small, as $0_n \in \mathrm{locmin}(P_1)$. Now use the linear approximation formula

   $$f_0(t\xi) = f_0(0_n) + tf'_0(0_n) \cdot \xi + o(|t|),$$

   which holds as $f_0 \in \mathcal{D}^1(0_n)$. This leads to $f'_0(0_n) \cdot \xi = 0$. Therefore, $f'_0(0_n) = 0_n^T$, as $\xi \in X$ is arbitrary.

   (b) Proceed as above, but now the quadratic approximation formula

   $$f_0(t\xi) = f_0(0_n) + tf'_0(0_n) \cdot \xi + \frac{t^2}{2}f''_0(0_n)[\xi, \xi] + o(t^2)$$

holds, as $f_0 \in \mathcal{D}^2(0_n)$. This leads to $f_0''(0_n)[\xi, \xi] \geq 0$, using $f_0'(0_n) = 0_n^T$.

(c) Write the following quadratic approximation formula:

$$f_0(\xi) = f_0(0_n) + f_0'(0_n) \cdot \xi + \frac{1}{2} f_0''(0_n)[\xi, \xi] + o(|\xi|^2)$$

for all $\xi \in X$. The assumptions give that the second term of the right-hand side is zero, and that the third term is bounded below by the expression $K|\xi|^2$, for some constant $K > 0$ that does not depend on ξ (see appendix A). Therefore, the inequality $f_0(\xi) > f_0(0_n)$ holds for all $\xi \in X$ of sufficiently small modulus.

2. **Problems with equality constraints**

   (a) If the regularity assumption does not hold, then choose a nontrivial linear relation between the rows of $F'(0_n)$, say,

   $$\sum_{i=1}^m y_i f_i'(\widehat{x}) = 0_n^T.$$

   Then the Lagrange equations are seen to hold (with the choice $\lambda_0 = 0$). If the regularity assumption holds, then take an arbitrary nonzero $\xi \in \ker F'(0_n)$. Show that this is a direction vector of an admissible variation in the following sense. The tangent space theorem can be applied; it gives that $t\xi + r(t\xi)$ is a solution of $F(x) = 0_m$ for $|t|$ sufficiently small, and that $r(t\xi) = o(|t|)$. Therefore, it is natural to say that

   $$(t\xi + r(t\xi))_{t \in \mathbb{R}}$$

   is an admissible variation with direction vector ξ. Observe that

   $$f_0(t\xi + r(t\xi)) \geq f_0(0_n)$$

   for $|t|$ sufficiently small, as $0_n \in \text{locmin}(P_2)$ and $t\xi + r(t\xi)$ is admissible. Moreover, $r(t\xi) = o(|t|)$. Now use the linear approximation formula

   $$f_0(t\xi + r(t\xi)) = f_0(0_n) + t f_0'(0_n) \cdot \xi + o(|t|),$$

   which holds as $f_0 \in \mathcal{D}^1(0_n)$. This leads to $f_0'(0_n) \cdot \xi = 0$. Therefore, $f_0'(0_n)$ is a linear combination of the vectors

$f_1'(0_n), \ldots, f_m'(0_n)$, the rows of $F'(0_n)$, as ξ is an arbitrary solution of $F'(0_n)\xi = 0_m$ (here the theory of linear equations is used). That is, the Lagrange equations hold (with $\lambda_0 = 1$).

(b) Choose a selection of Lagrange multipliers $\lambda = (1, y)$ for which
$$\mathcal{L}_x(0_n, \lambda) = 0_n^T.$$
Proceed as above, but now rewrite the inequality
$$f_0(t\xi + r(t\xi)) \geq f_0(0_n)$$
as
$$\mathcal{L}(t\xi + r(t\xi), \lambda) \geq \mathcal{L}(0_n, \lambda),$$
and use the quadratic approximation formula
$$\mathcal{L}(t\xi + r(t\xi), \lambda)$$
$$= \mathcal{L}(0_n, \lambda) + \mathcal{L}_x(0_n, \lambda) \cdot (t\xi + r(t\xi))$$
$$+ \frac{1}{2}\mathcal{L}_{xx}(0_n, \lambda)[t\xi + r(t\xi), t\xi + r(t\xi)] + o(t^2),$$
which holds as $f_0 \in \mathcal{D}^1(0_n)$ and $F \in \mathcal{D}^1(0_n, \mathbb{R}^m)$. This leads to
$$\mathcal{L}_{xx}(0_n, \lambda)[\xi, \xi] \geq 0,$$
using $\mathcal{L}_x(0_n, \lambda) = 0_m$ and
$$\mathcal{L}_{xx}(0_n, \lambda)[t\xi + r(t), t\xi + r(t\xi)] = t^2 \mathcal{L}_{xx}(0_n, \lambda)[\xi, \xi] + o(t^2).$$

(c) Write the following quadratic approximation formula:
$$\mathcal{L}(\xi + r(\xi), \lambda)$$
$$= \mathcal{L}(0_n, \lambda) + \mathcal{L}_x(0_n, \lambda)(\xi + r(\xi))$$
$$+ \frac{1}{2}\mathcal{L}_{xx}(0_n, \lambda)[\xi + r(\xi), \xi + r(\xi)] + o(|\xi|^2)$$

for all $\xi \in \ker F'(0_n)$. The assumptions give that the second term of the right-hand side is zero and that the third term is bounded below by $K|\xi|^2$ for some constant $K > 0$ that does not depend on ξ (see appendix A). Therefore,

$$\mathcal{L}(\xi + r(\xi), \lambda) > \mathcal{L}(0_n, \lambda)$$

for all $\xi \in \ker F'(0_n)$ of sufficiently small modulus. This inequality can be rewritten as $f_0(\xi + r(\xi)) > f_0(0_n)$. This gives $f_0(x) > f_0(0_n)$ for all admissible x of problem (P_2) that are close enough to 0_n, by virtue of the tangent space theorem.

3. Problems with equality and inequality constraints

(a) **Convex case**

i. Assume $f_0(0_n) = 0$; this does not restrict the generality of the argument. Consider the set

$$\mathcal{C} = \{z \in \mathbb{R}^{m+1} : \exists x \in A,\ z_i \geq f_i(x),\ 0 \leq i \leq k,$$

$$F(x) = (z_{k+1}, \ldots, z_m)^T\}.$$

This set has the following properties:

- $\mathcal{C}$ is a convex set (if x is "good" for z and x' is "good" for z', then

$$x_\alpha = (1 - \alpha)x + \alpha x'$$

is "good" for

$$z_\alpha = (1 - \alpha)z + \alpha z'$$

for all α with $0 \leq \alpha \leq 1$).

- $0_{m+1} \in \partial \mathcal{C}$, as $0_{m+1} \in \mathcal{C}$ and $0_{m+1} \notin \text{int}\,\mathcal{C}$

(if $0_{m+1} \in \text{int}\,\mathcal{C}$, then $(-\varepsilon, 0, \ldots, 0)^T \in \mathcal{C}$ for sufficiently small $\varepsilon > 0$; this would contradict the assumption $0_n \in \min(P_3)$).

Therefore, the supporting hyperplane theorem can be applied. This gives the existence of a nonzero vector $\lambda \in (\mathbb{R}^{m+1})'$ such that

$$\lambda \cdot z \geq 0 \text{ for all } z \in \mathcal{C}. \qquad (*)$$

Substitute in $(*)$ suitable elements of $\mathcal{C}$, and write $\lambda = (\lambda_0, \mu, y)$ with $\lambda_0 \in \mathbb{R}_+$, $\mu \in (\mathbb{R}^k)'$, $y \in (\mathbb{R}^{m-k})'$. This gives $\lambda_0 \geq 0$ and the KKT conditions:

1) Each $z \in \mathbb{R}^{m+1}$ with $z_i \geq 0$, $0 \leq i \leq k$, and $z_i = 0$, $k+1 \leq i \leq m$, is an element of $\mathcal{C}$ (as 0_n is "good" for such an element z). This gives $\lambda_0 \geq 0$ and the nonnegativity conditions.

2) If $f_i(0_n) < 0$ for some $i \in \{1, \ldots, k\}$, then

$$z = (0, \ldots, 0, f_i(0_n), 0, \ldots, 0)^T \in \mathcal{C} \overset{\lambda_i \geq 0}{\Rightarrow} \lambda_i = 0.$$

This gives the complementarity conditions.

3) The set

$$\{z \in \mathbb{R}^{m+1} : \exists x \in A, \ z_i = f_i(x), \ 0 \leq i \leq k,$$

$$F(x) = (z_{k+1}, \ldots, z_m)^T\}$$

is contained in $\mathcal{C}$

$$\Rightarrow \mathcal{L}(x, \lambda) \geq 0 = \mathcal{L}(0_n, \lambda) \ \forall x \in A.$$

This gives the minimum condition.

ii. Let x be an admissible point. Then, refering to the three KKT conditions—A, B and C—where they are used,

$$f_0(x) \overset{B}{\geq} \mathcal{L}(x, \lambda) \overset{A}{\geq} \mathcal{L}(0_n, \lambda) \overset{C}{=} f_0(0_n).$$

(b) **Differentiable case**

The proof will use not only the main theorem of differential calculus ("the tangent space theorem"), but also the main theorem of the convex calculus ("the supporting hyperplane theorem"). The convexity of the problem comes from the inequality constraints.

i. We distinguish two cases.

   A. **Nonregular case**
   If $\operatorname{rank} F'(0_n) < m - k$, then choose a nontrivial linear relation between $r_i = f'_i(0_n)$, $1 \leq j \leq m-k$, the rows of $F'(0_n)$, say,
   $$\sum_{i=1}^{m-k} y_i r_i = 0_n^T;$$
   then the first order conditions are seen to hold (with $\lambda_0 = 0$ and $\mu_j = 0$, $1 \leq j \leq k$). If the regularity assumption $\operatorname{rank} F'(0_n) = m - k$ holds, but the other regularity assumption—from statement (ii) for the differentiable case—does not hold, that is,
   $$\not\exists \xi \in \mathbb{R}^n : f'_i(0_n) \cdot \xi < 0, \ \forall i \in I, \ F'(0_n)\xi = 0_{m-k},$$
   then argue as follows. Choose a subset $J \subseteq I$ and an element $\bar{i} \in I \setminus J$ for which the following properties hold true:
   $$\exists \xi \in \mathbb{R}^n : f'_i(0_n) \cdot \xi < 0, \ \forall i \in J, \ F'(0_n)\xi = 0_{m-k}, \tag{$*$}$$
   $$\not\exists \xi \in \mathbb{R}^n : f'_i(0_n) \cdot \xi < 0, \ \forall i \in J \cup \{\bar{i}\},$$
   $$F'(0_n)\xi = 0_{m-k}.$$
   Then the following implication holds true:
   $$f'_i(0_n) \cdot \xi < 0 \ \forall i \in J,$$
   $$F'(0_n)\xi = 0_{m-k} \Rightarrow f'_{\bar{i}}(0_n) \cdot \xi \geq 0.$$
   This implies the following statement: the linear—and so convex—problem
   $$f'_{\bar{i}}(0_n) \cdot \xi \to \min, \ f'_i(0_n) \cdot \xi \leq 0, \ i \in J,$$
   $$F'(0_n)\xi = 0_{m-k}$$

has value 0 and solution $\widehat{\xi} = 0_n$; moreover, it has a Slater point, by (∗). Apply the first order necessary conditions for convex problems, which have been proved above. Use the resulting selection of Lagrange multipliers

$$\tilde{\lambda} = (1, (\tilde{\mu}_i)_{i \in J}, \tilde{y})$$

to define for problem (P_3) the following selection of Lagrange multipliers $\lambda = (\lambda_0, \mu, y)$:

$$\lambda_0 = 0, \ \mu_{\bar{i}} = 1, \ \mu_i = \tilde{\mu}_i, \ i \in J,$$

$$\mu_i = 0, \ i \notin J \cup \{\bar{i}\}, \ i \geq 1, \ y = \tilde{y}.$$

Then the first order conditions for smooth problems (P_3) hold for this selection of multipliers.

B. **Regular case**
Take a vector $\xi \in \mathbb{R}^n$ for which $f'_i(0_n) \cdot \xi < 0 \ \forall i \in I$ and $F'(0_n)\xi = 0_{m-k}$. Apply the tangent space theorem to F. It follows that $t\xi + r(t\xi)$ is admissible for (P_3) for all sufficiently small $t > 0$. To see this, it remains to verify that the inequality constraints

$$f_i(t\xi + r(t\xi)) < 0$$

hold for all $i \in I$ (for $i \notin I$ they hold, as $f_i(0_n)$ is negative).
For each $i \notin I$ we have

$$f_i(0_n) < 0 \text{ and so } f_i(t\xi + r(t\xi)) < 0$$

for all sufficiently small $t > 0$. For each $i \in I$ we have $f_i(0_n) = 0$ and $f'_i(0_n) \cdot \xi < 0$ and so

$$f_i(t\xi + r(t\xi)) = f_i(0_n) + tf'_i(0_n) \cdot \xi + o(t) < 0$$

for all sufficiently small $t > 0$.
Therefore, one has

$$f_0(t\xi + r(t\xi)) \geq f_0(0_n)$$

for each such ξ and each sufficiently small $t > 0$, using $0_n \in \text{locmin}(P_3)$. This implies $f'_0(0_n) \cdot \xi \geq 0$, using

$$f_0(t\xi + r(t\xi)) = f_0(0_n) + tf'_0(0_n) \cdot \xi + o(t).$$

This proves the implication

$$f'_i(\hat{x}) \cdot \xi < 0 \;\forall i \in I, \; F'(0_n)\xi = 0_{m-k} \Rightarrow$$

$$f'_0(0_n) \cdot \xi \geq 0.$$

This implies the following statement: the linear—and so convex—problem

$$f'_0(0_n) \cdot \xi \to \min,$$

$$f'_i(0_n) \cdot \xi \leq 0, \; i \in I, \; F'(0_n)\xi = 0_{m-k}$$

has value 0 and solution $\widehat{\xi} = 0_n$; moreover, it has a Slater point by the second regularity assumption. Apply the first order necessary conditions for convex problems. Extend the resulting selection of Lagrange multipliers

$$\tilde{\lambda} = (1, (\mu_i)_{i \in I}, y)$$

by defining $\mu_i = 0$ for all $i \in \{1, \ldots, k\} \setminus I$; then the first order conditions for (P_3) hold for the resulting selection of Lagrange multiplier $\lambda = (\lambda_0, \mu, y)$.

ii. Choose a selection of Lagrange multipliers $\lambda = (1, \mu, y)$ for which the first order conditions hold. Rewrite the inequality

$$f_0(t\xi + r(t\xi)) \geq f_0(0_n)$$

as

$$\mathcal{L}(t\xi + r(t\xi), \lambda) \geq \mathcal{L}(0_n, \lambda)$$

and use the following quadratic approximation formula:

$$\mathcal{L}(tv + r(tv), \lambda)$$

$$= \mathcal{L}(0_n, \lambda) + \mathcal{L}_x(0_n, \lambda) \cdot (t\xi + r(t\xi))$$

$$+ \frac{1}{2} \mathcal{L}_{xx}(0_n, \lambda)[t\xi + r(t\xi), t\xi + r(t\xi)] + o(t^2),$$

which holds as $f_i \in \mathcal{D}^2(0_n)$, $1 \leq i \leq k$,
and $F \in \mathcal{D}^2(0_n, \mathbb{R}^{m-k})$. This leads to the inequality

$\mathcal{L}_{xx}[\xi, \xi] \geq 0$, using the first order necessary conditions, and

$$\mathcal{L}_{xx}(0_n, \lambda)[t\xi + r(t\xi), t\xi + r(t\xi)]$$

$$= t^2 \mathcal{L}_{xx}(0_n, \lambda)[\xi, \xi] + o(t^2).$$

iii. Write the quadratic approximation formula

$$\mathcal{L}(\xi + r(\xi), \lambda)$$

$$= \mathcal{L}(0_n, \lambda) + \mathcal{L}_x(0_n, \lambda) \cdot \xi + \frac{1}{2} \mathcal{L}_{xx}(0_n, \lambda)[\xi, \xi] + o(|\xi|^2)$$

for all ξ under consideration, which holds as

$$f_i \in \mathcal{D}^2(0_n), \ 1 \leq i \leq k, \ \text{and} \ F \in \mathcal{D}^2(0_n, \mathbb{R}^{m-k}).$$

The assumptions give that the second term on the right-hand side is zero, and that the third term is bounded below by $K|\xi|^2$ for some constant $K > 0$, for all ξ under consideration. Therefore,

$$\mathcal{L}(\xi + r(\xi), \lambda) > \mathcal{L}(0_n, \lambda)$$

for all ξ under consideration that are of sufficiently small modulus. This inequality implies the following one:

$$f_0(\xi + r(\xi)) > f_0(0_n).$$

This gives $f_0(x) > f_0(0_n)$ for all admissible x close enough to 0_n, by virtue of the following lemma.

For each admissible point x of (P_3), we define $d(x)$ to be the distance between x and the cone of admissible directions,

$$\{\xi \in X : f_i'(0_n) \cdot \xi \leq 0, \ 1 \leq i \leq k, \ F'(0_n)\xi = 0_{m-k}\}.$$

Lemma F.10 $d(x) = o(|x|)$.

This lemma can be derived by making use of the tangent space theorem.

4. Conic smooth-convex problems

(a) Distinction of two cases

i. **Nonregular case.** Write $F = (f_1, \ldots, f_m)^T$. If $\operatorname{rank} F'(\widehat{x}) < m - k$, then choose a nontrivial linear relation among the vectors $f_i'(\widehat{x})$, $1 \leq i \leq m$; this gives that the first order conditions hold (with $\lambda_0 = 0$). Now assume

$$\nexists \xi \in X : F'(\widehat{x})\xi = 0_m, \ \widehat{x} + \xi \in \operatorname{int} C.$$

Then choose a subset $J \subseteq \{1, \ldots, m\}$ and $\bar{i} \notin J$, $\bar{i} \geq 1$ for which the following properties hold true:

$$\exists \xi \in X : f_i'(\widehat{x}) \cdot \xi = 0, \ i \in J, \ \widehat{x} + \xi \in \operatorname{int} C,$$

$$\nexists \xi \in X : f_i'(\widehat{x}) \cdot \xi = 0, \ i \in J \cup \{\bar{i}\}, \ \widehat{x} + \xi \in \operatorname{int} C.$$

This implies the following statement: the linear—and so convex—problem

$$f_{\bar{i}}'(\widehat{x}) \cdot \xi \to \min, \ f_i'(\widehat{x}) \cdot \xi = 0 \ \forall i \in J, \ \widehat{x} + \xi \in C$$

has value zero and solution $\widehat{\xi} = 0$, and it has a Slater point. Apply the theorem for convex problems. This gives the existence of numbers λ_i, $i \in J$, for which

$$\min_{\xi \in C}(f_{\bar{i}}'(\widehat{x}) + \sum_{i \in J}\lambda_i f_i'(\widehat{x}))\xi = 0.$$

This completes the proof of the theorem in the degenerate case: it remains to put

$$\lambda_0 = 0, \ \lambda_{\bar{i}} = 1, \ \lambda_i = 0, \ i \notin J \cup \{\bar{i}\},$$

and to note that the first order conditions of (P_3) hold for the resulting enlarged selection of multiplier λ.

ii. **Regular case.** Apply the tangent space theorem to the vector function F. Choose an arbitrary $\xi \in \ker F'(\widehat{x})$ for which $\widehat{x} + \xi \in \operatorname{int} C$. Then

$$F(\widehat{x} + t\xi + r(t\xi)) = 0_m, \ r(t\xi) = o(t), \ \widehat{x} + t\xi + r(t\xi) \in C,$$

using the tangent space theorem. Therefore, one has
$$f_0(\widehat{x} + t\xi + r(t\xi)) \geq f_0(\widehat{x})$$
for each such ξ and each sufficiently small $t > 0$, using $\widehat{x} \in \text{locmin}(P_4)$. This implies $f_0'(\widehat{x}) \cdot \xi \geq 0$, using
$$f_0(\widehat{x} + t\xi) = f_0(\widehat{x}) + tf_0'(\widehat{x}) \cdot \xi + o(t).$$
This proves the implication
$$F'(\widehat{x})\xi = 0, \ \widehat{x} + \xi \in \text{int} C \Rightarrow f_0'(\widehat{x}) \cdot \xi \geq 0.$$
This implies the following statement: the linear—and so convex—problem
$$f_0'(\widehat{x}) \cdot w \to \min, \ F'(\widehat{x})w = 0, \ w \in C$$
has value zero and solution $\widehat{w} = 0$; moreover, it has a Slater point v by the regularity assumption. We apply the theorem for convex problems. It remains to note that the first order conditions of (P_4) hold for the resulting selection of Lagrange multiplier λ and that this selection has $\lambda_0 \neq 0$.

(b) Choose a selection of Lagrange multipliers $\lambda = (1, \mu, y)$ for which the first order conditions hold. Then rewrite the inequality
$$f_0(\widehat{x} + t\xi + r(t\xi)) \geq f_0(\widehat{x})$$
as
$$\mathcal{L}(\widehat{x} + t\xi + r(t\xi), \lambda) \geq \mathcal{L}(\widehat{x})$$
and use the following quadratic approximation formula:
$$\begin{aligned}\mathcal{L}(\widehat{x} &+ t\xi + r(t\xi), \lambda) \\ &= \mathcal{L}(\widehat{x}, \lambda) + \mathcal{L}_x(\widehat{x}, \lambda) \cdot (t\xi + r(t\xi)) \\ &+ \tfrac{1}{2}\mathcal{L}_{xx}(\widehat{x}, \lambda)[t\xi + r(t\xi), t\xi + r(t\xi)] + o(|t|^2),\end{aligned}$$
which holds as $f_0 \in \mathcal{D}^2(\widehat{x})$ and $F \in \mathcal{D}^2(\widehat{x}, \mathbb{R}^{m-k})$. This leads to
$$\mathcal{L}_{xx}(\widehat{x}, \lambda)[\xi, \xi] \geq 0,$$
using the first order necessary conditions as well as
$$\mathcal{L}_{xx}(\widehat{x}, \lambda)[t\xi + r(t), t\xi + r(t\xi)] = t^2 \mathcal{L}_{xx}(\widehat{x}, \lambda)[\xi, \xi] + o(|t|^2).$$

(c) Write the following quadratic approximation formula:

$$\mathcal{L}(\widehat{x}+\xi+r(\xi),\lambda)$$

$$=\mathcal{L}(\widehat{x},\lambda)+\mathcal{L}_x(\widehat{x},\lambda)\cdot\xi+\frac{1}{2}\mathcal{L}_{xx}(\widehat{x},\lambda)[\xi,\xi]+o(|\xi|^2)$$

for all $\xi \in \ker F'(\widehat{x})$. The assumptions give that the second term of the right-hand side is zero and that the third term is bounded below by $K|\xi|^2$ for some constant $K > 0$. Therefore,

$$\mathcal{L}(\widehat{x}+\xi+r(\xi),\lambda) > \mathcal{L}(\widehat{x},\lambda)$$

for all $\xi \in \ker F'(\widehat{x})$ of sufficiently small modulus, with $\xi \in \mathrm{int} C$. This inequality can be rewritten as

$$f_0(\widehat{x}+\xi+r(\xi)) > f_0(\widehat{x}).$$

This gives $f_0(x) > f_0(\widehat{x})$ for all admissible x sufficiently close to $\widehat{x}$, by virtue of the following lemma.

For each admissible point x of (P_4) we define $d(x)$ to be the distance between x and the cone of admissible directions,

$$\{t\xi \in X : t \geq 0, \ F'(0_n)\xi = 0_m, \ \widehat{x}+\xi \in C\}.$$

Lemma F.11 $d(x) = o(|x|)$.

This is a consequence of the tangent space theorem.

Appendix G

Conditions of Extremum from Fermat to Pontryagin

Plan. This text is meant for experts and advanced readers. Short proofs, from an advanced point of view, are given of all first order necessary conditions that are used in the solution of concrete finite- and infinite-dimensional extremal problems. Some proofs are novel. For example, we use Brouwer's fixed point theorem to prove the Lagrange multiplier rule under weak assumptions. Moreover, we use the theory of ordinary differential equations to give a straightforward proof of the formidable Pontryagin maximum principle.

G.1 NECESSARY FIRST ORDER CONDITIONS FROM FERMAT TO PONTRYAGIN

Notation:

1. $\mathbb{R}^n = \left\{ x = \begin{pmatrix} x_1 \\ \vdots \\ x_n \end{pmatrix} \mid x_i \in \mathbb{R},\ 1 \leq i \leq n \right\}$;

2. $x \in \mathbb{R}^n \Rightarrow x^T = (x_1, \ldots, x_n)$;

3. $(\mathbb{R}^n)' = \{y = (y_1, \ldots, y_n) \mid y_i \in \mathbb{R},\ 1 \leq i \leq n\}$;

4. If $y \in (\mathbb{R}^n)'$ and $x \in \mathbb{R}^n$ then $y \cdot x = \sum_{i=1}^n y_i x_i$;

5. $|x| = (x^T \cdot x)^{1/2}$;

6. $V \in \mathcal{O}(\mathbb{R}^n)$ means that V is an open set in $\mathbb{R}^n$;

7. $V \in \mathcal{O}(\widehat{x}, \mathbb{R}^n)$ means that V is a neighborhood of $\widehat{x}$ in $\mathbb{R}^n$, that is, $x \in V$ and $V \in \mathcal{O}(\mathbb{R}^n)$.

G.1.1 Finite-dimensional problems without constraints

Let $V \in \mathcal{O}(\mathbb{R}^n)$ and $f : V \to \mathbb{R}$. The problem

$$f(x) \to \min \qquad (P_1)$$

is called *the problem without constraints*.

Definition 1. A point $\widehat{x} \in V$ is called *a local minimum of* (P_1) if there exists $\varepsilon > 0$ such that $f(x) \geq f(\widehat{x}) \ \forall x \in V$ st $|x - \widehat{x}| \leq \varepsilon$. In this case we write $\widehat{x} \in \mathrm{locmin}(P_1)$.

Let $V \in \mathcal{O}(\widehat{x}, \mathbb{R}^n)$ and $f : V \to \mathbb{R}$. Then $f \in D^1(\widehat{x})$ means that there exists $a \in (\mathbb{R}^n)'$ such that $f(\widehat{x} + x) = f(\widehat{x}) + a \cdot x + o(x)$. The vector a is called *the derivative of f at the point $\widehat{x}$*; it is denoted $f'(\widehat{x})$.

Theorem 1 *Fermat theorem. Let $V \in \mathcal{O}(\widehat{x}, \mathbb{R}^n)$, $f : V \to \mathbb{R}$, $f \in D^1(\widehat{x})$ and $\widehat{x} \in \mathrm{locmin}(P_1)$. Then*

$$f'(\widehat{x}) = 0. \tag{1}$$

Throughout, we always use the symbol 0 without any qualifications to denote the zero element that is appropriate in the context. In theorem 1 it denotes $0_n^T = (0, \ldots, 0) \in (\mathbb{R}^n)'$.

Proof. Let $x \in \mathbb{R}^n$ and $g(\alpha) = f(\widehat{x} + \alpha x)$. From definition 1 it follows that $0 \in \mathrm{locmin}\, g$; from the condition $f \in D^1(\widehat{x})$ one has $g \in D^1(0)$. Hence from Fermat and the superposition theorem (also called the chain rule) we obtain that $0 = g'(0) = f'(\widehat{x}) \cdot x$ for all $x \Rightarrow f'(\widehat{x}) = 0$.

G.1.2 The simplest problem of the calculus of variations

Let $L : \mathbb{R}^3 \to \mathbb{R}$ ($L = L(t, x, \dot{x})$) be a continuous function of three variables.

The problem

$$J(x(\cdot)) = \int_{t_0}^{t_1} L(t, x(t), \dot{x}(t)) dt \to \min, \quad x(t_i) = x_i, \ i = 0, 1 \quad (P_2)$$

is called *the simplest problem of the calculus of variations*.

The problem (P_2) will be considered in the space $C^1([t_0, t_1])$, equipped with the C^1-norm $\|\cdot\|_{C^1([t_0,t_1])}$, defined by

$$\|x(\cdot)\|_{C^1([t_0,t_1])} = \max_{t \in [t_0,t_1]} \max(|x(t)|, |\dot{x}(t)|).$$

Definition 2. A local minimum $\widehat{x}(\cdot)$ of (P_2) in the space $C^1([t_0, t_1])$ is called *a weak local minimum* of (P_2). This means that there exists $\varepsilon > 0$ such that $J(x(\cdot)) \geq J(\widehat{x}(\cdot))$ for all $x(\cdot) \in C^1([t_0, t_1])$ with $x(t_i) = x_i$, $i = 0, 1$, and $\|x(\cdot) - \widehat{x}(\cdot)\|_{C^1([t_0,t_1])} \leq \varepsilon$.

Theorem 2 *Euler equation. Let a function $\widehat{x}(\cdot)$ be admissible in the problem (P_2)—this means that $\widehat{x}(\cdot) \in C^1([t_0, t_1])$ and $\widehat{x}(t_i) = x_i$, $i = 0, 1$—and let $L, L_x, L_{\dot{x}}$ be continuous in the neighborhood of the curve (in $\mathbb{R}^3$) $t \mapsto (t, \widehat{x}(t), \dot{\widehat{x}}(t))$, $t \in [t_0, t_1]$. If the function $\widehat{x}(\cdot)$ is* a weak local minimum in the problem (P_2), then the function $\widehat{L}_{\dot{x}}(\cdot)$ is continuously differentiable and the equation

$$-\frac{d}{dt}\widehat{L}_{\dot{x}}(t) + \widehat{L}_x(t) = 0 \qquad (2)$$

is satisfied.

(We have used the following abbreviations: $\widehat{L}_{\dot{x}}(t) = L_{\dot{x}}(t, \widehat{x}(t), \dot{\widehat{x}}(t))$, $\widehat{L}_x(t) = L_x(t, \widehat{x}(t), \dot{\widehat{x}}(t))$. Similar abbreviations will appear further on.)

Proof. 1) *Construction of a variation.* Write $\widehat{u}(\cdot) = \dot{\widehat{x}}(\cdot)$ and let $u(\cdot)$ be an arbitrary function from $C([t_0, t_1])$ for which $\int_{t_0}^{t_1} u(t)dt = 0$; write $u_\alpha(\cdot) = \widehat{u}(\cdot) + \alpha u(\cdot)$. Define $x_\alpha(\cdot) = \widehat{x}(\cdot) + \alpha y(\cdot)$ where $y(\cdot)$ is defined from the equalities $\dot{y} = u(t)$, $y(t_0) = 0$. It is clear that $x_\alpha(\cdot)$ is an admissible function in (P_2). Denote $g(\alpha) = g(\alpha; u(\cdot)) = J(x_\alpha(\cdot))$. From the definition of the weak local minimum we obtain that $0 \in$ locmin g.

2) *Differentiation of g and transformation of the derivative.* From the theorem of classical analysis about differentiation of an integral depending on parameters, it follows that $g \in D^1(0)$ and

$$g'(0; u(\cdot)) = \frac{d}{d\alpha} g(\alpha; u(\cdot))|_{\alpha=0} = \int_{t_0}^{t_1} (\widehat{L}_{\dot{x}}(t)u(t) + \widehat{L}_x(t)y(t))dt. \quad (i)$$

From the Fermat theorem we obtain that $g'(0; u(\cdot)) = 0$ (if $\int_{t_0}^{t_1} u(t)dt = 0$).

3) Changing in (i) $\widehat{L}_x(t)$ by $\dot{p}_0(\cdot)$, where $p_0(\cdot)$ is defined by the equalities $\dot{p}_0(t) = \widehat{L}_x(t)$, $p_0(t_1) = 0$, and integrating by parts, we have

$$\int_{t_0}^{t_1} (\widehat{L}_{\dot{x}}(t) - p_0(t))u(t)dt = 0 \quad \text{if} \quad \int_{t_0}^{t_1} u(t)dt = 0. \qquad (ii)$$

Let c be the constant for which $\int_{t_0}^{t_1} (\widehat{L}_{\dot{x}}(t) - p_0(t) - c)dt = 0$. Substituting the function $\bar{u}(\cdot) = \widehat{L}_{\dot{x}}(t) - p_0(t) - c$ into (ii) and denoting $p(\cdot) = p_0(\cdot) + c$, we come to the equality

$$\int_{t_0}^{t_1} (\widehat{L}_{\dot{x}}(t) - p(t))^2 dt = 0.$$

This proves the equality $p(t) = \widehat{L}_{\dot{x}}(t)$; besides, $\dot{p}(t) = \widehat{L}_x(t)$. Hence the function $\widehat{x}(\cdot)$ satisfies the Euler equation: $-\frac{d}{dt}\widehat{L}_{\dot{x}}(t) + \widehat{L}_x(t) = 0$.

G.1.3 Problems with equality and inequality constraints

Let $V \in \mathcal{O}(\mathbb{R}^n)$, $f_i : V \to \mathbb{R}$, $0 \leq i \leq m$. The problem

$$f_0(x) \to \min, \quad f_i(x) = 0, \quad 1 \leq i \leq m \qquad (P_{3.1})$$

is called *the problem with equality constraints*.

Let $V \in \mathcal{O}(\mathbb{R}^n)$, $f_i : V \cap \mathbb{R}_+^n \to \mathbb{R}$, $0 \leq i \leq m$. The problem

$$f_0(x) \to \min, \quad f_i(x) = 0, \quad 1 \leq i \leq m, \quad x \geq 0 \qquad (P_{3.2})$$

is called *the simplest problem with inequality constraints*. Note that all problems with equality and inequality constraints can be put into this form by the introduction of additional variables.

The function $\mathcal{L}(x, \lambda) = \sum_{i=0}^m \lambda_i f_i(x)$ is called *the Lagrange function* (for both problems $(P_{3.1})$ and $(P_{3.2})$); the λ_i are called *Lagrange multipliers* of these problems. An admissible point in an optimization problem is defined to be a point that satisfies all constraints of the problem.

We display for both problems $(P_{3.1})$ and $(P_{3.2})$ the definitions of a local minimum in the space $\mathbb{R}^n$, equipped with the Euclidean norm $|\cdot|$.

Definition 3.1. A point $\widehat{x} \in V$ is called *a local minimum of* $(P_{3.1})$ if there exists $\varepsilon > 0$ such that $f_0(x) \geq f_0(\widehat{x})$ for all admissible x in $(P_{3.1})$ with $|x - \widehat{x}| \leq \varepsilon$. In this case we write $\widehat{x} \in \text{locmin}(P_{3.1})$.

Definition 3.2. A point $\widehat{x} \in V \cap \mathbb{R}_+^n$ is called *a local minimum of* $(P_{3.2})$ if there exists $\varepsilon > 0$ such that $f_0(x) \geq f_0(\widehat{x})$ for all admissible x in $(P_{3.2})$ with $|x - \widehat{x}| \leq \varepsilon$. In this case we write $\widehat{x} \in \text{locmin}(P_{3.2})$.

Let $V \in \mathcal{O}(0, \mathbb{R}^n)$ and $f : V \cap \mathbb{R}_+^n \to \mathbb{R}$. Then $f \in D^1(+0)$ means that there exists $a \in (\mathbb{R}^n)'$ such that $f(x) = f(0) + a \cdot x + o(x) \ \forall x \in \mathbb{R}_+^n$. The vector a is called *the positive-side derivative of f at the point* 0; it is denoted $f'(+0)$.

Theorem 3.1 *Lagrange multiplier rule.* Let $V \in \mathcal{O}(\widehat{x}, \mathbb{R}^n)$, $f_i : V \to \mathbb{R}$, $f_i \in D^1(\widehat{x}) \cap C(V)$, $0 \leq i \leq m$. If $\widehat{x} \in \text{locmin}(P_{3.1})$, then there exists a nonzero vector $\lambda = (\lambda_0, \ldots, \lambda_m)$ of Lagrange multipliers with $\lambda_0 \geq 0$ such that

$$\mathcal{L}_x(\widehat{x}, \lambda) = \sum_{i=0}^m \lambda_i f_i'(\widehat{x}) = 0. \qquad (3.1)$$

The following result is essentially the John theorem for problems with equality and inequality constraints (under weak assumptions of differentiability).

Theorem 3.2 *Lagrange multiplier rule for problem* $(P_{3.2})$. *Let* $V \in \mathcal{O}(0, \mathbb{R}^n)$, $f_i : V \cap \mathbb{R}^n_+ \to \mathbb{R}$, $f_i \in D^1(+0) \cap C(V \cap \mathbb{R}^n_+)$, $0 \leq i \leq m$. *If* $0 \in \text{locmin}(P_{3.2})$, *then there exists a nonzero vector* $\lambda = (\lambda_0, \ldots, \lambda_m)$ *of Lagrange multipliers with* $\lambda_0 \geq 0$ *such that*

$$\mathcal{L}_x(0, \lambda) = \sum_{i=0}^{m} \lambda_i f'_i(+0) \geq 0. \tag{3.2}$$

Proof of theorem 3.1. Without loss of generality we can assume that $\hat{x} = 0$ and $f_0(0) = 0$.

1. Let $F(x) = \begin{pmatrix} f_0(x) \\ \vdots \\ f_m(x) \end{pmatrix}$ and so $F(0) = 0$. We consider the alternatives:

   (a) $F'(0)\mathbb{R}^n \neq \mathbb{R}^{m+1}$ or (b) $F'(0)\mathbb{R}^n = \mathbb{R}^{m+1}$.

In the first case the vectors $f'_i(0)$, $0 \leq i \leq m$ are linearly dependent and (3.1) follows from the definition of this fact.

Let us show that b) leads to a contradiction; this will establish the theorem.

2. At first we choose vectors $\{g_i\}_{i=0}^{m} \in \mathbb{R}^n$ such that $F'(0)g_i = e_i$, $0 \leq i \leq m$, where $\{e_i\}_{i=0}^{m}$ is some chosen basis in $\mathbb{R}^{m+1}$. We will construct a right inverse mapping $R : \mathbb{R}^{m+1} \to \mathbb{R}^n$ of the linear operator $F'(0) : \mathbb{R}^n \to \mathbb{R}^{m+1}$, that is, $F'(0)R(y) = y \; \forall y$. For $y = \sum_{i=0}^{m} \alpha_i e_i$, we put $R(y) = \sum_{i=0}^{m} \alpha_i g_i$. One can trivially show that $F'(0)R(y) = y \; \forall y \in \mathbb{R}^{m+1}$ and that there exists a constant $C > 0$ for which $|R(y)| \leq C|y| \; \forall y$. Hence $G(y) = F(R(y)) = y + \rho(y)$, $\rho(y) = o(y)$.

3. Let us take any $\delta > 0$ small enough that $|y - G(y)| \leq \frac{|y|}{2}$ for all y for which $|y| \leq \delta$. Take the vector $\eta = \begin{pmatrix} -\delta/2 \\ 0 \\ \vdots \\ 0 \end{pmatrix} \in \mathbb{R}^{m+1}$ and define the following continuous mapping $\Phi_\eta : B_{m+1}(0, \delta) \to \mathbb{R}^{m+1}$, where $B_{m+1}(0, \delta) = \{y \in \mathbb{R}^{m+1} \mid |y| \leq \delta\}$:

$$\Phi_\eta(y) = \eta + y - G(y).$$

We have: $|\Phi_\eta(y)| \leq |\eta| + |y - G(y)| \leq \delta$; hence Φ_η maps $B_{m+1}(0, \delta)$ into itself. From the Brouwer fixed point theorem (cf. [52]) it follows that there exists $\bar{y} \in B_{m+1}(0,\delta)$ for which $\bar{y} = \Phi_\eta(\bar{y}) \Rightarrow F(R(\bar{y})) = G(\bar{y}) = \eta$. That is, for each sufficiently small $\delta > 0$, we have shown the existence of a point $R(\bar{y}) \in \mathbb{R}^n$ near $0 \in \mathbb{R}^n - |R(\bar{y})| \le C\delta$ — that is admissible in $(P_{3.1})$, and with $f_0(R(\bar{y})) = -\delta/2$ smaller than $f_0(0) = 0$. Therefore $0 \notin \text{locmin}(P_{3.1})$. This contradiction leads to the theorem.

Proof of theorem 3.2. 1. Let F be the mapping constructed above. Here we consider the alternatives:

$$\text{(a) } F'(+0)\mathbb{R}^n_+ \ne \mathbb{R}^{m+1} \text{ or (b) } F'(+0)\mathbb{R}^n_+ = \mathbb{R}^{m+1}.$$

In the first case the image $F'(+0)\mathbb{R}^n_+$ is a convex set in $\mathbb{R}^{m+1}$ and the origin in $\mathbb{R}^{m+1}$ is a boundary point of it. By the separation theorem this set can be separated from the origin. That is, there exists a nonzero vector $\lambda = (\lambda_0, \ldots, \lambda_m)$ such that $\sum_{i=0}^m \lambda_i f_i'(0)x \ge 0 \,\forall x \ge 0 \Rightarrow$ (3.2).

Let us show that b) leads to a contradiction; this will establish the theorem.

2. At first we choose vectors $\{g_i\}_{i=0}^{m+1} \in \mathbb{R}^n_+$ such that $F'(0)g_i = e_i$, $0 \le i \le m+1$, where $\{e_i\}_{i=0}^{m+1}$ is some chosen set of vectors in $\mathbb{R}^{m+1}$ that span $\mathbb{R}^{m+1}$ as a convex cone. Here we will construct a right inverse mapping $R : \mathbb{R}^{m+1} \to \mathbb{R}^n_+$ of $F'(0) : \mathbb{R}^n_+ \to \mathbb{R}^{m+1}$. Every $y \in \mathbb{R}^{m+1}$ has a *unique* representation as a conic combination (= linear combination with nonnegative coefficients) of $\{e_i\}_{i=0}^{i=m}$ if we demand that at least one of the coefficients is zero: $y = \sum_{i=0}^{m+1} \alpha_i e_i$, $\alpha_i \ge 0$. Then we put $R(y) = \sum_{i=0}^{m+1} \alpha_i g_i \in \mathbb{R}^n_+$. The final part of the proof is the same as in theorem 3.1.

G.1.4 The Lagrange problem in the calculus of variations and the problem of optimal control

Let U be a subset of $\mathbb{R}^r$, $f : [t_0, t_1] \times \mathbb{R}^n \times U \to \mathbb{R}$ be a continuous function $(f = f(t, x, u))$, and $\varphi : [t_0, t_1] \times \mathbb{R}^n \times U \to \mathbb{R}^n$ be a continuous mapping $(\varphi = \varphi(t, x, u))$.

If $U = \mathbb{R}^r$, then the problem

$$J((x(\cdot), u(\cdot)) = \int_{t_0}^{t_1} f(t, x(t), u(t))dt \to \min, \; \dot{x} = \varphi(t, x, u),$$

$$x(t_i) = x_i, \; i = 0, 1 \qquad (P_{4.1})$$

is called *the Lagrange problem of the calculus of variations* (in Pontryagin's form). This problem we consider in the space

$$X_1 = C^1([t_0, t_1], \mathbb{R}^n) \times C([t_0, t_1], \mathbb{R}^r),$$

equipped with the norm $\|\cdot\|_{X_1}$ defined by

$$\|(x(\cdot), u(\cdot))\|_{X_1} = \max(\|x(\cdot)\|_{C^1([t_0,t_1],\mathbb{R}^n)}, \|u(\cdot)\|_{C([t_0,t_1],\mathbb{R}^r)}).$$

For an arbitrary subset U of $\mathbb{R}^r$, the problem

$$J((x(\cdot), u(\cdot))) = \int_{t_0}^{t_1} f(t, x(t), u(t))dt \to \min, \quad \dot{x} = \varphi(t, x, u),$$

$$(u(t) \in U \ \forall t), \quad x(t_i) = x_i, \quad i = 0, 1 \qquad (P_{4.2})$$

is called *the problem of optimal control*.

Problem $(P_{4.2})$ we consider in the space

$$X_2 = PC^1([t_0, t_1], \mathbb{R}^n) \times PC([t_0, t_1], \mathbb{R}^r),$$

equipped with the norm $\|\cdot\|_{C([t_0,t_1],\mathbb{R}^n)}$ on $PC^1([t_0, t_1], \mathbb{R}^n)$ (that is, $x(\cdot)$ is a piecewise continuously differentiable vector function and $u(\cdot)$ is a piecewise continuous vector function).

A pair of functions $(x(\cdot), u(\cdot))$ that satisfies the constraints of problem $(P_{4.1})$ or $(P_{4.2})$ is called an admissible one for this problem.

Definition 4.1. A local minimum of problem $(P_{4.1})$ in the space X_1 is called *a weak local minimum of the Lagrange problem* $(P_{4.1})$.

Definition 4.2. We say that an admissible pair $(\widehat{x}(\cdot), \widehat{u}(\cdot))$ in $(P_{4.2})$ is *a strong local minimum of the problem of optimal control* $(P_{4.2})$ if there exists $\varepsilon > 0$ such that $J(x(\cdot), u(\cdot)) \geq J(\widehat{x}(\cdot), \widehat{u}(\cdot))$ for all admissible pairs $(x(\cdot), u(\cdot))$ in $(P_{4.2})$ with

$$\|(x(\cdot) - \widehat{x}(\cdot))\|_{C([t_0,t_1],\mathbb{R}^n)} \leq \varepsilon.$$

The function

$$\mathcal{L}((x(\cdot), u(\cdot)), \lambda) = \int_{t_0}^{t_1} L(t, x(t), \dot{x}(t), u(t))dt,$$

where

$$L(t, x, \dot{x}, u) = \lambda_0 f(t, x, u) + p(t) \cdot (\dot{x} - \varphi(t, x, u)),$$

is called *the Lagrange function* for both problems $(P_{4.1})$ and $(P_{4.2})$.

A selection of Lagrange multipliers in these cases is a pair $(p(\cdot), \lambda_0)$, consisting of a vector function $p : [t_0, t_1] \to (\mathbb{R}^n)'$ and a nonnegative number λ_0.

Theorem 4.1 Euler-Lagrange equations for problem ($P_{4.1}$). *Let the pair $(\widehat{x}(\cdot), \widehat{u}(\cdot))$ be admissible in $(P_{4.1})$ and f, f_x, f_u, φ, φ_x, φ_u be continuous in a neighborhood of the curve $t \mapsto (t, \widehat{x}(t), \widehat{u}(t))$, $t \in [t_0, t_1]$ in $\mathbb{R} \times \mathbb{R}^n \times \mathbb{R}^r$. If the pair $(\widehat{x}(\cdot), \widehat{u}(\cdot))$ is a weak local minimum in the problem $(P_{4.1})$, then there exist a number λ_0 and a vector function $p(\cdot) \in C^1([t_0, t_1], (\mathbb{R}^n)')$, not both zero, such that the Euler equations over x:*

$$-\frac{d}{dt}\widehat{L}_{\dot{x}}(t) + \widehat{L}_x(t) = 0 \Leftrightarrow \dot{p} = -p\widehat{\varphi}_x(t) + \lambda_0 \widehat{f}_x(t) \qquad (4.1_a)$$

and over u:

$$\widehat{L}_u(t) = 0 \Leftrightarrow p(t)\widehat{\varphi}_u(t) = \lambda_0 \widehat{f}_u(t) \qquad (4.1_b)$$

are satisfied.

The formulae (4.1_a) and (4.1_b) are called *the Euler-Lagrange equations* for the Lagrange problem $(P_{4.1})$. The following result is known as the *Pontryagin maximum principle*.

Theorem 4.2 Necessary conditions for problem ($P_{4.2}$). *Let the pair $(\widehat{x}(\cdot), \widehat{u}(\cdot))$ be admissible in $(P_{4.2})$ and let f, f_x, φ, φ_x be continuous in a neighborhood of the curve $t \mapsto (t, \widehat{x}(t), \widehat{u}(t))$, $t \in [t_0, t_1]$ in $\mathbb{R} \times \mathbb{R}^n \times U$. If the pair $(\widehat{x}(\cdot), \widehat{u}(\cdot))$ is a strong local minimum in the problem $(P_{4.2})$, then there exist a nonnegative number λ_0 and a vector function*

$$p(\cdot) \in PC^1([t_0, t_1], (\mathbb{R}^n)'),$$

not both zero, such that the Euler equations over x:

$$-\frac{d}{dt}\widehat{L}_{\dot{x}}(t) + \widehat{L}_x(t) = 0 \Leftrightarrow \dot{p} = -p\widehat{\varphi}_x(t) + \lambda_0 \widehat{f}_x(t), \qquad (4.2_a)$$

and the following minimum condition over u:

$$\min_u L(t, \widehat{x}(t), \dot{\widehat{x}}(t), u) = \widehat{L}(t), \qquad (4.2_b)$$

for each point t of continuity of $u(\cdot)$ are satisfied.

Proof. Both proofs begin in the same way as the proof of theorem 2.

1) *Construction of families of variations.* For the Lagrange problem we choose an arbitrary finite auxiliary collection

$$\mathcal{U}_1 = \{u_i(\cdot)\}_{i=1}^N, \text{ where } u_i(\cdot) \in C([t_0, t_1], \mathbb{R}^r), \ 1 \le i \le N.$$

Define, for all $\alpha \in A_1 = \mathbb{R}^N$,

$$u_\alpha(\cdot;\mathcal{U}_1) = \widehat{u}(\cdot) + \sum_{i=1}^{N} \alpha_i u_i(\cdot).$$

For the problem of optimal control we choose an arbitrary finite auxiliary collection $\mathcal{U}_2 = \{(\tau_i, v_i)\}_{i=1}^{N}$ of mutually different pairs (τ_i, v_i) with $\tau_i \in (t_0, t_1]$ and $v_i \in U$, $1 \leq i \leq N$. Define, for all $\alpha \in A_2 = \mathbb{R}_+^N$ of sufficiently small norm, $u_\alpha(\cdot;\mathcal{U}_2)$ to be the function you get from $\widehat{u}(\cdot)$ by placing the following N "needles": place a needle of height v_i on the subinterval $[\tau_i - \alpha_i, \tau_i]$ for $1 \leq i \leq N$. That is, the resulting function takes on the subinterval $[\tau_i - \alpha_i, \tau_i]$ the value v_i for $1 \leq i \leq N$, and outside these intervals it coincides with $\widehat{u}(\cdot)$.

Let now $x_\alpha(\cdot;\mathcal{U}_j)$ be the solution of the Cauchy problem

$$\dot{x} = \varphi(t, x, u_\alpha(\cdot;\mathcal{U}_j)),$$

with the initial data $x(t_0) = x_0$ for $j = 1, 2$, and with $\alpha \in A_j$. Denote

$$g_0(\alpha;\mathcal{U}_j) = \mathcal{J}(x_\alpha(\cdot;\mathcal{U}_j), u_\alpha(\cdot;\mathcal{U}_j)) = \int_{t_0}^{t_1} f(t, x_\alpha(t;\mathcal{U}_j), u_\alpha(t;\mathcal{U}_j))dt$$

for $j = 1, 2$ and $\alpha \in A_j$.

2) *Differentiation of $g_0(\cdot;\mathcal{U}_j)$ and transformation of the derivatives.* Now we will use the theory of ordinary differential equations (on existence and uniqueness of solutions of a Cauchy problem and their continuous and differentiable dependence on parameters and initial data). This gives $g_0(\cdot;\mathcal{U}_j) \in D^1(0)$. Moreover, this leads in both cases to explicit formulas for the derivative (resp. positive-side derivative):

$$\frac{\partial}{\partial \alpha_i} g_0(0;\mathcal{U}_1) = \int_{t_0}^{t_1} (\widehat{f}_x(t) \cdot y(t; u_i(\cdot)) + \widehat{f}_u(t) \cdot u_i(t))dt, \qquad (i.1)$$

where $y(\cdot; u_i(\cdot)) : \dot{y} = \widehat{\varphi}_x(t)y + \widehat{\varphi}_u(t)u_i(t)$, $y(t_0; u_i(\cdot)) = 0$, respectively

$$\frac{\partial}{\partial \alpha_i} g(+0;\mathcal{U}_2) = \Delta_{\tau_i v_i} f + \int_{\tau_i}^{t_1} \widehat{f}_x(t) \cdot y(t; \tau_i, v_i)dt, \qquad (i.2)$$

where $y(\cdot; \tau_i, v_i) : \dot{y} = \widehat{\varphi}_x(t) \cdot y$, $y(\tau_i; \tau_i, v_i) = \Delta_{\tau_i v_i} \varphi$; moreover, we have twice used the abbreviation

$$\Delta_{\tau v} g = g(\tau, \widehat{x}(\tau), v) - \widehat{g}(\tau).$$

3) Let $g_k(\alpha;\mathcal{U}_j)$ be the k-th coordinate of the vector $x_\alpha(t_1;\mathcal{U}_j) - x_1$, $1 \leq k \leq n$, $j = 1, 2$.

Consider the problems

$$g_0(\alpha; \mathcal{U}_j) \to \min, \ \alpha \in A_j, \ g_k(\alpha; \mathcal{U}_j) = 0, \ 1 \le k \le n, \quad (P_{\mathcal{U}_j}),$$

$j = 1, 2$.

Let

$$\mathcal{L}(\alpha, \lambda(\mathcal{U}_j)) = \sum_{k=0}^{n} \lambda_k(\mathcal{U}_j) g_k(\alpha; \mathcal{U}_j))$$

be the Lagrange function of the problem $(P_{\mathcal{U}_j})$, $j = 1, 2$, $\alpha \in A_j$.

From definitions 4.1 and 4.2 it follows that 0 is a local minimum of these problems.

4) From theorems 3.1 and 3.2 we obtain that there exist vectors

$$\lambda(\mathcal{U}_j) = (\lambda_0(\mathcal{U}_j), \lambda_1(\mathcal{U}_j), \ldots, \lambda_n(\mathcal{U}_j)) = (\lambda_0(\mathcal{U}_j), \lambda'(\mathcal{U}_j)), \ |\lambda(\mathcal{U}_j)| = 1$$

such that $\mathcal{L}_\alpha(0, \lambda(\mathcal{U}_j)) = 0$ for $j = 1$ and ≥ 0 for $j = 2$. This leads in both cases to explicit conditions, with $1 \le i \le N$: for $j = 1$,

$$0 = \lambda_0(\mathcal{U}_1) \left(\int_{t_0}^{t_1} (\widehat{f}_x(t) \cdot y(t, u_i(\cdot)) + \widehat{f}_u(t) \cdot u_i(t)) dt \right) + \lambda'(\mathcal{U}_1) \cdot y(t_1; u_i(\cdot)),$$

and for $j = 2$,

$$0 \le \lambda_0(\mathcal{U}_2) \left(\int_{\tau_i}^{t_1} (\widehat{f}_x(t) \cdot y(t; \tau_i, v_i)) dt \right) + \lambda'(\mathcal{U}_2) \cdot y(t_1; \tau_i, v_i),$$

where $y(\cdot; \tau_i, v_i) : \dot{y} = \varphi_x(t)y, \ y(\tau_i; \tau_i, v_i) = \Delta_{\tau_i v_i} \varphi$.

One can easily understand that for each choice of auxiliary collection $\mathcal{U}_j$, $j = 1, 2$, the set $\Lambda(\mathcal{U}_j)$ of Lagrange multipliers of modulus one for which the first order necessary conditions for problem $(P_{\mathcal{U}_j})$ above hold is non-empty and compact. Clearly, if $N \le \bar{N}$ and $\mathcal{U}_j$ (resp. $\bar{\mathcal{U}}_j$) is an auxiliary collection of N (resp. $\bar{N}$) elements, with $\mathcal{U}_j \supseteq \bar{\mathcal{U}}_j$, then $\Lambda(\mathcal{U}_j) \subseteq \Lambda(\bar{\mathcal{U}}_j)$. Hence, by the defining property of compactness, it follows that the sets $\Lambda(\mathcal{U}_j)$ have a nonempty intersection, where $\mathcal{U}_j$ runs over all auxiliary sets for problem $(P_{4.j})$, $j = 1, 2$. That is, there exists for the problem $(P_{4.j})$ a vector $(\widehat{\lambda}_0(j), \widehat{\lambda}'(j))$ of Lagrange multipliers such that, for $j = 1$:

$$0 = \widehat{\lambda}_0(1) \left(\int_{t_0}^{t_1} (\widehat{f}_x(t) y(t, u(\cdot)) + \widehat{f}_u(t) u(t)) dt \right) + \lambda'(1) \cdot y(t_1; u(\cdot)) \ (ii.1)$$

$\forall u(\cdot)$ and, for $j = 2$:

$$0 \leq \widehat{\lambda}_0(2) \left(\int_{\tau_i}^{t_1} (\widehat{f}_x(t)y(t;\tau,v)dt \right) + \lambda'(2) \cdot y(t_1;\tau,v) \forall (\tau,v). \quad (ii.2)$$

5) Denote by $p(\cdot, j)$ the solution of the linear differential equation

$$-\dot{p} = p\widehat{\varphi}_x(t) - \widehat{\lambda}_0(j)\widehat{f}_x(t), \quad p_0(t_1) = -\lambda'(j), \quad j = 1, 2.$$

Then (4.1_a) and (4.2_a) are satisfied.

Substituting the expression $\dot{p}(t;j) + p(t;j)\widehat{\varphi}_x(t)$ for $\widehat{f}_x(t)$ into $(ii.j)$ for $j = 1, 2$ (in the cases $\widehat{\lambda}_0(j) = 1$), and integrating by parts, we obtain (4.1_b) and (4.2_b). The cases $\widehat{\lambda}_0(j) = 0$ are trivial.

G.2 CONDITIONS OF EXTREMUM OF THE SECOND ORDER

Each of the first order necessary conditions above can be supplemented by second order necessary and sufficient conditions. The point of this is that these necessary and sufficient conditions are almost equal. This gives the following insight. The second order necessary conditions are essentially the only other obstruction to local optimality besides the first order conditions. In this section we present for most of problem types above, the second order necessary conditions, and for some of these the second order sufficient conditions as well.

G.2.1 Necessary conditions of the second order and sufficient conditions

G.2.1.1 Finite-dimensional problems without constraints

Let us consider again the problem without constraints

$$f(x) \to \min. \qquad (P_1)$$

Let $V \in \mathcal{O}(\widehat{x}, \mathbb{R}^n)$ and $f : V \to \mathbb{R}$. One writes $f \in D^2(\widehat{x})$ if $f \in D^1(\widehat{x})$ and if moreover, there exists a symmetric $n \times n$ matrix A such that

$$f(\widehat{x} + x) = f(\widehat{x}) + f'(\widehat{x}) \cdot x + \frac{1}{2}x^T \cdot Ax + r(x)$$

with $r(x) = o(|x|^2)$. Then this matrix A is unique and it is denoted as $f''(\widehat{x})$. We write $A \succeq 0$ (resp. $A \succ 0$) if $x^T \cdot Ax \geq 0 \ \forall x \in \mathbb{R}^n$ (resp.

$x^T \cdot Ax > 0 \, \forall x \in \mathbb{R}^n \setminus \{0\}$). The following result is a supplement to theorem 1 (Fermat theorem).

Theorem 1' a) Necessary conditions of the second order.
Let $V \in \mathcal{O}(\widehat{x}, \mathbb{R}^n)$, $f : V \to \mathbb{R}$, $f \in D^2(\widehat{x})$, and $\widehat{x} \in \mathrm{locmin}(P_1)$. Then

$$f'(\widehat{x}) = 0 \quad \text{and} \quad f''(\widehat{x}) \succeq 0. \tag{1a}$$

b) Sufficient conditions of the second order. *If $f \in D^2(\widehat{x})$, $f'(\widehat{x}) = 0$ and $f''(\widehat{x}) \succ 0$, then $\widehat{x} \in \mathrm{locmin}(P_1)$.*

Proof. a) The equality $f'(\widehat{x}) = 0$ follows from theorem 1. If there exists $\bar{x} \neq 0$ such that $\bar{x}^T \cdot A\bar{x} < 0$, then (from the definition of $f \in D^2(\widehat{x})$) the function $\alpha \mapsto g(\alpha) = f(\widehat{x} + \alpha \bar{x})$ does not have a local minimum at 0, hence $\widehat{x} \notin \mathrm{locmin}(P_1)$.

b) As $f''(\widehat{x}) \succ 0$, there exists $\varepsilon > 0$ such that $x^T \cdot A\bar{x} \geq \varepsilon |x|^2 \, \forall x \in \mathbb{R}^n$. Therefore, for $\delta > 0$ chosen such that $|r(x)| \leq \frac{\varepsilon}{2}|x|^2$ if $|x| \leq \delta$, one has $f(\widehat{x} + x) \geq f(\widehat{x}) + \frac{\varepsilon}{2}|x|^2$ if $|x| \leq \delta$. Therefore, $\widehat{x} \in \mathrm{locmin}(P_1)$

G.2.1.2 The simplest problem of the calculus of variations

Consider now the simplest problem of the calculus of variations,

$$J(x(\cdot)) = \int_{t_0}^{t_1} L(t, x(t), \dot{x}(t)) dt \to \min, \quad x(t_i) = x_i, \; i = 0, 1. \quad (P_2)$$

The following result is a supplement to theorem 2 (Euler theorem). Here we present only the *necessary* conditions.

Theorem 2'. Necessary conditions of the second order.
Let a function $\widehat{x}(\cdot)$ be admissible in problem (P_2) and let L be twice continuously differentiable in a neighborhood of the curve

$$t \mapsto (t, \widehat{x}(t), \dot{\widehat{x}}(t)) \; t \in [t_0, t_1]$$

in $\mathbb{R}^3$. If the function $\widehat{x}(\cdot) \in C^2([t_0, t_1])$ is a weak minimum in problem (P_2), then the following conditions hold:

(1) Euler equation:

$$-\frac{d}{dt}\widehat{L}_{\dot{x}}(t) + \widehat{L}_x(t) = 0.$$

(2) Legendre condition:

$$A(t) = \widehat{L}_{\dot{x}\dot{x}}(t) \geq 0 \; \forall t \in [t_0, t_1].$$

(3) *Jacobi condition:* if $A(\cdot) \in C^1([t_0, t_1])$ and the strict Legendre condition $A(t) > 0 \; \forall t \in [t_0, t_1]$ is satisfied, then the interval (t_0, t_1) contains no zeros of the solution of the Cauchy problem consisting of the Jacobi equation $-\frac{d}{dt}(A(t)\dot{h}) + B(t)h = 0$, where $B(t) = \widehat{L}_{xx}(t) - \frac{d}{dt}\widehat{L}_{\dot{x}x}(t)$, and the initial conditions $h(t_0) = 0$, $\dot{h}(t_0) = 1$.

If $\widehat{x}(\cdot)$ is a strong minimum in (P_2), then, besides the three conditions, the following condition holds:

(4) *Weierstrass condition:*

$$\mathcal{E}\left(t, \widehat{x}(t), \dot{\widehat{x}}(t), u\right) \geq 0, \quad \forall u \in \mathbb{R}, \; t \in [t_0, t_1],$$

where $\mathcal{E}(t, x, \dot{x}, u) = L(t, x, u) - L(t, x, \dot{x}) - \langle L_{\dot{x}}(t), u - \dot{x}\rangle$.

The zeros of the Jacobi equation with the initial conditions above are called *conjugate points*, and the function $\mathcal{E}$ is called *the Weierstrass function*.

Proof. We begin with the necessary conditions of a strong minimum. These are direct corollaries of the maximum principle. Let us reformulate (P_2) as a problem of optimal control:

$$\int_{t_0}^{t_1} L(t, x, u)\, dt \to \min, \quad \dot{x} = u \in \mathbb{R}, \quad x(t_0) = x_0, \quad x(t_1) = x_1.$$
$$(P_2')$$

If $\widehat{x}(\cdot)$ is a strong minimum of (P_2), then by definition the pair $(\widehat{x}(\cdot), \dot{\widehat{x}}(\cdot))$ is an optimal process in (P_2'). The Lagrange function of problem (P_2') is

$$\mathcal{L}(x(\cdot), u(\cdot), p(\cdot)) = \int_{t_0}^{t_1} (L(t, x(t), u(t)) + p(t)(\dot{x}(t) - u(t)))\, dt.$$

From the necessary conditions for problem (P_2') it follows that, if $(\widehat{x}(\cdot), \dot{\widehat{x}}(\cdot))$ is an optimal process in (P_2'), then the Euler equation

$$-\dot{p}(t) + \widehat{L}_x(t) = 0 \qquad (i)$$

and the minimum condition

$$\min_{u \in \mathbb{R}}(L(t, \widehat{x}(t), u) - p(t)u) = \widehat{L}(t) - p(t)\dot{\widehat{x}}(t) \qquad (ii)$$

are satisfied. We see that the function $u \mapsto l(u) = L(t, \widehat{x}(t), u) - p(t)u$ attains its absolute minimum at the point $\dot{\widehat{x}}(t)$, and hence

$$l'(\dot{\widehat{x}}(t)) = 0,$$

$$l''(\dot{\widehat{x}}(t)) \geq 0. \qquad (iii).$$

From (iii) we see that the equality $p(t) = \widehat{L}_{\dot{x}}(t)$ holds. This equality together with (i) leads to the Euler equation. The inequality (iii) is equivalent to the Legendre condition: $\widehat{L}_{\dot{x}\dot{x}}(t) = A(t) \geq 0 \; \forall t \in [t_0, t_1]$. Now we prove that the Legendre condition is a necessary condition for a strong minimum. The condition (ii), being written in the form of the inequality

$$L(t, \widehat{x}(t), u) - \widehat{L}_{\dot{x}}(t)u \geq \widehat{L}(t) - \widehat{L}_{\dot{x}}(t)\dot{\widehat{x}}(t) \quad u \in \mathbb{R}^n,$$

is equivalent to the inequality $\mathcal{E}(t, \widehat{x}(t), \dot{\widehat{x}}(t), u) \geq 0 \; \forall u \in \mathbb{R}^n$. This means that the Weierstrass condition is a necessary condition for a strong minimum of (P_2).

Now we have to prove that the Legendre and Jacobi conditions are necessary conditions for a weak minimum.

Let us take $h(\cdot) \in C^1([t_0, t_1])$, $h(t_0) = h(t_1) = 0$. Denoting $g_{h(\cdot)}(\alpha) = J(\widehat{x}(\cdot) + \alpha h(\cdot))$, differentiating $g_{h(\cdot)}(\cdot)$ twice, and substituting $\alpha = 0$, we have

$$\mathcal{K}(h(\cdot)) := f''_{h(\cdot)}(0)$$

$$= \int_{t_0}^{t_1} (\widehat{L}_{\dot{x}\dot{x}}(t)\dot{h}^2(t) + 2\widehat{L}_{x\dot{x}}(t)\dot{h}(t)h(t) + \widehat{L}_{xx}(t)h^2(t))\, dt \geq 0 \qquad (iv)$$

(because $\widehat{x}(\cdot)$ is a weak minimum of (P_2)).

After integrating by parts, we obtain that the function $\bar{x}(t) \equiv 0$ is an absolute minimum in the problem

$$\mathcal{K}(x(\cdot)) = \int_{t_0}^{t_1} \left(A(t)\dot{x}^2(t) + B(t)x^2(t)\right) dt \to \min, \quad x(t_0) = x(t_1) = 0,$$
$$(v)$$

when $x(\cdot) \in C^1([t_0, t_1])$, and hence when $x(\cdot) \in PC^1([t_0, t_1])$, because a function from $PC^1([t_0, t_1])$ can be smoothed such that the value of $\mathcal{K}$ changes as little as desired. Hence the function $\bar{x}(t) \equiv 0$ satisfies the minimum condition from the necessary conditions for the problem (v), which consists of the inequality $A(t)u^2 \geq 0 \; \forall u \in \mathbb{R}$. Therefore, $A(t) \geq 0$ is a necessary condition for a weak minimum.

Now we prove the Jacobi condition. Let the solution $h(\cdot)$ of the Jacobi equation with the initial conditions given above have a zero at a point τ, $t_0 < \tau < t_1$. Then, integrating by parts, we have

$$\int_{t_0}^{\tau} \left(A(t)\dot{\widehat{h}}^2(t) + B(t)\widehat{h}^2(t) \right) dt$$

$$= \int_{t_0}^{\tau} \widehat{h}(t) \left(-\frac{d}{dt} A(t)\dot{\widehat{h}}(t) + B(t)\widehat{h}(t) \right) dt = 0.$$

Let us take

$$\bar{h}(t) = \begin{cases} \widehat{h}(t), & t \in [t_0, \tau], \\ 0, & t \in [\tau, t_1]. \end{cases}$$

We see that the piecewise function $\bar{h}(\cdot)$ attains the absolute minimum of the problem (v). From the Pontryagin maximum principle it follows that there exists a piecewise continuously differentiable function

$$\bar{p}(\cdot) = \begin{cases} A(t)\dot{\widehat{h}}(t), & t \leq \tau, \\ 0, & t \geq \tau. \end{cases}$$

From $\dot{\bar{h}}(\tau+0) = 0$ and from the continuity of $\bar{p}(\cdot)$ it follows that $\dot{\bar{h}}(\tau) = 0$. The equality $\bar{h}(\tau) = 0$ holds by the Jacobi condition; therefore, from the uniqueness of the solution of the Cauchy problem for a linear equation, one has $\bar{h}(t) \equiv 0$, but this contradicts the conditions $\bar{h}(0) = \dot{\widehat{h}}(0) = 1$.

G.2.2 Necessary conditions of the second order in smooth problems with equality constraints

Consider problem $(P_{3.1})$ from section G.1.3:

$$f_0(x) \to \min, \quad f_i(x) = 0, \ 1 \leq i \leq m, \quad (P_{3.1})$$

where $V \in \mathcal{O}(\widehat{x}, \mathbb{R}^n)$, $f_i : V \to \mathbb{R}$, $0 \leq i \leq m$ and $\widehat{x}$ is admissible. Denote by F the mapping $x \mapsto F(x) = \begin{pmatrix} f_1(x) \\ \vdots \\ f_n(x) \end{pmatrix}$. Write $F'(\widehat{x}) = A$, $L = \operatorname{Ker} A$.

The following result is a supplement to theorem 3.1 (Lagrange multiplier rule).

Theorem 3.1′ Conditions of the second order for problem $(P_{3.1})$. *Assume in problem* $(P_{3.1})$ *that* $f_i \in D^2(\widehat{x}) \cap C(V)$, $0 \leq i \leq m$, *and* $F'(\widehat{x})\mathbb{R}^n = \mathbb{R}^m$.

a) Necessary conditions. *If* $\widehat{x} \in \operatorname{locmin}(P_{3.1})$, *then there exists a vector* $\lambda = (1, \bar\lambda)$, $\bar\lambda = (\lambda_1, \ldots, \lambda_m)$ *such that*

$$(i)\ \mathcal{L}_x(\widehat{x}, \lambda) = 0, \quad (ii)\ x^T \mathcal{L}_{xx}(\widehat{x}, \lambda) x \geq 0 \quad \forall x \in L.$$

b) Sufficient conditions. *If for some* $\lambda = (1, \bar\lambda)$, $\bar\lambda = (\lambda_1, \ldots, \lambda_m)$ *the conditions*

$$(i)\ \mathcal{L}_x(\widehat{x}, \lambda) = 0, \quad (ii)\ x^T \mathcal{L}_{xx}(\widehat{x}, \lambda) x > 0 \quad \forall x \in L \setminus \{0\}$$

hold,

then $\widehat{x} \in \operatorname{strict}\ \operatorname{locmin}(P_{3.1})$

Proof. a) Condition (i) follows from Theorem 3.1 and the regularity assumption $F'(\widehat{x})\mathbb{R}^n = \mathbb{R}^m$. We have to prove (ii).

From Brouwer's fixed point theorem the following result can be deduced: *there exists a mapping* $r: L \to \mathbb{R}^n$ *with* $r(x) = o(|x|)$ *such that* $F(\widehat{x} + x + r(x)) = 0$ *for all* $x \in L$ (this result is called the tangent space theorem). Choose an arbitrary $x \in L$. Then, for all $\alpha \in \mathbb{R}$,

$$f_0(\widehat{x} + \alpha x + r(\alpha x))$$

$$= f_0(\widehat{x} + \alpha x + r(\alpha x)) + \bar\lambda \cdot F(\widehat{x} + \alpha x + r(\alpha x)) = \mathcal{L}(\widehat{x} + \alpha x + r(\alpha x), \lambda)$$

$$= \mathcal{L}(\widehat{x}, \lambda) + \mathcal{L}_x(\widehat{x}, \lambda) \cdot (\alpha x + r(\alpha x))$$

$$+ \frac{1}{2}(\alpha x + r(\alpha x))^T \mathcal{L}_{xx}(\widehat{x}, \lambda)(\alpha x + r(\alpha x)) + o(\alpha^2)$$

$$= f_0(\widehat{x}) + \frac{1}{2}\alpha^2 x^T \mathcal{L}_{xx}(\widehat{x}, \lambda) x + o(\alpha^2).$$

Moreover, $f_0(\widehat{x}) \leq f_0(\widehat{x} + \alpha x + r(\alpha x))$ for $|\alpha|$ sufficiently small, by the admissibility of $\widehat{x} + \alpha x + r(\alpha x)$ for problem $(P_{3.1})$ and the local minimality of $\widehat{x}$ for problem $(P_{3.1})$. Therefore,

$$f_0(\widehat{x}) \leq f_0(\widehat{x}) + \frac{1}{2}\alpha^2 x^T \mathcal{L}_{xx}(\widehat{x}) x + o(\alpha^2)$$

for $|\alpha|$ sufficiently small. This gives $x^T \mathcal{L}_{xx}(\widehat{x}) x \geq 0$.

b) We will argue by contradiction. Assume that conditions (i) and (ii) hold for some $\lambda = (1, \bar{\lambda})$, but that $\hat{x} \notin$ strict locmin$(P_{3.1})$. Choose $\alpha > 0$ such that the inequality $x^T \mathcal{L}_{xx}(\hat{x}, \lambda) x \geq \alpha |x|^2$ holds for all $x \in L$. Then there exists a sequence $\{x_k\}_{k \in \mathbb{R}}$ such that

$$x_k \neq 0 \ \forall k, \ \lim_{k \to \infty} x_k = 0, \ f_0(\hat{x}+x_k) \leq f_0(\hat{x}), \ f_i(\hat{x}+x_k) = 0, \ 1 \leq i \leq m.$$

Denoting $\varepsilon_k = |x_k|$ and choosing a subsequence $\frac{x_{k_l}}{|x_{k_l}|} \to \bar{x}$, we obtain that

$$x_{k_l} = \varepsilon_{k_l}(\bar{x} + \tilde{x}_{k_l}), \ \varepsilon_{k_l} \to 0, \ |\tilde{x}_{k_l}| \to 0, \ |\bar{x}| = 1.$$

Then

$$\mathcal{L}(\hat{x} + x_{k_l}) = \mathcal{L}(\hat{x}) + \mathcal{L}_x(\hat{x}, \lambda) \cdot x_{k_l} + \frac{\varepsilon_{k_l}^2}{2} \bar{x}^T \mathcal{L}_{xx}(\hat{x}, \lambda) \bar{x} + o(\varepsilon_{k_l}^2)$$

$$\geq f_0(\hat{x}) + \frac{1}{2}\alpha(\varepsilon_{k_l}|\bar{x}|)^2 + o(\varepsilon_{k_l}^2).$$

On the other hand,

$$\mathcal{L}(\hat{x} + x_{k_l}) = f_0(\hat{x} + x_{k_l}) \leq f_0(\hat{x})$$

for all l. Therefore,

$$f_0(\hat{x}) + \frac{1}{2}\alpha(\varepsilon_{k_l}|\bar{x}|)^2 + o(\varepsilon_{k_l}^2) \leq f_0(\hat{x})$$

for all l. Subtracting $f_0(\hat{x})$ on both sides, dividing by $\varepsilon_{k_l}^2$, and letting l tend to infinity, we are led to $|\bar{x}| = 0$. Contradiction.

Conditions of the second order for problems with equality and inequality constraints can be proved in an analogous way.

Appendix H

Solutions of Exercises of Chapters 1–4

> Sed haec aliaque artificia practica, quae ex usu multo facilius quam ex praeceptis ediscuntur, hic tradere non est instituti nostri.
>
> But it is not our intention to treat of these details or of other practical artifices that can be learned more easily by usage than by precept.
>
> C. F. Gauss

Chapter one
1.4

1. A global minimum $\hat{x}$ exists, as f is coercive ($f(x) \approx |x| - \frac{1}{2}x$ for $|x|$ large).

2. Fermat: $f'(x) = 0 \Rightarrow \frac{1}{2}(x^2+1)^{-\frac{1}{2}} \cdot 2x - \frac{1}{2} = 0$.

3. $4x^2 = x^2 + 1$, $x \geq 0$ (x cannot be negative, 'as square roots are nonnegative') $\Rightarrow x = \frac{1}{3}\sqrt{3}$.

4. $\hat{x} = \frac{1}{3}\sqrt{3}$.

1.5 Eliminate: $x_1 = t$; $x_2 = 1 - t$. This gives

1. $g(t) = e^{t-t^2} \to$ extr.

   A global maximum $\hat{x}_{\max}$ exists, but a global minimum does not exist; the value of the minimization problem $S_{\min}$ is zero, but this value is not assumed (note that $g(t) > 0$ for all t and $g(t) \to 0$ for $|t| \to +\infty$, and apply Weierstrass).

2. Fermat: $g'(t) = 0 \Rightarrow e^{t-t^2}(1 - 2t) = 0$.

3. $t = \frac{1}{2}$.

4. $\hat{x}_{\max} = (\frac{1}{2}, \frac{1}{2})^T$, the set of global minima—also called absolute minima—absmin is empty, and $S_{\min} = 0$.

1.6 Clear from Figure 1.16.

1.7 We define x to be the distance between the following two points of the river: the point that the caring neighbor visits, and the point that is the orthogonal projection onto the river of the house of the caring neighbor.

1. $f(x) = \sqrt{50^2 + x^2} + a\sqrt{100^2 + (300 - x)^2} \to \min$ (Fig. 1.17);

   for $a = 1$ (resp. $a = 2$) this is the problem to minimize length (resp. time). Existence of a global solution $\hat{x}(a)$ follows from the coercivity of f for all $a > 0$.

2. Fermat: $f'(x) = 0 \Rightarrow$
$$x/\sqrt{50^2 + x^2} = a(300 - x)/\sqrt{100^2 + (300 - x)^2}.$$

3. $\sin\alpha = a\sin\beta$. In the case $a = 1$ we get $\alpha = \beta$, and so the two triangles in Figure 1.17 are similar, $\Rightarrow (300 - x)/100 = x/50 \Rightarrow x = 100$.

4. $\hat{x}(1) = 100$ and $\hat{x}(2)$ can only be characterized by the equation $\sin\alpha = 2\sin\beta$.

1.8 Done.

1.9 The light is refracted (Fig. 1.21). The speed of light in air, v_1, is greater than the speed of light in water, v_2, and so, by Snellius, $\sin\alpha > \sin\beta$ and so $\alpha > \beta$. Therefore, you should not stretch out your hand in the direction where you see the fish, but in a more vertical direction.

1.10 Snellius:
$$\Rightarrow (x/\sqrt{x^2 + 50^2}) : 1 = 2 : 5 \Rightarrow 5x = 2\sqrt{x^2 + 50^2};$$

this is indeed the same equation as we obtained by the Fermat theorem (up to the same factor 2 on both sides).

1.11 The earth's atmosphere is less dense at the top and more dense at the bottom. Light travels more slowly in air than it does in vacuum, and so the light of the sun can get to a point beyond the horizon more quickly if, instead of just going in a straight line, it avoids the dense regions—where it goes slowly—by getting through them at a steeper slant. Therefore, when it appears to go below the horizon, it is already well below the horizon.

1.12 What you really see is the sky light "reflected" on the road: light from the sky, heading for the road at a small angle, can bend up before it reaches the ground and then it can reach the eye. Why? The air is very hot just above the road but cooler up higher. Hotter air is less dense than cooler air, and this decreases the speed of light less than does cooler air. That is to say, light goes faster in the hot region than in the cool region. Therefore, instead of the light "deciding to come" in the straightforward way, it also has a least-time path by which it goes into the region where it goes faster for awhile, in order to save time. So, it can go in a curve.

1.13 The fata morgana mirage occurs when there are alternating warm and cold layers near the ground. Instead of traveling straight through these layers, light is bent toward the colder, denser air. This is in accordance with the variational principle of Fermat, and the phenomenon is seen to be related to the one from the previous exercise, "mirages on hot roads." The result can be a rather complicated light path and a strange image of a distant object. The images may undergo rapid changes as the air layers move slightly up and down.

1.14

1. $f(x) = (10^7 + \frac{1}{4}10^6 x + 10^4 x \cdot x)/x = 10^7 x^{-1} + \frac{1}{4}10^6 + 10^4 x \to$ min, $x > 0$. Existence of a global solution $\hat{x}$ follows from Weierstrass, as $f(0+) = +\infty$ and $f(+\infty) = +\infty$.

2. Fermat: $f'(x) = 0 \Rightarrow -10^7 x^{-2} + 10^4 = 0$.

3. $x = \sqrt{1000}$.

4. $\hat{x} = \sqrt{1000}$.

Finally, we will find the integral solution. We have $31 < \sqrt{1000} < 32$. Therefore, the integral solution is 31 or 32, using that $f''(x) = \frac{2 \cdot 10^7}{x^3} > 0$ and so f' is monotonically increasing. A comparison shows $f(31) > f(32)$.

Conclusion: the integral solution is 32 floors.

1.15

1. $f(x) = \sqrt[x]{x} = e^{(\ln x)/x} \to$ extr, $x > 0$.

   A global maximum $\hat{x}_{\max}$ exists by Weierstrass, but a global minimum does not exist, although the value of the maximization problem $S_{\min}$ is finite: it is equal to zero (we have $f(0+) = 0$, $f(+\infty) = 1$, $f(x) > 0$ for all x, and $f(x) > 1$ for all $x > 1$).

2. Fermat: $f'(x) = 0 \Rightarrow e^{(\ln x)/x}(1 - \ln x)/x^2 = 0$.

3. $x = e$.

4. $\widehat{x}_{\max} = e$ the set of global minima absmin is empty, and $S_{\min} = 0$.

Finally we give the integral solutions: the global minimum is assumed at $x = 1$ and the global maximum is assumed at 3 (as $f(2) = \sqrt[2]{2} < \sqrt[3]{3} = f(3)$); here we use that f is monotonic increasing on $(0, e)$ and monotonic decreasing on $(e, +\infty)$, as can be seen from the course of the sign of the derivative f'.

1.16 To begin with, we ignore the constraint that the solution has to be an integer.

1.

1. $f(x) = 50x^3 - 35x^2 - 12x + 4 \to \min$, $x \geq 0$.

   Existence of a global solution $\widehat{x}$ follows from Weierstrass, as $f(+\infty) = +\infty$.

2. Fermat: $f'(x) = 0 \Rightarrow 150x^2 - 70x - 12 = 0$.

3. $(5x-3)(30x+4) = 0$, $x > 0 \Rightarrow x = 0.6$. The left-hand boundary point 0 can be excluded: f is decreasing at 0, as $f'(0) = -12 < 0$.

4. $\widehat{x} = 0.6$.

Finally, we will find the integral solution. One has $0 < 0.6 < 1$ and so the integral solution is either 0 or 1, using that f is monotonic decreasing on $(0, 0.6)$ and monotonic increasing on $(0.6, +\infty)$: indeed $f'(x) < 0$ for $0 \leq x < 0.6$ and $f'(x) > 0$ for $x > 0.6$. Comparison gives $f(0) < f(1)$.

Conclusion: 0 is the global solution (note that rounding off 0.6 to the nearest integer would lead to 1).

Note the subtlety of the logic of initially ignoring the integrality constraint: at first glance it might seem paradoxical that in the beginning we discard 0, while at the end 0 turns out to be the solution.

2. Let $\widehat{y}$ be the point of local maximum of f. Then $f'(x) = 3(x - \widehat{x})(x - \widehat{y})$ and so $f(x) = x^3 - \frac{3}{2}\widehat{x}x^2 - \frac{3}{2}\widehat{y}x^2 + 3\widehat{x}\widehat{y}x + c$ for some constant c. Assume that $n = 0$; this does not restrict the generality of the argument. Then $0 < \widehat{x} < 1$ and $\widehat{y} < 0$.

Then

$$f(1) - f(0) = 1 - \frac{3}{2}\widehat{x} - \frac{3}{2}\widehat{y} + 3\widehat{x}\widehat{y}.$$

This implies the following facts, as can be easily seen from a figure of the hyperbola $1 - \frac{3}{2}x - \frac{3}{2}y + 3xy = 0$:

- If $\widehat{x} < \frac{1}{2}$, we get $f(1) - f(0) > 0$, so the integer nearest to $\widehat{x}$ is 0 and this has smaller f-value than 1.

- If $\widehat{x} > \frac{2}{3}$, we get $f(1) - f(0) < 0$, so the integer nearest to $\widehat{x}$ is 1 and this has smaller f-value than 0.

1.17 1. Consider, to begin with, that 100 is written as a sum of n positive numbers for a fixed n. We consider the problem

$$\prod_{i=1}^{n} x_i \to \max, \quad \sum_{i=1}^{n} x_i = 100, \quad x_i \geq 0.$$

It is readily seen, for example, by repeated application of the result of the problem of Fermat (problem 1.4.1 and the subsequent remark) that it is optimal to have all x_i equal, $x_i = \frac{100}{n}$ for $1 \leq i \leq n$. This gives product $(\frac{100}{n})^n$. Now we let n vary. In order to find the number n which leads to a maximal product, we consider the problem

1. $f(x) = (\frac{100}{x})^x \to \max$, $x > 0$, where we allow the variable x to be nonintegral.

   Existence of a global solution follows from Weierstrass, as $f(0+) = 1$, $f(+\infty) = 0$ and $f(1) = 100 > 1$.

2. Fermat: $f'(x) = 0 \Rightarrow f(x)(\ln 100 - \ln x - 1) = 0$.

3. $x = 100/e \approx 36.79$.

4. $\widehat{x} = 100/e \approx 36.79$.

2. We now consider the additional requirement that the solution be an integer. The derivative $f'(x)$ is positive (resp. negative) to the left (resp. right) of $100/e$. Therefore, $f(x)$ is monotonic increasing (resp. decreasing) for x to the left (resp. right) of $100/e$. It follows that the integer n for which $f(n)$ is maximal is either 36 or 37. It remains to compare them:

$$f(36) \approx 9.40 \times 10^{15} \quad \text{and} \quad f(37) \approx 9.47 \times 10^{15}.$$

Conclusion: it is optimal to write 100 as a sum of 37 equal numbers, $100/37 \approx 2.70$.

A systematic solution of this problem will be given in problem 11.2 ("Wine on the wedding of the crown prince").

1.18 The additional benefit is 300 dollars, much more than the marginal cost, which is a few dollars for a drink and a snack. This marginal analysis shows why the passenger is taken. The optimization problem here is to make the choice giving the highest profit between two alternatives: to take the passenger or not.

1.19 The columns lead to the following qualitative conclusions: higher output gives higher total revenue, but also higher total cost. Moreover, for low output marginal revenue is larger than marginal cost, and for high output it is the other way round. We see that marginal revenue and cost are about equal for output level 4: indeed, then, an increase of output by one unit gives marginal revenue 60 - 51 = 9 and marginal cost 42 - 31 = 11, so an increase of output would lead to a smaller profit; for a decrease of output by one unit, these marginal values are 10 and 9, so a decrease of output would lead to a smaller profit as well. A calculation of the profits for each output level confirms that the optimal output level is 4. The optimization problem here is to choose, among the seven possible output levels, the one giving the highest profit.

1.20 1. If the price in dollars is 24, the profit will be $(24 - 5) \cdot 10000 = 190000$. Doing the same calculations for all prices and comparing results gives that the optimal price is 16 dollars and that this gives profit 550,000 dollars.

2. The optimal profit for the company as a function of t is the pointwise maximum of the functions $t \to (P-5-t)D$, where (P, D) ranges over all six pairs of possible prices P and corresponding demands D. It is straightforward to draw the graph of this function. This allows us to determine the profit optimizing price as a function of t, and this gives, in turn, the royalties for the artist as a function of t. We do not display the graphs, but only give the conclusion.

The artist's agent will advise him to ask slightly less than 11 euros; then the artist will get almost 330,000 dollars.

3. Selling for a fixed amount does not alter the incentives of the company for setting the price; if you ask slightly less than 550,000, then the record company will still make a—small—profit. Conclusion: selling for a fixed amount is better, provided the record company accepts all profitable offers.

4. The marketing department has clearly based itself on the estimate that the demand D for the dvd will be the excess of 26 over the dvd price P, multiplied by 5000. That is, $D = 5000(26 - P)$. Therefore, price P will lead to a profit

$$(P - 5)D = 5000(P - 5)(26 - P),$$

provided the artist does not ask for a recording fee. This is optimal for $P = 15\frac{1}{2}$.

(Here and in the remainder of the solution we use that a quadratic polynomial with roots at r_1 and r_2, has its extremum at $(r_1 + r_2)/2$, or equivalently, a parabola has its extremum precisely in between its roots.)

That is, the optimal price is $15\frac{1}{2}$ dollars and this gives a profit of 551,250 dollars, a slight improvement on the profit found in **1**. If the artist asks for a recording fee t, then the price P will lead to a profit

$$(P - 5 - t)D = 5000(P - 5 - t)(26 - P).$$

This is optimal for

$$P = \frac{(5+t) + 26}{2} = 15\frac{1}{2} + \frac{t}{2}.$$

Then the artist gets

$$tD = 5000t(26 - P) = 5000t \left(10\frac{1}{2} - \frac{1}{2}t\right).$$

This is optimal for $t = (0 + 21)/2 = 10\frac{1}{2}$. Therefore, the agent will advise the artist to ask $10\frac{1}{2}$ dollars; then the artist will get $275,625$ dollars. In passing, we note that the profit of the company is half this amount.

1.21

1. $f(L) = \frac{1}{3}\ln(wL) + \frac{2}{3}\ln(24 - L) \to \max$, $0 < L < 24$.

   Existence of a global maximum follows from Weierstrass ($f(0+) = -\infty$ and $f(24-) = -\infty$).

2. Fermat: $f'(L) = 0 \Rightarrow 1/(3L) - 2/(3(24 - L)) = 0$.

3. $3(24 - L) = 6L \Rightarrow L = 8$.

4. It is optimal to work precisely 8 hours, *whatever the wage rate is*.

1.22 For given m_1, m_2, n_1, n_2, p, r, the probability that from the group of people who get the placebo (resp. the medicine) precisely m_1 (resp. n_1) show an improvement of their condition is—by virtue of the formulas of the binomial distribution—

$$\frac{m!}{m_1!m_2!}p^{m_1}(1-p)^{m_2} \text{ (resp. } \frac{n!}{n_1!n_2!}(rp)^{n_1}(1-rp)^{n_2}).$$

We consider the problem of finding the values of p and r which maximize these two probabilities. For the people who get a placebo, the problem can be written as follows, after taking the logarithm of the first probability-function above, and omitting the logarithm of $\frac{m!}{m_1!m_2!}$ as we may—omitting a constant term from an objective function does not change the location of the extrema—

1. $f(p) = m_1 \ln p + m_2 \ln(1-p) \to \max$, $0 < p < 1$. Existence of a global maximum $\widehat{p}$ follows from the Weierstrass theorem, as $f(0+) = -\infty$ and $f(1-) = -\infty$.

2. Fermat: $f'(p) = 0 \Rightarrow \frac{m_1}{p} - \frac{m_2}{1-p} = 0$.

3. $p = \frac{m_1}{m_1+m_2} = \frac{m_1}{m}$.

4. $\widehat{p} = \frac{m_1}{m_1+m_2} = \frac{m_1}{m}$.

We take this as estimate for p. In the same way we are led to the estimate $\frac{n_1}{n}$ for pr. Taking the quotient of these two estimates gives the desired estimate, $r = \frac{mn_1}{m_1 n}$.

1.23 The remaining optimization problem is

$$\frac{1}{2}\pi + 2r - \frac{1}{2}\pi r^2 \to \max, \ 0 \le r \le 1,$$

where r is the radius of the inner half-circle. Using our knowledge of parabolas or applying Fermat we get that the global maximum is reached for $r = 2/\pi$, which is admissible. Therefore, this is the optimal radius for the inner half-circle. This leads to the optimal area $\pi/2 + 2/\pi \approx 2.207$. Thus, the area of the optimal sofa is slightly more than 40% larger than the area of the half-circle table.

1.24

1. $f(x) = x(8-x)(8-2x) = 2x^3 - 24x^2 + 64x \to \max$, $0 \le x \le 4$. Existence of a global solution $\widehat{x}$ follows from Weierstrass.

2. Fermat: $f'(x) = 0 \Rightarrow 6x^2 - 48x + 64 = 0$.

3. This quadratic equation has only one root in the admissible interval $[0,4]$, namely $x = 4 - \frac{4}{3}\sqrt{3}$. Compare values : $f(0) = f(4) = 0$ and $f(4 - \frac{4}{3}\sqrt{3}) > 0$.

4. $\widehat{x} = 4 - \frac{4}{3}\sqrt{3} \approx 1.69$.

1.25

1. The maximum (resp. minimum) value of the problem is $+\infty$ (resp. $-\infty$), as one sees by letting x tend to $+\infty$ and to $-\infty$, observing that $ax^3 + bx^2 + cx + d \approx ax^3$ for $|x|$ large. Therefore, there are no global extrema.

SOLUTIONS OF EXERCISES OF CHAPTERS 1–4   609

2. Fermat: $f'(x) = 0 \Rightarrow 3ax^2 + 2bx + c = 0$.

3. Distinguish three cases.

   (a) $4b^2 - 12ac > 0$. Then the stationary points are
   $$x_1 = (-2b - \sqrt{4b^2 - 12ac})/(6a)$$
   and $x_2 = (-2b + \sqrt{4b^2 - 12ac})/(6a)$;

   moreover, $a > 0 \Rightarrow x_1 = \text{locmax}$, $x_2 = \text{locmin}$, and $a < 0 \Rightarrow x_1 = \text{locmin}$, $x_2 = \text{locmax}$.

   (b) $4b^2 - 12ac < 0$. Then there are no stationary points and so no local extrema.

   (c) $4b^2 - 12ac = 0$. Then there is only one stationary point and it is not a local extremum.

4. There are no global extrema. There are local extrema precisely if $4b^2 - 12ac > 0$. Assume $a > 0$ (the other case, $a < 0$, is similar). Then there are a unique local maximum
   $$x_{\max} = (-2b - \sqrt{4b^2 - 12ac})/6a$$
   and a unique local minimum
   $$x_{\min} = (-2b + \sqrt{4b^2 - 12ac})/6a.$$

1.26 We assume that the triangle is a right triangle (with right angle at A) and equilateral (we may achieve this by a linear change of coordinates). This reduces the problem to one which we have solved already: maximizing the product of two nonnegative numbers with given sum (problem 1.6.1 and the subsequent remark).

Conclusion: the optimal choice of D (resp. E) is the midpoint of AB (resp. BC).

1.27 We have to consider the problem
$$1/(a_1 a_2) \to \text{extr}, \quad a_1 + a_2 = 3, \quad 1 \leq a_1 \leq \frac{3}{2}$$

by virtue of the hint, using also that one of the two numbers a_1, a_2 is less than or equal to $\frac{3}{2}$. Eliminate: $a_1 = t$, $a_2 = 3 - t$.

1. This gives $f(t) = (3t - t^2)^{-1} \to \text{extr}$, $1 \leq t \leq \frac{3}{2}$.

   Existence of a global minimum $\widehat{t}_{\min}$ and a global maximum $\widehat{t}_{\max}$ follows from Weierstrass.

2. Fermat: $f'(t) = 0 \Rightarrow (3 - 2t)/(3t - t^2)^2 = 0$.

3. $t = \frac{3}{2}$. Thus, the candidate extrema are 1 and $\frac{3}{2}$.
   Compare: $f(1) = \frac{1}{2}$, $f(\frac{3}{2}) = \frac{4}{9}$.

4. $\widehat{t}_{\min} = \frac{3}{2}$, $S_{\min} = \frac{4}{9}$ and $\widehat{t}_{\max} = 1$, $S_{\max} = \frac{1}{2}$.

Conclusion: for each line through the center of gravity of a given triangle, the ratio of the area of one of the two sides into which the line divides the triangle and the area of the given triangle is a number in the interval $[\frac{4}{9}, 1 - \frac{4}{9}]$, as required. The extreme values are attained when the line is parallel to one of the sides of the triangle.

1.28 Specialize the calculations in the proof of theorem 6.11 to the case $n = 2$.

1.29 Let R be the radius of the given ball. The problem to be solved is

$$h \cdot \pi(b/2)^2 \to \max, \quad b^2 + h^2 = (2R)^2, \quad b, h \geq 0,$$

where h and b are variables and R is given. Eliminate: $h = x$, $b^2 = 4R^2 - x^2$. This gives, omitting the constant factor $\frac{1}{4}\pi$ in the objective function, the problem

1. $f(x) = x(4R^2 - x^2) \to \max$, $0 \leq x \leq 2R$.

   Existence of a global solution $\widehat{x}$ follows from the Weierstrass theorem.

2. Fermat: $f'(x) = 0 \Rightarrow 4R^2 - 3x^2 = 0$.

3. $x = \frac{2}{3}R\sqrt{3}$. Thus, the candidate extrema are $0, 2R$ and $\frac{2}{3}R\sqrt{3}$.
   Compare: $f(0) = f(2R) = 0$ and $f(\frac{2}{3}R\sqrt{3}) > 0$.

4. $\widehat{x} = \frac{2}{3}R\sqrt{3}$. Conclusion: the optimal ratio is $\frac{\widehat{b}}{h} = \sqrt{2}$.

1.30 We assume that the given area A equals 1, as we may without restriction of the generality of the argument: we can achieve this by scaling, and scaling does not affect the problem. The given formulas lead to the problem

$$\pi h^2(R - h/3) \to \max, \quad 2\pi Rh = 1, \quad h, R > 0,$$

where the variables are h as well as R. Eliminate R: $h = x$ and $R = 1/(2\pi x)$.

SOLUTIONS OF EXERCISES OF CHAPTERS 1–4

1. This gives the problem $f(x) = \pi(x/(2\pi) - x^3/3) \to$ max, $x \geq 0$. Here we have allowed x to be zero in order to facilitate the proof of the existence of a global solution; this will be justified in hindsight, if the solution turns out to be nonzero.

   Existence of a global solution $\hat{x}$ follows from Weierstrass as $f(+\infty) = -\infty$.

2. Fermat: $f'(x) = 0 \Rightarrow 1/2 - \pi x^2 = 0$.

3. $x = 1/\sqrt{2\pi}$. Compare: $f(0) = 0$ and $f(1/\sqrt{2\pi}) > 0$.

4. $\hat{x} = 1/\sqrt{2\pi}$ and this is nonzero, as required. Conclusion: the height of the spherical segment and the radius of the sphere should both be $1/\sqrt{2\pi}$.

1.31 The problem can be modelled as follows:

$$\frac{1}{2}(2\cos\varphi)(2\pi\sin 2\phi) \to \text{max}, \ 0 \leq \varphi \leq \frac{1}{2}\pi$$

where φ is half the angle at the top of the cone in Figure 1.29 and with unit of length the radius of the ball. Writing $x = \sin\varphi$ this problem can be rewritten as follows:

1. $f(x) = x - x^3 \to$ max, $0 \leq x \leq 1$. Existence of a global solution $\hat{x}$ follows from Weierstrass.

2. Fermat: $f'(x) = 0 \Rightarrow 1 - 3x^2 = 0$.

3. $x = \frac{1}{3}\sqrt{3}$. Thus the candidate maxima are 0, 1 and $\frac{1}{3}\sqrt{3}$. Compare $f(0) = f(1) = 0$, and $f(\frac{1}{3}\sqrt{3}) > 0$.

4. Conclusion: $\hat{x} = \frac{1}{3}\sqrt{3}$.

Conclusion: the given problem has solution $\varphi = \arcsin(\frac{1}{3}\sqrt{3})$.

1.32 The two basic facts which underlie the solution of the previous exercise can be extended in a straightforward way.

- The $(n-1)$-dimensional volume of the boundary of an n-dimensional ball with radius r equals $w_n(r) = c(n)r^{n-1}$ for a positive constant $c(n)$ which depends only on n and the explicit form of which is not needed here.

- Consider a truncated cone in $\mathbb{R}^n$ for which the basis is an $(n-1)$-dimensional ball with radius r and for which the distance from the top to a point on the boundary of this ball is l.

The $(n-1)$-dimensional volume of the lateral boundary of this truncated cone is equal to

$$\frac{1}{n-1} l w_{n-1}(r).$$

These two facts imply by the same argument as in the previous exercise that the problem can be modeled as

$$\cos^{n-1}\varphi \sin^{n-2}\varphi \to \max,\ 0\le\varphi\le\frac{\pi}{2}.$$

1. This problem is equivalent to

$$f(x) = (1-x^2)^{\frac{n-1}{2}} x^{n-2} \to \max,\ 0\le x\le 1,$$

if n is odd, writing $x = \sin\varphi$ (resp.

$$f(x) = x^{n-1}(1-x^2)^{\frac{n-2}{2}} \to \max,\ 0\le x\le 1,$$

if n is even, writing $x = \cos\varphi$).

Existence of a global solution $\hat{x}$ follows from Weierstrass.

2. Fermat: $f'(x) = 0 \Rightarrow$

$$\frac{n-1}{2}(1-x^2)^{\frac{n-1}{2}-1}(-2x)x^{n-2} + (1-x^2)^{\frac{n-1}{2}}(n-2)x^{n-3} = 0$$

if n is odd (resp.

$$(n-1)x^{n-2}(1-x^2)^{\frac{n-2}{2}} + x^{n-1}\frac{n-2}{2}(1-x^2)^{\frac{n-2}{2}-1}(-2x) = 0$$

if n is even).

3. $x = \sqrt{\frac{n-2}{2n-3}}$ if n is odd (resp. $x = \sqrt{\frac{n-1}{2n-3}}$ if n is even).

4. The cone is optimal if half its top angle equals $\arcsin\sqrt{\frac{n-2}{2n-3}}$ if n is odd (resp. $\arccos\sqrt{\frac{n-1}{2n-3}}$ if n is even).

1.33 The idea of the solution of both parts is the same; here we present the solution of the first part. The average distance of x to the numbers $0, 1, 3$ an 10 is $\frac{1}{4}(|x| + |x-1| + |x-3| + |x-10|)$. We omit the factor $\frac{1}{4}$ as we may: omitting a positive factor does not change the location of the minima.

SOLUTIONS OF EXERCISES OF CHAPTERS 1–4   613

1. $f(x) = |x| + |x-1| + |x-3| + |x-10| \to \min$.

   Existence of a global solution $\widehat{x}$ follows from the coercivity of f (note that $f(x) \approx |x|$ for $|x|$ large).

2. Fermat: $f'(x) = 0 \Rightarrow \frac{x}{|x|} + \frac{x-1}{|x-1|} + \frac{x-3}{|x-3|} + \frac{x-10}{|x-10|} = 0$.

3. Note that if the sum of four numbers that are all $+1$ or -1 is equal to 0, then two of them have to be $+1$ and two -1. This gives $1 < x < 3$. Thus the candidate minima are the points of non-differentiability 1, 3, the boundary points 0, 10, and the stationary points $(1,3) = \{x : 1 < x < 3\}$. Comparison of the candidates shows that f takes at the points 0 and 10 values that are greater than the constant value that it takes at the closed interval $[1,3]$.

4. absmin $= [1,3]$.

1.34 It is immediately clear that the problem $x \to \min$, $x \geq 0$ has the unique solution $\widehat{x} = 0$. Now we consider the other problem.

1. $f_c(x) = x - c \ln x \to \min$, $x > 0$.

   Existence of a global solution $\widehat{x}_c$ follows from Weierstrass as $f(0+) = \infty$ and $f(+\infty) = +\infty$.

2. Fermat: $f'(x) = 0 \Rightarrow 1 - \frac{c}{x} = 0$.

3. $x = c$.

4. $\widehat{x}_c = c$. One has $\widehat{x} = 0$, clearly. Therefore $\lim_{c \downarrow 0} \widehat{x}_c = \widehat{x}$.

1.35 We have already established $x - \frac{1}{6}x^3 \leq \sin x \leq x$. Integrating gives
$$-1 + \frac{1}{2}x^2 - \frac{1}{24}x^4 \leq -\cos x \leq -1 + \frac{1}{2}x^2.$$
Integrating again gives
$$-x + \frac{1}{6}x^3 - \frac{1}{120}x^5 \leq -\sin x \leq -x + \frac{1}{6}x^3.$$
That is,
$$x - \frac{1}{6}x^3 \leq \sin x \leq x - \frac{1}{6}x^3 + \frac{1}{120}x^5.$$

Substituting $x = \frac{\pi}{180}$ in $\frac{1}{120}x^5$ gives that the error in approximating $\sin 1°$ by $\frac{\pi}{180} - \frac{1}{6}(\frac{\pi}{180})^3$ is less than 10^{-9}. Therefore, we get about nine decimals accuracy for $\sin 1°$.

1.36 We start with $1 \leq e^x \leq 3$ for $0 \leq x \leq 1$. This follows from the given approximation $1 \leq e \leq 3$ and the fact that e^x is increasing for $0 \leq x \leq 1$. Integrating and using $e^0 = 1$ gives $1 + x \leq e^x \leq 1 + 3x$. Integrating again gives

$$1 + x + \frac{1}{2}x^2 \leq e^x \leq 1 + x + \frac{3}{2}x^2.$$

We continue in this way till the difference between right and left is smaller than $\frac{1}{2}10^{-5}$. This gives

$$e \approx 1 + 1 + \frac{1}{2!} + \cdots + \frac{1}{9!}$$

up to five decimals. To compute $\sqrt{e}$ up to this accuracy we need fewer steps. We get

$$\sqrt{e} \approx 1 + \frac{1}{2} + \frac{1}{2!}\frac{1}{2^2} + \cdots + \frac{1}{7!}\frac{1}{2^7}$$

up to five decimals.

1.37 The difference between right and left tends to zero very slowly and for five decimals, you need to do about half a million steps of the approximation procedure, and the approximation is the sum of about half a million terms. If—instead—you use the approximation procedure to compute $\ln \frac{1}{2}$, as we may, by virtue of $\ln \frac{1}{2} = -\ln 2$, then you need only about twenty steps and the approximation is the sum of about twenty terms.

1.38 This exercise is essentially equivalent to the initial observation of the geologist problem (example 1.6). We let time run from $t = 0$ to $t = T$ and let $s_1(t)$ (resp. $s_2(t)$) be the distance covered after t seconds. The assumptions are

$$s_1(0) = s_2(0), \quad s_1'(t) \geq s_2'(t) \text{ for all } t \in [0, T].$$

Integrating the inequality from 0 to t gives $s_1(t) \geq s_2(t)$ for all $t \in [0, T]$, as required.

1.39 Integrate the inequality $C \leq f^{(n)}(x) \leq D$ from 0 to x. This gives

$$f^{(n-1)}(0) + Cx \leq f^{(n-1)}(x) \leq f^{(n-1)}(0) + Dx.$$

Continuing in this way, one gets, after integrating in total n times, the required inequality.

1.40 We only consider the case $f(a) > 0$—and so $f(b) < 0$. The other case is similar.

1. $g(t) = \int_a^t f(\tau)d\tau \to \min$, $t \in [a,b]$. Existence of a global solution $\hat{t}$ follows from Weierstrass.

2. Fermat: $g'(t) = 0 \Rightarrow f(t) = 0$.

3. The minimum is not attained at the boundary points, as $g'(a) = f(a) > 0$ and $g'(b) = f(b) < 0$.

4. The equation $f(t) = 0$ has a solution, $t = \hat{t}$.

Chapter two
2.1

1. $f(x) = x_1 x_2 + 50/x_1 + 20/x_2 \to \min$, $x_1, x_2 > 0$.

   Existence follows from the "coercivity" of f:

   - $x_1 \downarrow 0$: $f(x) \geq 50/x_1$ and this function of x_1 tends to $+\infty$ for $x_1 \downarrow 0$.
   - $x_1 \to +\infty$: for $x_2 \geq 1$, one has $f(x) \geq x_1$, and this function of x_1 tends to $+\infty$ for $x_1 \to +\infty$.
   - $x_2 \downarrow 0$: $f(x) \geq 20/x_2 \to \infty$ for $x_2 \downarrow 0$.
   - $x_2 \to +\infty$: for $x_1 \geq 1$, one has $f(x) \geq x_2 \to +\infty$ for $x_2 \to +\infty$.

2. Fermat: $f'(x) = 0_2 \Rightarrow$
   $$\frac{\partial f}{\partial x_1} = x_2 - \frac{50}{x_1^2} = 0, \quad \frac{\partial f}{\partial x_2} = x_1 - \frac{20}{x_2^2} = 0.$$

3. $x_1 = 20/x_2^2 = (20/50^2)x_1^4$. This gives $x_1^3 = 125$, and so $x_1 = 5$ and $x_2 = 50/5^2 = 2$.

4. $\hat{x} = (5,2)^T$.

2.2

1. $f(x) = x_1^2 + x_2^2 + x_3^2 - x_1 x_2 + x_1 - 2x_3 \to \min$.

   Existence of a unique global solution $\hat{x}$ follows from problem 2.4.1 (recall the matrix notation for quadratic functions (cf. the remark preceding problem 2.2.1), and its quadratic part is positive definite, as it can be written as $\frac{1}{2}(x_1 - x_2)^2 + \frac{1}{2}(x_1^2 + x_2^2) + x_3^2$).

2. Fermat: $f'(x) = 0_3^T \Rightarrow$

$$\frac{\partial f}{\partial x_1} = 2x_1 - x_2 + 1 = 0, \quad \frac{\partial f}{\partial x_2} = 2x_2 - x_1 = 0, \quad \frac{\partial f}{\partial x_3} = 2x_3 - 2 = 0.$$

3. $x = (-\frac{2}{3}, -\frac{1}{3}, 1)^T$.

4. $\widehat{x} = (-\frac{2}{3}, -\frac{1}{3}, 1)^T$.

2.3

1. $f(x) = 2x_1^4 + x_2^4 - x_1^2 - 2x_2^2 \to \min$.

   Existence of a global solution $\widehat{x}$ follows from Weierstrass, as
   $$f(x) = x_1^2(2x_1^2 - 1) + x_2^2(x_2^2 - 2) \to +\infty \quad \text{for} \quad |x| \to +\infty.$$

2. Fermat: $f'(x) = 0_2^T \Rightarrow$

   $$\frac{\partial f}{\partial x_1} = 8x_1^3 - 2x_1 = 0, \quad \frac{\partial f}{\partial x_2} = 4x_2^3 - 4x_2 = 0.$$

3. Nine stationary points: $x_1 = \pm\frac{1}{2}$, $x_1 = 0$ and $x_2 = \pm 1$, $x_2 = 0$. Comparison of values shows that four stationary points have lowest f-value: $x_1 = \pm\frac{1}{2}$, $x_2 = \pm 1$.

4. $\widehat{x} = (\pm\frac{1}{2}, \pm 1)$.

   Finally, $(0,0)^T$ is a local maximum, as can be seen by viewing f as the sum of $x_1^2(2x_1^2 - 1)$ and $x_2^2(x_2^2 - 2)$.

2.4 (i) Profit equals $P_1 Q_1 + P_2 Q_2 - TC =$

$$P_1(500 - P_1) + P_2(240 - \frac{2}{3}P_2) - (50000 + 20((500 - P_1) + (240 - \frac{2}{3}P_2))).$$

We go over to minimization.

1. $f(P) = P_1^2 + \frac{2}{3}P_2^2 - 520 P_1 - 253\frac{1}{3} P_2 + 64800 \to \min$, $P \in \mathbb{R}^2$.

   A unique global solution $\widehat{P}$ exists: this follows by problem 2.2 from the positive definiteness of the 2×2 diagonal matrix with on the diagonal the positive numbers $+1$ and $+\frac{2}{3}$.

2. Fermat: $f'(P) = 0 \Rightarrow$

   $$\frac{\partial f}{\partial P_1} = 2P_1 - 520 = 0, \quad \frac{\partial f}{\partial P_2} = \frac{4}{3}P_2 - 253\frac{1}{3} = 0.$$

SOLUTIONS OF EXERCISES OF CHAPTERS 1–4   617

3. $P = (260, 190)^T$.

4. $\widehat{P} = (260, 190)^T$ and maximal profit is $-S_{\min} = 26866\frac{2}{3}$.

(ii) With $P_1 = P_2 = p$, profit equals

$$P_1 Q_1 + P_2 Q_2 - (50000 + 20Q_1 + 20Q_2)$$

$$= p(500 - p) + p(240 - \frac{2}{3}p) - (50000 + 20(500 - p) + 20(240 - \frac{2}{3}p)).$$

1. $g(p) = -\frac{5}{3}p^2 + 773\frac{1}{3}p - 64800 \to \max, \ p \in \mathbb{R}$.

   Existence of a global maximum $\widehat{p}$ follows from coercivity of g (or from knowledge of parabolas, noting that the coefficient of p^2 is negative).

2. Fermat: $g'(p) = 0 \Rightarrow -\frac{10}{3}p + 773\frac{1}{3} = 0$.

3. $p = 232$.

4. $\widehat{p} = 232$. This gives maximal profit—after introduction of the law—$24906\frac{2}{3}$. Therefore, the loss of profit caused by the law is 1960.

(iii) Profit equals $P_1 Q_1 + P_2 Q_2 - TC =$

$$(500 - Q_1 - t)Q_1 + \left(360 - \frac{3}{2}Q_2\right)Q_2 - (50000 + 20Q_1 + 20Q_2).$$

1. $h(Q) = (500 - Q_1 - t)Q_1 + (360 - \frac{3}{2}Q_2)Q_2 - (50000 + 20Q_1 + 20Q_2) \to \min$.

   Existence of a unique solution follows from problem 2.2.

2. Fermat: $h'(Q) = 0_2^T \Rightarrow$

$$\frac{\partial h}{\partial Q_1} = 500 - t - 2Q_1 - 20 = 0, \quad \frac{\partial h}{\partial Q_2} = 360 - 3Q_2 - 20 = 0.$$

3. $Q_1 = 240 - \frac{1}{2}t, \ Q_2 = 113\frac{1}{3}$.

4. $\widehat{Q} = (240 - \frac{1}{2}t, 113\frac{1}{3})^T$. This shows that the tax has no influence on the sales in the industrial market and that the amount sold in the domestic market is lowered. The price in the domestic market will be

$$P_1 = 500 - (240 - \frac{1}{2}t) = 260 + \frac{1}{2}t,$$

so this price goes up. A calculation shows that

$$\text{profit}_{\text{old}} - \text{profit}_{\text{new}} = t^2/2 > 0 \text{ for } t > 0,$$

so the tax causes a deadweight loss.

2.5

1. Total profits are $\pi_c + \pi_p = p_c q_c - (5 + q_c) + p_p q_p - (3 + 2q_p)$. We go over to minimization.

$$f(p) = 5p_c^2 - 8p_p p_c + 6p_p^2 - 26p_c - 24p_p + 69 \to \min,$$

$$p = (p_c, p_p)^T \in \mathbb{R}^2.$$

Existence of a global solution follows, using problem 2.2 and the fact that the degree two part of $f(p)$ is

$$5p_c^2 - 8p_p p_c + 6p_p^2 = p_c^2 + 4(p_c - p_p)^2 + 2p_p^2 > 0$$

for all (p_p, p_c). Therefore, writing

$$A = \begin{pmatrix} -5 & 4 \\ 4 & -6 \end{pmatrix} \text{ and } b = (-13 \ -12)^T,$$

the given problem has a unique point of global minimum

$$(p_c, p_p)^T = -A^{-1}b = (9 \ 8)^T,$$

as a matrix calculation shows. This leads to the following values under monopoly: for Coca $p_c = 9$, $q_c = 16$, $\pi_c = 123$ and for Pepsi $p_p = 8$, $q_p = 4$, $\pi_p = 21$.

2. We consider Coca's (resp. Pepsi's) problem to choose the price p_c (resp. p_p) that maximizes its own profit π_c (resp. π_p) for a given fixed price p_p (resp. p_c) of its competitor. This is a one-dimensional optimization problem, which turns out to have a unique solution $p_c^{\text{new}} = \frac{2}{5}p_p + 3\frac{2}{5}$ (resp. $p_p^{\text{new}} = \frac{1}{3}p_c + 2\frac{1}{3}$). The equilibrium solution is $\bar{p}_c = 5$, $\bar{p}_p = 4$. This suggests writing the reaction relations above as follows:

$$p_c^{\text{new}} - 5 = \frac{2}{5}(p_p^{\text{old}} - 4),$$

$$p_p^{\text{new}} - 4 = \frac{1}{3}(p_c^{\text{old}} - 5).$$

SOLUTIONS OF EXERCISES OF CHAPTERS 1–4   619

This shows that, if Coca and Pepsi have each optimized $2k$ times, we have that $p_c - 5$ and $p_p - 4$ are both equal to $4(\frac{2}{15})^k$. In particular, we get convergence to the equilibrium solution. In the same way production levels and profits converge to the equilibrium levels.

3. We have, in the case of monopoly,
$$(p_c, p_p, q_c, q_p, \pi_c, \pi_p) = (9, 8, 16, 4, 123, 21),$$
and, in the case of duopoly,
$$(p_c, p_p, q_c, q_p, \pi_c, \pi_p) = (5, 4, 20, 12, 75, 21).$$

We see that competition has two beneficial effects: higher production levels and much lower prices.

2.6

1. We consider the problem $(40X - X^2) + (10Y - (Y^2 + 0.05X)) \to$ max. This problem has the unique solution $X = 19.975$ and $Y = 5$.

2. We consider the problems $40X - X^2 \to$ max and $10Y - Y^2 \to$ max; the reason that we have omitted the term $0.05X$ in the optimization problem of the laundry is that the laundry cannot influence X. The solutions of these problems are $X = 20$ and $Y = 5$. Therefore, the answer to the question is: no.

3. The per-unit tax should be 0.05.

2.7

1. $f(x) = \sum_{i=1}^{3} m_i |x - x_i| \to$ min, for given positive numbers m_i and given vectors $x_i \in \mathbb{R}^3$, where $1 \leq i \leq 3$. Existence of a global solution $\hat{x}$ follows from the coercivity of f ($f(x) \approx 3|x|$ for $|x|$ sufficiently large).

2. Fermat: $f'(x) = 0_3^T \Rightarrow \sum_{i=1}^{3} m_i \frac{x - x_i}{|x - x_i|} = 0_3$.

3. If there exists a—necessarily—unique triangle with sides m_1, m_2, and m_3, and $\angle x_i x_j x_k = \alpha_j$ if $\{i, j, k\} = \{1, 2, 3\}$, then we get an interior solution. Otherwise, the global solution is found by comparing the f-values at x_1, x_2, x_3.

2.8 $|Ax - b| \to$ min. Taking the square of the objective function and expanding the brackets gives the quadratic problem

$$f(x) = x^T A^T A x - 2b^T A x + b^T b \to \min,$$

using the following rules for taking transposes: $(CD)^T = D^T C^T$, $C^{TT} = C$, and $c^T = c$ for a number c (viewed as a 1×1 matrix). The matrix $A^T A$ is positive definite, as $x^T A^T A x = |Ax|^2 > 0$ for all $x \neq 0_n$, using that $Ax \neq 0_m$ as the rows of A are linearly independent. It follows, on using problem 2.2, that the given problem has a unique solution

$$\widehat{x} = (A^T A)^{-1} A^T b.$$

2.9 Formulation. Each of the two parts into which a given simplex in $\mathbb{R}^n$—the simplex, with vertices $x_1, \ldots, x_{n+1} \in \mathbb{R}^n$ is the set

$$\{\lambda_1 x_1 + \cdots + \lambda_{n+1} x_{n+1} : \lambda_1 + \cdots + \lambda_m = 1, \ \lambda_i \geq 0, \ 1 \leq i \leq m\}$$

—is divided by a hyperplane through its center of gravity has n-dimensional volume between $(n/(n+1))^n$ and $1 - (n/(n+1))^n$ times the n-dimensional volume of the given simplex. These constants cannot be improved. If n tends to infinity, then $(n/(n+1))^n$ tends to the positive constant e^{-1}, by the well-known limit $\lim_{n \to \infty}(1 + (1/n))^n = e$.

Proof. The problem can be modeled as an optimization problem in the same way as was done in the case $n = 2$: see the hint for exercise 1.27. That is, we consider the simplex in $\mathbb{R}^n$ with vertices $0_n, e_1, \ldots, e_n$ and write the equation of a hyperplane through the center of gravity $(1/n - 1, \ldots, 1/n - 1)^T$ as $a_1 x_1 + \cdots + a_n x_n = 1$. This leads to the following optimization problem

$$g(a) = \left(\prod_{i=1}^n \frac{n}{a_i} \right) \to \min, \ \sum_{i=1}^n a_i = n, \ a_i > 0, \ 1 \leq i \leq n.$$

This is readily found to have a unique global solution $a_i = 1$, $1 \leq i \leq n$, and so the optimal value is $(n/(n+1))^n$. It remains to use again the well-known limit $e = \lim_{n \to \infty}(1 + \frac{1}{n})^n$.

2.10 See theorem 6.15 and its proof.

2.11 A calculation of the integral brings the problem into the following form:

$$f(x) = 2x_1^2 + \frac{2}{3}x_2^2 - \frac{4}{3}x_1 + \frac{2}{5} \to \min.$$

SOLUTIONS OF EXERCISES OF CHAPTERS 1–4 621

This has a unique solution, $x_1 = \frac{1}{3}$, $x_2 = 0$. That is, the optimal polynomial is $t^2 - \frac{1}{3}$. Note that this agrees with the Rodrigues formula from theorem 2.12.

2.12 In the same way as in the previous exercise, the optimal polynomial is found to be the one given by the Rodrigues formula, $t^3 - \frac{3}{5}t$.

2.14 We have

$$f(\widehat{x} + \alpha \bar{x}) = f(\widehat{x}) + f'(\widehat{x}) \cdot (\alpha \bar{x}) + o(\alpha)$$

by the definition of the derivative. This equals $f(\widehat{x}) - \alpha + o(\alpha)$ by the definition of $\bar{x}$. Therefore, $f(x_\alpha) < f(\widehat{x})$ for sufficiently small positive α and this proves that $\widehat{x}$ is not a point of local minimum.

Chapter three We will not display the—easy—justifications for putting $\lambda_0 = 1$, except in the first two exercises.

3.1

1. $f_0(x) = e^{x_1 x_2} \to$ max, $f_1(x) = x_1^3 + x_2^3 - 1 = 0$.

   Existence of a global maximum $\widehat{x}$ follows by coercivity $(x_1 \sqrt[3]{1 - x_1^3} \approx -x_1^2$ for $|x_1|$ sufficiently large, and so $e^{x_1 \sqrt[3]{(1-x_1^3)}} \to 0$ for $|x_1| \to \infty$).

2. Lagrange function: $\mathcal{L} = \lambda_0 e^{x_1 x_2} + \lambda_1 (x_1^3 + x_2^3 - 1)$.

   Lagrange: $\mathcal{L}_x = 0_2^T \Rightarrow$

   $$\frac{\partial \mathcal{L}}{\partial x_1} = \lambda_0 x_2 e^{x_1 x_2} + 3\lambda_1 x_1^2 = 0, \quad \frac{\partial \mathcal{L}}{\partial x_2} = \lambda_0 x_1 e^{x_1 x_2} + 3\lambda_1 x_2^2 = 0.$$

   We put $\lambda_0 = 1$, as we may: if $\lambda_0 = 0$, then $\lambda_1 \neq 0$ and then the Lagrange equations would give $x_1 = x_2 = 0$, contradicting the equality constraint.

3. Eliminate λ_1: $x_1^3 = x_2^3 \Rightarrow x_1 = x_2$. Therefore, $x_1^3 = x_2^3 = 1/2$, using the equality constraint, and so $x_1 = x_2 = \frac{1}{2}\sqrt[3]{4}$.

4. $\widehat{x} = (\frac{1}{2}\sqrt[3]{4}, \frac{1}{2}\sqrt[3]{4})^T$.

3.2

1. $f_0(x) = x_1^2 + 12 x_1 x_2 + 2 x_2^2 \to$ extr, $f_1(x) = 4x_1^2 + x_2^2 - 25 = 0$.
   Global extrema exist by Weierstrass.

2. Lagrange function: $\mathcal{L} = \lambda_0 (x_1^2 + 12 x_1 x_2 + 2 x_2^2) + \lambda_1 (4x_1^2 + x_2^2 - 25)$,

Lagrange: $\mathcal{L}_x = 0_2^T \Rightarrow$

$$\frac{\partial \mathcal{L}}{\partial x_1} = \lambda_0(2x_1 + 12x_2) + \lambda_1(8x_1) = 0,$$

$$\frac{\partial \mathcal{L}}{\partial x_2} = \lambda_0(12x_1 + 4x_2) + \lambda_1(2x_2) = 0.$$

We put $\lambda_0 = 1$, as we may: if $\lambda_0 = 0$, then $\lambda_1 \neq 0$ and $x_1 = x_2 = 0$, contradicting the constraint.

3. Eliminate λ_1:

$$x_1 x_2 + 6x_2^2 = 24x_1^2 + 8x_1 x_2.$$

This can be rewritten as

$$6\left(\frac{x_2}{x_1}\right)^2 - 7\left(\frac{x_2}{x_1}\right) - 24 = 0,$$

provided $x_1 \neq 0$. This gives $x_2 = \frac{8}{3}x_1$ or $x_2 = -\frac{3}{2}x_1$. In the first (resp. second) case we get $x_1 = \pm\frac{3}{2}$ and so $x_2 = \pm 4$ (resp. $x_1 = \pm 2$ and so $x_2 = \mp 3$; we use the sign $\mp$ as x_2 has a different sign than x_1), using the equality constraint.

Compare: $f_0(2, -3) = f_0(-2, 3) = -50$ and

$$f_0(\frac{3}{2}, 4) = f_0(-\frac{3}{2}, -4) = 106\frac{1}{4}.$$

4. $(2, -3)$ and $(-2, 3)$ are global minima and $(\frac{3}{2}, 4)$ and $(-\frac{3}{2}, -4)$ global maxima.

3.3

1. $f_0(x) = x_1 x_2^2 x_3^3$, $f_1(x) = x_1^2 + x_2^2 + x_3^2 - 1 = 0$, $x \in \mathbb{R}_+^3$.

   Existence of global extrema follows from Weierstrass.

2. Lagrange function: $\mathcal{L} = \lambda_0 x_1 x_2^2 x_3^3 + \lambda_1(x_1^2 + x_2^2 + x_3^2 - 1)$.
   Lagrange: $\mathcal{L}_x = 0_3^T \Rightarrow$

$$\lambda_0 x_2^2 x_3^3 + 2\lambda_1 x_1 = 0, \ 2\lambda_0 x_1 x_2 x_3^3 + 2\lambda_1 x_2 = 0,$$

$$3\lambda_0 x_1 x_2^2 x_3^2 + 2\lambda_1 x_3 = 0.$$

We put $\lambda_0 = 1$, as we may.

3. We eliminate λ_1 from the stationarity equations: this gives

$$2x_1^2 = x_2^2 = \frac{2}{3}x_3^2$$

(we may ignore the possibility that x_1, x_2, or x_3 is zero; such points have f_0-value zero and so cannot be global extrema, clearly). This leads to the following stationary points: the points (x_1, x_2, x_3) with $x_1^2 = \frac{1}{6}$ and so $x_2^2 = \frac{1}{3}$, $x_3^2 = \frac{1}{2}$, using the equality constraint. Compare: the eight stationary points we have found all have the same value (up to sign).

4. There are eight global extrema $(\pm 1/\sqrt{6}, \pm 1/\sqrt{3}, \pm 1/\sqrt{2})$. Such a point is a point of global minimum (resp. maximum) if x_1 and x_3 have different sign (resp. the same sign).

3.4

1. $f_0(x) = x_3 \to$ extr,

$$f_1(x) = x_1^2 + x_2^2 + x_3^2 - x_1 x_3 - x_2 x_3 + 2x_1 + 2x_2 + 2x_3 - 2 = 0.$$

Existence of global extrema $\widehat{x}_{\max}$ and $\widehat{x}_{\min}$ follows from Weierstrass, as the quadratic part of the quadratic polynomial f_1 is positive definite—note that

$$f_1(x) = \frac{1}{2}(x_1 - x_3)^2 + \frac{1}{2}(x_2 - x_3)^2 + \frac{1}{2}x_1^2 + \frac{1}{2}x_2^2$$

—and so coercive. Therefore, the solution set of $f_1(x) = 0$ is bounded.

2. Lagrange function:

$$\mathcal{L} = \lambda_0 x_3 + \lambda_1(x_1^2 + x_2^2 + x_3^2 - x_1 x_3 - x_2 x_3 + 2x_1 + 2x_2 + 2x_3 - 2),$$

Lagrange: $\mathcal{L}_x = 0_3^T \Rightarrow$

$$\lambda_1(2x_1 - x_3 + 2) = 0, \; \lambda_1(2x_2 - x_3 + 2) = 0,$$

$$\lambda_0 + \lambda_1(2x_3 - x_1 - x_2 + 2) = 0.$$

We put $\lambda_0 = 1$, as we may.

3. The third Lagrange equation implies $\lambda_1 \neq 0$. Then the other Lagrange equations give $x_2 = x_1$ and $x_3 = 2x_1 + 2$. Substitute in the equality constraint: $2x_1^2 + 12x_1 + 6 = 0 \Rightarrow x_1 = -3 \pm \sqrt{6} \Rightarrow$

$$x = (-3 + \sqrt{6}, -3 + \sqrt{6}, -4 + 2\sqrt{6})^T$$

or $x = (-3 - \sqrt{6}, -3 - \sqrt{6}, -4 - 2\sqrt{6})^T$.

Compare: $f((-3 + \sqrt{6}, -3 + \sqrt{6}, -4 + 2\sqrt{6})^T)$
$> f((-3 - \sqrt{6}, -3 - \sqrt{6}, -4 - 2\sqrt{6})^T)$.

4. $x_{\max} = (-3 + \sqrt{6}, -3 + \sqrt{6}, -4 + 2\sqrt{6})^T$,
$\hat{x}_{\min} = (-3 - \sqrt{6}, -3 - \sqrt{6}, -4 - 2\sqrt{6})^T$.

3.5

1. Existence of global extrema follows from the Weierstrass theorem: the constraint $\sum_{i=1}^{5} x_i^2 = 4$ implies that the admissible set is bounded and the other assumptions are even simpler to verify.

2. Lagrange function:

$$\mathcal{L} = \lambda_0 \left(\sum_{i=1}^{5} x_i^4 \right) + \lambda_1 \left(\sum_{i=1}^{5} x_i \right)$$

$$+ \lambda_2 \left(\sum_{i=1}^{5} x_i^3 \right) + \lambda_3 \left(\left(\sum_{i=1}^{5} x_i^2 \right) - 4 \right),$$

Lagrange: $\mathcal{L}_x = 0_5^T \Rightarrow$

$$\frac{\partial \mathcal{L}}{\partial x_i} = 4\lambda_0 x_i^3 + \lambda_1 + 3\lambda_2 x_i^2 + 2\lambda_3 x_i = 0, \ 1 \le i \le 5.$$

3. How to use these equations? Well, the Lagrange equations give the following information: the coordinates of x are solutions of one polynomial equation of degree at most 3! Therefore, either three coordinates are equal, say $x_1 = x_2 = x_3$, or two pairs of coordinates are equal, say $x_1 = x_2$, $x_3 = x_4$.

    (a) In the first case, we get

    $$3x_1 + x_4 + x_5 = 0, \ 3x_1^3 + x_4^3 + x_5^3 = 0, \ 3x_1^2 + x_4^2 + x_5^2 = 4.$$

    We eliminate x_1 from the first two equations: this gives

    $$-(x_4 + x_5)^3 + 9x_4^3 + 9x_5^3 = 0.$$

    We divide by x_5^3 and expand; this gives that $\frac{x_4}{x_5}$ is a solution of the equation

    $$8y^3 - 3y^2 - 3y + 8 = 0.$$

The left-hand side of this equation is the product of $y+1$ and $8y^2-11y+8$, which has a negative discriminant. Therefore, it follows that $\frac{x_4}{x_5} = -1$. This leads to the following stationary points:

- $v_1 = (0,0,0,\sqrt{2},-\sqrt{2})$,
- $v_2 = (0,0,0,-\sqrt{2},\sqrt{2})$.

(b) In the second case, we get

$$2x_1+2x_3+x_5 = 0, \quad 2x_1^3+2x_3^3+x_5^3 = 0, \quad 2x_1^2+2x_3^2+x_5^2 = 4.$$

We proceed in a similar way as in the first case, now beginning with elimination of x_5 from the first two equations. This gives that $\frac{x_3}{x_1}$ is a solution of the equation

$$(y+1)(y^2+3y+1) = 0,$$

and so $\frac{x_3}{x_1}$ equals

$$-1, \quad -\frac{3}{2}+\frac{1}{2}\sqrt{5} \quad \text{or} \quad -\frac{3}{2}-\frac{1}{2}\sqrt{5}.$$

Substituting each of these three possibilities into the remaining equality constraint gives the following values for x_1:

$$\pm 1, \quad \pm(\frac{1}{10}\sqrt{10}+\frac{1}{2}\sqrt{2}), \quad \pm(\frac{1}{10}\sqrt{10}-\frac{1}{2}\sqrt{2}).$$

This involves conquering an obstacle: one has to calculate the square root of $\frac{3}{5}+\frac{1}{5}\sqrt{5}$. This equals $\frac{1}{10}\sqrt{10}+\frac{1}{2}\sqrt{2}$ and this can be written more compactly as $10^{-\frac{1}{2}}+2^{-\frac{1}{2}}$. It is very easy to check this outcome, but how to discover it is another matter: this requires methods from the theory of numbers. This leads to the following additional stationary points:

- $v_3 = (1,1,-1,-1,0)$,
- $v_4 = (-1,-1,1,1,0)$,
- $v_5 = (10^{-\frac{1}{2}}+2^{-\frac{1}{2}}, 10^{-\frac{1}{2}}+2^{-\frac{1}{2}}, 10^{-\frac{1}{2}}+2^{-\frac{1}{2}}, 10^{-\frac{1}{2}}+2^{-\frac{1}{2}}, \frac{2}{5}\sqrt{10})$,
- $v_6 = (-10^{-\frac{1}{2}}-2^{-\frac{1}{2}}, -10^{-\frac{1}{2}}-2^{-\frac{1}{2}}, -10^{-\frac{1}{2}}-2^{-\frac{1}{2}}, -10^{-\frac{1}{2}}-2^{-\frac{1}{2}}, -\frac{2}{5}\sqrt{10})$,
- $v_7 = (10^{-\frac{1}{2}}-2^{-\frac{1}{2}}, 10^{-\frac{1}{2}}-2^{-\frac{1}{2}}, 10^{-\frac{1}{2}}-2^{-\frac{1}{2}}, 10^{-\frac{1}{2}}-2^{-\frac{1}{2}}, \frac{2}{5}\sqrt{10})$,
- $v_8 = (-10^{-\frac{1}{2}}+2^{-\frac{1}{2}}, -10^{-\frac{1}{2}}+2^{-\frac{1}{2}}, -10^{-\frac{1}{2}}+2^{-\frac{1}{2}}, -10^{-\frac{1}{2}}+2^{-\frac{1}{2}}, -\frac{2}{5}\sqrt{10})$.

4. Comparison of all eight stationary points shows that the points of global minimum (resp. maximum) are v_5, v_6, v_7, v_8 (resp. v_1, v_2).

3.6 The problem can be modeled as follows, writing $x_1 = x$ and $x_2 = y$:

1. $f_0(x) = 3\ln x_1 + 2\ln x_2 \to$ max, $f_1(x) = \frac{1}{2}x_1 + x_2 - 5 = 0$, $x_1, x_2 > 0$.

   Existence of a global maximum $\widehat{x}$ follows, as $3\ln x_1 + 2\ln(5 - \frac{1}{2}x_1) \to -\infty$ for $x_1 \downarrow 0$ and for $x_1 \uparrow 10$.

2. Lagrange function: $\mathcal{L} = \lambda_0(3\ln x_1 + 2\ln x_2) + \lambda_1(\frac{1}{2}x_1 + x_2 - 5)$.
   Lagrange: $\mathcal{L}_x = 0_2^T$
   $$\Rightarrow \frac{\partial \mathcal{L}}{\partial x_1} = \frac{3\lambda_0}{x_1} + \frac{\lambda_1}{2} = 0, \quad \frac{\partial \mathcal{L}}{\partial x_2} = \frac{2\lambda_0}{x_2} + \lambda_1 = 0.$$

   We put $\lambda_0 = 1$, as we may.

3. Eliminate λ_1: $x_1 = 3x_2$. Use constraint: $x = (6,2)^T$.

4. It is optimal to play six games and to eat two ice creams.

3.7 We consider the general problem $3\ln x_1 + 2\ln x_2 \to$ max, $ax_1 + bx_2 = c$, $x_1, x_2 > 0$ for given positive constants a, b, c. We could solve this problem in a similar way as in the special case $a = \frac{1}{2}, b = 1, c = 5$ (see the previous exercise). This would give that there is a unique solution $(x_1, x_2) = (3c/5a, 2c/5b)$. This formula would show the dependence of the solution on a, b, and c if x_1 and x_2 were allowed to be fractional numbers. However, this does not make sense and the best we can say at this point is that the formula gives an approximation, from which the integral solution can be found by a non-straightforward "rounding off procedure."

3.8 (i) To simplify the calculations, we reformulate the constraint
$$10X^{1/2}Y^{1/2} = 90$$
as $\ln X + \ln Y = \ln 81$, by taking the logarithm. The question is whether the solution $(\widehat{X}, \widehat{Y})$ of the following problem has $\widehat{X} = \widehat{Y}$:

1. $8X + 16Y \to$ min, $\ln X + \ln Y = \ln 81$, $X, Y > 0$.

   Existence of a global solution follows from coercivity.

SOLUTIONS OF EXERCISES OF CHAPTERS 1–4

2. Lagrange function: $\mathcal{L} = \lambda_0(8X + 16Y) + \lambda_1(\ln X + \ln Y - \ln 81)$.
Lagrange: $\frac{\partial \mathcal{L}}{\partial X} = \frac{\partial \mathcal{L}}{\partial Y} = 0 \Rightarrow$

$$8\lambda_0 + \frac{\lambda_1}{X} = 0, \quad 16\lambda_0 + \frac{\lambda_1}{Y} = 0.$$

We put $\lambda_0 = 1$, as we may.

3. We eliminate λ_1: this gives $2Y = X$.

4. It is optimal to use twice as much units of input X than input Y. In particular, the proposal to use equal amounts is not optimal.

(ii) Let p (q) be the—positive—price of X (Y) and let C be a given constant. Then the problem is

$$pX + qY \to \min, \quad \ln X + \ln Y = C, \quad X, Y > 0.$$

Proceeding in a similar way as in (i), we get $pX = qY$. That is, it is optimal to spend equal amounts of money on input X and input Y. In particular, the input combination $X = 9$, $Y = 9$ is efficient if the prices of the two inputs are equal.

(iii) Using the result of (ii) we should have $X = 2Y$ in the optimal situation. This would give that there is a unique point of minimum, $(X, Y) = (40\sqrt{2}, 20\sqrt{2})$, if it made sense to have fractional solutions. Therefore, a precise analysis of this problem should include a rounding off procedure.

3.9 (i)

1. We consider the problem

$$10\sqrt{L_X} + 5\sqrt{L_Y} \to \max, \quad L_X + L_Y = 100, \quad L_X, L_Y \geq 0.$$

Existence of a global solution $(\widehat{L}_X, \widehat{L}_Y)$ follows from Weierstrass.

2. Lagrange function: $\mathcal{L} = \lambda_0(10\sqrt{L_X} + 5\sqrt{L_Y}) + \lambda_1(L_X + L_Y - 100)$.
Lagrange: $\frac{\partial \mathcal{L}}{\partial L_X} = \frac{\partial \mathcal{L}}{\partial L_Y} = 0 \Rightarrow$

$$\frac{10\lambda_0}{2\sqrt{L_X}} + \lambda_1 = 0, \quad \frac{5\lambda_0}{2\sqrt{L_Y}} + \lambda_1 = 0.$$

We put $\lambda_0 = 1$, as we may.

3. Eliminate λ_1: $2\sqrt{L_Y} = \sqrt{L_X}$ and so $4L_Y = L_X$. Use constraint: $L_Y = 20, L_X = 80$. It follows that

$$(\hat{L}_X, \hat{L}_Y) \in \{(80, 20), (100, 0), (0, 100)\}.$$

Comparison shows $(\hat{L}_X, \hat{L}_Y) = (80, 20)$.

4. Conclusion: if the worker wants to maximize profit, then it is optimal to allocate 80 hours to wallpapering and 20 hours to painting.

(ii) We consider the problem

$$10\sqrt{XY} \to \max, \ L_X + L_Y = 100, \ X = \sqrt{L_X}, \ Y = \sqrt{L_Y}, \ L_X, L_Y \geq 0.$$

This is seen to have the unique solution $L_X = L_Y = 50$. Conclusion: if the worker wants to maximize utility of consumption, then it is optimal to allocate 50 hours to wallpapering as well as to painting.

(iii) We consider the problem

$$10\sqrt{\tilde{X}\tilde{Y}} \to \max, \ 10\tilde{X} + 5\tilde{Y} = 10X + 5Y, \ X = \sqrt{L_X}, \ Y = \sqrt{L_Y},$$

$$L_X + L_Y = 100, \ L_X, L_Y \geq 0, \ \tilde{X}, \tilde{Y} \geq 0.$$

Here L_X (resp. L_Y) is the time allocated to wallpapering (resp. painting), X (resp. Y) is the total amount of wallpapering (resp. painting) the worker is doing in this time, and $\tilde{X}$ (resp. $\tilde{Y}$) is the total amount of wallpapering (resp. painting) that is done in his own house by himself and others. This problem can be reduced to the following one:

1.

$$f(\tilde{X}, \tilde{Y}, L_X, L_Y) = \ln \tilde{X} + \ln \tilde{Y} \to \max, \ L_X + L_Y = 100,$$

$$10\tilde{X} + 5\tilde{Y} = 10\sqrt{L_X} + 5\sqrt{L_Y}, \ L_X, L_Y > 0.$$

Existence of a global solution follows from Weierstrass.

2. Lagrange function: $\mathcal{L} =$

$$\lambda_0(\ln \tilde{X} + \ln \tilde{Y}) + \lambda_1(L_X + L_Y - 100)$$

$$+ \lambda_2(10\tilde{X} + 5\tilde{Y} - (10\sqrt{L_X} + 5\sqrt{L_Y})).$$

Lagrange: $\frac{\partial \mathcal{L}}{\partial \tilde{X}} = \frac{\partial \mathcal{L}}{\partial \tilde{Y}} = \frac{\partial \mathcal{L}}{\partial L_X} = \frac{\partial \mathcal{L}}{\partial L_Y} = 0 \Rightarrow$ (we have immediately put $\lambda_0 = 1$)

$$\tilde{X}^{-1} + 10\lambda_2 = 0, \ \tilde{Y}^{-1} + 5\lambda_2 = 0,$$

$$\lambda_1 - \lambda_2 10/\sqrt{L_X} = 0, \ \lambda_1 - \lambda_2 5/(2\sqrt{L_Y}) = 0.$$

3. We eliminate the other Lagrange multipliers. This gives $\tilde{Y} = 2\tilde{X}$ and $L_X = 4L_Y$. Therefore, $L_X = 80, \ L_Y = 20, \ \tilde{X} = \frac{5}{2}\sqrt{5}, \ \tilde{Y} = 5\sqrt{5}$, using the equality constraints.

4. Conclusion: if the worker wants to optimize utility, and trade is possible, then the following is optimal. To begin with, 80 hours should be allocated to wallpapering and 20 hours to painting. Moreover, he should do an amount of wallpapering $\frac{5}{2}\sqrt{5}$ in his own house and an amount $4\sqrt{5} - \frac{5}{2}\sqrt{5} = \frac{3}{2}\sqrt{5}$ for other people; thus he will be a net supplier of wallpapering. Finally, he should do an amount of painting $2\sqrt{5}$ in his own house and he should let someone else do an additional amount of $5\sqrt{5} - 2\sqrt{5} = 3\sqrt{5}$ in his house; thus he is a net demander of painting.

(iv) Profit maximization leads to larger production (measured in money) than utility maximization without trade: indeed $50\sqrt{5} > 75\sqrt{2}$. Moreover, utility maximization with trade leads to the same allocation as profit maximization and to a higher utility than utility maximization without trade: indeed $50\sqrt{\frac{5}{2}} > 50\sqrt{2}$.

3.10 (i)

1. The problem can be modeled as follows:

$$4\ln(l_1)+\ln(t_1)+5\ln(l_2)+20\ln(t_2) \to \max, \ l_1+l_2 = 13, \ t_1+t_2 = 22,$$

$$l_1, l_2, t_1, t_2 > 0.$$

Existence of a global maximum $(\widehat{l}_1, \widehat{t}_1, \widehat{l}_2, \widehat{t}_2)$ follows from "coercivity" for maximization (elimination of l_2 and t_2 makes this clear).

2. Lagrange function: $\mathcal{L} =$

$$\lambda_0(4\ln(l_1) + \ln(t_1) + 5\ln(l_2)$$

$$+20\ln(t_2)) + \lambda_1(l_1 + l_2 - 13) + \lambda_2(t_1 + t_2 - 22).$$

Lagrange: $\frac{\partial \mathcal{L}}{\partial l_1} = \frac{\partial \mathcal{L}}{\partial t_1} = \frac{\partial \mathcal{L}}{\partial l_2} = \frac{\partial \mathcal{L}}{\partial t_2} = 0 \Rightarrow$

$$\frac{4\lambda_0}{l_1} + \lambda_1 = 0, \quad \frac{\lambda_0}{t_1} + \lambda_2 = 0, \quad \frac{5\lambda_0}{l_2} + \lambda_1 = 0, \quad \frac{20\lambda_0}{t_2} + \lambda_2 = 0.$$

We put $\lambda_0 = 1$, as we may.

3. Eliminate the Lagrange multipliers: $5l_1 = 4l_2$, $20t_1 = t_2$. Substitute into the constraints: $(l_1, t_1, l_2, t_2) = (5\frac{7}{9}, 1\frac{1}{21}, 7\frac{2}{9}, 20\frac{20}{21})$.

4. $(\widehat{l_1}, \widehat{t_1}, \widehat{l_2}, \widehat{t_2}) = (5\frac{7}{9}, 1\frac{1}{21}, 7\frac{2}{9}, 20\frac{20}{21})$.

However, a nonintegral solution is not allowed, as we do not want to tear the albums apart. Rounding off to the nearest integers leads to $(\widehat{l_1}, \widehat{t_1}, \widehat{l_2}, \widehat{t_2}) = (6, 1, 7, 21)$ This happens to be the optimal integral point.

Optional. For this problem we display the verification that the integral point we have found is the optimal one. We will carry out the transition to an integral solution for the optimal distribution of the Spider-Man comic books, and then for the optimal distribution of the X-Men comic books. We can eliminate l_2, using the constraints, $l_2 = 13 - l_1$. Thus the Spider-Man comic books problem can be written as an unconstrained problem

$$g(l_1) = 4 \ln l_1 + 5 \ln(13 - l_1) \to \max.$$

The derivative $g'(l_1) = \frac{4}{l_1} - \frac{5}{13 - l_1}$ is seen to be a monotonic decreasing function. Therefore, $g'_1(l_1) > 0$ for $l_1 < 5\frac{7}{9}$ and $g'_1(l_1) < 0$ for $l_1 > 5\frac{7}{9}$. This shows that the optimal integral solution can be obtained by comparing $l_1 = 5$ and $l_1 = 6$: $4 \ln 5 + 5 \ln 8 = 16.83$ and $4 \ln 6 + 5 \ln 7 = 16.90$, and so $(l_1, l_2) = (6, 7)$ is the optimal "integral" way to distribute the Spider-Man comic books. The same method can be used for the X-Men comic book problem, and this leads to a comparison of $t_1 = 1$ and $t_1 = 2$—the two possibilities of rounding off $1\frac{1}{21}$: $\ln 1 + 20 \ln 21 = 60.89$ and $\ln 2 + 20 \ln 20 = 60.61$, and so $(t_1, t_2) = (1, 21)$ is the optimal "integral" way to distribute the X-Men comic books. Conclusion: it is optimal that the X-men fan gives away 1 of his comic books and the Spider-Man fan 7 of his.

(ii) The distribution might seem unfair to the Spider-Man fan: he gets only one of the 22 X-Men comic books, and he has to give away 7, that is, more than half of his 13 Spider-Man comic books.

(iii) We compare $(l_1, l_2) = (6, 8)$ and $(l_1, l_2) = (7, 7)$: $4 \ln 6 + 5 \ln 8 > 4 \ln 7 + 5 \ln 7$, and so it is optimal to give the additional Spider-Man comic book to the X-Men fan.

SOLUTIONS OF EXERCISES OF CHAPTERS 1-4   631

3.11 (i) On each of the n periods between successive moments of ordering, the graph of the inventory level decreases linearly from level x to level 0.

(ii) The problem can be modeled as follows

1. $f(x, n) = C_h x/2 + C_0 n \to \min$, $nx = A$, $n, A \geq 0$. Existence of a global solution $(\hat{x}, \hat{n})$ follows from Weierstrass.

2. Lagrange function: $\mathcal{L} = \lambda_0(\frac{C_h x}{2} + C_0 n) + \lambda_1(nx - A)$.

   Lagrange: $\frac{\partial \mathcal{L}}{\partial x} = \frac{\partial \mathcal{L}}{\partial n} = 0 \Rightarrow$

   $$C_h/2 + \lambda_1 n = 0, \quad \lambda_0 C_0 + \lambda_1 x = 0.$$

   We put $\lambda_0 = 1$, as we may.

3. Eliminate λ_1: $C_h x = 2 C_0 n$. Use constraint:

   $$(x, n) = (\sqrt{\frac{2 C_0 A}{C_h}}, \sqrt{\frac{C_h A}{2 C_0}}).$$

4. $(\hat{x}, \hat{n}) = (\sqrt{\frac{2 C_0 A}{C_h}}, \sqrt{\frac{C_h A}{2 C_0}})$.

It remains to take into account that n has to be an integer. To this end the value of the objective function has to be compared at two admissible points, corresponding to whether we round up or down. Finally we consider the second question. The Lagrange multiplier measures how sensitive the optimal costs are for the size of the annual requirement for the commodity.

3.12 The first error is that a point of local extremum of the constrained problem $f(x, y) \to$ extr, $g(x, y) = 0$ satisfies the same necessary conditions as a point of local extremum of the unconstrained problem $f(x, y) \to$ extr. This is "usually" not the case. The second error is the suggestion that the motivation for the Lagrange method is that it is more convenient than solving the system of equations consisting of the necessary conditions, $f_x = f_y = 0$, of the unconstrained problem, and the equality constraint $g = 0$ of the constrained problem $f(x, y) \to$ extr, $g(x, y) = 0$. However, the correct motivation is that the Lagrange method gives a system of equations satisfied by all local extrema of the given constrained problem and that the other system does not have this property.

3.13

1. $f_0(x) = 2x_1 + 3x_2 \to \max$, $f_1(x) = \sqrt{x_1} + \sqrt{x_2} - 5 = 0$. Existence of a solution follows from the Weierstrass theorem.

2. Lagrange function: $\mathcal{L} = \lambda_0(2x_1 + 3x_2) + \lambda_1(\sqrt{x_1} + \sqrt{x_2} - 5)$.
   Lagrange: $\mathcal{L}_x = 0_2^T \Rightarrow$

   $$\frac{\partial \mathcal{L}}{\partial x_1} = 2\lambda_0 + \lambda_1/(2\sqrt{x_1}) = 0, \quad \frac{\partial \mathcal{L}}{\partial x_2} = 3\lambda_0 + \lambda_1/(2\sqrt{x_2}) = 0.$$

   We put $\lambda_0 = 1$, as we may.

3. Eliminate λ_1: $2\sqrt{x_1} = 3\sqrt{x_2} \Rightarrow 4x_1 = 9x_2$. Use constraint: $x = (9, 4)^T$.

4. This seems to imply that the problem has a unique point of global maximum, $(x_1, x_2) = (9, 4)$. However, this point is not the solution of the given problem: for example, the objective function has a much higher value at the admissible point $(0, 25)$, which is in fact the point of global maximum. What is the explanation of this paradox? The Lagrange equations hold only at local extremal points where the following smoothness assumptions hold: the function f_0 is differentiable and the function f_1 is continuously differentiable. The function $\sqrt{x_1} + \sqrt{x_2} - 5$ is not differentiable at the point $(0, 25)$.

3.14 Done.

3.15

1. To prove this inequality, it suffices to show that the optimal value of the problem

   $$f_0(x) = D(x_1, \ldots, x_n) \to \max, \quad f_j(x) = |x_j|^2 - 1 = 0, \ 1 \leq j \leq n,$$

   is 1. Here $x_1, \ldots, x_n$ run over $\mathbb{R}^n$ and $D(x_1, \ldots, x_n)$ denotes the determinant of the $n \times n$ matrix X with columns $x_1, \ldots, x_n$.

   Existence of a global solution follows from Weierstrass.

2. Lagrange function: $\mathcal{L} = \lambda_0 D(x_1, \ldots, x_n) + \sum_{j=1}^n \lambda_j(|x_j|^2 - 1)$.
   Lagrange: $\mathcal{L}_x = 0_n^T \Rightarrow \frac{\partial \mathcal{L}}{\partial x_{ij}} = \lambda_0 \frac{\partial D}{\partial x_{ij}} + 2\lambda_j x_{ij} = 0$.
   We put $\lambda_0 = 2$, as we may.

3. We will derive from this that $XX^T = I_n$. Multiplying the (i, j)-Lagrange equation with x_{ij} and summing over all i gives

   $$\lambda_j = -\sum_{i=1}^n \frac{\partial D}{\partial x_{ij}} x_{ij};$$

the right-hand side equals $-D$, using that $D = D(x_1,\ldots,x_n)$ is linear in x_j. We recall from matrix theory that the product of the matrix X with the $n \times n$ matrix that has at the (j,i)-place $\frac{\partial D}{\partial x_{ij}}$—the companion matrix of X—equals DI_n. Combining this with the Lagrange equations and using that all multipliers are equal to $-D$, we get that $XX^T = I_n$, as required. Taking determinants gives $D^2 = 1$, and so $D = 1$ or $D = -1$.

4. This shows that the optimal value of the problem under consideration is 1; an example of a point of maximum is I_n.

3.16 We give a solution which is a straightforward extension of the solution of example 3.11.

1. In order to establish the inequalities, it suffices to show that the minimal (resp. maximal) value of the problem

$$f(x) = x_1^p + \cdots + x_n^p \to \text{extr}, \quad f_1(x) = x_1^q + \cdots + x_n^q - 1 = 0, \quad x_i \geq 0,$$

$$1 \leq i \leq n,$$

equals 1 (resp. $n^{1-\frac{p}{q}}$) if $p < q$. Existence of global extrema follows from the Weierstrass theorem. Note here that the admissible set is bounded as all nonnegative solutions of the equality constraint $x_1^q + \cdots + x_n^q = 1$ satisfy $0 \leq x_i \leq 1$ for all $1 \leq i \leq n$.

2. Lagrange function: $\mathcal{L} = \lambda_0(x_1^p + \cdots + x_n^p) + \lambda_1(x_1^q + \cdots + x_n^q - 1)$.
Lagrange: $\mathcal{L}_x = 0_n^T \Rightarrow$

$$\frac{\partial \mathcal{L}}{\partial x_i} \lambda_0 p x_i^{p-1} + \lambda_1 q x_i^{q-1} = 0, \ 1 \leq i \leq n.$$

We put $\lambda_0 = 1$, as we may.

3. The Lagrange equations are seen to give the following information, taking into account that all x_i are nonnegative: the variables x_i that are nonzero are all equal. This implies that they are all equal to $m^{-\frac{1}{q}}$, using the equality constraint; here m denotes the number of i for which x_i is nonzero. Substitution in the objective function gives the value $m(m^{-\frac{1}{q}})^p = m^{1-\frac{p}{q}}$. It follows that the minimal (resp. maximal) value of the given problem is the smallest (resp. largest) of the numbers $m^{1-\frac{p}{q}}$, where $m = 1,\ldots,n$.

4. Therefore, this minimal (resp. maximal) value is 1 (resp. $n^{1-\frac{p}{q}}$), using that $1 - \frac{p}{q} > 0$ as $p < q$.

3.17
3.18–3.21 Solutions are similar to the solution of exercise 3.4.
3.22–3.25 Solutions are similar to the solutions of exercise 3.5.

Chapter four.
4.1 At first sight it seems that we have to write down the KKT conditions and to distinguish $2^4 = 16$ cases. However, this is not necessary; we can make use of the special structure of this problem. To begin with, the problem has a unique point of global minimum $(x_1, x_2, x_3) = (0, 0, 0)$, by Weierstrass and strict convexity. Moreover, we make the educated guess that the points of global maximum of this problem are the three points that can be obtained from the admissible point $(12, 0, 0)$ by permutation of the coordinates. We verify this guess:

$$x_1^2 + x_2^2 + x_3^2 \leq (x_1 + x_2 + x_3)^2 = 12^2,$$

which is the value of the objective function at the point $(12, 0, 0)$. Moreover, the inequality is an equality precisely if two of the three coordinates of x are equal to zero.

4.2 The objective function is strictly convex, as it can be written as the sum of functions which are convex for obvious reasons,

$$\frac{1}{2}(x_1^2 + x_2^2) + \frac{1}{2}(x_1 - x_2)^2 + 3|x_1 - x_2 - 2|,$$

and one of which, $\frac{1}{4}(x_1^2 + x_2^2)$, is even strictly convex. We distinguish two cases.

Case (i): there is a solution for which $x_1 - x_2 - 2 = 0$. Then, substituting $x_1 = x_2 + 2$ into the objective function of the given problem, we get

$$(x_2 + 2)^2 - (x_2 + 2)x_2 + x_2^2 = x_2^2 + 2x_2 + 4;$$

putting the derivative equal to zero, gives $2x_2 + 2 = 0$, and so $x_2 = -1$—and $x_1 = 1$. Now it turns out to be useful to compare the objective function with the function that arises from it by replacing the absolute values by ordinary brackets. Calculating the derivative of

$$x_1^2 - x_1 x_2 + x_2^2 - 3(x_1 - x_2 - 2)$$

SOLUTIONS OF EXERCISES OF CHAPTERS 1–4 635

at $x = (1, -1)^T$ gives $(0, 0)$. This shows that $x = (1, -1)$ is a point of global minimum of the convex function

$$x_1^2 - x_1 x_2 + x_2^2 - 3(x_1 - x_2 - 2).$$

Using that $x = (1, -1)$ is a solution of $x_1 - x_2 - 2 = 0$, it follows that $(1, -1)$ is a global minimum of

$$x_1^2 - x_1 x_2 + x_2^2 + 3|x_1 - x_2 - 2|.$$

As the objective function is strictly convex, this is the unique point of global minimum. In particular, we do not have to look into the second case, $x_1 - x_2 - 2 \neq 0$.

4.3

1. $f_0(x) = e^{x_1 - x_2} - x_1 - x_2 \to \min$, $f_1(x) = x_1 + x_2 - 1 \leq 0$, $f_2(x) = -x_1 \leq 0$, $f_3(x) = -x_2 \leq 0$. This is a convex problem with strict convex objective function: this follows from the strict convexity of the function $t \mapsto e^t$.

2. Lagrange function $L(x, \lambda) = \lambda_0(e^{x_1 - x_2} - x_1 - x_2) + \lambda_1(x_1 + x_2 - 1) + \lambda_2(-x_1) + \lambda_3(-x_2)$.

   Karush-Kuhn-Tucker:

   - $L_{x_1} = L_{x_2} = 0 \Rightarrow \lambda_0(e^{x_1 - x_2} - 1) + \lambda_1 - \lambda_2 = \lambda_0(-e^{x_1 - x_2} - 1) + \lambda_1 - \lambda_3 = 0$,
   - $\lambda_i \geq 0$, $i = 1, 2, 3$,
   - $\lambda_1(x_1 + x_2 - 1) = \lambda_2(-x_1) = \lambda_3(-x_2) = 0$.

3. We select as the most promising case that the first two inequality constraints are tight: $x_1 + x_2 - 1 \leq 0$, $-x_1 \leq 0$, and that λ_0 is not zero, say $\lambda_0 = 1$. Then $x_1 = 0$, $x_2 = 1$, $\lambda_3 = 0$, and this leads to $\lambda_1 = 1 + e^{-1} \geq 0$ and $\lambda_2 = 2e^{-1} \geq 0$. That is, all KKT conditions hold.

4. The problem has the unique solution $(0, 1)$.

4.4 *Geometrical solution:* one should make the unique choice of points that have the property that the connecting line is orthogonal to each of the two given lines. We give a construction that leads to this choice: given two lines in space, assume that they do not lie in one plane, as this case is obvious. Take through each of these lines the plane that is parallel to the other line. Thus we get two different planes, each containing a line; we may assume for convenience that these planes are horizontal. Now take the orthogonal projection of each of these lines on the "other" plane. Thus we get two parallel planes and on each of them two intersecting lines. The two points of intersection form the solution of the given problem. One can derive that such a choice is possible and unique from the following facts (we will not display this derivation):

1. The point on a given line which is closest to a given point outside the line is the orthogonal projection of the given point on the given line.

2. The shortest distance between two parallel planes in (three-dimensional) space is the distance between a point on one plane and its orthogonal projection on the other plane.

Analytical solution: let l and m be two given lines in space. We exclude again the easy case that the lines lie in one plane. The problem can be modeled as

$$f(x,y) = |x - y| \to \min, x \in l, \ y \in m.$$

The function is readily seen to be convex; one can even check that it is strictly convex, using that l and m do not lie in one plane. Moreover, the function is seen to be coercive. This proves the existence and uniqueness of the solution $(\widehat{x}, \widehat{y})$. In order to write the stationarity conditions, let v (w) denote a direction vector of l (m). Then $(\alpha, \beta) = (0, 0)$ is the unique solution of the problem

$$f(\alpha, \beta) = |(\widehat{x} + \alpha v) - (\widehat{y} + \beta w)|^2 \to \min.$$

The stationarity conditions give that $\widehat{x} - \widehat{y}$ is orthogonal to v and w. One can use the stationarity conditions as well to give an explicit formula for the solution. Let $r + \alpha v$ ($s + \beta w$) be a parameter representation of the line l (m). Then the characterization of the solution $(r + \widehat{\alpha} v, s + \widehat{\beta} w)$ by the property that $(r + \widehat{\alpha} v) - (s + \widehat{\beta} w)$ is orthogonal to v and w leads to a system of two linear equations in $\widehat{\alpha}$ and $\widehat{\beta}$ and so to an explicit formula for the given problem, which we do not display.

4.5

1. $f(x) = \frac{1}{n+1}(|x - e_1| + \cdots + |x - e_n| + |x|) \to \min, \ x \in \mathbb{R}^n$.

   For $n = 1$ there are infinitely many solutions: all numbers of the closed interval $[0, 1]$. Assume $n \geq 2$.

   The existence of a point of global minimum follows from the coercivity of f: indeed, $f(x) \approx |x|$ if $|x|$ is sufficiently large and so $f(x) \to +\infty$ for $|x| \to +\infty$.

We will prove that there is precisely one point of global minimum $\widehat{x}$. To prove this, it suffices to show that f is strictly convex. To begin with, the problem is seen to be a convex problem, using the convexity of the modulus function. Moreover, the objective function is even strictly convex. Otherwise there would be a line in $\mathbb{R}^n$ on which all functions $|x - e_i|$, $1 \leq i \leq n$, and $|x|$ are not strictly convex. This would lead to the absurd conclusion that this line would have to contain the points e_i, $1 \leq i \leq n$, and 0_n, using the fact that the only lines on which the modulus function is not strictly convex are the lines containing the origin 0_n.

2. Fermat: $f'(x) = 0_n \Rightarrow$

$$\frac{x - e_1}{|x - e_1|} + \cdots + \frac{x - e_n}{|x - e_n|} + \frac{x}{|x|} = 0_n.$$

3. It turns out that the solution $\widehat{x}$ can be calculated exactly, by virtue of the knowledge that it is unique. To begin with, it must be a scalar multiple of the vector $e = (1, \ldots, 1)^T$, by the symmetry in the problem combined with the uniqueness of the solution. Indeed, if two of the coordinates of a solution were unequal, then interchanging them would give another solution. We write $\widehat{x} = te$ with $t \in \mathbb{R}$. Substitution of $x = te$ into the stationarity equation, leads to the equation

$$(1 - nt)/(\sqrt{nt^2 - 2t + 1}) = 1/\sqrt{n}.$$

This implies that $1 - nt > 0$, that is, $t < 1/n$. Moreover, it implies

$$(1 - 2nt + n^2t^2)/(nt^2 - 2t + 1) = 1/n,$$

and this gives the quadratic equation

$$(n^3 - n)t^2 - 2(n^2 - 1)t + (n - 1) = 0.$$

This has the following two solutions $(1 \pm 1/\sqrt{n+1})/n$. As $t < \frac{1}{n}$, we get that the unique solution of the problem is $\widehat{x} = (1 - 1/\sqrt{n+1})e/n$.

4. $\widehat{x} = (1 - 1/\sqrt{n+1})e/n$.

4.6 Taking this possibility into account from the beginning would not lead to another decision. You can either realize this without any calculation or see it formally by a comparison of the necessary and sufficient conditions of both optimization problems.

4.7 The problem with short-selling can be modeled as follows:

1. $f_0(x) = 200x_1^2 + 400x_2^2 + 100x_1x_2 + 899x_3^2 + 200x_2x_3 \to \min$,

$$f_1(x) = 0.12 - (0.1x_1 + 0.1x_2 + 0.15x_3) \leq 0,$$

$$f_2(x) = x_1 + x_2 + x_3 - 1 = 0.$$

This is a convex problem, as f_0 and f_1 are convex—for f_0 one can use one of the following three methods for showing that the quadratic part of f_0: the minor criteria, the symmetric sweeping method, or by giving an explicit description of this quadratic part as a sum of squares—and f_2 is affine(= linear plus constant).

2. Lagrange function: $\mathcal{L} =$

$$\lambda_0(200x_1^2 + 400x_2^2 + 100x_1x_2 + 899x_3^2 + 200x_2x_3)$$

$$+\lambda_1(0.12 - (0.1x_1 + 0.1x_2 + 0.15x_3)) + \lambda_2(x_1 + x_2 + x_3 - 1).$$

KKT:

- $\lambda_0(400x_1 + 100x_2) + \lambda_1(-0.1) + \lambda_2 = 0$,
  $\lambda_0(800x_2 + 100x_1 + 200x_3) + \lambda_1(-0.1) + \lambda_2 = 0$,
  $\lambda_0(1798x_3 + 200x_2) + \lambda_1(-0.15) + \lambda_2 = 0$,
- $\lambda_1 \geq 0$,
- $\lambda_1(0.12 - (0.1x_1 + 0.1x_2 + 0.15x_3)) = 0$.

We put $\lambda_0 = 1$, as we may.

3. We have to distinguish two cases: the inequality constraint is tight in the optimum or not. We consider the case that the inequality constraint is tight as the most promising one. Then we have a system of 5 linear equations in the 5 variables $x_1, x_2, x_3, \lambda_1, \lambda_2$. This turns out to have the following unique solution: $x_1 = 0.5$, $x_2 = 0.1$, $x_3 = 0.4$, $\lambda_1 = 10584$, $\lambda_2 = 1906.8$. This satisfies the remaining KKT condition, the nonnegativity of the multiplier λ_1 of the inequality constraint. As the KKT conditions are sufficient, it is not necessary to consider the case that the inequality constraint is not tight.

4. It follows that it is optimal to invest 50% of her savings in fund 1, 10% in fund 2 and 40% in fund 3.

SOLUTIONS OF EXERCISES OF CHAPTERS 1–4   639

In particular, we see that here it is not optimal to make use of the possibility to go short. Therefore, the problem with the additional constraint that it is forbidden to go short has the same solution.

4.8 The absence of an arbitrage opportunity implies that the positive orthant in $\mathbb{R}^n$ and the linear span of the basic exchange opportunities $x_1, \ldots, x_k$ have no point in common. Therefore we can apply the separation theorem. This gives the existence of a nonzero vector $p \in \mathbb{R}^n$ with $p^T \cdot x_i = 0$ for $1 \leq i \leq k$ and $p^T \cdot y \geq 0$ for all y in the positive orthant of $\mathbb{R}^n$. This is seen to imply that p is a nonzero, nonnegative vector with $p^T \cdot x_i = 0$ for $1 \leq i \leq k$, as required.

4.9 Application of the separation theorem to the first orthant in $\mathbb{R}^n$ and the convex cone spanned by the basic exchange opportunities x_i, $1 \leq i \leq k$, will do the job.

4.10 (i) We model the problem as a convex problem with strictly convex objective function.

1. $f_0(x) = -\ln x_1 - \ln x_2 - \ln x_3 \to \min$, $f_1(x) = x_1 + x_2 + x_3 - 4 = 0$, $f_2(x) = 2 - x_2 \leq 0$, $f_3(x) = x_1 - 1 \leq 0$, $x_i > 0$, $1 \leq i \leq 3$. Note the following details:

   (a) by taking minus the logarithm of the objective function, we have achieved that the problem is convex and that the objective function is strictly convex,

   (b) we have put into this model the following insights: it cannot be optimal to leave part of the income unused and it cannot be optimal that one of the commodities is not consumed.

2. Lagrange function: $\mathcal{L} =$
$$\lambda_0(-\ln x_1 - \ln x_2 - \ln x_3) + \lambda_1(x_1 + x_2 + x_3 - 4) + \lambda_2(2 - x_2) + \lambda_3(x_1 - 1).$$
   KKT:
   - $\mathcal{L}_x = 0_3^T \Rightarrow$
   $$-\lambda_0/x_1 + \lambda_1 + \lambda_3 = 0, \quad -\lambda_0/x_2 + \lambda_1 - \lambda_2 = 0, \quad -\lambda_0/x_3 + \lambda_1 = 0,$$
   - $\lambda_2, \lambda_3 \geq 0$,
   - $\lambda_2(2 - x_2) = 0$, $\lambda_3(x_1 - 1) = 0$. We put $\lambda_0 = 1$, as we may.

3. We begin with the one of the four cases that looks the most promising: $x_2 = 2$ and $x_1 = 1$—and so $x_3 = 1$. The "reason" for this is that without the constraints $x_1 \leq 1$, $x_2 \geq 2$, the solution is seen to be $x_1 = x_2 = x_3 = \frac{4}{3}$ and this violates both

constraints. The KKT conditions are reduced to $-1+\lambda_1+\lambda_3 = 0$, $-\frac{1}{2}+\lambda_1-\lambda_2 = 0$, $-1+\lambda_1 = 0$, $\lambda_2, \lambda_3 \geq 0$. Solving this gives $\lambda_1 = 1$, $\lambda_2 = \frac{1}{2}$, $\lambda_3 = 0$, and in particular the last two multipliers are nonnegative as required.

4. Conclusion: the problem has a unique solution—consume 1 unit of commodity 1, 2 units of commodity 2, and 1 unit of commodity 3.

(ii) In order to spot the most promising case we omit the constraints $x_2 \geq 2$ and $x_1 \leq 1$; then it is optimal to spend equal amounts on each commodity, $x_1 = 2x_2 = 3x_3$, and this leads to a solution that violates the omitted constraints. Therefore we consider the KKT conditions, with $\lambda_0 = 1$, for the case $x_1 = 1$ and $x_2 = 2$—and so $x_3 = \frac{1}{3}$:

$$-1+\lambda_1+\lambda_3 = 0, \quad -\frac{1}{2}+2\lambda_1-\lambda_2 = 0, \quad -3+3\lambda_1 = 0, \quad \lambda_2, \lambda_3 \geq 0.$$

Solving this gives $\lambda_1 = 1$, $\lambda_2 = 1\frac{1}{2}$, $\lambda_3 = 0$.

Conclusion: the problem has a unique solution—consume 1 unit of commodity 1, 2 units of commodity 2, and $\frac{1}{3}$ unit of commodity 3.

4.11 One can readily check that the objective function of the given problem is lower semicontinuous, differentiable, and strictly convex, and that it satisfies the assumption of theorem 4.18 (first we have to make a parallel shift to bring the origin into the admissible set). Therefore it has a unique solution, say $(p(x), d(x))$ and this solution can be characterized by the stationarity conditions. The explicit form of this condition is $p * d = x$. To do this exercise, it suffices to check this last statement.

We introduce the following notation: $v^r = (v_1^r, \ldots, v_n^r)$ for all positive vectors $v \in \mathbb{R}^n$ and all $r \in \mathbb{R}$. Moreover, let $\tilde{P}$ ($\tilde{D}$) be the linear subspace of $\mathbb{R}^n$ of vectors of the form $p_1 - p_2$, with $p_1, p_2 \in P$ ($d_1 - d_2$ with $d_1, d_2 \in D$).

The gradient of the function f_x, where we let p and d run over the entire space $\mathbb{R}^n$, is the vector $(d - x * p^{-1}, p - x * d^{-1})$, as one readily checks by partial differentiation with respect to the variables $p_1, \ldots, p_n, d_1, \ldots, d_n$.

Therefore, taking into account the constraints $p \in P$ and $d \in D$, it follows that the required stationarity conditions have the following explicit form:

$$(d - x * p^{-1})^T \cdot \tilde{p} = 0 \text{ for all } \tilde{p} \in \tilde{P} \text{ and } (p - x * d^{-1})^T \cdot \tilde{d} = 0 \text{ for all } \tilde{d} \in \tilde{D}.$$

SOLUTIONS OF EXERCISES OF CHAPTERS 1–4

These conditions can be rewritten as

$$(p^{\frac{1}{2}} * d^{\frac{1}{2}} - x * p^{-\frac{1}{2}} * d^{-\frac{1}{2}})^T \cdot p^{-\frac{1}{2}} * d^{\frac{1}{2}} * \tilde{p} = 0$$

for all $\tilde{p} \in \tilde{P}$, and

$$(p^{\frac{1}{2}} * d^{\frac{1}{2}} - x * p^{-\frac{1}{2}} * d^{-\frac{1}{2}})^T \cdot p^{\frac{1}{2}} * d^{-\frac{1}{2}} * \tilde{d} = 0$$

for all $\tilde{d} \in \tilde{D}$.

Note that $p^{-\frac{1}{2}} * d^{\frac{1}{2}} * \tilde{P}$ and $p^{\frac{1}{2}} * d^{-\frac{1}{2}} * \tilde{D}$ are orthogonal complements, as $\tilde{P}$ and $\tilde{D}$ are orthogonal complements. Therefore the conditions above are equivalent to

$$p^{\frac{1}{2}} * d^{\frac{1}{2}} - x * p^{-\frac{1}{2}} * d^{-\frac{1}{2}} = 0.$$

That is,

$$p * d = x.$$

4.12 The criterion for a tower of n coasters for not falling down ("admissibility") is that for each $k \leq n$ the center of gravity of the k highest coasters does not stick out more than the end of the $(k+1)$st coaster. It follows that for an optimal tower the criterion for optimality is that for each $k \leq n$, the center of gravity of the k highest coasters sticks out exactly as much as the end of the $(k+1)$st coaster. This allows us to compute recursively how far the optimal tower of n felts sticks out and it suggests a way to construct this tower.

The construction is as follows—the main idea is to build it starting from the top rather than from the bottom. Start with a vertical tower and move the top coaster as far as possible. Then move the top 2 coasters as far as possible without altering their relative positions. Continue with moving the top 3 coasters and so on. This leads to the optimal tower of n coasters.

It remains to compute how far the top coaster of the optimal tower of n coasters sticks out. Denote this length by $f(n)$, taking the length of a coaster as unit of length. We will translate the fact that the center of gravity of all n coasters of the optimal tower lies exactly above the edge of the table into a formula. We choose the vertical plane through the edge of the table as zero-level. The position of the center of gravity of the bottom coaster with respect to this plane is $f(n) - f(n-1) - \frac{1}{2}$ and that of the center of gravity of the top $n-1$ coasters is $f(n) - f(n-1)$.

Therefore the position of the center of gravity of all n coasters is the weighted average

$$\left(1 \cdot (f(n) - f(n-1)) - \frac{1}{2}\right) + (n-1)(f(n) - f(n-1))/n.$$

This has to be equal to 0. Rewriting the resulting equality gives the formula $f(n) = f(n-1) = 1/2n$. This formula holds for all n, and moreover, $f(0) = 0$. This leads to the formula

$$f(n) = \frac{1}{2} \sum_{k=1}^{n} \frac{1}{k}.$$

It is a well-known fact that this sum is approximately equal to $\ln n$. Therefore, the tower can in principle stick out as far as you want. However, in order to get very far you need an astronomical number of coasters.

4.14 We choose the constant sum of the distances of the ends of the lines to the bottom left corner of the paper to be the unit of length. We consider two neighboring lines from the drawing, say,

$$x_1/a_i + x_2/(1 - a_i) = 1$$

with $i = 1, 2$ and $a_1 \approx a_2$. We calculate their point of intersection. It turns out to be

$$((a_1 - a_2)/(a_2^{-1} - a_1^{-1}), (a_1 - a_2)/((1 - a_1)^{-1} - (1 - a_1)^{-1})).$$

We fix a_2 and let a_1 tend to a_2. Then the point of intersection tends to $(a_2^2, (1-a_2)^2)$. The coordinates of this point $x_1 = a_2^2$ and $x_2 = (1-a_2)^2$ satisfy the equation $\sqrt{x_1} + \sqrt{x_2} = 1$.

Conclusion: the curve on the child-drawing consists of the nonnegative solutions of the equation $\sqrt{x_1} + \sqrt{x_2} = 1$.

4.17

1. $f_0(x) \to \min$, $f_1(x) = a - x \leq 0$, $f_2(x) = x - b \leq 0$.

   Existence of a global solution $\hat{x}$ follows from the Weierstrass theorem.

2. Lagrange function: $\mathcal{L} = \lambda_0 f(x) + \lambda_1(a - x) + \lambda_2(x - b)$.

   KKT:

   - $\mathcal{L}_x = 0_n^T \Rightarrow \lambda_0 f'(x) - \lambda_1 + \lambda_2 = 0$,
   - $\lambda_1, \lambda_2 \geq 0$,

- $\lambda_1(a-x) = 0$, $\lambda_2(x-b) = 0$.

3. We distinguish three cases.

    (a) $a < x < b$. Then $\lambda_1 = \lambda_2 = 0$ and so $\lambda_0 > 0$. It follows that $f'(x) = 0$.

    (b) $x = a$. Then $\lambda_2(a-b) = 0$, and so $\lambda_2 = 0$. We may put $\lambda_0 = 1$: if $\lambda_0 = 0$, then $\lambda_1 \neq 0$, but this would contradict the stationarity condition. This gives $f'(x) - \lambda_1 = 0$, $\lambda_1 \geq 0$. Therefore, $f'(x) \geq 0$.

    (c) $\widehat{x} = b$. Similar to previous case.

4. Conclusion: $f'(\widehat{x}) = 0$ if $a < x < b$, $f'(\widehat{x}) \geq 0$ if $\widehat{x} = a$, and $f'(\widehat{x}) \leq 0$ if $\widehat{x} = b$.

Bibliography

[1] V.M. Alekseev, V.M. Tikhomirov, and S.V. Fomin, *Optimal Control*, Consultants Bureau, New York (1987).

[2] V.I. Arnold, *Huygens and Barrow, Newton and Hooke, Pioneers in mathematical analysis and catastrophe theory from evolvents to quasicrystals*, Birkhäuser Verlag, Basel (1990).

[3] M.S. Bazaraa, H.D. Sherali, and C.M. Shetty, *Nonlinear Programming: Theory and Algorithms*, John Wiley and Sons, New York (second edition) (1991).

[4] R. Bellman, *Dynamic Programming*, Princeton University Press, Princeton (1957).

[5] A. Ben-Tal and A. Nemirovski, *Lectures on Modern Convex Optimization: Analysis, Algorithms and Engineering Applications*, MPS/SIAM Series in Optimization 2, SIAM, Philadelphia (2001).

[6] D.P. Bertsekas, *Nonlinear Programming*, Athena Scientific, Belmont, Massachusetts (1995).

[7] D.P. Bertsekas, *Constrained Optimization and Lagrange Multiplier Methods*, Athena Scientific, Belmont, Massachusetts (1996).

[8] D.P. Bertsekas, *Convex Analysis and Optimization*, Athena Scientific, Belmont, Massachusetts (2003).

[9] O. Bolza, *Vorlesungen über Variationsrechnung*, Chelsea, New York (1962).

[10] S.E. Boyd and L. Vandenberghe, *Semidefinite programming*, SIAM Review 38(1)(1996):49-96.

[11] S.E. Boyd and L. Vandenberghe, *Convex Optimization*, Cambridge University Press, Cambridge (2004).

[12] J. Brinkhuis, *On the Fermat-Lagrange principle for mixed smooth convex problems* (Russian), Mat. Sbornik 192(5) (2001): 3-12; translation in Sb. Math. 102.

[13] J. Brinkhuis, *A Comprehensive View on Optimization: Reasonable Descent*, Econometric Institute Report EI 2005-23, Erasmus University Rotterdam (2005).

[14] F.H. Clarke, *Optimization and Nonsmooth Analysis*, Wiley-Interscience, New York (1983).

[15] V. Chvatal, *Linear Programming*, W.H. Freeman, New York (1983).

[16] R. Courant and D. Hilbert, *Methods of Mathematical Physics*, Vols. 1 and 2, Wiley, Berlin (1989).

[17] G.B. Dantzig, *Linear Programming and Extensions*, Princeton University Press, Princeton (1963).

[18] G.B. Dantzig, Linear programming. In J.K. Lenstra, A.H.G. Rinnooy Kan, and A. Schrijver, editors, *History of Mathematical Programming: A Collection of Personal Reminiscences*, CWI, North Holland, Amsterdam (1991).

[19] J.E. Dennis and R.B. Schnabel, *Numerical Methods for Unconstrained Optimization and Nonlinear equations*, SIAM Classics in Applied Mathematics 16, SIAM, Philadelphia (1996).

[20] M. Ehrgott, *Multicriteria Optimization*, Springer Lecture Notes in Economics and Mathematical Systems 491, Springer, New York (2000).

[21] W. Fenchel, *Convex Cones, Sets, and Functions*, Princeton University Press, Princeton (1951).

[22] W. Fenchel, *On conjugate convex functions*, Can. J. Math. 1, (1949): 73-77.

[23] P. de Fermat, *Oeuvres de Fermat*, vol. 1, Gauthier-Villars, Paris (1891).

[24] A.V. Fiacco and G.P. McCormick, *Nonlinear Programming: Sequential Unconstrained Minimization Techniques*, John Wiley and Sons, New York, (1968). Reprint as SIAM Classics in Appl. Math. 4, SIAM, Philadelphia (1990).

[25] R. Fletcher, *Practical Methods of Optimization*, John Wiley and Sons, New York (second edition) (2000).

[26] A.J. Goldman and A.W. Tucker, Theory of linear programming. In H.W. Kuhn and A.W. Tucker, editors, *Linear Inequalities and Related Systems*, Annals of Mathematical Studies 38, Princeton University Press, Princeton, (1956): 53–97.

[27] G.H. Golub and C.F. van Loan, *Matrix Computations*, John Hopkins University Press, Baltimore (1996).

[28] A. Grothendieck, *Recoltes et semailles*, Université des Sciences et Techniques de Languedoc, Montpellier (1985).

[29] O. Güler, C. Roos, T. Terlaky, and J.-Ph. Vial. Interior point approach to the theory of linear programming. Cahiers de Recherche 1992.3, Faculté des Sciences Économiques et Sociales, Université de Genève, Geneva (1992).

[30] G.H. Hardy, J.E. Littlewood, and G. Polya. *Inequalities*, Cambridge University Press, Cambridge (1934).

[31] C. Huygens, Horologium oscillatorium, sive, De motu pendulorum ad horologia aptato demonstrationes geometricae. Paris: F.Muguet, (1673).

[32] A.D. Ioffe and V.M. Tikhomirov, *Theory of Extremal Problems*, North-Holland, Amsterdam (1979).

[33] F. John, Extremum problems with inequalities as subsidiary conditions, in *Studies and Essays*, Courant Anniversary Volume, Interscience, New York (1948), pp. 187-204.

[34] L.V. Kantorovich, *Mathematical Methods of Organizing and Planning Production* [in Russian], Leningrad State University, Leningrad (1939).

[35] N. Karmarkar, *A New polynomial-time algorithm for linear programming*, Combinatorica 4 (1984): 373-395.

[36] W.E. Karush, *Minima of Functions of Several Variables with Inequalities as Side Conditions*, University of Chicago Press, Chicago (1939).

[37] J. Kepler, *Nova stereometria doliorum vinariorum*, in Johannes Kepler Gesammelte Werke, Munich (1937).

[38] L.G. Khachiyan, *A polynomial algorithm in linear programming*, Doklady Akademiia Nauk SSSR 244 (1979): 1093-1096, translation in English in Soviet Mathematics Doklady 20: 191-194.

[39] F.E. Kydland and E.C. Prescott, *Rules rather than discretion: the inconsistency of optimal plans*, Journal of Political Economy 85 (3) (June 1977): 473-492.

[40] H.W. Kuhn and A.W. Tucker (editors), *Linear Inequalities and Related Systems*, Ann. Math. Stud. 38 (1956), Princeton University Press, Princeton.

[41] H.W. Kuhn and A.W. Tucker, *Nonlinear programming*, in *Proceedings of the Second Berkeley Symposium*, University of California Press, Berkeley (1951): 481-492.

[42] J.L. Lagrange, *Mécanique analytique*, Paris (1788).

[43] J.L. Lagrange, *Théorie des fonctions analytiques*, Paris (1797).

[44] J.B.Lasserre, *Global optimization with polynomials and the problem of moments*, Siam J. Opt. Vol 11, No 3, (2001): 796-817.

[45] G.W. Leibniz, *Nova methodus pro maximis et minimis, itemque tangentibus, quae nec fractus, nec irrationales quantitates moratur et singulare pro illis calculi genus per G.G.L.* Acta eruditorum (1684): 467-473.

[46] D.G. Luenberger, *Optimization by Vector Space Methods*, Wiley-Interscience, New York (1997).

[47] D.G. Luenberger, *Linear and Nonlinear Programming*, Kluwer Academic, Boston, (2003).

[48] D.G. Luenberger, *Investment Science*, Oxford University Press, Cambridge (1998).

[49] G.G. Magaril-Il'yaev and V.M. Tikhomirov, *Convex Analysis: Theory and Applications*, Translations of Mathematical Monographs 222 American Mathematical Society, Providence (2003).

[50] O.L. Mangasarian, *Nonlinear Programming*, McGraw-Hill, New York (1969).

[51] P. Milgrom, *Putting Auction Theory to Work*, Cambridge University Press, Cambridge (2004).

[52] J.W. Milnor, *Topology from the Differential Viewpoint*, Princeton University Press, Princeton (1997).

[53] J. Mirrlees, *An Exploration in the Theory of Optimal Income Taxation*, Rev. Econ. Stud. 38: 175–208 (1971).

[54] R.B. Myerson, *Analysis of Conflict*, Harvard University Press, Cambridge (1991).

[55] J.F. Nash, Jr, *The bargaining problem*, Econometrica 18: 155-162 (1950).

[56] Y. Nesterov and A. Nemirovskii, *Interior-Point Polynomial Algorithms in Convex Programming*, SIAM Studies in Applied Mathematics 13, Philadelphia (1994).

[57] Y. Nesterov, *Introductory Lectures on Convex Programming: Basic Course*, Kluwer Academic Press, Boston (2003).

[58] L.W. Neustadt, *Optimization: A Theory of Necessary Conditions*, Princeton University Press, Princeton (1976).

[59] I. Newton, *Philosophiae naturalis principia mathematica*, London (1687).

[60] J. Nocedal and S.J. Wright, *Numerical Optimization*, Springer, New York (1998).

[61] L.S. Pontryagin, V.G. Boltyanskii, R.V. Gamkrelidze, and E.F. Mishchenko, *The Mathematical Theory of Optimal Processes*, Interscience, New York (1962).

[62] R.T. Rockafellar, *Convex Analysis*, Princeton University Press, Princeton (1970)

[63] C. Roos, T. Terlaky and J.-Ph. Vial, *Theory and Algorithms for Linear Optimization: An Interior Point Approach*, John Wiley and Sons, New York (1997).

[64] F. Ramsey, *A mathematical theory of savings*, Economic Journal (1928); reprinted in *Readings in Welfare Economics* K.J. Arrow and T. Scitovski, editors, Homewood, Ill.: Richard D. Irwin (1969).

[65] S. Rebello, *Long run policy and long run growth*, Journal of Political Economy 99 (1991): 500-521.

[66] P.M. Romer, *Increasing returns and long run growth*, Journal of Political Economy 94 (1986): 1002-1037.

[67] A. Rybakov, *Children of the Arbat*, "High School", Minsk, (in Russian), (1988).

[68] A. Schrijver, *Theory of Linear and Integer Programming*, John Wiley and Sons, New York (1986).

[69] S. Singh, *Fermat's Last Theorem*, Fourth Estate, London (1997).

[70] M. Spivak, *Calculus on Manifolds*, Benjamin/Cummings, New York (1965).

[71] M. Slater, *Lagrange multipliers revisited: a contribution to nonlinear programming*, Cowles Commission Discussion Paper, Math. 403 (1950).

[72] G.W. Stewart, *Afternotes goes to Graduate School: Lectures on Advanced Numerical Analysis*, SIAM, Philadelphia (1998).

[73] J.F. Sturm and S. Zhang, *Symmetric Primal-Dual Path Following Algorithms for Semidefinite Programming*, Applied Numerical Mathematics 29,(1999): 301–315,

[74] V.M. Tikhomirov, *Stories about Maxima and Minima*, Mathematical World 1, American Mathematical Society, Providence (1990).

[75] A. Wiles, *Modular elliptic curves and Fermat's Last Theorem*, Ann. of Math. 142 (1995): 443-551.

[76] L.C. Young, *Lectures on the Calculus of Variations and Optimal Control Theory*, W.B. Saunders, Philadelphia (1969).

Index

D^1, 10, 91
$D^1(\widehat{x}, \mathbb{R}^m)$, 525
$S_{\max}$, 520
$S_{\min}$, 520
$S_{\min}, S_{\max}$, 24
$U_n(\widehat{x}, \varepsilon)$, 90
e^π or π^e?, 391
l_p-norm, 118
n-variable Newton method, 300
n-variable modified Newton method, 301
$v(P)$, 520
$x \succ_C y$, 420
007 and Snellius, 65
007 and optimization, 35

active insect, 461
admissible, 141
aerodynamic problem, Newton's, 494
algorithm
    n-variable Newton method, 300
    n-variable modified Newton method, 301
    bisection method, 292
    center of gravity method, 304
    conjugate gradient method, 325
    definiteness, 511
    determination definiteness, 264
    ellipsoid method, 307
    Fibonacci method, 288
    golden section method, 291
    gradient method, 300
    grid method, 281
    method of steepest descent, 300
    modified Newton method, 297
    Newton method, 294
    primal-dual interior point methods for LP, 316
    self-concordant barrier methods, 335
    simplex method, 284

algorithms, and ball-coverings, 282
allocation, efficient and fair, 387
Apollonius, problem of, 164, 397
approximation of a function
    near point, 121
    on interval, 121
approximation theory, 118
approximation, measures of quality, 301
arbitrage
    benefits of, 252
    type A, 378
    type B, 382
arbitrage opportunity, 252, 378
arbitrage-free bounds for prices, 378
arbitrage-opportunities, criterion for absence, 252, 253
Archimedes
    defense of Syracuse, 77
    irritation of, 79
    problem of, 79
Archimedes's greatest achievement
    contemporary point of view, 77
    his own point of view, 78
    modern point of view, 77
astroid
    child drawing, 228
    pearl inside oyster, 165
Atlantis, 133
auction, 471
    highest bidder not winner, 363
    mechanism, 363

bad case, failure of multiplier rule, 139
ball-covering algorithm, 282
ball-coverings, and algorithms, 282
banks, independent central, 497
bargaining opportunity, 373
barrier methods

idea of, 53, 81
progress made by, 336
beach, selling ice cream, 102
Bellman, 441
Bellman equation, 445
bisection method, 292
  idea of, 45
  performance of, 302
Black and Scholes, formula of, 380
boundary, 527
boundary point, 527
bounded set, 539
  calculus, 542
brachistochrone, 475
brain teaser, 522

cake cutting, 293
calculations, confession of Newton, 13
calculus of variations, 476
cargo, maximum, 446
Cauchy sequence, 521
center of gravity, 32
  popular mistake, 74
center of gravity method, 304
central path, 323
certificate of nonsolvability, 395
chain rule, 525
  general case, 525
Chebyshev
  achievements of, 124
  and birth of approximation theory, 125
  passion of, 126
  steam machine, 125
  theory and practice, 125
Chebyshev polynomials, 411
child drawing, 226, 256
circle quadrangle, 162
classical analysis
  motivation for, 5
  two sentences of Newton, 26
closed loop form, 445
closed set, 527, 539
  calculus, 541
closure, 527
closure set, 231
coercive function, 543
  calculus, 544
  definition, 18

coercive, use of property, 19
coercivity for n variables, role of, 98
coincidence, explanation of, 34
coins, counterfeit, 473
cola war, 129
column vector, 506
Commander of the British Empire, 363
commitment, 497
  value of, 498
competition, beneficial effects of, 129
complementary slackness, 209, 237
  for LP problems, 216
completeness property, real numbers, 521
conditions of complementary slackness, 209
cone, 419
cone spanned by a set, 231
conic ordering, 420
conic problem, 423
conic smooth-convex problem, 418, 423
conjugate function, 532
conjugate gradient method, 325
consistency, time, 496
constraint qualifications, impracticibility of, 139
consumer theory, basic problem, 191, 253
continuous function, 539
  calculus, 540
  definition, 18
  geometric sense, 18
continuously differentiable function, 142
control, 445, 477
control problem, production and inventory, 462
convergent subsequence, 521
convex cone, 240, 419
  pointed, 420
  solid, 420
convex function, 210, 528
  criterion, 529
convex problems, efficient barrier methods, 335
convex set, 210, 528
convexity, 528

invertibility property, 202
  role in optimization, 45
  role of, 202
cost, of holding and ordering, 194
counterfeit, coins, 473
Cournot, 130
Cramer's rule, 507
crash course
  problem solving, 547
  theory, analytical style, 561
  theory, geometrical style, 553
cube law, and optimal speed of ships, 366
cutting plane method, 305
cycloid, 407, 475

Dantzig, invention of simplex method, 205
de L'Hospital, problem of, 80
definitions, modest role of, 10
derivative
  analytical definition, 7
  approximation definition, 29
  calculation of, 10, 523
  definition for n variables, 91
  geometrical definition, 27
  high school definition, 7
  not a number, 87
  physical definition, 26
  quadratic function, 93
  viewed as a velocity, 26
derivative for n variables
  in words, 92
  invention, 94
derivative vector function, 525
determinant, 508
differentiability
  for n variables, 91
  geometric sense of, 9
differentiable, 7
  strictly, 177, 560, 565
differential calculus, 10, 523
  geologists and power of, 11
dimension, 516
  solution set, 507
discount plane tickets, 219, 368
distribution, price buyer, 36
dot notation, 509
dual problem

idea of, 58
LP in standard form, 235
LP in symmetric form, 237
dual problems, 235
duality theory, 235
  idea of, 57
duopoly, 106
  and monopoly, 129
dynamic investment problem, 453
dynamic programming, 441
dynamic system, 445

elasticity of demand, 369
ellipsoid method, 307
  heart of matter, 75
  performance of, 314
envelope theorem
  n-variable case, 127
  equality constraints, 186
  idea of, 44
  one-variable case, 55
epigraph, 528
equimarginal rule, 30
equity and efficiency, 387
Euclid
  and the pharaoh, 73
  problem of, 73
Euclidean norm vector, 88
European call option, 380
exchange, 252, 253
existence of global solutions, 537
existence of solutions, 18
externalities
  and taxes, 131
  turned into internal costs, 131
extremum
  invention of sense of word, 14

family of problems, heart of the matter, 44
Farkas's lemma, 379, 396
fata morgana, explanation, 65
feedback form, 445
Fenchel transform, 532
Fermat
  method of, 5
  problem of, 152
  variational principle, 39
Fermat and Torricelli, problem of, 106

Fermat theorem
  economic formulation, 30
  for n variables, 97
  geometrical formulation, 27
  intuitive formulation, 5
  one-variable problems, 15
  physical formulation, 27
  prehistory of, 76
  rational decision maker, 29
Fermat-Lagrange condition, 427
Fermat-Lagrange theorem for perturbations, 428
Fermat-Torricelli point, 109
Fibonacci method, 288
first welfare theorem, 388
fish, illusion of position of, 64
flat, 236
flows of cargo, prediction of, 370
formula, solution set linear system, 507
four-step method
  basic form, 111
  userfriendly version, 20
fundamental theorem of algebra, optimization proof, 111
fundamental theorem of analysis, 535

Galilei
  tower of Pisa, 27
Galileo
  brachistochrone, 475
game, 383
game theory, 130
game with matches, 473
Gauss, advice of, 601
Gaussian elimination, 503
golden barriers and self-concordancy, 339
golden cones, 339
golden section method, 291
government regulations, need for, 102
Grünbaum, theorem of, 306
gradient, definition, 91
Gram, formula of, 409
grid method, 281
guess who?, 293

Hadamard product, 255
Hamilton-Jacobi-Bellman equation, 445

Hamiltonian function, 500
Hessian, 263
hidden convexity, 257
hinged quadrangle, 160
  paradox, 195
HJB equation, 445
Huygens, pendulum clock of, 406

ice cream cone, 339
identity matrix, 510
implicit function theorem, 168
incentives, may fail, 160
inconsistency, time, 496
inequality
  arithmetic-quadratic means, 156
  between l_p-norms, 157, 197
  Cauchy-Bunyakovskii-Schwarz, 89
  geometric-arithmetic means, 31
  geometric-quadratic means, 154
  Hadamard, 197
  triangle, 89
infimum, 520
inflation, low, 497
inner product, 509
insect climbing topaz, 284
interior, 527
interior point, 527
interior point methods, 316
  idea of, 53
intermediate value theorem, 83
inverse function theorem, 526
inverse matrix, 510
investment, 378
  dynamic problem, 453
  mean variance, 222
  risk minimization, 251
investor, 497
invisible hand, 387
  can fail, 102, 130
involute of curve, 401
iron dust experiment, 203
isochrone, 406

Jacobi matrix, 525
Jacobi methods, 113
John theorem, 248, 423

Kantorovich, invention of linear programming, 205

INDEX 655

Karmarkar, interior point algorithms LP, 285
Karush-Kuhn-Tucker conditions, 208
Karush-Kuhn-Tucker theorem, 210
Kepler
  problem of, 77, 153
  wine barrels, 75
kernel, 516
Khachian, polynomial algorithm LP, 285
KKT conditions, 208
  advantages, 211
KKT multipliers
  shadow price interpretation, 244
KKT theorem, 210
  convex problems, 247
  history, 216
  smooth problems, 248, 423
knapsack problem, 467
Kydland, 496

labor or leisure, 69
Lagrange function, 142, 208
Lagrange multiplier rule, 143
Lagrange multipliers, 142, 208
  embryonic form, 45
Lagrange, description of multiplier rule, 135
Landau
  small o, 29, 91
lazy insect, 460
leading minor, 512
least squares line, 109
least squares solution, system of linear equations, 131
Legendre
  and 70 assistants, 124
  and least squares method, 124
  main achievements, 123
lemma, centered systems, 522
length vector, 88
light, refraction of, 39
line search, 286
linear combination, 506
linear complementarity problem, 318
linear conic problems, 436
linear function on sphere, 267
linear independence, 516
linear operator, 515

linear programming, 204
linear span, 516
linear subspace, 516
Lipschitz function, 280
local minimum (maximum)
  motivation, 14
  definition, 14
Lorenz cone, 339
low inflation, 497
lower semicontinuous function, 249
LP, 204

main linear part, 92
market, Cournot's definition, 130
Marshall, and marginal analysis, 30
martingale, 379, 381
matrix, 505
matrix multiplication, 509
matrix norm, 113
maximum likelihood method, 70
maximum, linear function on sphere, 155, 217
mean variance investment, 222
median, 81
method
  of maximal entropy, 370
  of maximum likelihood, 370
method of steepest descent, 300
minimax theorem, of von Neumann, 384
minimization of a quadratic function, 101, 409
minimum(maximum)
  absolute, 15
  global, 14
  local, 14
  relative, 15
minor, 511
minor criteria, 511
mirages on hot roads, 65
mixed smooth-convex problems, 417
modified Newton method
  for n variables, 301
  for one variable, 297
modulus vector, 88
monopolist, competing over time, 455
monopoly, 455
  and duopoly, 129
monotonic convergence, 521

monotonic subsequence, 521
Morse theory, 133
multiplier rule, 143
   "carrot and stick" policy, 138
   counterintuitive, 138
   economic interpretation, 185
   power of, 160
   universality of, 138
multipliers
   essential detail, 146
   interpretation as shadow prices, 184

name of the game
   in economics, 30
   in mathematics, 30
Nash, 130, 373
Nash bargaining, 256, 373
Nash equilibrium, 106, 384
necessary conditions, for conic problems, 425
necessary conditions, higher order, 271
neighborhood, 526
   motivation, 526
Nemirovskii, 335
Nesterov, 335
Newton method
   for n variables, 300
   for one variable, 294
Newton method, idea of, 46
Newton-Leibniz formula, 535
Nim, 473
Nobel prize 2004, economics, 496
nonlinear programming problem, 206
nonnegative orthant, 339

open ball, 90
open set, 526
   motivation, 526
optimal control, 477
optimal seller's price, 36
optimality
   location, 107, 131
   market clearing prices, 60
   outcome game, 69
   path of light, 39
   position of political party, 102
   profit maximization, 105
   royalties, 68
   shape barrels, 76

skyskrapers, 65
social welfare, 193
sofa, 71
standby ticket, 67
optimization methods in a nutshell, 43
optimization problem solved by a conservation law, 38
option, 380
orthogonal diagonalization
   symmetric matrix, 112, 164, 269
orthogonal matrix, 112, 511
orthogonality of vectors, 88

Pareto efficient, 388
partial derivative, 523
   connection with derivative, 95
   role, 94
path of light, 39
path-following algorithm, 323
penalty, how to take, 382
pendulum clock, 406
perturbation, 426
   heart of the matter, 44
   smooth-convex, 418
playing games or eating ice cream, 191
pointed convex cone, 420
polynomials of least deviation, 118, 411
Pontryagin's maximum principle
   as realization of multiplier rule, 138
   discrete time, 477, 501
positive definite, 101, 393, 511
positive orthant
   exotic parametrization, 253
positive semidefinite, 511
positive semidefinite cone, 339
prediction, flows of cargo, 370
Prescott, 496
price discrimination, 128
   optimal discounts, 219, 368
   telecom market, 128
primal-dual interior point methods for LP, 316
principle of least time, 39
problem of Fermat, 30
problem with inequality constraint, 140

producer theory, basic problem, 192
product, row vector and column vector, 509
production and inventory, 462
profit maximization, not optimal, 105
proof multiplier rule, 180
Pythagorean theorem
n-dimensional, 88

quadratic function, 93

rank, 516
rank matrix, 507
RAS method, 370
ray, 419
real numbers, 519
reasonable descent, proof of necessary conditions, 134
refined tangent space theorem, 189
relation, two high school formulas, 27
relativity of time, 432
remainder term, 92
risk minimization investment, 251
Rodrigues
  as banker, 123
  as mathematician, 123
  as social reformer, 123
  formula of, 119
rose in a lemonade glass, optimal stability, 31
rounding off warning, 66
row vector, 506
royal road, xv, 73
  for n-variable problems, 87
  for basic algorithms, 278
  for continuous time dynamic optimization, 477
  for convex problems, 204
  for discrete time dynamic programming, 443
  for equality constraints, 138
  for inequality constraints, 204
  for one-variable problems, 6
  for second order conditions, 262

Saturday stayover, 219, 368
screwdriver, 496
second derivative, 263
second order conditions, 262
  for conic problems, 433
  motivation, 262
  quadratic function on sphere, 268
  without multipliers, 264
second welfare theorem, 386, 389
self-concordant barrier method, 335
self-concordant barriers, 339
semidefinite programming, 340, 418
separation of convex sets, 253
setting sun, illusion of, 65
shareholders versus stakeholders, 106, 158
short position, 378
simplex method, 284
  idea of, 284
sine, behind the button, 13
slackness, 237
Slater point, 210
smoothness, role in optimization, 45
Snellius, law of, 39
soap films, 107
social welfare, 61, 193
solid convex cone, 420
solution of optimization problem
  by meaningful equation, 42
  exact (by formula), 45
  numerical (by algorithm), 45
speed clause, 366
spider catching insect, 458
stakeholders, 106
  versus shareholders, 106, 158
state, 445
stationary but not optimal, 17
stationary points, 16
  of independent interest, 39, 164, 165
steepest descent method, 300
strictly convex function, 529
strictly differentiable, 177, 560, 565
strong minimum, 427
subdifferential, 531
subgradient, 532
superposition principle, general case, 525
supporting hyperplane theorem, 231, 566
  proof using Weierstrass, 231
  proof without Weierstrass, 558
supremum, 520
Sylvester, criterion of, 409

symmetric matrix, 93, 510
  basic property, 112, 164, 269
  basic property and optimization, 163
  orthogonal diagonalization, 112, 164, 269

table, three holes in, 107
tangent space theorem, 171, 566
  proof without Weierstrass, 557
Tartaglia, problem of, 72
tax rate, 498
Taylor polynomials
  and remainder term, 83
  discovery of, 13
  illustrations of, 82
testing medicines, 70
time charter contract, 366
Torricelli point, 109
tower of Pisa, 255, 451
towers of Hanoi, 449
trade, benefits of, 192
trade-off
  high profit or certain profit, 36
  high return or low risk, 199
  modeled by optimization problem, 16
  profit or revenue, 158
  quality or performance, 3
  speed or reliability, 53
transpose
  of a matrix, 513
  of a vector, 513
twice differentiable, 263

unimodular functions, 287
universal algorithm
  existence impossible, 280
  existence unlikely, 278
unsolved problem, classical Greek geometers, 160

value
  of a problem, 520
  of an optimization problem, 24
variational principle of Fermat, 39
vector notation, 512
  power of, 87
von Neumann, minimax theorem, 384

waist of three lines, and convexity, 223
waist of two lines, 250
Walrasian equilibrium, 388
wedding crown prince problem, 448
Weierstrass theorem, 537
  n-variable case, 98
  derivation, 544
  for convex problems, 250
  one-variable case, 18
Weierstrass, triple life of, 28

zero sweeping, 503

CPSIA information can be obtained at www.ICGtesting.com
Printed in the USA
BVOW011708091212

307450BV00006B/12/P

9 780691 102870